Harmonic Analysis on Symmetric Spaces—Higher Rank Spaces, Positive Definite Matrix Space and Generalizations

Audrey Terras

Harmonic Analysis on Symmetric Spaces—Higher Rank Spaces, Positive Definite Matrix Space and Generalizations

Second Edition

 Springer

Audrey Terras
Department of Mathematics
University of California at San Diego
La Jolla, CA, USA

ISBN 978-1-4939-8042-0   ISBN 978-1-4939-3408-9   (eBook)
DOI 10.1007/978-1-4939-3408-9

Printed on acid-free paper

This Springer imprint is published by Springer Nature
The registered company is Springer Science+Business Media LLC New York

To the Vulcans and other logical creatures

There's always some error, even with experts. There's got to be, since there are so many variables. Put it this way—the geometry of space is too complicated to handle and hyperspace compounds all these complications with a complexity of its own that we can't even pretend to understand.

From I. Asimov, *Foundation's Edge*, Doubleday & Co., Inc., NY, 1982, p. 162.

Preface to the First Edition

Well, finally, here it is—the long-promised *Revenge of the Higher Rank Symmetric Spaces and Their Fundamental Domains*. When I began work on it in 1977, I would probably have stopped immediately if someone had told me that 10 years would pass before I would declare it "finished." Yes, I am declaring it finished—though certainly not perfected. There is a large amount of work going on at the moment as the piles of preprints reach the ceiling. Nevertheless, it is summer and the ocean calls. So I am not going to spend another 10 years revising and polishing. But, gentle reader, do send me your corrections and even your preprints.

I said it all in the Preface to Volume I [612]. So I will try not to repeat myself here. Yes, the "recent trends" mentioned in that Preface are still just as recent. And there are newer, perhaps even more pernicious tendencies here in the USA. This is the age of the billion dollar "defense" funding of research, the "initiatives" to put more power and money in the hands of fewer and fewer, the boondoggles to spend huge sums on supercomputers and to bring space war movies into the university. Yes, and compartmentalized research is still in the ascendancy. But I do not feel much happier looking at the international mathematical community that just declared most female, minority, and third world mathematicians unfit to speak at the international congress in Berkeley last summer. Oh well, many fields were not represented either. But, for me, the best conference is one run democratically and covering a wide spectrum of viewpoints, a conference in which anyone who wishes can speak on their research. Infinite diversity in infinite combinations!

Well, so much for purple prose. Clearly I am hoping for some forgiving readers. I also need readers who are willing to work out lots of exercises on a large variety of topics. Yes, once more there are lots of exercises. But, isn't it boring to read other people's proofs? In Chapter 1 this can mean some rather complicated calculations on matrix space. In Section 2.1 of Chapter 2, this will require some familiarity with beginning differential geometry—tangent spaces, differentials, and the like. Some parts of Section 2.2 of Chapter 2 will demand a little knowledge of beginning algebraic number theory.

Perhaps I should repeat one thing from the Preface to Volume I—the warning that I am very bad at proofreading. And the formulas in Volume II are much worse than in Volume I. So please do remember this when a formula looks weird.

Thanks again to all who helped me write this. You know who you are, I hope. Live long and prosper!

Encinitas, CA, USA   Audrey Terras
August 1987

Preface to the Second Edition

It is marvelous and a bit scary to return to this garden after 17 years or so. Sadly many people such as Serge Lang, Hans Maass, and Atle Selberg are now gone. But there are many new flowers. Many more women are working in this field. Yeah! The books of Jorgenson and Lang [333], [334], as well as that of Elstrodt et al. [168] are welcome. So also is that of Goldfeld [230]. Thanks to younger people, there are even computations of automorphic forms for $GL(n, \mathbb{Z})$, when $n = 3$ and 4. See the website:

www.lmfdb.org

for many computations of higher rank modular forms and L-functions. There is much new work on random matrices. And, of course, there are many adelic books and papers.

I had hoped that this edition would be finished in the summer of 2013. It took more time than I expected to finish updating Volume II. There have been many changes since 1987 as I noted in the preface to Volume I and much lack of progress as well. I will not discuss adelic representation theory here either. I am still hoping that this volume is friendlier than works requiring adelic group representations. And, once again, I leave much to the reader. There is still no answer book for the exercises. Sorry.

I have added a few new sections, including one on Donald St. P. Richards' central limit theorem for $O(n)$-invariant random variables on the symmetric space of $GL(n, \mathbb{R})$, another on random matrix theory, and some discussions of mostly non-adelic advances in the theory of automorphic forms on arithmetic groups since 1987.

I should very belatedly thank Walter Kaufmann-Bühler for the translations of French and German that appear in the footnotes. We miss him very much. And, once more, I thank the Scientific Workplace people for allowing me to refuse to learn TEX and thus making the work on this book a much more pleasurable experience.

I am very grateful to Aloys Krieg and Anton Deitmar who bravely sent me long lists of errors at least 25 years ago. Sorry it took so long, but I hope that finally I managed to correct them all. I am also extremely grateful to Andrew Odlyzko and

Andrew Booker for sending me Figure 1.2 and Figure 1.32, respectively, which give evidence for some of the most interesting conjectures touched on in this volume.

There are lots of other people I should thank, especially my POSSLQ and my students.

When I refer to Volume I, now I will always refer to the new edition [612]. It should go without saying that it will be assumed that the reader has some acquaintance with Volume I.

Live long and prosper!

Encinitas, CA, USA   Audrey Terras
August 2015

Contents

List of Figures

Chapter 1
The Space $\mathcal{P}_n$ of Positive $n \times n$ Matrices

1.1 Geometry and Analysis on $\mathcal{P}_n$

The story so far:
In the beginning the Universe was created. This has made a lot of people very angry and been widely regarded as a bad move. Many races believe that it was created by some sort of god, though the Jatravartid people of Viltvodle VI believe that the entire Universe was in fact sneezed out of the nose of a being called the Great Green Arkleseizure.

From *The Restaurant at the End of the Universe*, by Douglas Adams, Harmony Books, NY, 1980.
Reprinted by permission of The Crown Publishing Group.

1.1.1 Introduction

In this chapter, our universe is a higher rank symmetric space $\mathcal{P}_n$, the **space of positive $n \times n$ real matrices**:

$$\mathcal{P}_n = \left\{ Y = (y_{ij})_{1 \leq i,j \leq n} \mid y_{ij} \quad \text{real}, \; {}^t Y = Y, \; Y \text{ positive definite} \right\}. \tag{1.1}$$

Here and throughout this book, "**Y is positive**" means that the quadratic form

$$Y[x] = {}^t x Y x = \sum_{i,j=1}^{n} x_i y_{ij} x_j > 0, \quad \text{if } x \in \mathbb{R}^n, \; x \neq 0. \tag{1.2}$$

We shall always write vectors x in $\mathbb{R}^n$ as column vectors and so the transpose is ${}^t x = (x_1, \ldots, x_n)$.

© Springer Science+Business Media New York 2016
A. Terras, *Harmonic Analysis on Symmetric Spaces—Higher Rank Spaces, Positive Definite Matrix Space and Generalizations*, DOI 10.1007/978-1-4939-3408-9_1

The universe which we now enter is one of dimension at least 6 and this means that we will have trouble drawing meaningful pictures, not to mention keeping our calculations on small pieces of paper. The reader is advised to get some big sheets of paper to do some of the exercises.

Exercise 1.1.1. Some books define a real $n \times n$ matrix X to be positive if all of its entries are positive (see Berman and Plemmons [46], Minc [452], and Pullman [495]). Show that this concept is totally different from (1.2) in the sense that neither implies the other when $n > 1$, even when X is symmetric.

We can lower the dimension of our space by one (and make the Lie group involved simple) by looking at the symmetric space of **special positive matrices (the determinant one surface in $\mathcal{P}_n$):**

$$\mathcal{SP}_n = \{Y \in \mathcal{P}_n \mid |Y| = \det Y = 1\} . \tag{1.3}$$

Recall that we can identify $\mathcal{SP}_2$ with the Poincaré upper half plane H (see Exercise 3.1.9 on p. 154 of Vol. I).

Our goal in this chapter is to extend to $\mathcal{P}_n$ as many of the results of Chapter 3, Vol. I, as possible. For example, we will study analogues of our favorite special functions—gamma, K-Bessel, and spherical. The last two functions will be eigenfunctions of the Laplacian on $\mathcal{P}_n$. They will display a bit more complicated structure than the functions we saw in Vol. I, but they and related functions have had many applications. For example, James [328–330] and others have used zonal polynomials and hypergeometric functions of matrix argument to good avail in multivariate statistics (see also Muirhead [468]).

There are, in fact, many applications of analysis on $\mathcal{P}_n$ in **multivariate statistics**, which is concerned with data on several aspects of the same individual or entity; e.g., reaction times of one subject to several stimuli. Multivariate statistics originated in the early part of the last century with Fisher and Pearson. In Section 1.1 we will consider a very simple application of one of our coordinate systems for $\mathcal{P}_n$ in the study of partial correlations, with an example from agriculture. Limit theorems for products of random matrices are also of interest; e.g., in demography (see Cohen [114]). One can say something about this subject as well, by making use of harmonic analysis on $\mathcal{P}_n$.

We will see that $\mathcal{P}_n$ is a **symmetric space**; i.e., a Riemannian manifold with a geodesic-reversing isometry at each point. Moreover it is a homogeneous space of the Lie group $GL(n, \mathbb{R})$—the **general linear group** of all $n \times n$ non-singular real matrices. The definition of Lie group is given around formula (2.1) of Chapter 2. By **homogeneous space**, we essentially refer to the identification of $\mathcal{P}_n$ with the quotient $K \backslash G$ in Exercise 1.1.5 below. Because $\mathcal{P}_n$ is a symmetric space, harmonic analysis on $\mathcal{P}_n$ will be rather similar to that on the spaces considered in Volume I—$\mathbb{R}^n$, S^2, and H. For example, the ring of G-invariant differential operators on $\mathcal{P}_n$ is easily shown to be commutative from the existence of a geodesic-reversing isometry. There are four main types of symmetric spaces: compact Lie groups, quotients of compact Lie groups, quotients of noncompact semisimple

Lie groups, and Euclidean spaces. E. Cartan classified these types further (see Helgason [273, 275, 278, 282]). We are mainly considering a special example of the third type here (if we restrict to the determinant 1 surface (1.3)). In the next chapter we will look at the more general theory.

Work of Harish-Chandra and Helgason allows us to extend the results of Section 3.2 of Volume I to $\mathcal{P}_n$, by obtaining an analogue of the Fourier or Mellin transform on $\mathcal{P}_n$ (see Section 1.3 or Helgason [275]). Of course this result is more complicated than those discussed in Volume I and the available tables of this transform are indeed very short. Still we will be able to generalize some of our earlier applications.

We will also study the fundamental domain in $\mathcal{P}_n$ for our favorite **modular group** $GL(n, \mathbb{Z})$, the discrete group of $n \times n$ matrices with integer entries and determinant $+1$ or -1. But our attempt to generalize Section 3.7 of Volume I and so obtain a theory of harmonic analysis on $\mathcal{P}_n/GL(n, \mathbb{Z})$ will not be anywhere near as satisfactory as it was for $n = 2$ in Chapter 3 of Volume I, although work of Langlands [392], Selberg [543], and others allows us to complete many of the foundational results.

One of the reasons for singling out the symmetric space $\mathcal{P}_n$ is that arguments for $\mathcal{P}_n$ can often be generalized to arbitrary symmetric spaces. One such example is the characterization of geodesics (see Chapter 2). This sort of approach is used by Mostow [465], for example.

There are also many applications in **physics**. Some are analogues of applications considered in Chapter 3 of Volume I. One can solve the heat and wave equations on $\mathcal{P}_n$ and consider central limit theorems. There are applications in quantum mechanics for systems of coherent states coming from higher rank symmetric spaces such as $\mathcal{P}_n$ (see Hurt [312, 313], Monastyrsky and Perelmonov [457]). It is possible to make a statistical study of eigenvalues of random symmetric matrices with implications for quantum mechanics (see Mehta [441] and Section 1.3.5). Recent work on quantum field theory involves function field analogues of some of the subjects that we will consider. See Frenkel [187] and Ooguri's interview [480] of Ed Witten.

Many of our applications will be in **number theory**. For example, in Section 1.4, we will begin with the work of Minkowski [453, Vol. II, pp. 53–100] on fundamental domains for $\mathcal{P}_n/GL(n, \mathbb{Z})$—work that is basic for algebraic number theory: e.g., in the discussions of the finiteness of the class number and Dirichlet's unit theorem (see Vol. I, Section 1.4). We will find that integrals over $\mathcal{P}_n/GL(n, \mathbb{Z})$ lead to some higher dimensional integral tests for the convergence of matrix series. We will also find that some of the integrals that we need for number-theoretic applications were computed independently by number theorists and statisticians in the early part of the last century. For example, one of Wishart's formulas will be very useful in applying our matrix integral test. Wishart was a statistician and his distribution is of central importance in multivariate statistics (see Section 1.2, Farrell [173, 174], and Muirhead [468]). It certainly came as a surprise to number theorists and physicists to learn that some of the integrals needed by physicists in their study of eigenvalues of random Hermitian matrices (see Mehta's 1st edition [441]) were

computed two decades before the physicists needed them in a Norwegian paper by
the number theorist Selberg [542]. Many number theorists are, in fact, interested
in such eigenvalue problems, thanks to possible connections with the Riemann
hypothesis (see Montgomery [458] and Hejhal [270]–as well as Section 1.3.5).

There are many sorts of fundamental domains for $GL(n, \mathbb{Z})$ besides that of
Minkowski. We will also consider the domain obtained by Grenier [241, 242]
using a generalization of the highest point method discussed for $n = 2$ in Vol.
I, Section 3.3. We discovered recently that this domain was also considered by
Hermite as well as Korkine and Zolotareff. The highest point method is similar
to that used by Siegel for the Siegel modular group $Sp(n, \mathbb{Z})$ (see Section 2.2,
Maass [426, p. 168], or Siegel [565]). The domain obtained by Grenier et al. has an
advantage over Minkowski's in that it has an exact box shape at infinity. This allows
one to use it to compute the integrals arising in the parabolic terms of the Selberg
trace formula, for example (see Vol. I, pp. 270 and 355 ff.). One would hope to be
able to generalize many of the other results in Volume I, Chapter 3, which depended
on knowing the explicit shape of the fundamental domain. There are many places to
look for applications of the study of the fundamental domain for $GL(n, \mathbb{Z})$:

(1) algorithms for class numbers and units in number fields,
(2) higher dimensional continued fraction algorithms,
(3) geometric interpretations of $\zeta(3)$,
(4) the study for $Y \in \mathcal{P}_n$ of the numbers $a_m(Y) = \#\{a \in \mathbb{Z}^n \mid Y[a] = {}^t aYa = m\}$,
    where m is a given integer. Here a is a column vector and ${}^t a$ denotes its
    transpose. In 1883, Smith and Minkowski won a prize for proving Eisenstein's
    formula on the subject. This work was greatly generalized by C.L. Siegel
    [565] and later reinterpreted adelically by Tamagawa and Weil [662, Vol. III,
    pp. 71–157].
(5) densest lattice packings of spheres in Euclidean space, Hilbert's 18th problem.
    This has applications in coding (see Section 1.4.1 or Sloane [568]).
(6) study of matrix analogues of the Riemann zeta function and corresponding
    automorphic forms generalizing the theta function. For example, we will
    investigate the following zeta function:

$$Z(s) = \sum_{A \in \mathbb{Z}^{n \times n} \text{rank } n / GL(n,\mathbb{Z})} |A|^{-s}, \quad \text{for} \quad s \in \mathbb{C} \quad \text{with} \quad \operatorname{Re} s > n, \quad (1.4)$$

where $|A| = $ determinant of A. Here the sum is over $n \times n$ integral matrices A
of rank n running through a complete set of representatives for the equivalence
relation

$$A \sim B \text{ iff } A = BU \quad \text{for some} \quad U \in GL(n, \mathbb{Z}).$$

This zeta function might be called the Dedekind zeta function attached to the
simple algebra $\mathbb{Q}^{n \times n}$. It turns out to have an analytic continuation and functional
equation and one of Riemann's original proofs can (with some difficulty) be

made to show this. But actually $Z(s)$ is a product of Riemann zeta functions and this leads to a formula for the volume of the fundamental domain for $GL(n, \mathbb{Z})$ involving a product of values of Riemann zeta functions at the integers from 2 to n.

Perhaps the main number theoretical application of harmonic analysis on $\mathcal{P}_n/GL(n, \mathbb{Z})$ is the study of L-functions corresponding to modular forms for $\Gamma = GL(n, \mathbb{Z})$ and analogues of Hecke's correspondence. By a **modular form** we mean a "nice" function on the symmetric space of interest with some invariance property under a modular group such as $GL(n, \mathbb{Z})$ or the Siegel modular group $Sp(n, \mathbb{Z})$. The modular form is thus a special case of an **automorphic form** with invariance under a general (perhaps non-arithmetic) discrete subgroup $\Gamma \subset G$ acting on the symmetric space G/K. Often, instead of calling these functions automorphic forms we will name them for one of the main mathematicians who studied them (e.g., Maass forms, Siegel modular forms). Or sometimes maybe we will call them modular forms, indicating that they are invariant under some kind of modular group like $GL(n, \mathbb{Z})$.

Results for L-functions corresponding to modular forms in the classical manner, using matrix Mellin transforms, can be found in Section 1.5 (see also Maass [419, 426], Goldfeld [230], Kaori Imai (Ota or also Ohta) [317], and Bump [83]). This sort of result goes back to Koecher [359], who studied matrix Mellin transforms (over $\mathcal{P}_n/GL(n, \mathbb{Z})$) of Siegel modular forms, but not enough complex variables were present to invert Koecher's transform. The L-function (1.4) above is an example of one of the functions considered in Koecher's theory. Kaori Imai Ota [317] and Weissauer [663] show how to obtain a converse theorem for such a transform. See Section 2.2. Solomon [570] as well as Bushnell and Reiner [93] obtain generalizations of such zeta functions and applications to the theory of algebras and combinatorics.

We will not discuss the adelic-representation theoretic view of Hecke theory on higher rank groups such as $GL(n)$—a theory which has been developed by Jacquet, Langlands, Piatetski-Shapiro, and others (see the Corvallis conference volume edited by Borel and Casselman [66]). Jacquet et al. [325] show that "roughly speaking, all infinite Euler products of degree 3 having a suitable analytic behavior are attached to automorphic representations of $GL(3)$." For $GL(n)$, $n \geq 4$, they must "twist" the Euler product by all automorphic cuspidal representations of $GL(j)$, $1 \leq j \leq n - 2$. The $GL(3)$ result is applied to attach automorphic representations to cubic number fields, in a similar way to that in which Maass [417] found non-holomorphic cusp forms for congruence subgroups of $SL(2, \mathbb{Z})$ (of level greater than one) corresponding to Hecke L-functions of real quadratic fields. The Langlands philosophy indicates that non-abelian Galois groups of number fields should have Artin L-functions coming from modular forms for matrix groups. More recent references are Bump [84], Goldfeld [230], and Goldfeld and Hundley [232].

One might also hope for higher rank analogues of the classical applications of harmonic analysis in number theory; e.g., in the proof of the prime number theorem (see, for example, Davenport [130], Grosswald [249], or Steven J. Miller and Ramin Takloo-Bighash [448]). There is not much in print in this direction. However, already Herz [293] had been motivated by the desire to understand the asymptotics of

$$\#\left\{T \in \mathbb{Z}^{k \times m} \mid A - {}^tTT \in \mathcal{P}_m\right\}, \quad \text{as } A \to \infty.$$

Herz hoped to use his theory of special functions on $\mathcal{P}_m$ to carry out this research, but he did not apparently manage to do this. We will be able to use the Poisson summation formula for $GL(n, \mathbb{Z})$ to study the related question of the asymptotics of

$$\#\left\{\gamma \in GL(n, \mathbb{Z}) \mid \text{Tr}\left({}^t\gamma\gamma\right) \leq x\right\}, \quad \text{as } x \to \infty,$$

where Tr denotes the trace of a matrix. Bartels [38] has obtained results on this question for very general discrete groups Γ acting on symmetric spaces. One goal of the theory developed in Sections 1.4 and 1.5 is the search for higher rank analogues of the work of Sarnak on units in real quadratic fields (see Vol. I, p. 363). Dorothy Wallace [642–650] gives results in this direction. One would also hope for higher degree analogues of results of Hejhal [271] on the distribution of solutions of quadratic congruences, results that were first proved by Hooley [305]. And there should be analogues of the work of Elstrodt et al. [165–168]. Bump [83], Bump and Goldfeld [89], Bump and Hoffstein [90], as well as Bump et al. [87], Goldfeld [230], and Goldfeld and Hundley [232] have generalized many classical results to $GL(n)$.

Warning to the Reader   This volume will be much more sketchy and demanding than the preceding volume [612]. The theory is still being developed. And it is not possible to say the final word. There are many research problems here. Comparison should be made with the work of Jorgenson and Lang [333] and [334] (who complain loudly about my "exercises"). See also Goldfeld [230].

I have chosen a point of view that I think best approximates that of Volume I, as well as that of Maass [424, 426] and Siegel [565, Vol. III, pp. 97–137]. Many experts might disagree with me here. And I'm sure that some would tell the reader to do everything adelically or in the language of group representations or both. I urge the reader to look at other references for a broader perspective on the subject.

I have sought to provide a large number of examples rather than general theorems. I ask the reader to consider special cases, look at examples using pencil and paper or even a computer. I have found that it is very helpful to use a computer to understand fundamental domains by drawing pictures of projections of them (see Section 1.4.3).

It is assumed that the reader has read about Haar or G-invariant measures on Lie groups G as well as G-invariant measures on quotient spaces G/H for closed subgroups H of G. Some references for this are Vol. I, Dieudonné [137], Helgason [275, 282], Lang [388], and Weil [659]. For example, Haar measure on the multiplicative abelian group

$$A = \{a \in \mathcal{P}_n \mid a \text{ is diagonal}\}$$

is given by

$$da = \prod_{j=1}^{n} a_j^{-1} da_j, \quad \text{if} \quad a = \begin{pmatrix} a_1 & 0 & \cdots & 0 \\ 0 & a_2 & \cdots & 0 \\ \vdots & \vdots & \ddots & \vdots \\ 0 & 0 & \cdots & a_n \end{pmatrix}. \tag{1.5}$$

Here the group operation on A is matrix multiplication and da_j is Lebesgue measure on the real line.

Throughout this book, we will use the following notation, for a square matrix A

$$\begin{aligned} |A| &= \det(A) = \text{determinant}(A) \quad \text{Tr}(A) = \text{trace}(A), \\ I_n &= n \times n \text{ identity matrix}, \quad {}^t A = \text{transpose}(A). \end{aligned} \tag{1.6}$$

Exercise 1.1.2 (A Nilpotent Lie Group). Define N to be the group of all upper triangular $n \times n$ real matrices with ones on the diagonal. Show that N forms a group under matrix multiplication and that this group is not abelian unless $n = 2$. Show that Haar measure on N is given by:

$$dn = \prod_{1 \le i < j \le n} dx_{ij} \quad \text{if} \quad n = \begin{pmatrix} 1 & x_{12} & x_{13} & \cdots & x_{1n} \\ 0 & 1 & x_{23} & \cdots & x_{2n} \\ 0 & 0 & 1 & \cdots & x_{3n} \\ \vdots & \vdots & \vdots & \cdots & \vdots \\ 0 & 0 & 0 & \ddots & x_{n-1,n} \\ 0 & 0 & 0 & \cdots & 1 \end{pmatrix}.$$

Here dx_{ij} is the Lebesgue measure on the real line. Show that right and left Haar measures on N are the same. Note that some people use the word **unipotent** rather than nilpotent for N.

Exercise 1.1.3. Show that the Haar measure on $G = GL(n, \mathbb{R})$ is given by

$$dg = \|g\|^{-n} \prod_{i,j} dg_{ij}.$$

Show that the right and left Haar measures are the same in this case.

From the theory of integrals on quotients G/H where H is a closed subgroup of the Lie group G, we will need an understanding of the formula:

$$\int_{g \in G} f(g)\, dg = \int_{gH = \bar{g} \in G/H} \left(\int_{h \in H} f(gh)\, dh \right) d\bar{g}. \qquad (1.7)$$

Here dg and dh are Haar measures on G, H, respectively. This formula really defines the G-invariant measure $d\bar{g}$ on the quotient space G/H. Such an integral is determined up to a positive constant. Formula (1.7) holds provided that, for example, both G and H are unimodular (i.e., left and right Haar measures are the same), and even more generally (see Helgason [282, p. 91]). If there is a **fundamental domain** S for H in G; i.e., a measurable subset S of G which can be identified with G/H, then the integral over G/H is the same as the integral over S and (1.7) is clear since G is a disjoint union of cosets sH over $s \in S$. It is formula (1.7) that leads to our integral test for the convergence of matrix series in Section 1.4.4. Helgason [282, pp. 139 ff.], uses this result to find some interesting formulas involving generalizations of the Radon transform. Weil [662, Vol. I, pp. 339–357] employed (1.7) to prove an integral formula of Siegel, which leads to the existence of dense lattice packings of spheres in Euclidean spaces of high dimensions—a result that we will discuss in Section 1.4.4.

It would also be useful for the reader to make herself or himself comfortable with calculus on manifolds (see Yvonne Choquet-Bruhat, Cécile CeWitt-Morette, and Margaret Dillard-Bleick [106], Dieudonné [137], or Lang [388], for example). We will need to be able to change variables in Laplace operators, in particular. Courant and Hilbert [125, p. 224] give an easy method to do this using the calculus of variations. Suppose that the old variables are $x \in \mathbb{R}^n$ and the new ones are $y \in \mathbb{R}^n$. Let the Jacobian matrix of the change of variables be

$$J = (\partial x_i / \partial y_j).$$

Then we have the following formulas for the arc length, volume, and Laplacian in the new coordinates:

$$
\left.
\begin{aligned}
ds^2 &= \sum_{j=1}^{n} dx_j^2 = \sum_{i,j=1}^{n} g_{ij} dy_i dy_j, \quad \text{where } G = {}^t J J, \\[2mm]
dv &= \prod_{j=1}^{n} dx_j = g^{1/2} \prod_{j=1}^{n} dy_j, \quad \text{if } g = |G| = \det G, \\[2mm]
\Delta &= g^{-1/2} \sum_{i=1}^{n} \frac{\partial}{\partial y_i} g^{1/2} \sum_{j=1}^{n} g^{ij} \frac{\partial}{\partial y_j}, \quad \text{for } G^{-1} = (g^{ij}).
\end{aligned}
\right\} \qquad (1.8)
$$

Exercise 1.1.4. Prove (1.8) using the calculus of variations, as in Courant and Hilbert [125, p. 224].

Hint. See formulas (2.2)–(2.6) in Vol. I. The calculus of variations says that if u minimizes the integral

$$F(u) = \int \sum_{i=1}^{n} u_{x_i}^2 \, dx,$$

subject to the constraint

$$K(u) = \int u^2 dx = \text{constant},$$

then u must satisfy the Euler-Lagrange equation:

$$\Delta u = \sum_{i=1}^{n} u_{x_i x_i} = \lambda u;$$

i.e., u is an eigenfunction for the Euclidean Laplacian.

Most of the rest of this section concerns the basic facts about $\mathcal{P}_n$. There are lots of exercises in advanced (or not so advanced) calculus on matrix space. We have tried to give extensive hints. The summary §1.1.6 includes tables of formulas which put these exercises into a nutshell. Many readers may want to skip to the summary, as these exercises can be time consuming. But we do this at our peril, of course.

1.1.2 Elementary Results

Now we begin the detailed consideration of the geometry and analysis of $\mathcal{P}_n$, the space of positive $n \times n$ real matrices (1.1). Other references for this section are Helgason [275–282], Hua [308], Maass [426], and Selberg [543].

Now $\mathcal{P}_n$ is the symmetric space of the **general linear group** $G = GL(n,\mathbb{R})$ of non-singular $n \times n$ real matrices. The **action** of $g \in GL(n, \mathbb{R})$ on $Y \in \mathcal{P}_n$ is given by

$$Y[g] = {}^t g Y g, \text{ where } {}^t g \text{ denotes the transpose of } g. \tag{1.9}$$

Let $K = O(n)$ be the **orthogonal group** of matrices $k \in G = GL(n, \mathbb{R})$ such that ${}^t k k = I$. Perhaps we should write $O(n, \mathbb{R})$ rather than $O(n)$, but we will not do this in this chapter as here the group will almost always consist of real matrices. If it is necessary, we will add the field to the notation.

Consider the map

$$K \backslash G \to \mathcal{P}_n$$

$$Kg \mapsto I[g]. \tag{1.10}$$

This map provides an identification of the two spaces as homogeneous spaces of G. By a **homogeneous space** X of G, we mean that X is a differentiable manifold with transitive differentiable **right G-action**

$$G \times X \to X$$

$$(g, x) \mapsto x \circ g$$

with the following two properties:

$$x \circ e = x, \text{ for } e = \text{ identity of } G \text{ and all } x \in X,$$
$$x \circ (gh) = (x \circ g) \circ h, \text{ for all } g, h \in G \text{ and } x \in X.$$

Exercise 1.1.5. Prove that the map (1.10) identifies the two spaces as homogeneous spaces of $G = GL(n, \mathbb{R})$. Make sure that you show that the mapping preserves the group actions. The action of $g \in G$ on a coset $Kh \in K \backslash G$ is $\sigma_g (Kh) = Khg$.
Hint. You need to use the general result which says that, having fixed a point p in a homogeneous space X of the group G, then, defining the **isotropy group** $G_p = \{g \in G \mid p \circ g = p\}$, it follows that the map

$$G_p \backslash G \to X$$

$$G_p g \mapsto p \circ g$$

is a diffeomorphism onto. There is a proof in Broecker and tom Dieck [81, p. 35].
    In order to see that the map (1.10) is onto, you can use the spectral theorem for positive matrices (Exercise 1.1.7 below).

Exercise 1.1.6 (Criteria for the Positivity of a Real Symmetric Matrix). Show that if $Y = {}^tY$ is a real $n \times n$ matrix, then Y is positive definite (as in (1.2)) if and only if one of the following equivalent conditions holds:

(a) All of the eigenvalues of Y are positive.
(b) If

$$Y = \begin{pmatrix} Y_k & * \\ * & * \end{pmatrix}, \text{ with } Y \in \mathbb{R}^{k \times k},$$

then $|Y_k| > 0$, for all $k = 1, 2, \ldots, n$.

Hint. It is helpful to use the partial Iwasawa decomposition discussed in Exercise 1.1.11 below.

Exercise 1.1.7 (Spectral Theorem for a Symmetric Matrix). Suppose $Y = {}^tY$ is an $n \times n$ real matrix. Then there is an orthogonal matrix k in $O(n)$ such that $Y = {}^tkDk = D[k]$, where D is the diagonal matrix

$$D = \begin{pmatrix} d_1 & \cdots & 0 \\ \vdots & \ddots & \vdots \\ 0 & \cdots & d_n \end{pmatrix}$$

and d_j is real and equal to the jth eigenvalue of Y. We can characterize the d_j by a sequence of maximum problems:

$$d_j = \max \left\{ {}^t xYx \ \middle| \ x \in \mathbb{R}^n, \ \|x\| = \sqrt{{}^t xx} = 1, \ {}^t xx_i = 0, \ i = 1, \ldots, j-1 \right\} = {}^t x_j Yx_j,$$

with $Yx_j = d_j x_j$. Here we maximize over x in $\mathbb{R}^n$ of length 1 and orthogonal to all the preceding $j - 1$ eigenvectors of Y. Then $k = (x_1 \ x_2 \ \cdots \ x_n)$, i.e., the matrix whose j^{th} column is x_j.

Hint. See Courant and Hilbert [125, Chapter 1].

There are many familiar applications of Exercise 1.1.7; e.g., to the discussion of normal modes of vibrating systems (see Courant and Hilbert [125]). Exercise 1.1.7 also leads to a numerical method for the solution of partial differential equations known as the Rayleigh-Ritz method (see Arfken [20, pp. 800–803]), a method which has developed into the modern finite element method (see Strang and Fix [582]).

Exercise 1.1.8 (Mini-Max Principle for Eigenvalues of $Y = {}^t Y$).

(a) Let Y and d_j be as in Exercise 1.1.7. Show that

$$d_j = \min_{w_1, \ldots, w_{j-1} \in \mathbb{R}^n} \left(\max_{\substack{x \in \mathbb{R}^n, \ \|x\|=1 \\ {}^t x \cdot w_i = 0, \ \forall \ i = 1, \ldots, j-1}} {}^t xYx \right).$$

(b) As an application of part (a) show that if

$$Y = \begin{pmatrix} Y_1 & * \\ * & * \end{pmatrix}$$

where Y is as in Exercise 1.1.7 and Y_1 is a $k \times k$ symmetric matrix, then

(jth eigenvalue of Y_1) $\leq$ (jth eigenvalue of Y) $\leq$ ((j $-$ 1)st eigenvalue of Y_1).

The mini-max principle can be generalized greatly and leads to many qualitative results about solutions of boundary value problems of mathematical physics. For example, it can be used to compare fundamental tones for vibrating strings. The string of larger density will have the lower fundamental tones (see Courant and Hilbert [125, Ch. VI]).

Exercise 1.1.9. Given two $n \times n$ real symmetric matrices P and Y with $P \in \mathcal{P}_n$, show that there is a matrix g in $GL(n, \mathbb{R})$ such that both $Y[g]$ and $P[g]$ are diagonal.
Hint. Use Exercise 1.1.7.

The space of symmetric real $n \times n$ matrices is easily seen to be identifiable with $\mathbb{R}^m$, where $m = n(n+1)/2$. And it follows from Exercise 1.1.9 that $\mathcal{P}_n$ is an **open cone** in $\mathbb{R}^m$, $m = n(n+1)/2$; i.e., $X, Y \in \mathcal{P}_n$ and $t > 0$ implies that $X + Y$ and tX are also in $\mathcal{P}_n$.

1.1.3   Geodesics and Arc Length

In order to view the space $\mathcal{P}_n$ as an analogue of the Poincaré upper half plane which was studied in Chapter 3 of Volume I, we need a notion of arc length. This will turn the space into a **Riemannian manifold**. Now $\mathcal{P}_n$ is an open set in $n(n+1)/2$-dimensional Euclidean space as can be seen from Exercise 1.1.6 in the last subsection. Thus it is indeed a differentiable manifold of the easiest sort to consider. The choice of coordinates is not a problem. To define the Riemannian structure, we need to define an arc length element (thinking as physicists). Differential geometers would consider a bilinear form on the tangent space or a continuous 2-covariant tensor field, or whatever. We want to take the point of view of a student of advanced calculus, however. But duty impels us to note that Lie groups people would say that it all comes from the Killing form on the Lie algebra. We will consider that approach in Chapter 2.

Well, anyway, suppose we just define the **arc length** element ds by the formula:

$$ds^2 = \text{Tr}\,((Y^{-1}dY)^2), \quad \text{where } Y = (y_{ij})_{1 \leq i,j \leq n}$$
$$\text{and } dY = (dy_{ij})_{1 \leq i,j \leq n}, \ \text{Tr} = \text{Trace}. \tag{1.11}$$

It is easy to show that ds is invariant under the group action of $g \in GL(n, \mathbb{R})$ on $Y \in \mathcal{P}_n$ given by (1.9). For $W = Y[g]$ is a linear function of Y. Recall from advanced calculus that a linear map is its own differential (see for example Lang [388, Chapter 5]). Therefore $dW = dY[g]$ and it follows that, upon plugging into the definition of ds, we obtain:

$$ds^2 = \text{Tr}\left((Y^{-1}dY)^2\right) = \text{Tr}\left(\left(g\left(W^{-1}\right){}^t g\left({}^t g^{-1}\right) dW\, g^{-1}\right)^2\right) = \text{Tr}\left((W^{-1}dW)^2\right).$$

Thus we have shown that our arc length on $\mathcal{P}_n$ behaves like that on the Poincaré upper half plane as far as group invariance properties go. Moreover, the arc length element is positive definite, since it is positive definite at $Y = I$ and the action of $GL(n, \mathbb{R})$ on $\mathcal{P}_n$ is transitive (see Exercise 1.1.5).

Recall that in the case of the Poincaré upper half plane H, the geodesics through the point gi, for g in $SL(2, \mathbb{R})$, have the form

$$g \begin{pmatrix} \exp(at) & 0 \\ 0 & \exp(-at) \end{pmatrix} i, \quad \text{for } t \in \mathbb{R},$$

where the action of the 2×2 matrix on i is by fractional linear transformation (see Volume I, page 151). Here a is any real number. Thus it will not come as a surprise to you that in the case of $\mathcal{P}_n$ the geodesics through the point $I[g]$ for $g \in GL(n, \mathbb{R})$ are given by:

$$
\begin{pmatrix}
\exp(a_1 t) & \cdots & 0 \\
\vdots & \ddots & \vdots \\
0 & \cdots & \exp(a_n t)
\end{pmatrix} [g], \text{ for } t \in \mathbb{R},
$$

where the numbers a_i are arbitrary real numbers. It may not surprise you either if we reveal that the method of proof of this fact (Theorem 1.1.1 below) is just a rather straightforward generalization of the proof that works in the Euclidean case and in the upper half plane case. The only problem will be the computation of a rather nasty Jacobian. Mercifully Hans Maass [426] computes the Jacobian for us.

It is perhaps more convenient to view a geodesic through the identity in $\mathcal{P}_n$ as the **matrix exponential**:

$$
\exp(tX) = \sum_{n \geq 0} \frac{(tX)^n}{n!}, \text{ for } t \in \mathbb{R}, \ X \in \mathbb{R}^{n \times n}, \ {}^t X = X. \tag{1.12}
$$

The following exercise reviews the basic properties of exp. There is an analogue for any Riemannian manifold (see Helgason [273, 278]). In Chapter 2 we will consider the analogue for Lie groups G.

Exercise 1.1.10 (The Matrix Exponential).

(a) Show that the matrix power series below converges absolutely for all $A \in \mathbb{R}^{n \times n}$

$$
\exp(A) = \sum_{n \geq 0} \frac{A^n}{n!}.
$$

(b) Show that if $AB = BA$, for A, B in $\mathbb{R}^{n \times n}$, then

$$
\exp(A + B) = \exp(A) \exp(B).
$$

Show that this equality may fail if $AB \neq BA$.

(c) Show that if U is in $GL(n, \mathbb{R})$, then $\exp(U^{-1}AU) = U^{-1} \exp(A)U$.

(d) Using the notation (1.6), show that

$$
|\exp(A)| = \exp(\mathrm{Tr}(A)).
$$

Hint. See Lang [385, pp. 295–296].

There are applications of the matrix exponential to the solution of ordinary differential equations (see Apostol [17, Vol. II, pp. 213–214]).

In order to prove the theorem characterizing the geodesics in $\mathcal{P}_n$, we need to have an analogue of rectangular coordinates in the Poincaré upper half plane. This analogue is what I call **partial Iwasawa coordinates** defined as follows for $Y \in \mathcal{P}_n$, with the notation (1.9):

$$Y = \begin{pmatrix} V & 0 \\ 0 & W \end{pmatrix} \begin{bmatrix} I_p & 0 \\ X & I_q \end{bmatrix}, \text{ for } V \in \mathcal{P}_p, \ W \in \mathcal{P}_q, \ X \in \mathbb{R}^{q \times p}, \ n = p + q. \quad (1.13)$$

Here I_p denotes the $p \times p$ identity matrix. When $n = 2$, Exercise 3.1.9 of Volume I shows that these coordinates (or those from Exercise 1.1.11(b) below) correspond to rectangular coordinates in the Poincaré upper half plane. There are as many versions of these partial Iwasawa coordinates as there are partitions of n into two parts. We feel impelled to say that somehow all these different coordinate systems measure ways of approaching the boundary of the symmetric space. Thus if we manage to try to do the analysis of behavior of functions like Eisenstein series for $GL(n)$, we will have to make serious use of these coordinates.

Partial Iwasawa coordinates generalize the well-known technique from high school algebra known as "completing the square," since if $a \in \mathbb{R}^p$ and $b \in \mathbb{R}^q$,

$$Y \begin{bmatrix} a \\ b \end{bmatrix} = V[a] + W[Xa + b],$$

which is indeed a sum of squares if $n = 2$ and $p = q = 1$.

At the end of this section, we will see an application of partial Iwasawa coordinates in multivariate statistics—with results going back to Pearson in 1896 and Yule in 1897 (see also T.W. Anderson [8, p. 27]).

There is also another type of partial Iwasawa coordinates, as the next exercise shows.

Exercise 1.1.11 (Partial Iwasawa Coordinates).

(a) Show that the equation (1.13) can always be solved uniquely for V, W, X, once Y in $\mathcal{P}_n$ is given, along with $p, q \in \mathbb{Z}^+$, such that $n = p + q$.

(b) Obtain a second set of partial Iwasawa coordinates, using the equality:

$$Y = \begin{pmatrix} F & 0 \\ 0 & G \end{pmatrix} \begin{bmatrix} I_p & H \\ 0 & I_q \end{bmatrix}, \text{ with } F \in \mathcal{P}_p, \ G \in \mathcal{P}_q, \ H \in \mathbb{R}^{p \times q}, \ n = p + q.$$

We have seen that there are many sorts of partial Iwasawa coordinates. In fact, you can continue the idea to decompose Y into an analogous product, with any number (between 2 and n) of block matrices along the diagonal. This gives coordinate systems associated with any partition of n. Such coordinate systems will be needed in our study of Eisenstein series in Section 1.3. But let's restrain ourselves since we do not need these more general decompositions for our goal of characterizing the geodesics in $\mathcal{P}_n$. However, later in this section, we will need the full Iwasawa decomposition corresponding to the partition of n given by

$$n = 1 + \cdots + 1$$

—a decomposition which is studied in the following exercise.

Exercise 1.1.12 (The Full Iwasawa Decomposition). Show that repetition of the partial Iwasawa decomposition in Exercise 1.1.11(b) leads to the full Iwasawa decomposition: $Y = a[n]$, where the matrix a is positive diagonal and the matrix n is upper triangular with 1's on the diagonal.[1]

The coordinate system of Exercise 1.1.12 corresponds to the full Iwasawa decomposition on the group $G = GL(n, \mathbb{R})$ (see Helgason [275–278, 282], and Iwasawa [322] as well as Chapter 2). Siegel [562, p. 29] calls the full Iwasawa decomposition the "Jacobi transformation." Weil [658, p. 7] calls it "Babylonian reduction." Numerical analysts call it the "LDU factorization" and obtain it by Gaussian elimination (see Strang [581]). The Gram-Schmidt orthogonalization process can also be used to derive the full Iwasawa decomposition.

Exercise 1.1.13. Suppose that Y_k is the upper left-hand $k \times k$ corner of the matrix Y. For

$$1 \le i_1 \le \cdots \le i_h \le n \text{ and } 1 \le j_1 \le \cdots \le j_h \le n,$$

set

$$Y(i_1, \ldots, i_h \mid j_1, \ldots, j_h) = \det(Y_{i_a j_b})_{1 \le a,b \le h}.$$

Then show that the matrices a and n from Exercise 1.1.12 are given by:

$$a = \begin{pmatrix} a_1 & \cdots & 0 \\ \vdots & \ddots & \vdots \\ 0 & \cdots & a_n \end{pmatrix}, \quad n = \begin{pmatrix} 1 & \cdots & x_{ij} \\ \vdots & \ddots & \vdots \\ 0 & \cdots & 1 \end{pmatrix}, \quad \text{with}$$

$a_i = |Y_i|/|Y_{i-1}|$ and $x_{ij} = Y(1, \ldots, i-1, i | 1, \ldots, i-1, j)/|Y_i|$.
Hint. (See Minkowski [453, Vol. II, p. 55] and Hancock [260, Vol. II, pp. 536–539]). Note that if we choose ${}^t u = (u_1, \ldots, u_i, 0, 0, \ldots, 0, 1, 0, \ldots, 0)$ with the one in the jth place, then we can minimize $Y[u]$ over such vectors u. If $Y = a[n]$, then

$$Y[u] = a_1(u_1 + x_{12}u_2 + \cdots + x_{1n}u_n)^2 + a_2(u_2 + x_{23}u_3 + \cdots + x_{2n}u_n)^2 + \ldots + a_n u_n^2,$$

and thus to minimize $Y[u]$, we need

$$u_i + x_{ij} = 0,$$

---

[1] Hopefully it will not be too confusing to use n for matrix size as well as a matrix in the nilpotent group N.

$$2Yu = \text{grad } Y[u] = \left(\frac{\partial Y[u]}{\partial u_1}, \dots, \frac{\partial Y[u]}{\partial u_i} \right) = 0.$$

Cramer's rule completes the proof of the formula for x_{ij}. The formula for a_i is easily proved.

Now, in order to obtain the geodesics on $\mathcal{P}_n$, we must express the **arc length element in partial Iwasawa coordinates** (1.13):

$$ds_Y^2 = ds_V^2 + ds_W^2 + 2 \text{ Tr} \left(V^{-1} \left({}^t dX \right) W \, dX \right), \tag{1.14}$$

where ds_V is the element of arc length in $\mathcal{P}_p$, ds_W is the analogue for $W \in \mathcal{P}_q$, and dX is the matrix of differentials dx_{ij} if $X = (x_{ij}) \in \mathbb{R}^{q \times p}$.

To prove (1.14) write:

$$Y = \begin{pmatrix} V & 0 \\ 0 & W \end{pmatrix} \begin{bmatrix} I_p & 0 \\ X & I_q \end{bmatrix} = \begin{pmatrix} V + W[X] & {}^t X W \\ W X & W \end{pmatrix}.$$

Then

$$Y^{-1} = \begin{pmatrix} V^{-1} & -V^{-1} \left({}^t X \right) \\ -X V^{-1} & V^{-1}[{}^t X] + W^{-1} \end{pmatrix}$$

and

$$dY = \begin{pmatrix} dV & 0 \\ 0 & dW \end{pmatrix} \begin{bmatrix} I_p & 0 \\ X & I_q \end{bmatrix} + \begin{pmatrix} 0 & {}^t dX \\ 0 & 0 \end{pmatrix} \begin{pmatrix} V & 0 \\ 0 & W \end{pmatrix} \begin{pmatrix} I_p & 0 \\ X & I_q \end{pmatrix}$$

$$+ \begin{pmatrix} I_p & {}^t X \\ 0 & I_q \end{pmatrix} \begin{pmatrix} V & 0 \\ 0 & W \end{pmatrix} \begin{pmatrix} 0 & 0 \\ dX & 0 \end{pmatrix}.$$

This allows one to compute

$$dY \, Y^{-1} = \begin{pmatrix} L_0 & L_1 \\ L_2 & L_3 \end{pmatrix}.$$

We might wish for any symbol manipulation language and a nice friendly computer to help us out here.

Exercise 1.1.14 (Proof of Formula (1.14)).

(a) Prove that

$$L_0 = dV \, V^{-1} + {}^t X \, W \, dX \, V^{-1},$$

$$L_1 = -dV \, V^{-1} \left({}^t X \right) - {}^t X \, W \, dX \, V^{-1} \left({}^t X \right) + {}^t X \, dW \, W^{-1} + {}^t dX,$$

$$L_2 = W \, dX \, V^{-1},$$

$$L_3 = -W \, dX \, V^{-1} \left({}^t X \right) + dW \, W^{-1}.$$

(b) Use part (a) to show that

$$ds^2 = \mathrm{Tr}\left(L_0^2 + L_1 L_2\right) + \mathrm{Tr}\left(L_2 L_1 + L_3^2\right)$$
$$= \mathrm{Tr}\left((V^{-1}\,dV)^2\right) + \mathrm{Tr}\left((W^{-1}\,dW)^2\right) + 2\mathrm{Tr}\left(V^{-1}\,(^t dX)\,W\,dX\right).$$

(c) Find ds^2 for the other partial Iwasawa decomposition given in Exercise 1.1.11 (b).

Answer. $ds_Y^2 = \mathrm{Tr}\left((F^{-1}\,dF)^2\right) + \mathrm{Tr}\left((G^{-1}\,dG)^2\right) + 2\,\mathrm{Tr}\left(dH\,G^{-1}\,(^t dH)\,F\right)$, for

$$Y = \begin{pmatrix} F & 0 \\ 0 & G \end{pmatrix} \begin{bmatrix} I & H \\ 0 & I \end{bmatrix}.$$

Exercise 1.1.15. (a) Show that if $V = {}^t AA$, $W = {}^t BB$, $H = B\,dX\,A^{-1}$, for $A \in GL(p, \mathbb{R})$, $B \in GL(q, \mathbb{R})$, then

$$\mathrm{Tr}\left(V^{-1}\,(^t dX)\,W\,dX\right) = \mathrm{Tr}\left(H\,(^t H)\right) \geq 0.$$

(b) Conclude from Exercise 1.1.14 and part (a) that when seeking a curve $T(t) \in \mathcal{P}_n$, such that $T(0) = I$, $T(1) = D$, a diagonal matrix, the arc length will only be decreased by taking the X-coordinate in the partial Iwasawa decomposition of $T(t)$ in (1.13) to be zero.

(c) Use induction to conclude from part (b) that we only decrease the arc length by taking $T(t)$ to be diagonal for all $t \in [0, 1]$.

Theorem 1.1.1 (Geodesics in $\mathcal{P}_n$). *A geodesic segment $T(t)$ through I and Y in $\mathcal{P}_n$ has the form:*

$$T(t) = \exp(t\,A[U]), \quad 0 \leq t \leq 1,$$

where Y has the spectral decomposition:

$$Y = (\exp A)[U] = \exp(A[U]), \quad \text{for} \quad U \in O(n)$$

and

$$A = \begin{pmatrix} a_1 & 0 & \cdots & 0 \\ 0 & a_2 & \cdots & 0 \\ \vdots & \vdots & \ddots & \vdots \\ 0 & 0 & \cdots & a_n \end{pmatrix}, \quad a_j \in \mathbb{R}, \quad j = 1, 2, \ldots, n.$$

The length of the geodesic segment is:

$$\left(\sum_{j=1}^{n} a_j^2 \right)^{1/2}.$$

Proof. Use Exercise 1.1.7 to write $Y = D[U]$ with U in the orthogonal group $O(n)$ and

$$D = \begin{pmatrix} d_1 & 0 & \cdots & 0 \\ 0 & d_2 & \cdots & 0 \\ \vdots & \vdots & \ddots & \vdots \\ 0 & 0 & \cdots & d_n \end{pmatrix}, \quad d_j > 0, \ j = 1, 2, \ldots, n.$$

Then $D = \exp A$ with $d_j = \exp(a_j)$, $j = 1, \ldots, n$. From Exercise 1.1.10 we know that $Y = (\exp A)[U] = \exp(A[U])$.

Exercise 1.1.15 tells us that a geodesic $T(t)$ in $\mathcal{P}_n$ such that $T(0) = I$ and $T(1) = Y[U]$ has the form:

$$T(t) = \begin{pmatrix} d_1(t) & 0 & \cdots & 0 \\ 0 & d_2(t) & \cdots & 0 \\ \vdots & \vdots & \ddots & \vdots \\ 0 & 0 & \cdots & d_n(t) \end{pmatrix}[U], \quad \begin{array}{c} d_j(0) = 1, \quad d_j(1) = d_j, \\ j = 1, \ldots, n, \end{array}$$

where $d_j(t) = \exp(a_j(t))$. This is possible, since all the d_j's are positive. Using the definition (1.11) of arc length in $\mathcal{P}_n$, it is easily seen that the arc length of the curve $T(t)$ is:

$$\int_0^1 \left(\sum_{j=1}^{n} {a_j'}^2 \right)^{1/2} dt.$$

The curve $a(t) \in \mathbb{R}^n$ such that $a(0) = 0$ and $a(1) = (a_1, \ldots, a_n)$ which minimizes this distance is the straight line:

$$a_j(t) = t a_j, \ j = 1, \ldots, n.$$

This completes the proof of Theorem 1.1.1.   ∎

Exercise 1.1.16. Use Exercise 1.1.9 and Theorem 1.1.1 to find the geodesics through two arbitrary points in $\mathcal{P}_n$.

In Chapter 2 we will generalize Theorem 1.1.1 to an arbitrary noncompact symmetric space. For example, define the **symplectic group** $Sp(n, \mathbb{R})$ of all $(2n) \times (2n)$ real matrices g such that

$$^t gJg = J, \qquad \text{if} \quad J = \begin{pmatrix} 0 & I_n \\ -I_n & 0 \end{pmatrix}. \tag{1.15}$$

Then we can consider the **symmetric space** attached to $Sp(n, \mathbb{R})$ to be the subset of $\mathcal{P}_{2n}$ consisting of all positive $(2n) \times (2n)$ symplectic matrices.[2] In fact, this subset of $\mathcal{P}_{2n}$ will be seen to be a **totally geodesic submanifold** of $\mathcal{P}_{2n}$; i.e., the geodesics in the larger space which join two points in the smaller space must actually lie entirely in the smaller space.

Exercise 1.1.17 (The Determinant one Surface).

(a) As in Exercise 1.1.5, show that the determinant one surface $\mathcal{SP}_n$, defined in (1.3), can be identified with $K\backslash G$, for $G = SL(n, \mathbb{R})$, $K = SO(n)$. Here $SL(n, \mathbb{R})$ denotes the **special linear group** of all determinant one matrices in $GL(n, \mathbb{R})$ and $SO(n)$ denotes the special orthogonal group which is defined analogously as the determinant one matrices in $O(n)$.
(b) Show that $\mathcal{SP}_n$ is a totally geodesic submanifold of $\mathcal{P}_n$.

Hint. You can relate the geometric structures on $\mathcal{P}_n$ and $\mathcal{SP}_n$ using formula (1.8) and the substitution

$$Y = tW, \text{ for } t > 0 \text{ and } W \in \mathcal{SP}_n.$$

The group $SL(n, \mathbb{R})$ is a **simple** Lie group and thus certainly semisimple—a standard hypothesis for many theorems in the theory of group representations and harmonic analysis (see the next chapter). But $GL(n, \mathbb{R})$ has a center consisting of nonzero scalar matrices and is thus not semisimple. $GL(n, \mathbb{R})$ is only **reductive**. But this should not make us abandon $GL(n)$ for its "simple" relative. For many theorems about simple and semisimple Lie groups are proved by first settling the case of $GL(n, \mathbb{R})$ and then using the Adjoint representation. For examples of this phenomenon, see the next chapter and Helgason [273, pp. 234–237]. Moreover, it is often true that calculations are easier on $GL(n, \mathbb{R})$ than $SL(n, \mathbb{R})$. Thus we will usually consider the general rather than special linear group.

Exercise 1.1.18 (A Geodesic-Reversing Isometry). Show that the map $\sigma_I(Y) = Y^{-1}$ fixes the point I, preserves arc length, and reverses geodesics through the point I.
Hint. You could use the fact that $W = Y^{-1}$ implies that $\quad Y\, dW + dY\, W = 0$.

---

[2] Some authors write $Sp(2n, \mathbb{R})$ instead of $Sp(n, \mathbb{R})$. Confusing!

Note that one can easily find a geodesic-reversing isometry of $\mathcal{P}_n$ at each point Y in $\mathcal{P}_n$ by translating the result of Exercise 1.1.18 using the action of $GL(n, \mathbb{R})$. It is the existence of these **geodesic-reversing isometries** that turns the Riemannian manifold $\mathcal{P}_n$ into a **symmetric space**. Such spaces were classified by E. Cartan in the 1920s (see Cartan [94, 95], Helgason [275–278], and Chapter 2). The geodesic-reversing isometry is used, for example, in the proof that the G-invariant differential operators on $\mathcal{P}_n$ form a commutative algebra (see part 2 of Lemma 1.1.2 in Section 1.1.5). Selberg [543, p. 51] notes that for this commutativity, it would suffice to assume that the Riemannian manifold S is only a **weakly symmetric space**. That is, S is assumed to have a group G of isometries such that G is locally compact and transitive on S and having a fixed isometry μ (which may not lie in G) such that $\mu G \mu^{-1} = G$, and $\mu^2 \in G$. And for S to be weakly symmetric it is assumed that for every pair $x, y \in S$, there is an element $m \in G$ such that $mx = \mu y$ and $my = \mu x$. Ernest Vinberg has proved that it is not necessary to assume $\mu^2 \in G$. Selberg showed that weakly symmetric spaces give rise to Gelfand pairs. A group G with a subgroup K is called a **Gelfand pair** (G, K) if $L^1 (K \backslash G / K)$ is a commutative algebra under convolution on G.

The **locally symmetric spaces** are quotients of a symmetric space G/K with a discrete group $\Gamma \subset G$; thus of the form $K \backslash G / \Gamma$.

1.1.4   Measure and Integration on $\mathcal{P}_n$

Now that we have completed our study of geodesics on $\mathcal{P}_n$, it is time to consider the $GL(n, \mathbb{R})$-invariant volume element and integration on $\mathcal{P}_n$. For $Y = (y_{ij}) \in \mathcal{P}_n$, let $dy_{ij} =$ Lebesgue measure on $\mathbb{R}$. Then we define the $GL(n, \mathbb{R})$-**invariant volume element** $d\mu$ on $\mathcal{P}_n$ by:

$$d\mu = d\mu_n(Y) = |Y|^{-(n+1)/2} \prod_{1 \leq i \leq j \leq n} dy_{ij}. \qquad (1.16)$$

One can show that $d\mu_n$ in (1.16) is the volume element associated by (1.8) to the Riemannian metric defined by (1.11).

To prove that the **volume element given in formula (1.16) is invariant under $GL(n, \mathbb{R})$**, suppose that $g \in GL(n, \mathbb{R})$ and $W = Y[g]$. We follow Maass [426, p. 23]. The **Jacobian** of the mapping $Y \mapsto W$ is:

$$J(g) = \|dW/dY\| = \|g\|^{n+1}, \text{ for } \|g\| = \text{absolute value of } \det(g). \qquad (1.17)$$

To prove this, note that $J(g)$ is multiplicative; i.e.,

$$J(gh) = J(g)J(h), \quad \text{for all } g, \ h \in G.$$

And $J(g)$ is clearly a polynomial in the entries of g. So we can assume that g lies in the set of matrices whose eigenvalues are pairwise distinct. Then g has the form

$U^{-1}DU$, with D diagonal. Suppose that D has j^{th} diagonal entry d_j. Then $Y[D] = (d_i y_{ij} d_j)$, which implies that

$$J(U^{-1}DU) = J(D) = \prod_{1 \le i \le j \le n} |d_i d_j| = \|D\|^{n+1} = \|U^{-1}DU\|^{n+1}.$$

This completes the proof of formula (1.17). One quickly deduces from (1.17) the fact that the volume element in (1.16) is invariant under the action of $GL(n, \mathbb{R})$.

In the following Exercises we compute the Jacobian for various coordinate systems that we call partial Iwasawa and full Iwasawa coordinates.

Exercise 1.1.19 (The Jacobian for Partial Iwasawa Coordinates).

(a) Show that if we have the partial Iwasawa coordinate decomposition defined by:

$$Y = \begin{pmatrix} V & 0 \\ 0 & W \end{pmatrix} \begin{bmatrix} I_p & 0 \\ X & I_q \end{bmatrix}, \quad V \in \mathcal{P}_p, \ W \in \mathcal{P}_q, \ X \in \mathbb{R}^{q \times p}, \ n = p + q,$$

then

$$\left| \frac{\partial Y}{\partial (V, W, X)} \right| = |W|^p.$$

(b) If instead, we write an alternate partial Iwasawa decomposition:

$$Y = \begin{pmatrix} F & 0 \\ 0 & G \end{pmatrix} \begin{bmatrix} I_p & H \\ 0 & I_q \end{bmatrix}, \quad F \in \mathcal{P}_p, \ G \in \mathcal{P}_q, \ H \in \mathbb{R}^{p \times q}, \ n = p + q,$$

then

$$\left| \frac{\partial Y}{\partial (F, G, H)} \right| = |F|^q.$$

Hint. (a) Set

$$Y = \begin{pmatrix} A & B \\ {}^t B & C \end{pmatrix}, \quad \text{with } {}^t B = WX,$$

and note that $|\partial(WX)/\partial X| = |W|^p$—you get one $|W|$ for each column of X.

We can use Exercise 1.1.19 to obtain **the relation between invariant volumes in partial Iwasawa coordinates** (1.13):

$$d\mu_n(Y) = |V|^{-q/2} |W|^{p/2} \, d\mu_p(V) \, d\mu_q(W) \, dX, \tag{1.18}$$

where $d\mu_n$ is the invariant volume on $\mathcal{P}_n$ defined by (1.16) and dX is Lebesgue measure on $\mathbb{R}^{q \times p}$.

By iterating partial Iwasawa decompositions, one obtains the full Iwasawa decomposition of $Y = I[g] = I[an]$, where $g \in G = GL(n, \mathbb{R})$, $a \in G$ is diagonal with positive diagonal entries, $n \in G$ is upper triangular with one's on the diagonal. It follows that $g = kan$, with $k \in O(n)$. This is known as the **Iwasawa decomposition** of $g \in G$.

Exercise 1.1.20 (The Jacobian for the Full Iwasawa Decomposition of $GL(n,\mathbb{R})$).

(a) Define for $G = GL(n, \mathbb{R})$, $K = O(n)$,

$$A = \{a \in G \mid a \text{ is positive and diagonal}\},$$

$$N = \{n \in G \mid n \text{ is upper triangular with ones on the diagonal}\}.$$

Write for $x \in G$, the ANK-Iwasawa decomposition as $x = a(x)n(x)k(x)$, with $a(x) \in A$, $n(x) \in N$, and $k(x) \in K$. Show that the integral formula for the ANK-Iwasawa decomposition is:

$$\int_G f(x)dx = \int_A \int_N \int_K f(ank) \, dk \, dn \, da,$$

where all the measures are left $(=$ right) Haar measures.

(b) For $x \in G$, the KAN-Iwasawa decomposition of x can be written $x = K(x)A(x)N(x)$, with $K(x) \in K$, $A(x) \in A$, and $N(x) \in N$. Show that the **relation between the two Iwasawa decompositions** is: $A(x^{-1}) = a(x)^{-1}$, $K(x^{-1}) = k(x)^{-1}$, and $N(x^{-1}) = a(x)n(x)^{-1}a(x)^{-1}$.

(c) Show that if we set $n^a = ana^{-1}$ for $n \in N$, $a \in A$, then $n^a \in N$ and the Jacobian is

$$\alpha(a) \overset{\text{(defn)}}{=} |dn^a/dn| = \prod_{1 \leq i < j \leq n} \frac{a_i}{a_j} = \prod_{i=1}^{n} a_i^{n-2i+1}.$$

Here, if

$$n = \begin{pmatrix} 1 & \cdots & x_{ij} \\ \vdots & \ddots & \vdots \\ 0 & \cdots & 1 \end{pmatrix}, \quad \text{then} \quad dn = \prod dx_{ij} = \begin{cases} \text{the left } (= \text{right})\text{-invariant} \\ \text{Haar measure on } N, \end{cases}$$

by Exercise 1.1.2.

(d) Prove the integral formula for the KAN-Iwasawa decomposition:

$$\int_G f(x) \, dx = \int_K \int_A \int_N f(kan) \, \alpha(a) \, dn \, da \, dk, \quad \text{with } \alpha(a) \text{ as in (c)}.$$

Hints. (See Lang [387, pp. 37–40].) Note that one can normalize the left Haar measures on G and K along with the G-invariant measure on G/K to obtain as in (1.7):

$$\int_G f(x)dx = \int_{gK=\bar{g}\in G/K} \int_{k\in K} f(gk)\, dk\, d\bar{g}.$$

Now G/K can be identified with AN, and we need to only show that $\int_A \int_N f(an)\, dn\, da$ gives a left AN-invariant integral on AN. To see this, note that if $a_1 \in A$ and $n_1 \in N$, then we have the equality below for $n_2 = a^{-1}n_1a$:

$$\int_A \int_N f(a_1n_1\, an\,)\, dn\, da = \int_A \int_N f(a_1an_2n)dn\, da.$$

Then use the left invariance of da and dn, to complete the proof of (a).

It is also clear that if

$$n = \begin{pmatrix} 1 & \cdots & x_{ij} \\ \vdots & \ddots & \vdots \\ 0 & \cdots & 1 \end{pmatrix} \quad \text{and} \quad a = \begin{pmatrix} a_1 & \cdots & 0 \\ \vdots & \ddots & \vdots \\ 0 & \cdots & a_n \end{pmatrix},$$

then

$$ana^{-1} = \begin{pmatrix} 1 & \cdots & y_{ij} \\ \vdots & \ddots & \vdots \\ 0 & \cdots & 1 \end{pmatrix}, \quad \text{for } y_{ij} = a_ix_{ij}a_j^{-1}.$$

Now the left Haar measure on N is just $dn = \prod dx_{ij}$ (see Exercise 1.1.2), where dx_{ij} is just the usual Lebesgue measure on $\mathbb{R}$. The formula for the Jacobian in part (c) follows easily from these considerations. And it implies that

$$\int_N f(an)\, dn = \alpha^{-1}(a) \int_N f(na)\, dn.$$

Part (d) follows from this and part (a), along with the fact that all the groups A, N, K, G are unimodular (i.e., the right and left Haar measures are the same). See the discussion before Proposition 2.1.1 in Section 2.1 below. Thus

$$\int_G f(x)\, dx = \int_G f(x^{-1})\, dx = \int_N \int_A \int_K f(k^{-1}a^{-1}n^{-1})\, \alpha^{-1}(a)\, dk\, da\, dn.$$

Exercise 1.1.21 (Relation Between Measures on $\mathcal{P}_n$ and AN).

(a) Show that, if we write $da = \prod da_j/a_j$ and $dn = \prod dx_{ij}$, for

$$
a = \begin{pmatrix} a_1 & 0 & \cdots & 0 \\ 0 & a_2 & \cdots & 0 \\ \vdots & \vdots & \ddots & \vdots \\ 0 & 0 & \cdots & a_n \end{pmatrix} \quad \text{and} \quad n = \begin{pmatrix} 1 & x_{12} & \cdots & x_{1n} \\ 0 & 1 & \cdots & x_{2n} \\ \vdots & \vdots & \ddots & \vdots \\ 0 & 0 & \cdots & 1 \end{pmatrix},
$$

then

$$
\int_{\mathcal{P}_n} f(Y)\, d\mu_n(Y) = c \int_A \int_N f(I[(an)^{-1}])\, dn\, da = c \int_A \int_N f(I[an])\alpha(a)\, dn\, da.
$$

Use induction on Exercise 1.1.19(b) to see that $c = 2^n$.

(b) Another approach to this problem goes as follows. Make the change of variables from Y in $\mathcal{P}_n$ to an upper triangular matrix T via

$$
Y = {}^t T T, \quad T = \begin{pmatrix} t_1 & \cdots & t_{ij} \\ \vdots & \ddots & \vdots \\ 0 & \cdots & t_n \end{pmatrix}, \quad t_i > 0,\ t_{ij} \in \mathbb{R}.
$$

Show that the Jacobian of this change of variables is:

$$
\left| \frac{\partial Y}{\partial T} \right| = 2^n \prod_{j=1}^{n} t_j^{n-j+1}.
$$

Later (see § 1.3), we will need to consider integrals over the **boundary B of the symmetric space $\mathcal{P}_n$**, which is defined to be the compact space $B = K/M$, where M is the subgroup of K consisting of diagonal matrices with entries ± 1. Using the Iwasawa decomposition, we can identify the boundary B with G/MAN. This is done as follows:

$$
\left. \begin{array}{l} G/MAN \to K/M \\ gMAN \ \mapsto K(g)M. \end{array} \right\} \tag{1.19}
$$

Here $K(g)$ is the K-part of g in the Iwasawa decomposition $G = KAN$, as in Exercise 1.1.20. This mapping is well defined, as is easily checked, since $xN = Nx$, for $x \in MA$. Now an **element $g \in G$ acts on the boundary element kM** via:

$$
g(kM) = K(gk)M. \tag{1.20}
$$

More information on boundaries of symmetric spaces and compactifications can be found in Chapter 2 (see also Furstenberg [194], Gérardin [216], Helgason [275, 281], Koranyi [364], and Moore [459]). In the case of $SL(2, \mathbb{R})$, the symmetric space is the Poincaré upper half plane H, and the boundary can be identified with the circle, which can clearly be thought of as the compactified boundary of H.

Exercise 1.1.22 (The Jacobian of the Action of G on the Boundary B). Show that using the notation of Exercise 1.1.20 we

$$\int_B h(b)\, db = \int_B h(g(b))\alpha^{-1}(A(gk))\, db, \quad \text{if } b = kM \in B.$$

Hint. (See Helgason [275, pp. 50–51].) We want to use Exercise 1.1.20, and apply the integral formula for the KAN-Iwasawa decomposition to a function on the boundary of the form

$$\int_A \int_N f(kan)\, \alpha(a)\, dn\, da.$$

The whole theory of harmonic functions on the unit disc can be extended to symmetric spaces such as $\mathcal{P}_n$ (see Chapter 2 of this Volume or Helgason [275, 281] or [282, pp. 36 and 78 for the history]). Godement [222] defines a function u which is infinitely differentiable on a symmetric space G/K to be **harmonic** if $Lu = 0$ for all G-invariant differential operators L on G/K such that L annihilates constants. Furstenberg [194] shows that, when u is bounded, it suffices for the Laplacian of u to be zero in order for u to be harmonic. The **Poisson kernel** on $G/K \times B$ can be defined by:

$$P(gK, b) = d(g^{-1}(b))/db.$$

And Furstenberg [194] proved the **Poisson integral representation** for a bounded harmonic function u on G/K:

$$u(x) = \int_{b \in B} P(x, b)\, \mu(b)\, db,$$

where μ denotes some bounded measurable function on the boundary B of G/K. Helgason vastly generalized this result in 1970 (see his book [281, pp. 279–280]). See also Chapter 2.

Exercise 1.1.23 (Relation Between Invariant Measures on $\mathcal{P}_n$ and the Determinant One Surface $\mathcal{SP}_n$). Show that we can define an $SL(n, \mathbb{R})$-invariant measure dW on the symmetric space $\mathcal{SP}_n$ defined in (1.3) by setting, for $Y \in \mathcal{P}_n$,

$$\left. \begin{aligned} Y &= t^{1/n}W, \text{ with } t > 0 \text{ and } W \in \mathcal{SP}_n, \\ d\mu_n(Y) &= |Y|^{-(n+1)/2} \prod dy_{ij} = t^{-1}dt\, dW. \end{aligned} \right\} \qquad (1.21)$$

Another useful coordinate system on $\mathcal{P}_n$ is **polar coordinates**:

$$Y = a[k], \text{ for } a \in A \text{ and } k \in K, \qquad (1.22)$$

with A and K as in Exercise 1.1.20. The existence of this decomposition follows from the spectral theorem (Exercise 1.1.7). On the group level, formula (1.22) becomes $G = KAK$. Physicists often call this the **Euler angle decomposition** (see Wigner [668]). Numerical analysts (and now anyone who takes or teaches the linear algebra part of a calculus sequence) call it the **singular value decomposition** (see Strang [581, p. 139]). Polar coordinates have been very useful in multivariate statistics (see James [330] and Muirhead [468]). They are also the coordinates used by Harish-Chandra and Helgason to do harmonic analysis on $\mathcal{P}_n$ (see § 1.3).

Exercise 1.1.24 (The Invariant Volume Element in Polar Coordinates). Show that in polar coordinates $Y = a[k]$, for $a \in A$, $k \in K$, the invariant volume (1.16) is:

$$d\mu_n(Y) = c_n \prod_{j=1}^{n} a_j^{-(n-1)/2} \prod_{1 \le i < j \le n} |a_i - a_j|\, da\, dk,$$

where

$$a = \begin{pmatrix} a_1 & \cdots & 0 \\ \vdots & \ddots & \vdots \\ 0 & \cdots & a_n \end{pmatrix}, \qquad \begin{aligned} da &= \prod_{j=1}^{n} \frac{da_j}{a_j}, \text{ and} \\ dk &= \text{Haar measure on } K = O(n), \end{aligned}$$

with dk normalized so that $\int_K dk = 1$. The positive constant c_n will be determined in Proposition 1.2.3 of Section 1.2.1.
Hint. Note that if $Y = a[k]$, for $k \in K$ and $a \in A$, then

$$dY = da[k] + {}^t dk\, a\, k + {}^t k\, a\, dk,$$

and

$${}^t dk\, k + {}^t k\, dk = 0.$$

Thus

$$dY = \{da - (dk\, k^{-1})a + a(dk\, k^{-1})\}[k].$$

Now $X = dk\, k^{-1}$ is a skew-symmetric matrix. And $(aX - Xa)_{ij} = (a_i - a_j)x_{ij}$ if $X = (x_{ij})$.

Polar coordinates give a $(2^n n!)$-fold covering of $\mathcal{P}_n$, since the entries of a are the eigenvalues of $Y = a[k]$. Thus they are unique up to the action of the **Weyl group** W of permutations of the a_j, $j = 1, 2, \ldots, n$, as well as the action of the group M of orthogonal diagonal matrices. The latter are matrices which are diagonal with $+1$ or -1 as the entries.

Exercise 1.1.25 (The Arc Length in Polar Coordinates). Show that in polar coordinates (1.22), $Y = a[k]$, for $a \in A$, $k \in K$, the arc length (1.11) is:

$$ds_Y^2 = \sum_{j=1}^n a_j^{-2} da_j^2 + \text{Tr}\left(\left(a^{-1} dk\, k^{-1} a - dk\, k^{-1} \right)^2 \right).$$

Set $dk\, k^{-1} = X = $ a skew-symmetric matrix. Then show that you can rewrite the formula as:

$$ds_Y^2 = \sum_{j=1}^n a_j^{-2} da_j^2 + 2 \sum_{1 \le i < j \le n} \frac{(a_i - a_j)^2}{a_i a_j} x_{ij}^2, \text{ for } X = (x_{ij}) = dk\, k^{-1}.$$

You might want to write out the case $n = 2$ first. Then

$$k = \begin{pmatrix} \cos\theta & \sin\theta \\ -\sin\theta & \cos\theta \end{pmatrix}, \quad dk\, k^{-1} = \begin{pmatrix} 0 & d\theta \\ -d\theta & 0 \end{pmatrix}.$$

Hint. See the hint for Exercise 1.1.24 or see Muirhead [468, p. 241]. Thus

$$Y^{-1} dY = {}^t k a^{-1} \left(da - \left(dk\, k^{-1} \right) a + a \left(dk\, k^{-1} \right) \right) k$$

and if $X = dk\, k^{-1}$, we have

$$\text{Tr}((Y^{-1} dY)^2) = \text{Tr}\left(\left(a^{-1} da - a^{-1} X a + X \right)^2 \right)$$
$$= \text{Tr}\left((a^{-1} da)^2 \right) - 2\text{Tr}\left(X a^{-1} X a \right) + 2\text{Tr}\left(X^2 \right).$$

The formula of Exercise 1.1.24 was first proved in 1939 by three statisticians working independently (see Fisher [180], Hsu [307], and Roy [517]).

It is a familiar fact from calculus that integral operators are easier to deal with than differential operators. Integrals tend to make sequences converge while derivatives tend to make sequences diverge. Thus people use integral operators to study spectral theory for differential operators—the theory of Green's functions or resolvent kernels. See Courant and Hilbert [125], or Lang [388], for example. In fact, some mathematicians, perhaps motivated by the words of Hermann Weyl, in some of his early papers, have decided to throw out the differential operators altogether (cf. our remarks on p. 131 of Vol. I). In particular, there is a lot of work on spherical functions that does not discuss differential equations at all. For p-adic groups, this may be a good idea, but I don't think it is so clever to forget about differential operators entirely—even if they may be ugly and/or hairy. Anyway, in this section we want to discuss the integral operators that can be used to replace the differential operators. That brings us to our next topic—convolution operators, as in Section 3.7 of Volume I. Of course, these convolution operators will be of central importance for analysis on $\mathcal{P}_n$ and on fundamental domains thereof, just as in Chapter 3, Vol. I.

In order to define convolution, we must use Exercise 1.1.5 to identify $\mathcal{P}_n$ with the homogenous space $K \backslash G$ where $G = GL(n, \mathbb{R})$ and $K = O(n)$, and then we must think of functions on $\mathcal{P}_n$ as functions on G by writing:

$$f(x) = f(I[x]), \text{ for } x \in G. \tag{1.23}$$

Suppose that f and g are in $L^1(\mathcal{P}_n, d\mu_n)$ and define the **convolution** (splat) of f and g by:

$$(f * g)(a) = C_g f(a) = \int_{G = GL(n, \mathbb{R})} f(b) g(ab^{-1}) \, db. \tag{1.24}$$

Here db is the right or left (they are equal) Haar measure on G as in Exercise 1.1.3. The difference between the definitions given in formula (1.24) above and in formula (3.134) in Vol. I is due to the fact that we are thinking that $\mathcal{P}_n$ has a right G-action while the Poincaré upper half plane H has a left G-action.

Lemma 1.1.1 (Properties of Convolution Operators). *Throughout this list of properties we assume that g is a right $K = O(n)$-invariant function on $G = GL(n, \mathbb{R})$ which is infinitely differentiable with compact support (to be cautious) i.e., $g \in C_c^\infty(G/K)$. We will ultimately need to generalize this, however. And g will be convolved with integrable right K-invariant functions f.*

(1) The operator C_g defined by (1.24) commutes with the action of $c \in GL(n, \mathbb{R})$ on functions $f(a)$, $a \in G$, defined by

$$f^c(a) = f(ac).$$

Thus we say that C_g is a G-invariant integral operator.

(2) If $g(a) = g(a^{-1})$, for all $a \in G$ and if g is real-valued, then the convolution operator C_g is a self-adjoint operator with respect to the usual inner product on $L^2(\mathcal{P}_n, d\mu_n)$. Here $d\mu_n$ is the G-invariant volume element defined by (1.16).

(3) $C_{g*h}f = C_g C_h f$.

(4) The operators C_g commute for functions g which are K-bi-invariant (or radial in the sense of polar coordinates (1.22)); i.e., $g(kak') = g(a)$, for all $a \in G, k, k' \in K$. Such functions g must be symmetric functions of the diagonal elements of $a \in A$. Thus considering g as a function of Y in $\mathcal{P}_n$, we see that $g(Y)$ must be a function of $\mathrm{Tr}(Y^j)$, for $j = 1, 2, \ldots, n$.

(5) $C_g : L^2(\mathcal{P}_n, d\mu_n) \to C^\infty(\mathcal{P}_n)$.

Exercise 1.1.26. Prove Lemma 1.1.1 by imitating the proof of Lemma 3.7.2 in Volume I. You may have to modify some of the arguments slightly since the action of G on $\mathcal{P}_n$ is a right rather than a left action.

Selberg [543] and Maass [426] consider these convolution integral operators from a slightly different point of view. Given an integrable kernel $k(Y, W)$ on $\mathcal{P}_n \times \mathcal{P}_n$, they define an **invariant integral operator** L to be given on integrable functions f on $\mathcal{P}_n$ by:

$$Lf(Y) = \int_{W \in \mathcal{P}_n} k(Y, W) f(W) \, d\mu_n(W), \text{ for } Y, W \in \mathcal{P}_n,$$

provided that $L(f^a) = (Lf)^a$ for all $a \in G$. Here for $f : \mathcal{P}_n \to \mathbb{C}$ we define $f^a(Y) = f(Y[a])$, for all $Y \in \mathcal{P}_n$. Clearly L is an invariant operator if and only if the kernel k has the following property for $a \in G$:

$$k(Y[a], W[a]) = k(Y, W), \text{ for almost all } Y, W \in \mathcal{P}_n. \tag{1.25}$$

Kernels $k : \mathcal{P}_n \times \mathcal{P}_n \to \mathbb{C}$ which satisfy (1.25) are called **point-pair invariants** by Selberg and Maass. Note that for such k

$$k(Y, W) = k(I[a], I[b]) = k(I[ab^{-1}], I) \doteq g(ab^{-1}), \text{ when } a, b \in G. \tag{1.26}$$

Thus the invariant operator L is really a convolution operator. Moreover, the function g in (1.26) must be K bi-invariant, since $Y = I[k_1 a]$ and $W = I[k_2 b]$ for $k_1, k_2 \in K$ implies that

$$g(ab^{-1}) = g(k_1(ab^{-1})k_2^{-1}).$$

1.1.5 Differential Operators on $\mathcal{P}_n$

Given a Riemannian manifold, one always has a Laplacian (also known as the Laplace-Beltrami operator) Δ defined using (1.8) once one knows the arc length element. The goal of this chapter is the resolution of functions on $\mathcal{P}_n$ or $\mathcal{P}_n/GL(n, \mathbb{Z})$ in eigenfunctions of this Laplacian. These are analogues of the main results of Chapter 3 in Volume I. However, life is more complicated in $\mathcal{P}_n$. There are G-invariant differential operators which are not polynomials in the Laplacian. Here a differential operator L on $\mathcal{P}_n$ is said to be **invariant** with respect to $G = GL(n, \mathbb{R})$ if L commutes with the action of G; i.e., if for $a \in G$ and $f \in C^\infty(\mathcal{P}_n)$, defining

$$f^a(Y) = f(Y[a]), \text{ when } Y \in \mathcal{P}_n, \tag{1.27}$$

then we have $(Lf)^a = (Lf^a)$. And we define

$$D(\mathcal{P}_n) = \text{the algebra of } G\text{-invariant differential operators on } \mathcal{P}_n. \qquad (1.28)$$

Proceeding as in Maass [426], we can find examples of invariant differential operators on $\mathcal{P}_n$ using the total differential of $f \in C^\infty(\mathcal{P}_n)$ defined by:

$$\left. \begin{aligned} df &= \text{Tr}\left(dY \tfrac{\partial}{\partial Y} f\right) = \sum_{1 \le i \le j \le n} \tfrac{\partial f}{\partial y_{ij}} dy_{ij}, \quad \text{writing} \\ dY &= (dy_{ij})_{1 \le i,j \le n} \text{ and } \tfrac{\partial}{\partial Y} = \left(\tfrac{1}{2}(1 + \delta_{ij})\tfrac{\partial}{\partial y_{ij}}\right)_{1 \le i,j \le n}. \end{aligned} \right\} \qquad (1.29)$$

Here $\delta_{ij} = 1$ if $i = j$ and 0 otherwise. It follows that if $a \in G = GL(n, \mathbb{R})$ and $W = Y[a]$, then

$$\begin{aligned} df &= \text{Tr}\left(dW \frac{\partial}{\partial W}\right) f = \text{Tr}\left({}^t a \, dY \, a \, \frac{\partial}{\partial W}\right) f \\ &= \text{Tr}\left(dY \, a \frac{\partial}{\partial W} \, {}^t a\right) f = \text{Tr}\left(dY \frac{\partial}{\partial Y}\right) f. \end{aligned}$$

The transformation formula for the matrix differential operator in (1.29) is thus:

$$\frac{\partial}{\partial W} = a^{-1} \frac{\partial}{\partial Y} {}^t a^{-1}, \quad \text{if } W = Y[a], \quad \text{for } Y \in \mathcal{P}_n \text{ and } a \in G. \qquad (1.30)$$

From this formula, one can easily prove the G-invariance of the differential operators L_j defined by

$$L_j = \text{Tr}\left(\left(Y \frac{\partial}{\partial Y}\right)^j\right), \, j = 1, 2, 3, \ldots. \qquad (1.31)$$

Exercise 1.1.27. Prove this last statement. Show also that L_2 defined by (1.31) is the Laplacian on $\mathcal{P}_n$. You may want to postpone this last verification until we have discussed what happens to these differential operators when we express them in partial Iwasawa coordinates (see (1.33)).

Lemma 1.1.2. *(1) A differential operator L in $D(\mathcal{P}_n)$ is uniquely determined by its action on K bi-invariant or radial functions $f(Y) = f(Y[k])$ for all $Y \in \mathcal{P}_n$ and $k \in K$.*
(2) $D(\mathcal{P}_n)$ is a commutative algebra.

Proof. (1) Given $g \in C^\infty(\mathcal{P}_n)$, we can construct a radial function $g^\#$ as follows:

$$g^\#(Y) = \int_{k \in K} g(Y[k]) \, dk, \text{ where } dk = \text{Haar measure on } K, \text{ and } \int_{k \in K} dk = 1.$$

Then, for $L \in D(\mathcal{P}_n)$, we have $L(g^{\#}) = (Lg)^{\#}$, so that $L(g^{\#})(I) = (Lg)(I)$. Suppose that $L, M \in D(\mathcal{P}_n)$ are identical on radial or K bi-invariant functions. Then for any $g \in C^{\infty}(\mathcal{P}_n)$ and any a in G, we have

$$Lg(a) = (Lg^a)(I) = (Mg^a)(I) = Mg(a).$$

Thus $L = M$.

(2) In this proof, which follows Selberg [543], let us use the point-pair invariant kernel notation

$$f(ab^{-1}) = k(a, b).$$

Then, if $L \in D(\mathcal{P}_n)$, write $L_1 k$ when L acts on the first argument of k and $L_2 k$ when L acts on the second argument of k. Note that:

$$L_1 k(a, b) = L(f^{b^{-1}})(a) = (Lf)^{b^{-1}}(a).$$

Now we want to make use of the geodesic-reversing isometry of $\mathcal{P}_n$ at I which is given by $\sigma(Y) = Y^{-1}$ (see Exercise 1.1.18). So we set

$$f^{\sigma}(a) = f(a^{-1}) \text{ and } k^{\sigma}(a, b) = f^{\sigma}(ab^{-1}) = k(b, a).$$

Define, for $L \in D(\mathcal{P}_n)$, the differential operator L^{σ} by:

$$L^{\sigma} f = [(L(f^{\sigma^{-1}}))]^{\sigma}.$$

In Section 1.2, we will show that

$$L^{\sigma} = \overline{L^*} = \text{ the complex conjugate adjoint operator.}$$

Next we need to prove the following fact.

Claim.

$$L_1 k(a, b) = L_2^{\sigma} k(a, b).$$

Proof of Claim. To believe this claim, we need to use the fact that $\mathcal{P}_n$ is indeed a weakly symmetric space in the sense of Selberg; i.e., we need to know that for each $X, Y \in \mathcal{P}_n$, there is a matrix $g \in G$ such that

$$X[g] = Y^{-1} \text{ and } Y[g] = X^{-1}.$$

To see this, use the fact that there is a matrix $h \in G$ such that

$$X[h] = I \text{ and } Y[h] = D \text{ positive diagonal.}$$

If we replace g by $hg\,{}^t h$, the equations we seek to solve become:

$$X[hg\,{}^t h] = Y^{-1} \quad \text{and} \quad Y[hg\,{}^t h] = X^{-1}.$$

This means that we need to find g such that

$$I[g] = (Y[h])^{-1} = D^{-1},$$

and

$$D[g] = (X[h])^{-1} = I.$$

The solution is thus $g = D^{-1/2}$. Since D is a positive diagonal matrix, we can indeed take its square root.

Now to prove the claim, note the following sequence of equalities:

$$\begin{aligned}
L_1 k(X, Y) = k'(X, Y) &= k'(Y^{-1}[g^{-1}], X^{-1}[g^{-1}]) = k'(Y^{-1}, X^{-1}) \\
&= k'(Y, \sigma X)|_{Y \to \sigma Y} = (L_Y) k(Y, \sigma X)|_{Y \to \sigma Y} \\
&= (L_Y)^\sigma k(Y^{-1}, X^{-1}) = (L_Y)^\sigma k(X, Y).
\end{aligned}$$

This completes the proof of the claim.

Thanks to the claim, we see that if L and M are both in $D(\mathcal{P}_n)$, we can write

$$L_1 M_1 k = L_1 M_2^\sigma k = M_2^\sigma L_1 k = M_1 L_1 k,$$

since differential operators acting on different arguments certainly commute. The proof of (2) is thus accomplished. ∎

Theorem 1.1.2 (Structure of the Ring of G-invariant Differential Operators on $\mathcal{P}_n$). *The differential operators*

$$L_j = \mathrm{Tr}\left(\left(Y\frac{\partial}{\partial Y}\right)^j\right), \quad j = 1, 2, \ldots, n$$

form an algebraically independent basis for the ring $D(\mathcal{P}_n)$ of $GL(n, \mathbb{R})$-invariant differential operators on $\mathcal{P}_n$. Thus $D(\mathcal{P}_n)$ can be identified with $\mathbb{C}[X_1, \ldots, X_n] =$ the ring of polynomials in n indeterminates. In particular, $D(\mathcal{P}_n)$ is a commutative ring.

Proof (Of All but the Algebraic Independence of the Operators L_j). This is a result of Selberg [543, pp. 49–51, 57]. We follow the discussion given by Maass [426, pp. 64–67]. See also Jorgenson and Lang [334]. An invariant differential operator L on $\mathcal{P}_n$ has the form $L = L(Y, \partial/\partial Y)$. We want to show that L is a polynomial in $\mathrm{Tr}((Y\partial/\partial Y)^j)$, $j = 1, 2, \ldots, n$. The proof involves induction on the degree of $L(Y, X)$ considered as a polynomial in the entries of the matrix X.

The invariance of L implies that

$$L\left(Y[a], X\left[{}^t a^{-1}\right]\right) = L(Y, X), \quad \text{for all } a \in G.$$

So we may assume that $Y = I$ and X is diagonal with diagonal entries x_j, $j = 1, \ldots, n$. If $k \in K$, then $L(I, X) = L(I, X[k])$. Thus L must be a symmetric polynomial in $x_1, \ldots, x_n$, since we can take k to be any permutation matrix.

Then the fundamental theorem on symmetric polynomials says that L is a polynomial in $x_1^j + \cdots + x_n^j$, $j = 1, 2, \ldots, n$. Going back to the old variables, we find that there is a polynomial p with complex coefficients such that $L(Y, X) = p(\text{Tr}(YX), \text{Tr}((YX)^2), \ldots, \text{Tr}((YX)^n))$. Replacing X by $\partial/\partial Y$ from formula (1.29), we have

$$L(Y, \partial/\partial Y) = p(L_1, \ldots, L_n) + M, \quad \text{where } L_j = \text{Tr}\left((Y\partial/\partial Y)^j\right)$$

and $M \in D(\mathcal{P}_n)$ has lower degree than L, for the homogeneous terms of highest degree in two differential operators must commute. The induction hypothesis completes the proof that the L_j, $j = 1, 2, \ldots, n$, do indeed generate $D(\mathcal{P}_n)$.

See Maass [426, pp. 64–67], for the proof that the operators L_j, $j = 1, 2, \ldots, n$, are algebraically independent.   ∎

Theorem 1.1.2 is rather surprising since we have found a large number of differential operators which behave like the Laplacian for the Riemannian structure on $\mathcal{P}_n$ as in (1.8). One of them is the Laplacian, of course. But there are also others, including a differential operator of degree one. This result can also be generalized to arbitrary symmetric spaces (see Chapter 2 and Helgason [273, p. 432]). The generalization requires results of Harish-Chandra on the algebra of invariant differential operators for a semisimple Lie group as well as Chevalley's generalization of the fundamental theorem on symmetric polynomials. Chevalley's result can be found in Carter [97, Ch. 9].

The next exercise is useful in showing that the set of eigenfunctions for the invariant differential operators in $D(\mathcal{P}_n)$ is the same as that for the convolution integral operators in formula (1.24).

Exercise 1.1.28. Suppose that $g : \mathcal{P}_n \to \mathbb{C}$ is infinitely differentiable with compact support. Identify it with a function on $G = GL(n, \mathbb{R})$ as in Lemma 1.1.1. Suppose that $L \in D(\mathcal{P}_n)$. If $g(a) = g(a^{-1})$ for all $a \in G$, show that

$$LC_g = C_g L \quad \text{and} \quad LC_g = C_{Lg},$$

where C_g denotes the convolution integral operator defined by (1.24).
Hint. Recall Lemma 3.7.2 in Volume I.

Now consider what happens to the Laplace operator in the various coordinate systems which have been introduced. We begin with partial Iwasawa coordinates:

$$Y = \begin{pmatrix} V & 0 \\ 0 & w \end{pmatrix} \begin{bmatrix} I & x \\ 0 & 1 \end{bmatrix}, \quad V \in \mathcal{P}_{n-1}, \; w > 0, \; x \in \mathbb{R}^{n-1}. \tag{1.32}$$

We know from Exercise 1.1.14 that

$$ds_Y^2 = \mathrm{Tr}\left(\left(V^{-1}dV\right)^2\right) + (w^{-1}dw)^2 + 2w^{-1}V[dx]$$
$$= ds_V^2 + (w^{-1}dw)^2 + 2w^{-1}V[dx].$$

So the Riemannian metric tensor G_Y for $\mathcal{P}_n$ is:

$$G_Y = (g_{ij}) = \begin{pmatrix} G_V & 0 & 0 \\ 0 & 2w^{-1}V & 0 \\ 0 & 0 & w^{-2} \end{pmatrix},$$

where

$$|G_Y| = 2^{n-1}|V|^{1-n}w^{-n-1}$$

and G_V is the Riemannian metric tensor for $V \in \mathcal{P}_{n-1}$, $G_V^{-1} = (g_V^{ijkl})$. Thus, by formula (1.8), we find that if $V^{-1} = (v^{ij})$, then the **Laplacian in partial Iwasawa coordinates** (1.32) is:

$$\left. \begin{aligned} \Delta_Y &= w^{(n+1)/2}\frac{\partial}{\partial w}w^2 w^{(-n-1)/2}\frac{\partial}{\partial w} + \frac{1}{2}w\sum_{i,j=1}^{n-1} v^{ij}\frac{\partial^2}{\partial x_i \partial x_j} + L_V, \\[2ex] \text{where} \quad L_V &= |V|^{(n-1)/2}\sum_{\substack{i,j,k,l=1 \\ i\le j,\, k\le l}}^{n-1} \frac{\partial}{\partial v_{ij}}|V|^{(1-n)/2}g_V^{ijkl}\frac{\partial}{\partial v_{kl}}. \end{aligned} \right\} \tag{1.33}$$

Note that L_V is not the Laplacian Δ_V on $\mathcal{P}_{n-1}$, since:

$$\Delta_V = |V|^{n/2}\sum_{\substack{i,j,k,l=1 \\ i\le j,\, k\le l}}^{n-1} \frac{\partial}{\partial v_{ij}}|V|^{-n/2}g_V^{ijkl}\frac{\partial}{\partial v_{kl}}.$$

However, we can rewrite (1.33) as:

$$\Delta_Y = \left(w\frac{\partial}{\partial w}\right)^2 + \frac{1-n}{2}w\frac{\partial}{\partial w} + \frac{1}{2}w\sum_{i,j=1}^{n-1} v^{ij}\frac{\partial^2}{\partial x_i \partial x_j} + \Delta_V + \frac{1}{2}\mathrm{Tr}\left(V\frac{\partial}{\partial V}\right). \tag{1.34}$$

Exercise 1.1.29. Deduce formula (1.34) from formula (1.33).

Now let's consider what happens to the differential operator $\mathrm{Tr}((Y\partial/\partial Y)^2)$ in the partial Iwasawa decomposition (1.32). Let

$$Y = \begin{pmatrix} F & h \\ {}^t h & g \end{pmatrix} \quad \text{and} \quad \frac{\partial}{\partial Y} = \begin{pmatrix} \partial/\partial F & \frac{1}{2}\partial/\partial h \\ \frac{1}{2}\,{}^t(\partial/\partial h) & \partial/\partial g \end{pmatrix}.$$

Then

$$F = V, \quad g = w + V[x], \quad h = Vx, \quad dF = dV,$$
$$dg = dw + dV[x] + {}^t dx V x + {}^t x V dx, \quad dh = dV \cdot x + V dx.$$

Substitute this into the total differential (1.29) to obtain

$$\mathrm{Tr}\left(dY\frac{\partial}{\partial Y}\right) = \mathrm{Tr}\left(dF\frac{\partial}{\partial F}\right) + dg\frac{\partial}{\partial g} + {}^t dh\frac{\partial}{\partial h},$$

and compare the result with

$$\mathrm{Tr}\left(dY\frac{\partial}{\partial Y}\right) = \mathrm{Tr}\left(dV\frac{\partial}{\partial V}\right) + dw\frac{\partial}{\partial w} + {}^t dx\frac{\partial}{\partial x}.$$

This leads to the following formulas:

$$\frac{\partial}{\partial V} = \frac{\partial}{\partial F} + x\frac{\partial}{\partial g}{}^t x + \frac{1}{2}\left\{x\left(\frac{{}^t\partial}{\partial h}\right) + {}^t\left(x\left(\frac{{}^t\partial}{\partial h}\right)\right)\right\},$$

$$\frac{\partial}{\partial w} = \frac{\partial}{\partial g}, \quad \frac{\partial}{\partial x} = 2Vx\frac{\partial}{\partial g} + V\frac{\partial}{\partial h}.$$

It follows that, setting $\Omega = x\left(\frac{{}^t\partial}{\partial x}\right)$,

$$\frac{\partial}{\partial g} = \frac{\partial}{\partial w}, \quad \frac{\partial}{\partial h} = V^{-1}\frac{\partial}{\partial x} - 2x\frac{\partial}{\partial w},$$

$$\frac{\partial}{\partial F} = \frac{\partial}{\partial V} + x\frac{\partial}{\partial w}{}^t x - \frac{1}{2}\left\{\Omega V^{-1} + V^{-1}\,{}^t\Omega\right\}.$$

The preceding calculation is a little tricky since $\partial/\partial F$ must be symmetric. So you must put in

$$\frac{1}{2}\left(\Omega V^{-1} + V^{-1}\,{}^t\Omega\right) \quad \text{and not just } \Omega V^{-1} = x\left(\frac{{}^t\partial}{\partial x}\right)V^{-1}.$$

In the term ${}^t(x\,{}^t\partial/\partial x)$, the order is to differentiate first, then multiply.

Our calculation implies that, setting $\Omega = x \left(\frac{{}^t \partial}{\partial x} \right)$ again, we have

$$
Y \frac{\partial}{\partial Y} = \begin{pmatrix} V \frac{\partial}{\partial V} - \frac{1}{2} \, {}^t \Omega & \frac{1}{2} \frac{\partial}{\partial x} \\ {}^t x V \frac{\partial}{\partial V} - w \frac{\partial}{\partial w} \, {}^t x - \frac{1}{2} \, {}^t x \, {}^t \Omega + \frac{1}{2} w \left(\frac{{}^t \partial}{\partial x} \right) V^{-1} & \frac{1}{2} \, {}^t x \frac{\partial}{\partial x} + w \frac{\partial}{\partial w} \end{pmatrix}.
$$

If we square this matrix operator and take the trace, we find via mathematical induction that $\mathrm{Tr}((Y \partial / \partial Y)^2)$ has the same partial Iwasawa decomposition as that of Δ_Y which was given in (1.34). Therefore the two operators are indeed the same.

Next consider the Laplacian in polar coordinates (1.22). We saw in Exercises 1.1.24 and 1.1.25 that if $X = -{}^t X = dk\, k^{-1}$, $k \in K$, then

$$
ds_Y^2 = \sum_{j=1}^{n} \left(\frac{da_j}{a_j} \right)^2 + 2 \sum_{1 \le i < j \le n} \frac{(a_i - a_j)^2}{a_i a_j} x_{ij}^2.
$$

So the Riemannian metric tensor in (1.8) becomes:

$$
G = \begin{pmatrix} a_1^{-2} \\ & \ddots \\ & & a_n^{-2} \\ & & & 2(a_1 - a_2)^2/(a_1 a_2) & & & & & 0 \\ & & & & \ddots \\ & & & & & 2(a_1 - a_n)^2/(a_1 a_n) \\ & & 0 & & & & \ddots \\ & & & & & & & 2(a_{n-1} - a_n)^2/(a_{n-1} a_n) \end{pmatrix}
$$

with

$$
|G| = 2^{n(n-1)/2} \prod_{j=1}^{n} a_j^{-(n+1)} \prod_{1 \le i < j \le n} (a_i - a_j)^2.
$$

By formula (1.8), if $a_1 > a_2 > \cdots > a_n$, then the **Laplacian in polar coordinates** is:

$$
\Delta = \prod_{j=1}^{n} a_j^{(n+1)/2} \prod_{1 \le i < j \le n} (a_i - a_j^{-1}) \sum_{k=1}^{n} \frac{\partial}{\partial a_k} a_k^2 \prod_{j=1}^{n} a_j^{-(n+1)/2} \prod_{1 \le i < j \le n} (a_i - a_j) \frac{\partial}{\partial a_k}
$$

$$
+ \sum_{1 \le i < j \le n} \frac{a_i a_j}{2(a_i - a_j)^2} \frac{\partial^2}{\partial x_{ij}^2}.
$$

Clearly we can rewrite this as:

$$\Delta = \sum_{k=1}^{n}\left\{ a_k^2\frac{\partial^2}{\partial a_k^2} + \left(\sum_{\substack{j=1 \\ j\neq k}}^{n}\frac{a_k^2}{a_k - a_j} - \frac{n-3}{2}a_k \right)\frac{\partial}{\partial a_k}\right\} + \frac{1}{2}\sum_{1\leq i<j\leq n}\frac{a_i a_j}{(a_i - a_j)^2}\frac{\partial^2}{\partial x_{ij}^2}.$$

Therefore (cf. Muirhead [468, p. 242])

$$\Delta = \sum_{k=1}^{n}a_k^2\frac{\partial^2}{\partial a_k^2} - \frac{n-3}{2}\sum_{k=1}^{n}a_k\frac{\partial}{\partial a_k} + \frac{1}{2}\sum_{1\leq i<j\leq n}\frac{a_i a_i}{(a_i - a_j)^2}\frac{\partial^2}{\partial x_{ij}^2}$$

$$+ \sum_{k=1}^{n}\left(\sum_{\substack{j=1 \\ j\neq k}}^{n}\frac{a_k^2}{a_k - a_j} \right)\frac{\partial}{\partial a_k}. \tag{1.35}$$

1.1.6 A List of the Main Formulas Derived in Section 1.1

Now we can finally summarize our results. This will be convenient for future reference.

First **the results of changing to partial Iwasawa coordinates** are:

$$\left.\begin{array}{l} Y = \begin{pmatrix} V & 0 \\ 0 & w \end{pmatrix}\begin{bmatrix} I & x \\ 0 & 1 \end{bmatrix}, \text{ for } V \in \mathcal{P}_{n-1}, \ w > 0, \text{ and } x \in \mathbb{R}^{n-1}, \\[2ex] ds_Y^2 = \text{Tr}\left((Y^{-1}dY)^2 \right) = ds_V^2 + (w^{-1}dw)^2 + 2w^{-1}V[dx], \\[2ex] d\mu_n(Y) = |Y|^{-(n+1)/2}\prod_{1\leq i\leq j\leq n}dy_{ij} = w^{(1-n)/2}|V|^{1/2}d\mu_{n-1}(V)\frac{dw}{w}dx, \\[2ex] \Delta_Y = \text{Tr}\left((Y\partial/\partial Y)^2 \right) \\[2ex] \qquad = \Delta_V + \frac{1}{2}\text{Tr}(V\partial/\partial V) + \left(w\frac{\partial}{\partial w} \right)^2 - \frac{n-1}{2}w\frac{\partial}{\partial w} + \frac{1}{2}w\left(\frac{^t\partial}{\partial x} \right)V^{-1}\frac{\partial}{\partial x}. \end{array}\right\} \tag{1.36}$$

For discussions of these formulas, see Exercises 1.1.14, 1.1.19, and 1.1.29.

Then **the results of changing to polar coordinates** are:

$$Y = a[k], \quad a \in A, \ k \in K,$$

$$ds_Y^2 = \sum_{j=1}^{n} \left(a_j^{-1} da_j \right)^2 + 2 \sum_{1 \leq i < j \leq n} \frac{(a_i - a_j)^2}{a_i a_j} x_{ij}^2, \quad \text{for } X = dk \ k^{-1} = (x_{ij}),$$

$$d\mu_n(Y) = c_n \prod_{j=1}^{n} a_j^{-(n-1)/2} \prod_{1 \leq i < j \leq n} \left| a_i - a_j \right| \ da \ dk,$$

$$da = \prod_{j=1}^{n} \frac{da_j}{a_j}, \quad \int_K dk = 1, \quad dk = \text{Haar measure,}$$

$$\Delta = \sum_{i=1}^{n} a_i^2 \frac{\partial^2}{\partial a_i^2} - \frac{n-3}{2} \sum_{i=1}^{n} a_i \frac{\partial}{\partial a_i} + \frac{1}{2} \sum_{1 \leq i < j \leq n} \frac{a_i a_j}{(a_i - a_j)^2} \frac{\partial^2}{\partial x_{ij}^2} + \sum_{k=1}^{n} \left(\sum_{\substack{j=1 \\ j \neq k}}^{n} \frac{a_k^2}{a_k - a_j} \right) \frac{\partial}{\partial a_k}.$$

$$\tag{1.37}$$

Here $K = O(n)$ and A is the multiplicative group of positive $n \times n$ diagonal matrices. For discussions of these results, see Exercises 1.1.24 and 1.1.25 as well as the discussion before formula (1.35). The positive constant c_n will be determined in Section 1.2.1, where we will show that it is given by (1.40) below.

Next we list a few integral formulas. For $x \in G = GL(n, \mathbb{R})$, let dx denote a Haar measure (as in Exercise 1.1.3). For $k \in K = O(n)$, let dk be a Haar measure chosen so that the volume of K is one.

Let N be the nilpotent group of upper triangular matrices n of the form:

$$n = \begin{pmatrix} 1 & x_{12} & \cdots & x_{1n} \\ 0 & 1 & \cdots & x_{2n} \\ \vdots & \vdots & \ddots & \vdots \\ 0 & 0 & \cdots & 1 \end{pmatrix},$$

with Haar measure $dn = \prod dx_{ij}$, $dx_{ij} = $ Lebesgue measure on $\mathbb{R}$ (see Exercise 1.1.2). Let $A = A_n$ be the **positive diagonal group** of matrices a of the form

$$a = \begin{pmatrix} a_1 & 0 & \cdots & 0 \\ 0 & a_2 & \cdots & 0 \\ \vdots & \vdots & \ddots & \vdots \\ 0 & 0 & \cdots & a_n \end{pmatrix}, \quad a_j > 0,$$

with Haar measure $da = \prod da_i / a_i$.

Let T_n be the **triangular group** of matrices t of the form

$$t = \begin{pmatrix} t_{11} & \cdots & t_{1n} \\ \vdots & \ddots & \vdots \\ 0 & \cdots & t_{nn} : \end{pmatrix}, \quad t_{jj} > 0,$$

with measure $dt = \prod dt_{ij}$, $dt_{ij} = $ Lebesgue measure on $\mathbb{R}$. Note that we can identify both T_n and $\mathcal{P}_n$ with $K\backslash G$, $G = GL(n, \mathbb{R})$, $K = O(n)$, pointwise. However T_n is a group under matrix multiplication and $\mathcal{P}_n$ is not a group under matrix multiplication.

Let dW denote an $SL(n, \mathbb{R})$-invariant measure on $\mathcal{SP}_n$, the determinant one surface in $\mathcal{P}_n$ defined in (1.3). If M denotes the diagonal matrices in K, set $B = K/M$, the boundary of $\mathcal{P}_n$, with K-invariant volume db. For $b = kM$ in B and $g \in G$, define $A(g(b))$ to be the A-part of gk in its KAN-Iwasawa decomposition as in Exercise 1.1.20. Define

$$\left. \begin{aligned} \alpha(a) &= \prod_{i=1}^{n} a_i^{n-2i+1}, \quad \text{for } a \in A_n, \text{ as above;} \\ \beta(t) &= \prod_{i=1}^{n} t_{ii}^{-i}, \quad \text{for } t \in T_n, \text{ as above;} \\ \gamma(a) &= \prod_{i=1}^{n} a_i^{-(n-1)/2} \prod_{1 \le i < j \le n} |a_i - a_j|, \quad \text{for } a \in A_n, \text{ as above.} \end{aligned} \right\} \quad (1.38)$$

Then we have the following integral formulas (using the notation in (1.38)):

$$\left. \begin{aligned} \int_G f(x)dx &= \int_A \int_N \int_K f(ank) \, dk \, dn \, da = \int_K \int_A \int_N f(kan) \, \alpha(a) \, dn \, da \, dk; \\ \int_{\mathcal{P}_n} f(Y) \, d\mu_n(Y) &= 2^n \int_A \int_N f(I[(an)^{-1}]) \, dn \, da \\ &= 2^n \int_A \int_N f(I[an]) \, \alpha(a) \, dn \, da = 2^n \int_{T_n} f(I[t]) \, \beta(t) \, dt; \\ \int_{\mathcal{P}_n} f(Y) \, d\mu_n(Y) &= \int_{t>0} \int_{W \in \mathcal{SP}_n} f(t^{1/n}W) \, dW \, t^{-1} \, dt; \\ \int_B h(b) \, db &= \int_B h(g(b)) \, \alpha^{-1}(A(g(b))) \, db; \\ \int_{\mathcal{P}_n} f(Y) \, d\mu_n(Y) &= c_n \int_K \int_A f(a[k]) \, \gamma(a) \, da \, dk. \end{aligned} \right\}$$

$$(1.39)$$

The formulas in (1.39) are proved in Exercises 1.1.20–1.1.24. The positive constant c_n in the last formula in (1.39) will be determined in Section 1.2.1, where we will show that:

$$c_n^{-1} = \pi^{-(n^2+n)/4} \prod_{j=1}^{n} j\Gamma(j/2). \tag{1.40}$$

Here Γ denotes the gamma function. Hopefully it will not be confused with Γ the discrete subgroup of G.

Exercise 1.1.30. Show that right and left Haar measures are different for the multiplicative group $T_n = AN$ of upper triangular $n \times n$ matrices with positive diagonal entries. The group T_n is called a "solvable" Lie group.

Exercise 1.1.31. Is the Laplacian Δ a negative operator on the square integrable functions $f \in L^2(\mathcal{P}_n)$ such that $\Delta f \in L^2(\mathcal{SP}_n)$?

Exercise 1.1.32 (Grenier [241]). When Y lies in the determinant one surface $\mathcal{SP}_{n+1}$, write:

$$Y = \begin{pmatrix} v & 0 \\ 0 & v^{-1/n}W \end{pmatrix} \begin{bmatrix} 1 & {}^t x \\ 0 & I_n \end{bmatrix}, \quad v > 0, \ X \in \mathbb{R}^n, \ W \in \mathcal{SP}_n.$$

Show that if ds^2 is the arc length $\mathrm{Tr}((Y^{-1}dY)^2)$ on the determinant one surface, $d\mu_{n+1}$ is the G-invariant volume, and Δ_{n+1} is the corresponding Laplacian, we have the following expressions relating these quantities for rank $n + 1$ and those for the rank n case:

$$ds_Y^2 = \tfrac{n+1}{n} v^{-2} dv^2 + 2v^{(n+1)/n} \, W^{-1}[dx] + ds_W^2,$$

$$d\mu_{n+1}(Y) = v^{(n-1)/2} \, dv \, dx \, d\mu_n(W),$$

$$\Delta_{n+1} = \tfrac{n}{n+1} \{ v^2 \tfrac{\partial}{\partial v^2} + \tfrac{n+3}{2} v \tfrac{\partial}{\partial v} \} + \tfrac{1}{2} v^{-(n+1)/n} \, W\left[\tfrac{\partial}{\partial x}\right] + \Delta_n.$$

1.1.7   An Application to Multivariate Statistics

References for this application are Anderson [8], Morrison [464], and Muirhead [468].

A random variable X in $\mathbb{R}^n$ is **normal with mean** $\mu \in \mathbb{R}^n$ and covariance Σ in $\mathcal{P}_n$, which is written $N(\mu, \Sigma)$ if it has the probability density:

$$(2\pi)^{-n/2} |\Sigma|^{-1/2} \exp\left\{ -\frac{1}{2} \Sigma^{-1} [x - \mu] \right\}$$

(see Anderson [8, p. 17]). It follows that if X is a normal random variable in $\mathbb{R}^n$ distributed according to $N(\mu, \Sigma)$ and if $A \in \mathbb{R}^{m \times n}$, with $m \le n$, then $Y = AX$ is normal and distributed according to $N(A\mu, \Sigma[{}^t A])$.

Notions of partial and multiple correlation are quite important in the analysis of data. We can use the Iwasawa decomposition to aid in this analysis. Such results go back to Pearson and Yule in the late 1890s (see Anderson [8, pp. 27–28]). If we partition the random variable ${}^t X = ({}^t X_1, {}^t X_2)$, mean ${}^t \mu = ({}^t \mu_1, {}^t \mu_2)$, and covariance

$$\Sigma = \begin{pmatrix} \Sigma_{11} & \Sigma_{12} \\ {}^t \Sigma_{12} & \Sigma_{22} \end{pmatrix},$$

with $X_1 \in \mathbb{R}^p$, $X_2 \in \mathbb{R}^q$, $n = p + q$, $\Sigma_{11} \in \mathcal{P}_p$, then (using the result at the end of the last paragraph) we can see that X_i is normal and distributed according to $N(\mu_i, \Sigma_{ii})$, for $i = 1, 2$. The **conditional distribution** of X_1 holding $X_2 = x_2$ constant is **normally distributed** according to $N(\mu_1 + H(x_2 - \mu_2), V)$, where H and V are defined by the Iwasawa decomposition:

$$\Sigma = \begin{pmatrix} V & 0 \\ 0 & W \end{pmatrix} \begin{bmatrix} I & 0 \\ {}^t H & I \end{bmatrix}, \quad \text{for } V \in \mathcal{P}_p, \ W \in \mathcal{P}_q, \ n = p + q.$$

To see this, note that the conditional distribution of X_1 holding $X_2 = x_2$ constant is:

$$g(x_1 | x_2) = \frac{(2\pi)^{-n/2} |\Sigma|^{-1/2} \exp\left\{ -\frac{1}{2} \Sigma^{-1}[x - \mu] \right\}}{(2\pi)^{-q/2} |W|^{-1/2} \exp\left\{ -\frac{1}{2} W^{-1}[x_2 - \mu_2] \right\}}.$$

But we know that

$$\Sigma^{-1}[x] = V^{-1}[x_1 - Hx_2] + W^{-1}[x_2].$$

Therefore

$$g(x_1 | x_2) = (2\pi)^{-p/2} |V|^{-1/2} \exp\left\{ -\frac{1}{2} V^{-1}[x_1 - \mu_1 - H(x_2 - \mu_2)] \right\}.$$

Note that $V = \Sigma_{11} - \Sigma_{22}^{-1}[{}^t \Sigma_{12}]$ and $H = \Sigma_{12} \Sigma_{22}^{-1}$. The matrix H is called the **matrix of regression coefficients** of X_1 on x_2. The entries of V are called the **partial covariances**. The **partial correlation** between the ith entry of X_1 and the jth entry of X_1, holding $X_2 = x_2$ fixed is:

$$\rho_{ij} = \frac{v_{ij}}{\sqrt{v_{ii} v_{jj}}}.$$

A fundamental problem in statistics is the estimation of the mean and the covariance, after making N sample observations of the random variable X. Suppose X is distributed according to $N(\mu, \Sigma)$ and we have N observations $x_1, \ldots, x_N$, $N > n$. The **likelihood function** is

$$L = (2\pi)^{-nN/2} |\Sigma|^{-N/2} \exp\left\{ -\frac{1}{2} \sum_{i=1}^{N} \Sigma^{-1}[x_i - \mu] \right\}.$$

Maximizing L over Σ, μ gives the **maximum likelihood estimates** for Σ, μ and which are:

$$\widehat{\mu} = \bar{x} = \frac{1}{N} \sum_{k=1}^{N} x_k, \quad \widehat{\Sigma} = \frac{1}{N} \sum_{k=1}^{N} I[{}^t(x_i - \bar{x})]$$

(see Anderson [8, pp. 44–48]). Correlation coefficients can be estimated from $\widehat{\Sigma}$.

Now we want to consider an example discussed in 1907 by Hooker (see Anderson [8, p. 82]). Suppose that X_1 represents hay yield in hundredweights per acre, X_2 represents spring rainfall in inches, X_3 represents accumulated spring temperature over $42\,°F$, for a certain English region, measured over a period of 20 years. One looks at the data and uses maximum likelihood estimates for the mean, covariance, and correlation coefficients. The result of these calculations is:

$$\widehat{\mu} = \bar{x} = \begin{pmatrix} 28.02 \\ 4.91 \\ 594.00 \end{pmatrix}, \quad \begin{pmatrix} \widehat{\sigma}_1 \\ \widehat{\sigma}_2 \\ \widehat{\sigma}_3 \end{pmatrix} = \begin{pmatrix} 4.42 \\ 1.10 \\ 85.00 \end{pmatrix}, \quad \widehat{\sigma}_i^2 = \widehat{\sigma}_{ii}, \ \widehat{\rho}_{ij} = \widehat{\sigma}_{ij} / (\widehat{\sigma}_i \widehat{\sigma}_j),$$

$$\begin{pmatrix} 1 & \widehat{\rho}_{12} & \widehat{\rho}_{13} \\ \widehat{\rho}_{21} & 1 & \widehat{\rho}_{23} \\ \widehat{\rho}_{31} & \widehat{\rho}_{32} & 1 \end{pmatrix} = \begin{pmatrix} 1.00 & 0.80 & -0.40 \\ 0.80 & 1.00 & -0.56 \\ -0.40 & -0.56 & 1.00 \end{pmatrix}.$$

We can then ask: Is high temperature correlated with low yield or is high temperature correlated with low rainfall and thus with low yield? To answer this question, one estimates the partial correlation between X_1 and X_3 while holding $X_2 =$ constant. The Iwasawa decomposition method discussed above then leads to the result that this correlation is:

$$\frac{\widehat{\sigma}_1 \widehat{\sigma}_3 (\widehat{\rho}_{13} - \widehat{\rho}_{12}\widehat{\rho}_{23})}{\sqrt{\widehat{\sigma}_1^2 (1 - \widehat{\rho}_{12}^2) \widehat{\sigma}_3^2 (1 - \widehat{\rho}_{23}^2)}} = 0.0967.$$

Thus, if the effect of the rainfall is removed, yield and temperature are positively correlated. So both high temperature and high rainfall increase yield, but usually high rainfall occurs with low temperature.

Exercise 1.1.33. Imitate the preceding example with X_1=yearly global mean temperature deviation from average in C^o, X_2=yearly carbon dioxide emissions, X_3=number of your choice chosen from world population, sunspot numbers, number of species going extinct, sea level, area of glaciers, ...—measured each year from 1961 to 2015. Then compute partial correlations between the various X_i and X_j. Write an essay on climate change.

1.2 Special Functions on $\mathcal{P}_n$

Attempts have been made to generalize hypergeometric functions to the case of several variables, based on the construction of a many-dimensional analogue to the hypergeometric series [P. Appell & J. Kampé de Fériet, *Fonctions Hypergéométriques et Hypersphériques, Polynomes d'Hermite*, Gauthier-Villars, Paris, 1926]. However, this approach leads to functions which, in the opinion of the author, do not sufficiently fully reflect the multidimensionality of the domain. The present article is concerned with another approach to the theory of special functions for several variables. Special functions of a single variable can be expressed, as we know, in terms of elementary functions, viz., the power and exponential functions, by use of simple integral representations. It is precisely these integral representations that are taken as the pattern for the definition of the many-dimensional analogues of special functions.

From Gindikin [218, p. 1].

1.2.1 Power and Gamma Functions

This section concerns the matrix argument analogues of functions we encountered in Volume I—gamma, K-Bessel, and spherical functions. The approach is similar to that of Gindikin [218] quoted above—an approach that emphasizes integral representations for the functions. The main references for this section are Bengtson [42], Bump [83], Gindikin [218], Helgason [273–282], James [328–330], Maass [426], Muirhead [468], and Selberg [543]. See also Goldfeld [230], Gurarie [254], Hua [308], Vilenkin [635], and Wawrzyńczyk [657]. Of course, our chief concern will always be eigenfunctions of the ring $D(\mathcal{P}_n)$ of invariant differential operators whose structure was given in Theorem 1.1.2 of the preceding section.

A newer reference is the article of Richards on matrix argument special functions in the NIST Digital Library of Mathematical Functions (http://dlmf.nist.gov/). See F.W. Olver, D.W. Lozier, R.F. Boisvert, and C.W. Clark [481]. This replaces the old book by Milton Abramowitz and Irene Stegun [1]—a book so important to me when I first obtained a programmable calculator in 1973 or so that I needed 2 copies, one at home and one at UCSD. In those days my calculator (or even the UCSD computer I soon managed to use) did not know how to compute a K-Bessel function of 1 variable unless I could convince someone (namely, my ex) to write a program for it.

The most basic special function on $\mathcal{P}_n$ is a generalization of the power function y^s, $y \in \mathcal{P}_1 = \mathbb{R}^+$, $s \in \mathbb{C}$, appearing in the Mellin transform of Vol. I, Section 1.4. The **power function** $p_s(Y)$, for $Y \in \mathcal{P}_n$, and $s = (s_1, \ldots, s_n) \in \mathbb{C}^n$, is defined by:

$$p_s(Y) = \prod_{j=1}^{n} |Y_j|^{s_j}, \tag{1.41}$$

where $Y_j \in \mathcal{P}_j$ is the $j \times j$ upper left-hand corner in Y, $j = 1, 2, \ldots, n$. Note that when $n = 2$ and $Y \in \mathcal{SP}_2$, the power function can be identified with the function on the upper half plane defined by $p_s(x + iy) = y^s$ for $y > 0$ (see Exercise 3.1.9 of Volume I).

The power function (1.41) was introduced by Selberg [543, pp. 57–58]. Tamagawa [588, p. 369] calls it a right spherical function. In the language of Harish-Chandra and Helgason [275, p. 52], the power function is:

$$\exp[\lambda(H(gk))], \quad \text{for } g \in G, \ k \in K,$$

where $H(x) = \log A(x)$, if $x = K(x)A(x)N(x)$ is the Iwasawa decomposition of $x \in G$ (see Exercise 1.1.20 of Section 1.1.4). So H maps the group A of positive diagonal matrices into arbitrary diagonal matrices; i.e., into $\mathbb{R}^n$. Now λ is a linear functional on $\mathbb{R}^n$, which can be identified with an n-tuple of complex numbers. Thus the composition of exp, λ, and the H-function does indeed become a power function.

It is also possible to view the power function $p_s(Y)$ as a homomorphism of the **triangular group T_n of upper triangular matrices with positive diagonal entries**:

$$t = \begin{pmatrix} t_{11} & t_{12} & \cdots & t_{1n} \\ 0 & t_{22} & \cdots & t_{2n} \\ \vdots & \vdots & \ddots & \vdots \\ 0 & 0 & \cdots & t_{nn} \end{pmatrix}, \quad t_{jj} > 0. \tag{1.42}$$

If $r = (r_1, \ldots, r_n) \in \mathbb{C}^n$ and $t \in T_n$, define the **homomorphism** $\tau_r : T_n \to \mathbb{C} - 0$ by:

$$\tau_r(t) = \prod_{j=1}^{n} t_{jj}^{r_j}. \tag{1.43}$$

Clearly $\tau_r(t_1 t_2) = \tau_r(t_1)\tau_r(t_2)$. So you can think of τ_r as a homomorphism or, more significantly, as a **degree one representation** of the triangular group T_n. The following proposition relates (1.41) and (1.43).

Proposition 1.2.1 (Properties of the Power Function).

(1) Relation of p_s and τ_r.
   Suppose that $Y = I[t]$ for $t \in T_n$ as in (1.42). Then the power function from (1.41) and the group homomorphism (1.43) are related by

$$p_s(Y) = \tau_r(t), \quad \text{if } r_j = 2(s_j + \cdots + s_n)$$

and upon setting $r_{n+1} = 0$, we have

$$s_j = (r_j - r_{j+1})/2.$$

(2) Action of T_n on p_s.
   If $Y \in \mathcal{P}_n$ and $t \in T_n$, then $p_s(Y[t]) = p_s(Y)p_s(I[t])$.
(3) Power Functions Are Eigenfunctions of Invariant Differential Operators.
   If $L \in D(\mathcal{P}_n)$, then $Lp_s = \lambda_L(s)p_s$; i.e., p_s is an eigenfunction of L with eigenvalue $\lambda_L(s) = Lp_s(I)$.
(4) A Symmetry.
   Set $s = (s_1, \ldots, s_n)$, $s^* = (s_{n-1}, \ldots, s_2, s_1, -(s_1 + \cdots + s_n))$, and

$$\omega = \begin{pmatrix} 0 & & & 1 \\ & & \cdot & \\ & \cdot & & \\ 1 & & & 0 \end{pmatrix}.$$

   Then $p_s(Y^{-1}[\omega]) = p_{s^*}(Y)$, for all $Y \in \mathcal{P}_n$. Also $\omega^2 = I$ and $s^{**} = s$.

Proof. (1) Note that $|Y_j| = t_1^2 \cdots t_j^2$. The result follows easily.
(2) This follows from part (1) and the fact that τ_r is a homomorphism.
(3) Set $W = Y[t]$ for $Y \in \mathcal{P}_n$, $t \in T_n$. Then if $L \in D(\mathcal{P}_n)$, write L_W for L acting on the W-variable. We have the following equalities, if we make use of the G-invariance of L as well as part (2):

$$L_W p_s(W) = L_Y p_s(Y[t]) = p_s(I[t])L_Y p_s(Y).$$

   Then set $Y = I$ to obtain $W = I[t]$ and $Lp_s = \lambda_L(s)p_s$, with the eigenvalue stated in the proposition.
(4) See Exercise 1.2.1 below.   ∎

Note. The formula for the eigenvalue of the invariant differential operator L acting on the power function p_s is not a very useful one. In the case of the Poincaré upper half plane we had $\Delta y^s = s(s-1)y^s$ and it was clear that we could find powers $s \in \mathbb{C}$ to match any eigenvalue $\lambda = s(s-1)$, using high school algebra. To generalize this, we need a better formula for the eigenvalues $\lambda_L(s)$ than that in part (3). We will do better soon, with a fair amount of work, which was done for us by Maass [426].

Exercise 1.2.1. Show that if $t \in T_n$ and ω is as defined in part (4) of the preceding proposition, then

$$t[\omega] = \begin{pmatrix} t_{nn} & 0 & \cdots & 0 & 0 \\ * & t_{n-1,n-1} & \cdots & 0 & 0 \\ \vdots & \vdots & \ddots & \vdots & \vdots \\ * & * & \cdots & t_{22} & 0 \\ * & * & \cdots & * & t_{11} \end{pmatrix}.$$

Use this result to prove part (4) of the preceding proposition.

The power function is the appropriate kernel for the $\mathcal{P}_n$ analogue of the Mellin transform which we will call the Helgason–Fourier transform. We will see in Section 1.3 that the Helgason–Fourier transform does indeed have many of the properties of the usual Fourier and Mellin transforms. For example, the $\mathcal{P}_n$ Fourier transform does have an inversion formula (if one also includes a variable from $K = O(n)$ or the boundary K/M). It is also possible to consider an analogue of the Laplace transform on $\mathcal{P}_n$, a transform with a more elementary inversion formula, whose proof requires only ordinary Euclidean Fourier transforms and Cauchy's theorem in one variable as is seen in the next exercise.

Exercise 1.2.2 (The Laplace Transform). Define the **Laplace transform** of f : $\mathcal{P}_n \to \mathbb{C}$ at the symmetric matrix $Z \in \mathbb{C}^{n \times n}$ by:

$$\mathcal{L}f(Z) = \int_{Y \in \mathcal{P}_n} f(Y) \exp[-\mathrm{Tr}(YZ)] \, dY, \text{ where } dY = \prod_{1 \le i \le j \le n} dy_{ij}.$$

For sufficiently nice functions f, the integral above converges in a right half plane, $\mathrm{Re}\, Z > X_0$, meaning that $\mathrm{Re} Z - X_0 \in \mathcal{P}_n$. Show that the inversion formula for this transform is:

$$(2\pi i)^{-n(n+1)/2} \int_{\mathrm{Re}\, Z = X_0} \mathcal{L}f(Z) \exp[\mathrm{Tr}(YZ)] dZ = \begin{cases} f(Y), & \text{for } Y \in \mathcal{P}_n, \\ 0, & \text{otherwise.} \end{cases}$$

Here $dZ = \prod dz_{ij}$. And the integral is over symmetric matrices Z with fixed real part (in the domain of absolute convergence).

Hints. This result is discussed by Bochner [55, pp. 686–702], Bochner and Martin [56, pp. 90–92, 113–132], Herz [293, pp. 479–480], and Muirhead [468, p. 252]. You can use inversion of the Euclidean Fourier transform on the space of symmetric $n \times n$ real matrices and imitate the proof that worked for $n = 1$ (see Exercise 1.2.18 of Vol. I).

The most basic example of a Helgason–Fourier transform (or of a matrix Laplace transform) on $\mathcal{P}_n$ is the **gamma function** for $\mathcal{P}_n$ defined by:

$$\Gamma_n(s) = \int\limits_{Y \in \mathcal{P}_n} p_s(Y) \exp[-\text{Tr}(Y)] \, d\mu_n(Y), \qquad (1.44)$$

for $s \in \mathbb{C}^n$ with Re s_j sufficiently large; that is,

$$\text{Re}(s_j + \cdots + s_n) > (j-1)/2, \quad j = 1, \ldots, n.$$

In fact, we can use the Iwasawa decomposition to write $\Gamma_n(s)$ as a product of ordinary gamma functions $\Gamma_1 = \Gamma$:

$$\Gamma_n(s) = \pi^{n(n-1)/4} \prod_{j=1}^{n} \Gamma\left(s_j + \cdots + s_n - \frac{j-1}{2} \right). \qquad (1.45)$$

Exercise 1.2.3. Prove formula (1.45) by making the change of variables $Y = I[t]$, for $t \in T_n$ defined by (1.42), using formulas (1.38) and (1.39) from Section 1.1.6 to get

$$\Gamma_n(s) = 2^n \int\limits_{T_n} \exp\{-\text{Tr}(I[t])\} \, \tau_r(t) \prod_{j=1}^{n} t_{jj}^{-j} \prod_{1 \leq i \leq j \leq n} dt_{ij}.$$

Exercise 1.2.4. Show that

$$\int\limits_{Y \in \mathcal{P}_n} p_s(Y) \exp\left\{-\text{Tr}\left(YX^{-1}\right)\right\} \, d\mu_n(Y) = p_s(X) \Gamma_n(s).$$

Exercise 1.2.4 will be useful in the study of the algebra $D(\mathcal{P}_n)$ of invariant differential operators on $\mathcal{P}_n$—a study which was begun in § 1.1.5. It will also be necessary when we consider analogues of Hecke's correspondence between modular forms and Dirichlet series in later sections of this chapter and the next.

A special case of the product formula (1.45) for Γ_n was found in 1928 by the statistician Wishart [670]. A more general result is due to Ingham [320]. Later Siegel needed this special case of (1.45) in his work on quadratic forms (see [565, Vol. I, pp. 326–405]). Such gamma functions for $\mathcal{P}_n$ and more general domains of positivity are considered by Gindikin [218]. More general integrals of this type appear in quantum statistical mechanics (see the first edition of the random matrix book of Mehta [441, p. 40]) where a conjecture is given for the value of

$$\int_{\mathbb{R}^n} \exp\{-k^t xx\} \prod_{i<j} |x_i - x_j|^s \, dx. \qquad (1.46)$$

Selberg had already proved the conjecture 23 years earlier in [542]—a paper in Norwegian. Macdonald and Dyson have generalized the conjecture to arbitrary groups. See also Greg Anderson, Alice Guionnet, and Ofer Zeitouni [7]. Regev [502] has used a result of Bechner to prove Macdonald's conjecture for the main types of simple Lie groups. See Macdonald [427]. There is a special case of Selberg's formula in Exercise 1.2.5. We state the general formula in our discussion of quantum chaos in Section 1.3.5.

Next we compute the constant in the integral formula for polar coordinates (see formulas (1.37), (1.39), and (1.40)) from Section 1.1.6.

Proposition 1.2.2 (Volume of $O(n)$). Let $dk\ k^{-1} = (dh_{ij}(k))_{1 \le i < j \le n}$ and set $dh(k) = \prod_{1 \le i < j \le n} dh_{ij}(k)$. Then

$$\mathrm{Vol}(O(n)) = \int_{k \in K = O(n)} dh(k) = \frac{2^n \pi^{n^2/2}}{\Gamma_n \left(0, \ldots, 0, \frac{n}{2}\right)}.$$

Proof (From Muirhead [468, pp. 63–71]). First note that

$$\int_{X \in \mathbb{R}^{n \times n}} \exp[-\mathrm{Tr}(X\ {}^tX)]\, dX = \pi^{n^2/2}.$$

Now change variables via $X = kt$, for $k \in K = O(n)$ and $t \in T_n$ (the upper triangular matrices with positive diagonal entries). This is possible by the Gram-Schmidt orthogonalization process. Then

$$k^{-1} dX = dt + (k^{-1} dk)t,$$

and one finds the Jacobian of this change of variables to be:

$$\prod_{i=1}^{n} t_{ii}^{n-i}.$$

If we use the formula for changing variables from $Y \in \mathcal{P}_n$ to $t \in T_n$ via $Y = I[t]$ in Exercise 1.2.3, then we obtain

$$\pi^{n^2/2} = \mathrm{Vol}(K) \int_{T_n} \exp[-\mathrm{Tr}(I[t])] \prod_{i=1}^{n} t_{ii}^{n-i}\, dt = \mathrm{Vol}(K) 2^{-n} \Gamma_n \left(0, \ldots, 0, \frac{n}{2}\right).$$

This completes the proof. ∎

Note that the measure $dh(k)$ in Proposition 1.2.2 is an unnormalized Haar measure for K. However, in the integral formula for polar coordinates (last line of (1.39) and (1.40)) of Section 1.1.6, we normalized the Haar measure on K to obtain:

$$\int_{k \in K} dk = 1.$$

Now we can compute the constant in this integral formula.

Proposition 1.2.3 (The Constant in the Integral Formula for Polar Coordinates). *Let dk denote Haar measure on $K = O(n)$, normalized so that $\int_{k \in K} dk = 1$, and let*

$$\gamma(a) = \prod_{i=1}^{n} a_i^{-(n-1)/2} \prod_{1 \le i < j \le n} |a_i - a_j|, \quad \text{if } a = \begin{pmatrix} a_1 & \cdots & 0 \\ \vdots & \ddots & \vdots \\ 0 & \cdots & a_n \end{pmatrix} \in A,$$

with $a_i > 0$ and $da = \prod_{i=1}^{n} da_i/a_i$. Then the integral formula for polar coordinates, given in the last line of (1.39) and (1.40) of Section 1.1.6, is:

$$\int_{Y \in \mathcal{P}_n} f(Y) d\mu_n(Y) = c_n \int_{a \in A} \int_{k \in K} f(a[k]) \gamma(a) da \, dk,$$

with constant c_n given by

$$c_n^{-1} = \pi^{-n^2/2} n! \, \Gamma_n\left(0, \dots, 0, \frac{n}{2}\right) = \pi^{-(n^2+n)/4} \prod_{j=1}^{n} j \, \Gamma(j/2).$$

Proof. It suffices to prove the result for K bi-invariant functions f; i.e., $f(a) = f(a[k])$, for $a \in A$ and $k \in K$. Then, from Exercise 1.1.24 of § 1.1.4, we have

$$\int_{Y \in \mathcal{P}_n} f(Y) d\mu_n(Y) = \frac{\text{Vol}(K)}{2^n \, n!} \int_{a \in A} f(a) \gamma(a) da,$$

where the volume of K is computed as in Proposition 1.2.2. So the formula for this volume which is given in Proposition 1.2.2 completes the proof of Proposition 1.2.3.

∎

Exercise 1.2.5 (Evaluation of a Special Case of Selberg's Integral). Set

$$D(a) = \prod_{1 \le i < j \le n} |a_i - a_j|, \quad \text{for } a \in A = \text{positive diagonal matrices.}$$

Then a limiting case of Selberg's integral [542] is:

$$S(p, z) = \int_{a \in A} |a|^p D(a)^{2z} \exp[-\mathrm{Tr}(a)] \, da$$

$$= \prod_{k=1}^{n} \Gamma(1 + kz)\Gamma(p + (k-1)z)/\Gamma(1 + z).$$

Check that the formula (1.45) for $\Gamma_n(s)$ and Proposition 1.2.3 give Selberg's result in the case $z = \frac{1}{2}$. Selberg [542] gives evaluations of integrals which appear in random matrix theory (see Mehta [441] and Section 1.3.5).

Now we want to return to our study of $D(\mathcal{P}_n)$, the G-invariant differential operators on $\mathcal{P}_n$ (see Section 1.1.5). First we need a few definitions. We define the **adjoint** L^* of a differential operator L in $D(\mathcal{P}_n)$ by the following formula, assuming that f, g are such that the integrals converge:

$$\int_{Y \in \mathcal{P}_n} (Lf)(Y)\overline{g(Y)} \, d\mu_n(Y) = \int_{Y \in \mathcal{P}_n} f(Y)\overline{(L^*g)(Y)} \, d\mu_n(Y). \tag{1.47}$$

The geodesic-reversing isometry σ of $\mathcal{P}_n$ at the identity is $\sigma(Y) = Y^{-1}$ and we can define L^σ for $L \in D(\mathcal{P}_n)$ by:

$$L^\sigma f = L(f \circ \sigma^{-1}) \circ \sigma, \quad \text{where "} \circ \text{" denotes composition of functions.} \tag{1.48}$$

In Theorem 1.2.1 we will show that $L^\sigma = \overline{L}^*$. In order to do this, we will need a result about the eigenvalues of the invariant differential operators acting on power functions. To put this result in its best form, we need the **proper normalization of variables in the power function** (Selberg [543, p. 57]) defined on the triangular group T_n by:

$$\varphi_r(t) = \prod_{i=1}^{n} t_{ii}^{2r_i + i - (n+1)/2}, \quad \text{for } t \in T_n, \; r \in \mathbb{C}^n. \tag{1.49}$$

Exercise 1.2.6. Find $s = s(r)$ such that $p_s(I[t]) = \varphi_r(t)$.

Selberg [543, p. 58] states the following theorem.

Theorem 1.2.1 (Normalized Power Functions and Invariant Differential Operators).

(1) If $L_j = \mathrm{Tr}((Y\partial/\partial Y)^j)$, for $j = 1, 2, \ldots, n$, as in Theorem 1.1.2 of Section 1.1.5, and φ_r is the normalized power function from formula (1.49), then

$$L_i \varphi_r = \lambda_i(r)\varphi_r$$

and $\lambda_i(r)$ is a symmetric polynomial in r_j of degree i and having the form:

$$\lambda_i(r_1, \ldots, r_n) = r_1^i + \cdots + r_n^i + \text{terms of lower degree.}$$

(2) The effect of $L \in D(\mathcal{P}_n)$ on power functions $p_s(Y)$ determines L uniquely.
(3) For $L \in D(\mathcal{P}_n)$, using the notation (1.47) and (1.48), we have

$$L^\sigma = \overline{L}^*,$$

with $\overline{}$ denoting complex conjugation.

Proof (Maass [426, pp. 70–76]).

(1) We shall use induction on n. Recall that we showed in Section 1.1.5 that if $Y \in \mathcal{P}_n$ has partial Iwasawa decomposition

$$Y = \begin{pmatrix} V & 0 \\ 0 & w \end{pmatrix} \begin{bmatrix} I & x \\ 0 & 1 \end{bmatrix}, \text{ for } V \in \mathcal{P}_{n-1}, \ w > 0, \ x \in \mathbb{R}^{n-1}, \tag{1.50}$$

then, setting $\Omega = x \frac{{}^t \partial}{\partial x}$,

$$Y \frac{\partial}{\partial Y} = \begin{pmatrix} V \frac{\partial}{\partial V} - \frac{1}{2}\,{}^t\Omega & \frac{1}{2}\frac{\partial}{\partial x} \\ {}^t x V \frac{\partial}{\partial V} - w \frac{\partial}{\partial w}\,{}^t x - \frac{1}{2}\,{}^t x\,{}^t\Omega + \frac{1}{2} w \frac{{}^t \partial}{\partial x} V^{-1} & \frac{1}{2}\,{}^t x \frac{\partial}{\partial x} + w \frac{\partial}{\partial w} \end{pmatrix}. \tag{1.51}$$

If L_1 and L_2 are matrix differential operators, we will write $L_1 \sim L_2$ if they agree on functions on $\mathcal{P}_n$ which are independent of the x-variable in (1.50).

It can be proved inductively that if

$$A_h = \left(V \frac{\partial}{\partial V} + \frac{1}{2} I \right)^h - \frac{1}{2} \sum_{j=0}^{h-1} \left(V \frac{\partial}{\partial V} + \frac{1}{2} I \right)^j \left(w \frac{\partial}{\partial w} \right)^{h-1-j}, \tag{1.52}$$

then

$$\left(Y \frac{\partial}{\partial Y} \right)^h \sim \begin{pmatrix} A_h & 0 \\ {}^t x A_h - \left(w \frac{\partial}{\partial w} \right)^h {}^t x & \left(w \frac{\partial}{\partial w} \right)^h \end{pmatrix}. \tag{1.53}$$

Exercise 1.2.7. Prove formula (1.53), using (1.52) to define A_h and induction.
Hint. Note that if c is a real variable which does not depend on x, but may depend on the other variables, then

$$\frac{\partial}{\partial x} \left(c\,{}^t x \right) = cI,$$

where I denotes the $(n-1) \times (n-1)$ identity matrix.

It follows from (1.53) that

$$\mathrm{Tr}\left(\left(Y\tfrac{\partial}{\partial Y}\right)^h\right) \sim \mathrm{Tr}\left(\left(V\tfrac{\partial}{\partial V} + \tfrac{1}{2}I\right)^h\right) + \left(w\tfrac{\partial}{\partial w}\right)^h$$
$$-\tfrac{1}{2}\sum_{j=0}^{h-1}\mathrm{Tr}\left(\left(V\tfrac{\partial}{\partial V}+\tfrac{1}{2}I\right)^j\right)\left(w\tfrac{\partial}{\partial w}\right)^{h-1-j}. \tag{1.54}$$

This formula is peculiarly unsymmetrical.

Next, set up the notation, $Y = I[t]$, for $t \in T_n$,

$$V = \begin{pmatrix} t_{11}^2 & & 0 \\ & \ddots & \\ 0 & & t_{n-1,n-1}^2 \end{pmatrix}\begin{bmatrix} 1 & & * \\ & \ddots & \\ 0 & & 1 \end{bmatrix}, \quad w = t_{nn}^2,$$

$$\varphi_r(Y) = \prod_{j=1}^{n} t_{jj}^{2r_j+j-(n+1)/2}, \quad \varphi_a(V) = \prod_{j=1}^{n-1} t_{jj}^{2a_j+j-n/2},$$

$$r = (a,b),\ a \in \mathbb{C}^{n-1},\ b \in \mathbb{C}.$$

Then

$$\varphi_r(Y) = |V|^{-1/4}w^{b+(n-1)/4}\varphi_a(V). \tag{1.55}$$

Now we need the following exercise.

Exercise 1.2.8. Show that

$$\mathrm{Tr}\left(\left(V\tfrac{\partial}{\partial V}\right)^h\right)|V|^m = |V|^m\,\mathrm{Tr}\,(m^h I),$$

where I denotes the $(n-1) \times (n-1)$ identity matrix.

Hint. Use expansion of $|V|$ by minors to find $\partial|V|/\partial v_{ij}$ is the i,j cofactor of V, that is,

$$N_{ij} = (-1)^{i+j}|M_{ij}|,$$

where M_{ij} is the matrix obtained from V by crossing out the ith row and the jth column of V. Then

$$|V| = \sum_{i=1}^{n} v_{ij}|N_{ij}|$$

and $V^{-1} = |V|^{-1}\,{}^t(N_{ij})$. Note that we need $\tfrac{1}{2}$ in the off-diagonal entries of $\partial/\partial V$ in order to make this exercise work for symmetric matrices.

Putting together formulas (1.54), (1.55) and Exercise 1.2.8, we see that

$$|V|^{\frac{1}{4}} \, w^{\frac{1-n}{4}} \, \mathrm{Tr} \left(\left(Y \tfrac{\partial}{\partial Y} \right)^h \right) \varphi_r(Y) = \left\{ \mathrm{Tr} \left(\left(V \tfrac{\partial}{\partial V} + \tfrac{1}{4}I \right)^h \right) + \left(w \tfrac{\partial}{\partial w} + \tfrac{n-1}{4} \right)^h \right.$$

$$\left. - \tfrac{1}{2} \sum_{j=0}^{h-1} \mathrm{Tr} \left(\left(V \tfrac{\partial}{\partial V} + \tfrac{1}{4}I \right)^j \right) \left(w \tfrac{\partial}{\partial w} + \tfrac{n-1}{4} \right)^{h-1-j} \right\} \varphi_a(V) w^b. \qquad (1.56)$$

Now the eigenvalue of interest is:

$$\lambda_h(r) = \left[\mathrm{Tr} \left(\left(Y \frac{\partial}{\partial Y} \right)^h \right) \varphi_r(Y) \right]_{Y=I}. \qquad (1.57)$$

Exercise 1.2.9. Check that when $n = 2$ the eigenvalue defined by formula (1.57), with $Y \in \mathcal{P}_2$, is a symmetric polynomial in r_1 and r_2 having highest degree term

$$r_1^h + r_2^h, \quad h = 1, 2, 3, \dots.$$

The complete polynomial is:

$$\left(r_1 + \frac{1}{4} \right)^h + \left(r_2 + \frac{1}{4} \right)^h - \frac{1}{2} \sum_{j=0}^{h-1} \left(r_1 + \frac{1}{4} \right)^j \left(r_2 + \frac{1}{4} \right)^{h-1-j}.$$

Now suppose $n \geq 3$ and proceed by induction, assuming that

$$\mathrm{Tr} \left(\left(V \frac{\partial}{\partial V} + \frac{1}{4}I \right)^h \right) \varphi_a(V) = \xi_h(a)\varphi_a(V), \quad \text{where } a_1 = r_1, \dots, a_{n-1} = r_{n-1}$$

and $\xi_h(a)$ is a symmetric polynomial in a with highest degree term

$$r_1^h + \cdots + r_{n-1}^h.$$

Clearly

$$\left(w \frac{\partial}{\partial w} + \frac{n-1}{4} \right)^h w^b = v_h(b)w^b, \quad b = r_n, \quad v_h(b) = \left(b + \frac{n-1}{4} \right)^h.$$

Then, by formula (1.56), $\lambda_h(r)$ is invariant under permutations of $r_1, \dots, r_{n-1}$ and $\lambda_h(r)$ has highest degree term $r_1^h + \cdots + r_{n-1}^h + r_n^h$, for

$$\lambda_h(r) = \xi_h(a) + \nu_h(b) - \frac{1}{2} \sum_{j=0}^{h-1} \xi_j(a)\nu_{h-1-j}(b).$$

In order to see that λ_h is symmetric in the last variable too, we have to note that you could also do this expansion with

$$Y = \begin{pmatrix} v & 0 \\ 0 & W \end{pmatrix} \begin{bmatrix} 1 & {}^t x \\ 0 & I \end{bmatrix}, \quad v > 0, \quad W \in \mathcal{P}_{n-1}, \quad x \in \mathbb{R}^{n-1}.$$

Then you would find $\lambda_h(r)$ to be symmetric in $r_2, \ldots, r_n$, by the induction assumption. In fact, you can make the same argument with an arbitrary partial Iwasawa decomposition as in Exercise 1.2.11 below. This completes the proof of part (1), since we have checked the case $n = 2$, by Exercise 1.2.9.

(2) Suppose $L = p(L_1, \ldots, L_n)$, where p is a polynomial in n indeterminates. and $L_j = \mathrm{Tr}((Y\partial/\partial Y_j))$, for $j = 1, 2, \ldots, n$. Then $L\varphi_r = 0$ implies that $p(\lambda_1(r), \ldots, \lambda_n(r))$ vanishes for all $r \in \mathbb{C}^n$. But then p must vanish identically, since the eigenvalues $\lambda_j(r)$ form a basis for the symmetric polynomials in $r_1, \ldots, r_n$ by part (1). Thus the mapping from $\mathbb{C}^n$ to $\mathbb{C}^n$ which takes r to $(\lambda_1(r), \ldots, \lambda_n(r))$ is onto.

Exercise 1.2.10. Suppose that $\lambda_1(r), \ldots, \lambda_n(r)$ form a basis for all the symmetric polynomials in $r_1, \ldots, r_n$.

(a) Prove that the mapping from $\mathbb{C}^n$ to $\mathbb{C}^n$ taking r to $(\lambda_1(r), \ldots, \lambda_n(r))$ is onto.
(b) Prove that $\lambda_j(r) = \lambda_j(r')$ for all $j = 1, \ldots, n$, implies that $r' = (r_{\sigma(1)}, \ldots, r_{\sigma(n)})$ for some permutation σ of n elements.

Hint. Let $u_j(r) = $ the jth elementary symmetric polynomial. Then

$$\sum_{j=0}^{n} u_j(s)x^j = \prod_{j=1}^{n} (x - s_j), \quad \text{for } s \in \mathbb{C}^n.$$

(3) First, note that $k(X, Y) = \exp(-\mathrm{Tr}(YX^{-1}))$ is a point-pair invariant leading to a convolution operator as in § 1.1.4. If $L \in D(\mathcal{P}_n)$, write $L_X k(X, Y)$ when L acts on the X-variable and $L_Y k(X, Y)$ when L acts on the Y-variable. Just as in the proof of part 2 of Lemma 1.1.2 in § 1.1.5, we use the fact that

$$L_X^\sigma k(X, Y) = L_Y k(X, Y).$$

This fact implies the second in the following sequence of equalities which stem from Exercise 1.2.4:

$$(L^\sigma p_s)(X)\Gamma_n(s) = \int_{Y \in \mathcal{P}_n} \{L_X^\sigma \exp(-\mathrm{Tr}(YX^{-1}))\} p_s(Y) \, d\mu_n(Y)$$

$$= \int_{Y \in \mathcal{P}_n} \{L_Y \exp(-\mathrm{Tr}(YX^{-1}))\} p_s(Y)\, d\mu_n(Y)$$

$$= \int_{Y \in \mathcal{P}_n} \exp(-\mathrm{Tr}(YX^{-1}))(\overline{L}^* p_s(Y))\, d\mu_n(Y)$$

$$= (\overline{L}^* p_s)(X) \Gamma_n(s).$$

Here we use the fact that the adjoint operator is also G-invariant. This completes the proof of Theorem 1.2.1. ∎

Exercise 1.2.11. Run through the preceding proof for a general Iwasawa decomposition

$$Y = \begin{pmatrix} V & 0 \\ 0 & W \end{pmatrix} \begin{bmatrix} I & X \\ 0 & I \end{bmatrix}, \quad \text{for} \quad V \in \mathcal{P}_r, \ W \in \mathcal{P}_{n-r}, \ X \in \mathbb{R}^{r \times (n-r)}.$$

Hint. (See Maass [426, pp. 70–76].)

Our study of the gamma function for $\mathcal{P}_n$ is now at an end. It will find applications in statistics at the end of the section. And these Γ-functions will also appear in functional equations of L-functions and Eisenstein series for $GL(n, \mathbb{Z})$ in Section 1.3.

One can also study **matrix incomplete gamma functions**. We saw an example of these incomplete gamma functions in Exercise 3.6.5 in Volume I when we obtained the analytic continuation of L-functions corresponding to Maass wave forms. These incomplete gamma functions appear in the analytic continuation of Dedekind zeta functions of number fields as well as in the analytic continuation of Eisenstein series for $GL(n, \mathbb{Z})$ (see Section 1.4 of Volume I and Section 1.5 which follows). More information on incomplete gamma functions can be found in Terras [602]. **Matrix beta functions** will arise in Section 1.3 as part of the computation of the Plancherel or spectral measure for inversion of the Helgason–Fourier transform on $\mathcal{P}_n$. See also Gindikin [218].

Exercise 1.2.12. (a) Consider the power function given by formula (1.49) and the operator

$$L_j = \mathrm{Tr}((Y\partial/\partial Y)^j), \quad \text{for} \ Y \in \mathcal{P}_n.$$

If $L_j \varphi_r(Y) = \lambda_j^n(r)\varphi_r(Y)$, show that the eigenvalue of L_1 is $\lambda_1^n(r) = r_1 + \cdots + r_n$ and the eigenvalue of $\Delta_Y = L_2$ is $\lambda_2^n(r) = r_1^2 + \cdots + r_n^2 + (n - n^3)/48$.

(b) Then show that if instead we consider the Laplacian on the determinant one surface $\mathcal{SP}_3$ as in Exercise 1.1.32 of Section 1.1.6, we find that

$$\Delta p_s(Y) = \left\{ \frac{2}{3}(s_1^2 + s_1 s_2 + s_2^2) + s_1 + s_2 \right\} p_s(Y).$$

Hint. (a) You can use formulas (1.55) and (1.56) or you can use formula (1.34) of Section 1.1.5 to see that

$$\Delta_Y \sim (w\partial/\partial w)^2 + ((1-n)/2)w\partial/\partial w + \Delta_V + \frac{1}{2}\text{Tr}(V\partial/\partial V).$$

Now formula (1.55) implies that:

$$\lambda_2^n(r) = [\Delta_Y |V|^{-1/4}\varphi_a(V)w^{b+(n-1)/4}]_{Y=I}$$

$$= \left(b + \tfrac{n-1}{4}\right)^2 + \tfrac{1-n}{2}\left(b + \tfrac{n-1}{4}\right) + \lambda_2^{n-1}\left(r_1 - \tfrac{1}{4}, \cdots, r_{n-1} - \tfrac{1}{4}\right)$$

$$+ \tfrac{1}{2}\lambda_1^{n-1}\left(r_1 - \tfrac{1}{4}, \cdots, r_{n-1} - \tfrac{1}{4}\right).$$

1.2.2 K-Bessel Functions

K-Bessel functions for $\mathcal{P}_n$ have been discussed by various authors with vastly different points of view. We will attempt to say a little more about some of the other developments at the end of this section. The closest references to our treatment are: Herz [293], Bengtson [42], Bump [83], Kaori Imai (Ota) and Terras [318], Maass [426, Ch. 18], and Terras [605–607]. Bessel functions analogous to the classical J-Bessel function are to be found in Bochner [55], Godement's article in Séminaire Cartan [547, exposé 9], Gelbart [207], and Gross et al. [246]. The classical Whittaker functions are confluent hypergeometric functions generalizing K-Bessel functions (see Lebedev [398]). Whittaker functions and Whittaker models for representations of real, complex, p-adic and adelic groups are discussed by Bump [83], Goldfeld [230], Hashizume [265], Jacquet [323], Piatetski-Shapiro in Borel and Casselman [66, Vol. I, pp. 209–212], Schiffman [534], Shalika [552], and Shintani [560]. Related references are Goodman [234], Goodman and Wallach [235], and Kostant [367]. Hypergeometric functions of matrix argument are also considered by Gindikin [218], Gross and Richards [248], Herz [293], James [328–330], Maass [426, Chapter 18], Muirhead [468], and Shimura [555].

Many of the preceding references are motivated by the number-theoretical problem of obtaining Fourier expansions of automorphic forms and this will be our main application (see Section 1.5 and the references mentioned there).

Others seek to solve statistical problems such as that of finding the noncentral Wishart distribution (see Herz [293], Muirhead [468], and the discussion at the end of this section). Still others seek uniqueness results about representations.

The K-Bessel functions which we study are not the most general of those mentioned above, but the suffice for our purposes and to give an introduction to the subject. Consideration of Kirillov's theory of the representations of the nilpotent group N (see Kirillov [348], Proskurin [494], and Moore [459]) serves to clarify the concepts. It is also useful to view Bessel and Whittaker functions in the light of the theory of the operators intertwining pairs of representations (see Dieudonné [137, Vol. VI], Hashizume [265], Kirillov [349], Mackey [430, pp. 363 ff.], and Vilenkin [635, Ch. VIII]). However, we will not delve into group representations in this volume.

A **character** of an abelian group G is a group homomorphism from G into $\mathbb{T}$, the multiplicative group of complex numbers of norm 1. Define the abelian group $N(m, n - m)$, for $1 \leq m < n$, and define the character χ_A of $N(m, n - m)$ for fixed $A \in \mathbb{R}^{m \times (n-m)}$ by:

$$
\begin{aligned}
N(m, n - m) &= \left\{ U = \begin{pmatrix} I & X \\ 0 & I \end{pmatrix} \middle| X \in \mathbb{R}^{m \times (n-m)} \right\}, \\
\chi_A \begin{pmatrix} I & X \\ 0 & I \end{pmatrix} &= \exp(2\pi i \, \mathrm{Tr}({}^t A \, X)), \quad \text{for } X \in \mathbb{R}^{m \times (n-m)}.
\end{aligned}
\tag{1.58}
$$

Using the notation (1.58), we will say that $f : \mathcal{P}_n \to \mathbb{C}$ is a **K-Bessel function** if, for some fixed $A \in \mathbb{R}^{m \times (n-m)}$, f has the following 3 properties:

(a) f transforms by $N(m, n - m)$ according to χ_A;
    i.e., $f(Y[U]) = \chi_A(U) f(Y), \quad \forall \, Y \in \mathcal{P}_n, \ U \in N(m, n - m)$;
(b) f is an eigenfunction for all the G-invariant differential operators $L \in D(\mathcal{P}_n)$;
(c) f grows at most like a power function at the boundary.

$$
\tag{1.59}
$$

This definition is analogous to (3.14) of Volume I. However, if one simply thinks of the behavior of $K_s(y)$, as y approaches infinity, one might think that the growth condition (c) is somewhat weak. But recall that in the case of $SL(2, \mathbb{R})$ we found that, for f as in (1.59),

$$
f\left(\begin{pmatrix} 1/y & 0 \\ 0 & y \end{pmatrix} \begin{bmatrix} 1 & x \\ 0 & 1 \end{bmatrix} \right) = c y^{1/2} K_{s-1/2}(2\pi |R| y),
$$

if $R \neq 0$ (see Exercise 3.2.1 of Vol. I). Here $K_s(y)$ denotes the ordinary K-Bessel function. As y approaches infinity, the function $K_s(y)$ approaches zero exponentially. But (c) in (1.59) is still O.K., since for $\mathrm{Re}\, s > 0$, as y approaches 0, $K_s(y)$ blows up like y^{-s} (see Exercise 3.2.2 in Vol. I). Moreover, if $R = 0$, then we obtain (for the case of $SL(2, \mathbb{R})$):

$$
f\left(\begin{pmatrix} 1/y & 0 \\ 0 & y \end{pmatrix} \begin{bmatrix} 1 & x \\ 0 & 1 \end{bmatrix} \right) = c y^s + d y^{1-s}.
$$

Just as in the case $n = 2$ (see Vol. I, Section 3.5) these K-Bessel functions appear in Fourier expansions of Maass forms for $GL(n, \mathbb{Z})$, that is, expansions with respect to the abelian groups $N(m, n - m)$ in (1.58) above (cf. Kaori Imai (Ota) and Terras [318], and Terras [605–607]). Such expansions are analogous to those used by Siegel in his study of Siegel modular forms (see Siegel [565, Vol. III, pp. 97–137], and Section 1.5).

Property (a) of (1.59) says that we are studying a special function corresponding to a representation of G induced from the character of $N(m, n-m)$ given by $\chi_A(U)$, for $U \in N(m, n-m)$, using the notation of (1.58).

Kirillov [348] shows that (up to equivalence) one obtains the infinite dimensional irreducible unitary representations of the nilpotent group N of upper triangular matrices with one on the diagonal by inducing the representations corresponding to a character of $N(m, n-m)$ with $m = [n/2]$. The finite dimensional (actually one-dimensional) irreducible unitary representations of N come from a different construction which we shall discuss at the end of this section in connection with Whittaker functions. It is only the infinite dimensional representations that contribute to the Plancherel formula for N.

In a sense, the K-Bessel functions considered here are analogous to the Eisenstein series for maximal parabolic subgroups of $GL(n)$ which will be discussed in Section 1.5. The Whittaker functions to be considered at the end of this section are similar to the Eisenstein series for minimal parabolic subgroups of $GL(n)$ which will also be studied in Section 1.5.

It is easy to give examples of functions satisfying the conditions in (1.59). In what follows we will find it natural to define two sorts of K-Bessel functions. To distinguish them, we use the capital "K" for the function in (1.61) below and the small "k" for the function in (1.60) below. Part (2) of Theorem 1.2.2 shows that the two functions are really essentially the same. Imitating formula (3.16) in Section 3.2 of Vol. I, we define **the first type of matrix k-Bessel function** to be:

$$k_{m,n-m}(s|Y,A) = \int_{X \in \mathbb{R}^{m \times (n-m)}} p_{-s}\left(Y^{-1}\begin{bmatrix} I & 0 \\ {}^t X & I \end{bmatrix}\right) \exp\{2\pi i \, \mathrm{Tr}({}^t AX)\} \, dX, \qquad (1.60)$$

for $s \in \mathbb{C}^n$ with coordinates restricted to suitable half planes, $Y \in \mathcal{P}_n$, $A \in \mathbb{R}^{m \times (n-m)}$, $1 \le m < n$. Here $p_s(Y)$ denotes the power function (1.41). Formula (1.60) is useful for demonstrating that $k_{m,n-m}$ satisfies (1.59), parts (a) and (b), since it is clearly an eigenfunction for any differential operator in $D(\mathcal{P}_n)$ and has the correct invariance property under transformation by elements of $N(m, n-m)$.

The **second type of K-Bessel function** is defined by:

$$K_m(s \mid V, W) = \int_{Y \in \mathcal{P}_m} p_s(Y) \exp\{-\mathrm{Tr}(VY + WY^{-1})\} \, d\mu_m(Y), \qquad (1.61)$$

for $V, W \in \mathcal{P}_m$, $s \in \mathbb{C}^m$, or W singular with $\mathrm{Re}\, s_j$ suitably restricted. The function (1.61) generalizes the formula in part (a) of Exercise 3.2.1 in Vol. I. This second type of K-Bessel function is useful in the study of convergence properties and analytic continuation in the s-variable. Herz [293, p. 506] considers the special case of (1.61) with $s_j = 0$ for $j \neq n$.

At this point, it is not clear how K_m is related to $k_{m,n-m}$. It will turn out that Bengtson's formula in Theorem 1.2.2 relates the two functions and thus gives a generalization of the result in Exercise 3.2.1 of Vol. I. We review this latter result in the next example.

Example 1.2.1 (The One Variable Case). When $m = 1$, formula (1.61) is the ordinary K-Bessel function K_s defined in Exercise 3.2.1 (a) of Volume I, since for $a, b > 0$, $s \in \mathbb{C}$:

$$K_1(s \mid a, b) = \int_0^\infty y^{s-1} \exp\{-(ay + b/y)\}\, dy = 2 \left(\frac{b}{a}\right)^{s/2} K_s\left(2\sqrt{ab}\right).$$

When $n = 2$ and $m = 1$, $a \in \mathbb{R}$, by part (a) of the same exercise, we have:

$$k_{1,1}\left(s, 0 \left| \begin{pmatrix} 1/y & 0 \\ 0 & y \end{pmatrix}, a \right.\right) = \int_{x \in \mathbb{R}} p_{-s}\left(\begin{pmatrix} y & 0 \\ 0 & 1/y \end{pmatrix}\begin{bmatrix} 1 & 0 \\ x & 1 \end{bmatrix}\right) \exp(2\pi i a x)\, dx$$

$$= y^s \int_{x \in \mathbb{R}} (y^2 + x^2)^{-s} \exp(2\pi i a x)\, dx$$

$$= \begin{cases} 2\pi^{1/2}\Gamma(s)^{-1}|\pi a|^{s-1/2} y^{1/2} K_{s-1/2}(2\pi |a|y), & \text{if } a \neq 0, \\ \Gamma(\tfrac{1}{2})\Gamma(s - \tfrac{1}{2})\Gamma(s)^{-1} y^{1-s}, & \text{if } a = 0. \end{cases}$$

In the next example, we see that our matrix argument k-Bessel functions can sometimes be factored into products of ordinary K-Bessel functions and Γ-functions. However, we must caution the reader that this does not seem to be a general phenomenon. Thus these Bessel functions differ greatly from the gamma functions considered in the last section.

Example 1.2.2 (A Factorization in a Special Case). Using the first remarks in the proof of part (5) in Theorem 1.2.2 below, we find that:

$$k_{2,1}(s_1, s_2, 0 | I, (a, 0)) = \int\int (1 + x_1^2)^{-s_1}(1 + x_1^2 + x_2^2)^{-s_2} \exp(2\pi i a x_1)\, dx_1 dx_2$$

$$= \int (1 + x_1^2)^{-s_1 - s_2 + 1/2} \exp(2\pi i a x_1)\, dx_1 \int (1 + y^2)^{-s_2}\, dy$$

$$= k_{1,1}\left(s_1 + s_2 - \frac{1}{2}, 0 \left| I, a\right.\right) k_{1,1}(s_2, 0 | I, 0)$$

$$= k_{1,1}\left(s_1 + s_2 - \frac{1}{2}, 0 \left| I, a\right.\right) B\left(\frac{1}{2}, s_2 - \frac{1}{2}\right),$$

where we have used the substitution $x_2 = (1 + x_1^2)^{1/2} y$ and the **beta function** is $B(p,q) = \Gamma(p)\Gamma(q)/\Gamma(p+q)$. The method of Example 1.2.2 extends to $k_{m,1}$ by part (5) of Theorem 1.2.2:

$$k_{m,1}(s_1, s_2, 0 \mid I_{m+1}, (a_1, 0)) = k_{m-1,1}(s_2, 0 \mid I_m, 0) \, k_{1,1}(-b, 0 \mid I, a_1),$$

where $s_1 \in \mathbb{C}$, $s_2 \in \mathbb{C}^{m-1}$, $a_1 \in \mathbb{R}$, $b = (m-1)/2 - \sum_{j=1}^{m} s_j$. However, when $a_2 \neq 0$, there does not appear to be such a factorization. Thus the k- and K-Bessel functions for $\mathcal{P}_n$ do not, in general, appear to factor into products of ordinary K-Bessel functions.

Exercise 1.2.13. Prove that the first matrix k-Bessel function $k_{m,n-m}(s|Y,A)$ in (1.60) is an eigenfunction for all the differential operators in $D(\mathcal{P}_n)$ when considered as a function of $Y \in \mathcal{P}_n$. And show that it has the invariance property (a) in (1.59) (again when considered as a function of Y).

The following exercise generalizes the first asymptotic result on the ordinary K-Bessel function in Exercise 3.2.2 in Vol. I.

Exercise 1.2.14. (a) Show that $K_m(s|I, 0) = \Gamma_m(s)$, where the Γ-function is defined in (1.44).
(b) Show that $K_m(s|A, B) \sim p_s(A^{-1})\Gamma_m(s)$, as $B \to 0$, for fixed $A \in \mathcal{P}_m$.

Exercise 1.2.15. Show that if $y > 0$ and $a \in \mathbb{R}$, then

$$k_{1,1}\left(s, 0 \,\left|\, \begin{pmatrix} 1/y & 0 \\ 0 & y \end{pmatrix}, a \right.\right) = y^{1-s} k_{1,1}(s, 0 \mid I_2, ay), \text{ for } s \in \mathbb{C}.$$

Exercise 1.2.16. Suppose that g is an element of the triangular group T_m defined in (1.42). Show that

$$K_m\left(s \mid V[{}^t g], W[g^{-1}]\right) p_s(I[g]) = K_m(s \mid V, W).$$

Exercise 1.2.16 shows that we can reduce one of the positive matrix arguments in K_m to the identity. However, it is convenient for our purposes to separate the arguments V and W An illustration of this convenience can be found, for example, in Exercise 1.2.14, where we see that we can treat the case that one of the arguments is singular.

The following theorem gives the main properties (known to the author) of these matrix argument K-Bessel functions. It is mainly due to Tom Bengtson [42].

Theorem 1.2.2 (Properties of Matrix K-Bessel Functions).

(1) Convergence and Decay at Infinity.
   Suppose that λ is the smallest element in the set of eigenvalues of V and W in $\mathcal{P}_m$. Then

$$K_m(s \mid V, W) = O(\lambda^{-m(m+1)/4} \exp(-2m\lambda)), \quad \text{as } \lambda \to \infty,$$

for fixed s. In particular, the integral (1.61) converges for all $s \in \mathbb{C}^n$ if $V, W \in \mathcal{P}_m$. And $K_m(s|V, W) \to 0$ exponentially as the eigenvalues of V and W all go to infinity.

(2) Bengtson's Formula Relating the Two Bessel Functions.
Let $s \in \mathbb{C}^m$,

$$s^\# = -s + (0, \ldots, 0, (n - m)/2) \quad \text{and} \quad s^* = (s_{m-1}, \ldots, s_1, -(s_1 + \cdots + s_m)).$$

Then, assuming that the coordinates of s are restricted to suitable half planes:

$$\Gamma_m(-s^*) k_{m,n-m} \left(s, 0 \left| \begin{pmatrix} V & 0 \\ 0 & W \end{pmatrix}, A \right. \right) = \pi^{m(n-m)/2} |W|^{m/2} K_m \left(s^\# \mid W[\pi\ {}^t A], V^{-1} \right).$$

(3) K-Bessel Functions with a Singular Argument Reduce to Lower Rank K-Bessel Functions.
For $\sigma_1 \in \mathbb{C}^m, \sigma_2 \in \mathbb{C}^{n-m}, V \in \mathcal{P}_m, W \in \mathcal{P}_{n-m}$, let

$$p_s \begin{pmatrix} V & 0 \\ 0 & W \end{pmatrix} = p_{\sigma_1}(V) p_{\sigma_2}(W) |W|^{m/2}.$$

Then

$$K_n \left(s \left| \begin{pmatrix} A & 0 \\ 0 & B \end{pmatrix} \begin{bmatrix} I & 0 \\ {}^t C & I \end{bmatrix}, \begin{pmatrix} 0 & 0 \\ 0 & D \end{pmatrix} \right. \right)$$

$$= \pi^{m(n-m)/2} |B|^{-m/2} p_{\sigma_1}(A^{-1}) \Gamma_m(\sigma_1) K_{n-m}(\sigma_2 | B, D).$$

We need to assume that $\operatorname{Re} \sigma_1$ is sufficiently large for the convergence of Γ_m.

(4) The Argument in $\mathcal{P}_n$ of the Matrix k-Bessel Function Can Be Reduced to I.
Let $V = g\ {}^t g$ for $g \in T_m$; i.e., g is upper triangular with positive diagonal. If $a \in \mathbb{R}^n, V \in \mathcal{P}_m, w > 0$, then

$$k_{m,1} \left(s, 0 \left| \begin{pmatrix} V & 0 \\ 0 & w \end{pmatrix}, a \right. \right) = p_{-s}(V^{-1}) |V|^{-1/2} w^{m/2} k_{m,1} \left(s, 0 \left| I_{m+1}, w^{1/2} g^{-1} a \right. \right).$$

Here $s \in \mathbb{C}^m$.

(5) An Inductive Formula for k-Bessel Functions.
For $s_1 \in \mathbb{C}, s_2 \in \mathbb{C}^{m-1}, a_1 \in \mathbb{R}, a_2 \in \mathbb{R}^{m-1}$, we have the following formula, if $a = (a_1, a_2)$ and $s = (s_1, s_2)$ are suitably restricted for convergence and $b = (m-1)/2 - \sum_{j=1}^{m} s_j$:

$$k_{m,1}(s,0|I_{m+1},a)$$
$$= \int_{u\in\mathbb{R}} (1+u^2)^b\, k_{m-1,1}\left(s_2,0\;\Big|\;I_m, a_2\sqrt{1+u^2}\right)\exp(2\pi i a_1 u)\,du.$$

Proof (Bengtson [42]).

(1) Since λ is the smallest element of the set of eigenvalues of V and W, $V[x] \geq \lambda\,{}^t xx$, for $x \in \mathbb{R}^m$ and $\mathrm{Tr}(V[X]) \geq \lambda\mathrm{Tr}(I[X])$ if $X \in \mathbb{R}^{m\times m}$. By the integral formula for the Iwasawa decomposition (see formulas (1.38) and (1.39) of § 1.1.6), we have, upon setting $Y = I[t]$, $t \in T_n$, the triangular group:

$$K_m(s|V,W) \leq 2^m \int_{t\in T_m} \exp\{-\lambda\,\mathrm{Tr}\left(I[t]+I[t^{-1}]\right)\}\prod_{i=1}^{m} t_{ii}^{\mathrm{Re}\,r_i-i}\prod_{1\leq i\leq j\leq m} dt_{ij}.$$

The variables $r \in \mathbb{C}^m$ are related to $s \in \mathbb{C}^m$ by the formula given in part (1) of Proposition 1.2.1.

Write

$$t^{-1} = \begin{pmatrix} t_{11}^{-1} & & t^{ij} \\ & \ddots & \\ 0 & & t_{mm}^{-1} \end{pmatrix}.$$

Then

$$K_m(s|V,W) \leq 2^m \prod_{j=1}^{m}\int_{t_{jj}>0} \exp\{-\lambda(t_{jj}^2+t_{jj}^{-2})\}\,t_{jj}^{\mathrm{Re}\,r_j-j}\,dt_{jj}$$
$$\times \prod_{1\leq i<j\leq m}\int_{t_{ij}\in\mathbb{R}} \exp\{-\lambda(t_{ij}^2+(t^{ij})^2)\}\,dt_{ij}$$
$$\leq \left(\tfrac{\pi}{\lambda}\right)^{m(m-1)/4}\prod_{j=1}^{m} K_{(\mathrm{Re}\,r_j-j+1)/2}(2\lambda),\; s\in\mathbb{R}.$$

For the final estimate, we need to know that $K_s(y) \leq (\pi/(2y))^{1/2}e^{-y}$, for $y > 0$, $s \in \mathbb{R}$ (see Lebedev [398]).

(2) Let Ξ denote the left-hand side of the equality that we are trying to prove. Then

$$
\Xi = \Gamma_m(-s^*)k_{m,n-m}\left(s,0 \left| \begin{pmatrix} V & 0 \\ 0 & W \end{pmatrix}, A\right.\right)
$$

$$
= \int_{Y\in\mathcal{P}_m} p_{-s^*}(Y)\exp\{-\mathrm{Tr}(Y)\}\,d\mu_m(Y)
$$

$$
\times \int_{X\in\mathbb{R}^{m\times(n-m)}} p_{-s}\left(V^{-1} + W^{-1}\left[{}^tX\right]\right)\exp\{2\pi i\,\mathrm{Tr}\left({}^tAX\right)\}dX.
$$

Now we want to use Exercise 1.2.4 of Section 1.2.1. In order to do this, we utilize another property of power functions from part (4) of Proposition 1.2.1 in Section 1.2.1 and obtain:

$$
p_{-s}\left(V^{-1}+W^{-1}\left[{}^tX\right]\right) = p_{-s^*}\left(\left(\left(V^{-1}+W^{-1}\left[{}^tX\right]\right)[\omega]\right)^{-1}\right),
$$

$$
\omega = \begin{pmatrix} 0 & & & 1 \\ & & \cdot & \\ & \cdot & & \\ & \cdot & & \\ 1 & & & 0 \end{pmatrix}.
$$

Then, by Exercise 1.2.4, we have the following equalities, letting $Z = Y[\omega]$:

$$
\Xi = \int_{Y\in\mathcal{P}_m}\int_{X\in\mathbb{R}^{m\times(n-m)}} p_{-s^*}(Y)\exp\left\{-\mathrm{Tr}\left(\left(V^{-1}+W^{-1}\left[{}^tX\right]\right)[\omega]Y\right)\right\}
$$

$$
\times \quad \exp\{2\pi i\mathrm{Tr}\left({}^tAX\right)\}\,dX\,d\mu_m(Y)
$$

$$
= \int_{Z\in\mathcal{P}_m}\int_{X\in\mathbb{R}^{m\times(n-m)}} p_{-s}(Z^{-1})\exp\left\{-\mathrm{Tr}\left(\left(V^{-1}+W^{-1}\left[{}^tX\right]\right)Z - 2\pi i\,{}^tAX\right)\right\}
$$

$$
\times \quad dX\,d\mu_m(Z).
$$

Now complete the square in the exponent. Let $Z = Y^2$ with $Y \in \mathcal{P}_m$, $W = Q^2$, $Q \in \mathcal{P}_{n-m}$, and change variables via $U = YXQ^{-1}$ to obtain:

$$
\Xi = |W|^{m/2}\int_{Z\in\mathcal{P}_m}|Z|^{-(n-m)/2}\int_{U\in\mathbb{R}^{m\times(n-m)}} p_{-s}(Z^{-1})
$$

$$
\times \exp\left\{-\mathrm{Tr}\left(V^{-1}Z + {}^tUU - 2\pi i\,{}^tAY^{-1}UQ\right)\right\}\,dU\,d\mu_m(Z).
$$

Let $C = i\pi Y^{-1}AQ$ and observe that

$$
\mathrm{Tr}\left({}^tUU - 2\pi i\,{}^t\left(Y^{-1}AQ\right)U\right) = \mathrm{Tr}\left(I[U - C] + W\left[\pi\,{}^tA\right]Z^{-1}\right).
$$

Thus

$$\Xi = |W|^{m/2} \int\limits_{Z \in \mathcal{P}_m} |Z|^{-(n-m)/2} p_{-s}\left(Z^{-1}\right) \exp\left\{-\mathrm{Tr}\left(V^{-1}Z + W\left[\pi\,{}^t A\right]Z^{-1}\right)\right\} d\mu_n(Z)$$

$$\times \int\limits_{U \in \mathbb{R}^{m \times (n-m)}} \exp\left\{-\mathrm{Tr}\left(I[U-C]\right)\right\}\,dU$$

$$= |W|^{m/2} \pi^{m(n-m)/2} K_m\left(s^\# \mid W\left[\pi\,{}^t A\right], V^{-1}\right).$$

(3) Set Ξ equal to the left-hand side of the equality we are trying to prove. By definition then

$$\Xi = \int\limits_{\mathcal{P}_n} p_s(Y) \exp\left\{-\mathrm{Tr}\left(\begin{pmatrix} A & 0 \\ 0 & B \end{pmatrix}\begin{bmatrix} I & 0 \\ {}^t C & I \end{bmatrix}Y + \begin{pmatrix} 0 & 0 \\ 0 & D \end{pmatrix}Y^{-1}\right)\right\}\,d\mu_n(Y).$$

Let Y be expressed according to the appropriate partial Iwasawa decomposition:

$$Y = \begin{pmatrix} V & 0 \\ 0 & W \end{pmatrix}\begin{bmatrix} I & X \\ 0 & I \end{bmatrix}.$$

By Exercise 1.1.19 of Section 1.1.4, Ξ is

$$\int\limits_{X \in \mathbb{R}^{m \times (n-m)}} \int\limits_{V \in \mathcal{P}_m} \int\limits_{W \in \mathcal{P}_{n-m}} p_{\sigma_1}(V)\,p_{\sigma_2}(W)\,|W|^{m/2}$$

$$\times \exp\left\{-\mathrm{Tr}\left(VA + V[X+C]B + WB + W^{-1}D\right)\right\}$$

$$\times |V|^{(n-2m-1)/2}|W|^{-(n+1)/2} \prod_{i,j} dw_{ij}\,dv_{ij}\,dx_{ij}$$

$$= \pi^{m(n-m)/2}|B|^{-m/2}p_{\sigma_1}(A^{-1})\Gamma_m(\sigma_1)K_{n-m}\left(\sigma_2 \mid B, D\right).$$

(4) The left-hand side of the equality that we wish to prove is:

$$\int\limits_{x \in \mathbb{R}^m} p_{-s}\left(V^{-1} + xw^{-1}\,{}^t x\right) \exp\left(2\pi i\,{}^t ax\right)\,dx$$

which equals:

$$w^{m/2}|V|^{-1/2}p_{-s}(V^{-1}) \int\limits_{u \in \mathbb{R}^m} p_{-s}\left(I + I\left[{}^t u\right]\right) \exp\left(2\pi i\,{}^t a\,w^{1/2}\left({}^t g^{-1}\right)u\right)\,du,$$

upon setting $u = {}^t g x w^{-1/2}$.

(5) First note that the upper left $j \times j$ corner of the matrix $(I + x\,{}^t x)$, with $x = a$ column vector in $\mathbb{R}^m$, is

$$\left(I_j + w_j\,{}^t w_j\right), \quad \text{where } {}^t w_i = (x_1 \cdots x_j).$$

And the matrix $\left(w_j\,{}^t w_j\right)$ is a $j \times j$ matrix of rank one. The unique nonzero eigenvalue of $\left(w_j\,{}^t w_j\right)$ is $\|w_j\|^2 = x_1^2 + \cdots + x_j^2$. We can therefore find $k \in O(j)$ such that

$$\left(I_j + w_j\,{}^t w_j\right)[k] = I_j + \begin{pmatrix} \|w_j\|^2 & 0 \\ 0 & 0 \end{pmatrix}.$$

Thus $\left|I_j + w_j\,{}^t w_j\right| = 1 + \|w_j\|^2$.

It follows from these considerations that

$$k_{m,1}\left(s \mid I_{m+1}, a\right) = \int\limits_{x \in \mathbb{R}^m} p_{-s}\left(I + x\,{}^t x\right) \exp\left(2\pi i\,{}^t ax\right)\, dx$$

$$= \int\limits_{x \in \mathbb{R}^m} \left(1 + x_1^2\right)^{-s_1} \left(1 + x_1^2 + x_2^2\right)^{-s_2} \cdots \left(1 + x_1^2 + \cdots + x_m^2\right)^{-s_m} \exp\left(2\pi i\,{}^t ax\right)\, dx.$$

Now make the change of variables $x_j^2 = \left(1 + x_1^2\right) u_j^2$, $j = 2, \ldots, m$, to complete the proof. ∎

Exercise 1.2.17. (a) Can you generalize property (4) of Theorem 1.2.2 to $k_{m,n-m}$?
(b) (Bengtson [42]). Show that $K_2\left(s \mid {}^t qq, I_2\right)$ converges for $s \in \mathbb{C}^2$, $q \in \mathbb{R}^2$, when

$$\begin{aligned}
&\text{Re } s_2 \text{ and } \text{Re}\,(s_1 + s_2) < -\tfrac{1}{2}, && \text{if } q = 0; \\
&\text{Re } s_2 < 0, && \text{if } q_2 = 0, \quad q_1 \neq 0; \\
&\text{Re}\,(s_1 + s_2) < 0, && \text{if } q_2 \neq 0.
\end{aligned}$$

Note that this is the function $k_{2,1}$ essentially.
(c) Obtain a functional equation for $k_{2,1}$.

Hint. (c) Use property (5) of Theorem 1.2.2 and the functional equation of the ordinary K-Bessel function.

Remaining Questions

(1) Concerning the K-Bessel Functions.

(a) Are these K-Bessel functions products of ordinary K-Bessel functions as was the case for the gamma function of matrix argument? The answer must be "No, except under very special circumstances, as in Example 1.2.2 above."
(b) Does (1.59) lead to a unique function? Here the answer appears to be "Yes" and " No." For many functions satisfying (1.59) can be constructed out of the same basic function; e.g.,

$$f(Y) = k_{m,n-m}\left(s \mid Y\begin{bmatrix} {}^{t}A^{-1} & 0 \\ 0 & {}^{t}B \end{bmatrix}, C\right),$$

with $A \in GL(m, \mathbb{R})$, $B \in GL(n - m, \mathbb{R})$, $C \in \mathbb{R}^{m \times (n-m)}$, such that $R = ACB$. For it is easily seen that

$$f\left(Y\begin{bmatrix} I_m & X \\ 0 & I_{n-m} \end{bmatrix}\right) = \exp\{2\pi i \mathrm{Tr}({}^{t}(ACB)X)\}f(Y).$$

(c) **Are there relations between $k_{m,n-m}$ and $k_{n-m,m}$? What functional equations do the $k_{m,n-m}$ satisfy?** See Exercise 1.2.17(c). The theory of Eisenstein series for maximal parabolic subgroups of $GL(n)$, which will be discussed in Section 1.5, leads us to expect that there is essentially only one functional equation (e.g., that of Exercise 1.2.17). Note that $m > n - m$ implies that $k_{m,n-m}(s, 0 \mid Y, A)$, $s \in \mathbb{C}^m$, which is the function related to $K_m\left(s^{\#} \mid W[\pi \, {}^{t}A], V^{-1}\right)$ by part (2) of Theorem 1.2.2, has more s-variables than the same function with m and $n - m$ interchanged. This means that you cannot use parts (2) and (3) of Theorem 1.2.2 to write $k_{m,n-m}$ as a product of lower rank functions.

(d) **Can one generalize the Kontorovich-Lebedev inversion formula** (3.19) of Section 3.2 in Vol. I to $\mathcal{P}_n$ and then obtain harmonic analysis on $\mathcal{P}_n$ in partial Iwasawa coordinates, thus generalizing Theorem 3.2.1 in Vol. I? This leads one to ask again: "What functional equations do matrix K-Bessel functions satisfy?" One is also led to attempt to generalize the Laplace transform relations between K-Bessel functions and spherical functions (Exercise 3.2.13 of Vol. I) to a matrix version involving the spherical functions for $\mathcal{P}_n$ to be considered in the next section.

Gelbart [207], Gross and Kunze [247], and Herz [293] generalize the Hankel transform to a transform involving matrix J-Bessel functions (which are operator valued in the first two references) and show that such a transform can be used to generalize Theorem 2.2.2 of Vol. I and decompose the Fourier transform on matrix space $\mathbb{R}^{k \times m}$ in polar coordinates for that space. Define, for $k \geq m$, the **compact Stiefel manifold**

$$V_{k,m} = \{X \in \mathbb{R}^{k \times m} \mid {}^{t}XX = I\} \cong O(k)/O(k - m).$$

Then polar coordinates for $X \in \mathbb{R}^{k \times m}$ are $R \in \mathcal{P}_m$ and $V \in V_{k,m}$ with $X = VR^{1/2}$. Of course, harmonic analysis on the Stiefel manifold involves representations of the orthogonal group. Thus one expects to see matrix-valued J-Bessel functions. In any case, this work on J-Bessel functions and inversion formulas for Hankel transforms certainly leads one to expect a similar theory for K-Bessel functions.

(2) More General Hypergeometric Functions for $\mathcal{P}_n$.

(a) Can one relate the K-Bessel functions with the J-Bessel functions considered by Gelbart [207], Gross et al. [246], Herz [293], and Muirhead [468, Ch. 10]? Gelbart [207] and Gross et al. [246] consider matrix-valued **J-Bessel functions** defined for an irreducible unitary representation λ of a compact Lie group U acting on a real finite dimensional inner product space X by orthogonal linear transformations via:

$$J_\lambda(w, z) = \int_U \exp\{i(w|uz)\}\lambda(u) \, du. \tag{1.62}$$

for $w, z \in X^{\mathbb{C}} = X \otimes_{\mathbb{R}} \mathbb{C}$, the complexification of X. Here $(w|z)$ is the complex bilinear form on $X^{\mathbb{C}}$ that uniquely extends the inner product on X. James (see Muirhead [468, p. 262]) defines a function $_0F_1$ which is the case $\lambda(u) \equiv 1$ identically in (1.62), with $(w|z) = \text{Tr}(wz)$.

Herz [293] considers an analogue of the J-Bessel function given by

$$A_\delta(M) = (2\pi i)^{-n} \int_{\text{Re } Z = X_0} \exp\left\{\text{Tr}\left(Z - MZ^{-1}\right)\right\} |Z|^{-\delta-p} \, dZ, \tag{1.63}$$

for $n = m(m + 1)/2$, $p = (m + 1)/2$, $\delta \in \mathbb{C}$ with $\text{Re } \delta > p - 1$, Z in the Siegel upper half space $\mathcal{H}_m$ with fixed real part $X_0 \in \mathcal{P}_m$. We will consider $\mathcal{H}_m$ more carefully in Chapter 2. Herz finds that this function is needed to express the noncentral Wishart distribution in multivariate statistics (see also Muirhead [468, Ch. 10]). In addition, such functions arise in summation formulas considered by Bochner [55] in his study of matrix analogues of the circle problem. But there do not appear to be good estimates for the error terms in these formulas. Such integrals also appear in Fourier coefficients of Eisenstein series for $Sp(n, \mathbb{Z})$. See Godement's article in Séminaire Cartan [547, Exposé 9].

How are (1.62) and (1.63) related? The answer is to be found in Herz [293, p. 493] and Muirhead [468, p. 262]. See also Gelbart [207] and the references indicated there. Gelbart applies his results to the construction of holomorphic discrete series representations of $Sp(n, \mathbb{R})$. Gross et al. [246] apply their results similarly to $U(n, n)$.

In the next section, we will develop an asymptotic relation between spherical functions for $\mathcal{P}_n$ and J-Bessel type functions (which are spherical functions for the Euclidean group of the tangent space to $\mathcal{P}_n$ at the point I). Such a relation comes from Lemma 4.3 in Helgason [279]. Actually, to be precise, we will relate spherical functions for $\mathcal{P}_n$ with James and Muirhead's $_0F_0$ function of 2 arguments for $\mathcal{P}_n$. Presumably this function is related to the $_0F_1$ for $\mathcal{P}_{n-1}$. We will use the same methods as Helgason [279] specialized to our case.

(b) **How do the *K*-Bessel functions relate to matrix argument confluent hyper-geometric functions** considered by Gindikin [218], Herz [293], James [328–330], and Muirhead [468, pp. 264, 472]? One can define a **matrix argument confluent hypergeometric function of the first kind** Φ_n (also known as $_1F_1$) by:

$$\Phi_n(a, c; X) = \frac{\Gamma_n(c)}{\Gamma_n(a)\Gamma_n(c-a)} \int\limits_{0 < Y < I} \exp[\mathrm{Tr}(XY)] \, |Y|^a \, |I-Y|^{c-a-(n+1)/2} \, d\mu_n(Y),$$

(1.64)

for $a, b \in \mathbb{C}$, with a, b suitably restricted and X a symmetric $n \times n$ matrix. The domain of integration is the subset of $Y \in \mathcal{P}_n$ such that $I - Y \in \mathcal{P}_n$. Muirhead [468, p. 447] shows that Φ_n gives the moment of the generalized variance of the noncentral Wishart distribution. This is due to Herz [293] and Constantine.

A **matrix confluent hypergeometric function of the second kind** can be defined for $a, c \in \mathbb{C}$ and $X \in \mathcal{P}_n$ by:

$$\Psi_n(a, c; X) = \Gamma_n(a)^{-1} \int\limits_{Y \in \mathcal{P}_n} \exp[-\mathrm{Tr}(XY)] \, |Y|^a \, |I + Y|^{c-a-(n+1)/2} \, d\mu_n(Y).$$

(1.65)

Muirhead [468, p. 474] uses this function to express certain statistical quantities coming from the T_0^2-statistic, which was proposed by Lawley in 1938 and Hotelling in 1947 in connection with a military problem—the air testing of bombsights.

If we do ask for a relation between our *K*-Bessel function (1.61) and the confluent hypergeometric function (1.65), we find that it is only clear for the case $n = 1$, when the functions are the classical ones considered by Lebedev [398].

One can show, for example, that the classical *K*-Bessel function is a special case of Ψ_1:

$$K_s(z) = \sqrt{\pi}(2z)^s e^{-z} \Psi_1(s + 1/2, 2s + 1; \, 2z).$$

This fact is proved in Lebedev [398, pp. 118 and 274]. The main results needed to prove it are:

$$\left. \begin{array}{l} \textbf{(i) } K_s(z) = \frac{1}{2} \displaystyle\int_0^\infty u^{-s-1} \exp\left[-\frac{z}{2}\left(u + \frac{1}{u}\right)\right] \, du; \\[2em] \textbf{(ii) } u^{-s-(1/2)} = \Gamma\left(s + \frac{1}{2}\right)^{-1} \displaystyle\int_0^\infty e^{-xu} x^{s-(1/2)} \, dx; \\[2em] \textbf{(iii) } K_{1/2}(z) = \left(\frac{\pi}{2z}\right)^{1/2} e^{-z}. \end{array} \right\}$$

(1.66)

Thus it is worthwhile generalizing (iii) in (1.66) to $\mathcal{P}_n$. We already have the analogues of (i) and (ii). The analogous relation between the J-Bessel type function in (1.63) and $_1F_1$ in (1.64) is proved by Muirhead [468, p. 262].

     Maass [426, Ch. 18] finds that the confluent hypergeometric functions of matrix argument occur in the Fourier coefficients of certain nonholomorphic automorphic forms for the Siegel modular group $Sp(n, \mathbb{Z})$. See also Shimura [555]. This suggests that we could relate Ψ and K by relating (nonholomorphic) Eisenstein series for $Sp(n, \mathbb{Z})$ and those for $GL(n, \mathbb{Z})$.

(c) **What is the connection between the K-Bessel functions and Whittaker functions?** Whittaker functions and Fourier expansions of automorphic forms as sums of these functions are discussed by Bump [83], Goldfeld [230], Jacquet [323], Jacquet et al. [325], Proskurin [494], Schiffman [534], and Shalika [552]. For $r \in \mathbb{R}^{m-1}$, $Y \in \mathcal{P}_m$, and $s \in \mathbb{C}^m$, with $\operatorname{Re} s$ suitably restricted for convergence, the **Whittaker function** can be defined by:

$$W(s \mid Y, r) = \int\limits_{n \in N} p_{-s}\left(Y^{-1}[{}^t n]\right) \exp\left(2\pi i \sum_{i=1}^{m-1} r_i x_{i,i+1}\right) dn, \qquad (1.67)$$

where N is the nilpotent group of real $m \times m$ upper triangular matrices with ones on the diagonal,

$$n = \begin{pmatrix} 1 & & x_{ij} \\ & \ddots & \\ 0 & & 1 \end{pmatrix}, \quad \text{and} \quad dn \text{ is found in Exercise 1.1.2.}$$

    The exponential appearing in the integral (1.67) is easily seen to be a one-dimensional character of N. The integral itself can easily be shown to converge wherever the numerator in the Harish-Chandra c-function of Section 1.3 converges (i.e., when b_m given by (1.154) in Section 1.3 below converges). One also sees easily that

$$W(s \mid Y[n], r) = \exp\left\{2\pi i \sum_{i=1}^{m-1} r_i x_{i,i+1}\right\} W(s \mid Y, r).$$

Thus the Whittaker function satisfies the analogues of properties (1.59) in which the abelian group $N(m, n-m)$ is replaced by the nilpotent group N. Kirillov [348] shows that the characters of N in the transformation formula above are (up to unitary equivalence) the only finite (actually one) dimensional irreducible unitary representations of N. There are also infinite dimensional irreducible unitary representations, as we mentioned earlier. It is possible to view $W(s \mid Y, r)$ as an analogue of Selberg's Eisenstein series $E_{(n)}$ (to be defined in formula (1.249) Section 1.5.1) with the largest possible number of complex variables $s \in \mathbb{C}^n$; i.e., the highest dimensional part of the spectrum of the

Laplacian. Thus one can follow ideas of Jacquet and use techniques developed by Selberg for Eisenstein series in order to obtain $n!$ functional equations for the Whittaker functions (see Bump [83], Goldfeld [230], and Jacquet [323]). The idea is to write the Whittaker function for $\mathcal{P}_n$ as an integral of Whittaker type functions of lower rank, such as the k-Bessel function (1.60). This is analogous to writing an Eisenstein series with n complex variables as a sum of Eisenstein series with a smaller number of complex variables (see Lemma 1.5.2 of Section 1.5.1, for example).

More explicitly, one can write the Whittaker function as a Fourier transform of a k-Bessel function (1.60). For example, when $n = 3$:

$$
W(s \mid Y, r) = \int_{x_{12} \in \mathbb{R}} k_{2,1} \left(s \,\middle|\, Y \begin{bmatrix} 1 & -x_{12} & 0 \\ 0 & 1 & 0 \\ 0 & 0 & 1 \end{bmatrix}, (0, r_2) \right) \exp(2\pi i r_1 x_{12}) \, dx_{12}.
$$

$$(1.68)$$

Then one can obtain properties of the Whittaker functions from those of the lower rank k-Bessel functions, and vice versa, since the k-Bessel function is also a Fourier transform of the Whittaker function. This same sort of idea relates the Fourier expansions of automorphic forms for $GL(n, \mathbb{Z})$ of Section 1.5.3 with those involving Whittaker functions. See also Section 1.5.4.

As we mentioned at the beginning of this section, there are many references on Whittaker functions, including those of Bump [83], Goldfeld [230], Goodman and Wallach [235], Hashizume [265], Jacquet [323], Kostant [367], and Shalika [552]. For example, Hashizume considers a Whittaker model to come from intertwining operators between admissible representations of G and representations induced from a nondegenerate unitary character of N. He proves some general multiplicity results. Kostant connects the theory of Whittaker functions and the theory of Toda lattices. He obtains the complete integrability of the corresponding geometrically quantized system. Bump uses Shalika's multiplicity one theorem to argue that Whittaker functions give the Fourier coefficients of automorphic forms for $GL(3)$ and notes that Kostant shows that the solution space of the differential equations for $W(s|Y, r)$ coming from the operators in $D(\mathcal{P}_n)$ has dimension the order of the Weyl group, which is $n!$. But the solutions of polynomial growth form a one-dimensional subspace (i.e., we have multiplicity one). See also Goldfeld [230].

Piatetski-Shapiro (in Borel and Casselman [66, Vol. I, pp. 209–212]) uses the uniqueness of Whittaker models for representations to show that if π is an irreducible smooth admissible representation of the *adelized* $GL(n)$, then the multiplicity of π in the space of cusp forms is one or zero.

It is possible to use Theorem 1.2.2 along with Propositions 1.2.2 and 1.2.3 of Section 1.2.1 to evaluate various special integrals.

Example 1.2.3. For $s \in \mathbb{C}$, define a vector $r(s) \in \mathbb{C}^{2m}$ by setting every entry of $r(s)$ equal to 0 except the mth entry which is set equal to s. Then, by part (2) of Theorem 1.2.2 and Exercise 1.2.14, we have:

$$k_{m,m}(r(s) \mid I, 0) = \pi^{m^2/2} \frac{\Gamma_m \left(0, \ldots, 0, \frac{m}{2} - s\right)}{\Gamma_m(0, \ldots, 0, s)}.$$

Note that as in the proof of Proposition 1.2.2 in Section 1.2.1, we have the following equalities, where T_n is the group of upper triangular $n \times n$ matrices with positive diagonal entries:

$$k_{m,m}(r(s) \mid I, 0) = \int\limits_{X \in \mathbb{R}^{m \times m}} |I + {}^t XX|^{-s} \, dX$$

$$= \mathrm{Vol}(K) \int\limits_{t \in T_m} |I + I[t]|^{-s} \prod_{i=1}^{m} t_{ii}^{m-i} \, dt$$

$$= \mathrm{Vol}(K) \, 2^{-m} \int\limits_{Y \in \mathcal{P}_m} |I + Y|^{-s} \, |Y|^{m/2} \, d\mu_m(Y)$$

$$= \pi^{m^2/2} \Gamma_m(0, \ldots, 0, m/2)^{-1} \int\limits_{Y \in \mathcal{P}_m} |I + Y|^{-s} \, |Y|^{m/2} \, d\mu_m(Y).$$

Let $A =$ the group of positive diagonal matrices, as usual, and define

$$D(a) = \prod_{1 \le i < j \le n} |a_i - a_j|, \quad da = \prod_{j=1}^{n} \frac{da_j}{a_j}, \quad \text{for } a \in A,$$

$$I(s) = \int\limits_{a \in A} |I + a|^{-s} \, |a|^{1/2} \, D(a) \, da = \frac{\Gamma_m(0, \ldots, 0, \frac{m}{2} - s) \Gamma_m(0, \ldots, 0, \frac{m}{2})}{c_m \, \Gamma_m(0, ., 0, s)}.$$

Here $c_m^{-1} = \pi^{-m^2/2} \, m! \, \Gamma_m(0, \ldots, 0, m/2)$. Thus

$$I(s) = \frac{m! \, \Gamma_m \left(0, \ldots, 0, \frac{m}{2} - s\right) \Gamma_m \left(0, \ldots, 0, \frac{m}{2}\right)^2}{\pi^{m^2/2} \, \Gamma_m(0, \ldots, 0, s)}.$$

Exercise 1.2.18 (Mellin Transforms of K-Bessel Functions—A Generalization of Exercise 3.6.4 of Vol. I).   Show that if $s, r \in \mathbb{C}^n$, then if $B \in \mathcal{P}_n$,

$$\int\limits_{A \in \mathcal{P}_n} p_s(A) \, K_n (r \mid B, A) \, d\mu_n(A) = \Gamma_n(s) \Gamma_n(s + r) p_{s+r}(B^{-1}).$$

Hint. (Bengtson [42].) Note that the left-hand side is:

$$\int\limits_{A \in \mathcal{P}_n} \int\limits_{Y \in \mathcal{P}_n} p_s(A) p_r(Y) \, \exp\left\{-\mathrm{Tr}\left(Y + AY^{-1}\right)\right\} \, d\mu_n(Y) \, d\mu_n(A).$$

Let $Y = I[t]$, $t \in T_n$, and change variables via $C = A[t^{-1}]$.

Bump [83, Ch. X] computes a Mellin-type transform of his Whittaker function for $SL(3, \mathbb{R})$ and obtains a quotient of six gammas over one gamma. Such results are useful in the study of L-functions corresponding to automorphic forms for $GL(n)$, as we saw already in Volume I, Section 3.6.2 when we studied properties of Dirichlet series corresponding to Maass wave forms. See also Goldfeld [230] and § 1.5.4.

Exercise 1.2.19 (A Functional Equation).

(a) Let ω and s^* be as defined in part (4) of Proposition 1.2.1 in § 1.2.1. Show that if $A, B \in \mathcal{P}_n$, then

$$K_n(s \mid A, B) = K_n(s^* \mid B[\omega], A[\omega]).$$

(b) Show that if $k \in K = O(n)$, then for $s \in \mathbb{C}$, we have

$$K_m(0, s \mid A[k], B[k]) = K_m(0, s \mid A, B).$$

Exercise 1.2.20 (Inductive Formula for K-Bessel Functions). Prove that

$$K_m\left(s \left| \begin{pmatrix} A & 0 \\ 0 & B \end{pmatrix} \begin{bmatrix} I & 0 \\ {}^t Q & I \end{bmatrix}, I \right.\right)$$
$$= \int\limits_{X \in \mathbb{R}^{m \times (n-m)}} K_m\left(r_1 \mid A + B\left[{}^t X + {}^t Q\right], I\right) K_{n-m}\left(r_2 \mid B, I + I[X]\right) \, dX.$$

Here for $s \in \mathbb{C}^n$, we have chosen $r_1 \in \mathbb{C}^m$, $r_2 \in \mathbb{C}^{n-m}$ such that:

$$p_s\begin{pmatrix} A & 0 \\ 0 & B \end{pmatrix} = p_{r_1}(A) \, |A|^{(m-n)/2} \, p_{r_2}(B) \, |B|^{m/2}.$$

Hint. See Terras [606].

Exercise 1.2.21 (Writing the Matrix K-Bessel Function as an Integral of the Ordinary K-Bessel Function). Show that if $r \in \mathbb{C}^{m-1}$, $s \in \mathbb{C}$,

$$K_m(r, s \mid A, B) = \frac{2}{m} \int\limits_{W \in S\mathcal{P}_m} p_r(W^{-1}) \, K_{ms}\left(2\sqrt{\mathrm{Tr}(AW)\mathrm{Tr}(BW^{-1})}\right) \left(\frac{\mathrm{Tr}(BW^{-1})}{\mathrm{Tr}(AW)}\right)^{ms/2} dW,$$

where the measure dW is chosen as in Exercise 1.1.23 of § 1.1.4.

In Terras [606], motivated by the study of Fourier expansions of modular forms, I consider more general K-Bessel type functions with the power function p_s replaced by other sorts of eigenfunctions for $D(\mathcal{P}_n)$. It also seems that we need to answer the question of whether it is possible to move the variable B over to be next to C in $K(s \mid W[C], V[B])$. Exercise 1.2.16 allows us to do something in this direction. But it does not seem to be exactly what we will need later. See the end of the proof of Theorem 1.5.3 in Section 1.5.3 below.

Thus we close this section on Bessel functions with too many questions unresolved. There has not been much progress since the last edition, as far as I know. Sorry. This will not be the last such section. See the quotation at the beginning of Section 1.5.

1.2.3 Spherical Functions

We want to find an analogue for $\mathcal{P}_n$ of the notion of spherical harmonic, the basic function for Fourier analysis on the sphere which was considered in Chapter 2 of Volume I. Of course the symmetric space under consideration is a higher rank analogue of the Poincaré upper half plane and thus it has spherical functions which generalize the Legendre or conical functions discussed in Section 3.2 of Volume I. We will see in the next section that we can use the spherical functions for $\mathcal{P}_n$ to obtain a Fourier transform on $\mathcal{P}_n$, with properties generalizing those of the Fourier transform on H in Theorem 3.2.3 of Vol. I.

The theory of spherical functions really goes back to the study of spherical harmonics by Legendre, Laplace, and Jacobi in the late 1700s (see Chapter 2 of Vol. I). In 1916–1918 Funk [192] and Hecke [268, pp. 208–214] developed their integral formula for spherical harmonics (see Theorem 2.1.2, of Vol. I). In 1929 and 1934 Cartan [96] and Weyl [666, Vol. III, pp. 386–399] began the modern theory with the study of spherical functions associated with compact symmetric spaces. The compactness hypothesis was dropped in the 1950s by Gelfand [212], Godement [222], Harish-Chandra [263], and others. Selberg [543] gives the basic theory of spherical functions for the case under consideration. Other references for the general theory include: Barut and Rączka [39] (who call spherical functions "harmonic functions" on p. 302), Berezin and Gelfand [44], Dieudonné [137, Vol. V, Ch. XXI, Vol. VI, Ch. XXII], Ehrenpreis and Mautner [155], Gangolli [198], Godement [224], Helgason [273–282], Maurin [437], Mautner [439], Richards' article in Olver et al. [481], Satake [532], Tamagawa [588], Warner [655], and Wawrzyńczyk [657].

Many of the authors listed above are motivated by the desire to understand the representation theory of Lie groups. Others are prompted by number theoretic applications; e.g., connections with Hecke operators (see Section 3.6 of Vol. I and Section 1.5.2 following). Still others are inspired by the appearance of spherical functions in various statistical problems (see Farrell [173, 174] James [328–330], and Muirhead [468]). Finally, some are motivated by physical applications.

The following references are indicative of some of the possibilities for applications of harmonic analysis on Lie groups in physics: Barut and Rączka [39], Mackey [429–431], and Menotti and Onofri [444].

We define a **spherical function** to be a function

$$h : \mathcal{P}_n \to \mathbb{C}$$

with the following properties:

(1) $h(Y[k]) = h(Y)$,   for all $Y \in \mathcal{P}_n$ and $k \in O(n)$;
(2) $Lh = \lambda_L h$, $(\lambda_L \in \mathbb{C})$ for all invariant differential operators $L \in D(\mathcal{P}_n)$;
(3) $h(I) = 1$.

$$(1.69)$$

That is, we are seeking a rotation-invariant eigenfunction for all the invariant differential operators, normalized to have the value 1 at the identity. This definition should be compared with (1.59) in the last section.

We should probably call the functions satisfying (1.69) "zonal spherical functions" or "spherical functions of class one," but we will not do that here (cf. Section 2.1 of Vol. I), since we do not intend to consider the more general spherical functions transforming according to a nontrivial representation of K (but see Vol. I, pp. 113–114, 172).

As we shall see in Theorem 1.2.3, there are many ways to characterize spherical functions other than (1.69). In fact, H. Weyl has remarked that "their property as eigenfunctions of Laplace operators is merely accidental" (see Maurin [437, pp. 224–225]). However, one appears to have some difficulty in making connections with applications if one insists on throwing out the differential equations.

Example 1.2.4 (Spherical Functions on the Poincaré Upper Half Plane). For the Poincaré upper half plane H, the spherical function is a standard special function—the Legendre or conical function discussed in Section 3.2 of Vol. I. It is unique because it solves a second order singular ODE whose second solution has a singularity at i in H.

It is easy to write down such a spherical function by integration over $K = SO(2)$:

$$h(s|z) = \frac{1}{2\pi} \int\limits_0^{2\pi} \mathrm{Im}(k_{-u}(z))^s du, \qquad \text{where } k_u = \begin{pmatrix} \cos u & \sin u \\ -\sin u & \cos u \end{pmatrix}.$$

Here we use the notation of Vol. I. It follows that

$$h(s|z) = P_{-s}(\cosh r) \text{ if } z = k_u \exp(-r)i,$$

where the action of $g \in SL(2, \mathbb{R})$ on $z \in H$ is by fractional linear transformation (see page 172, Section 3.2 of Vol. I). Here P_{-s} denotes the Legendre function which can be defined by the integral:

$$P_s(t) = \frac{1}{2\pi} \int_0^{2\pi} \left\{ t + \sqrt{t^2 - 1} \cos u \right\}^s du.$$

The other solution to the second order ODE satisfied by the Legendre function is called Q and it has the following asymptotic behavior as r approaches 0:

$$Q_{s-1}(\cosh r) \sim - \left(\frac{1}{2} \right) \log(\cosh r - 1), \text{ as } r \to 0$$

(see Vol. I, pp. 338–339).

Motivated by the preceding example and the construction of k-Bessel functions satisfying (1.59) via formula (1.60); i.e., by integration over the appropriate subgroup of G, we construct a **spherical function** by integrating the power function over K; i.e.,

$$h_s(Y) = \int_{k \in K} p_s(Y[k]) dk, \quad \text{for } Y \in \mathcal{P}_n, \ s \in \mathbb{C}^n. \tag{1.70}$$

Part (4) of Theorem 1.2.3 shows that these are the only spherical functions for $\mathcal{P}_n$.

In the following discussion (just as in (1.23) of Section 1.1.4), we will sometimes identify functions $f(Y)$, $Y \in \mathcal{P}_n$, with functions on $G = GL(n, \mathbb{R})$, by writing $Y = I[g]$, for $g \in G$. Such a function on G will be left K-invariant.

Next we will need to show that we could equivalently require the spherical functions to be eigenfunctions of convolution integral operators. In Section 3.7, Vol. I, the analogous result was proved for the Poincaré upper half plane.

Recall now the definition of **convolution operators** in (1.24) of Section 1.1.4:

$$(f * g)(a) = C_g f(a) = \int_G f(b) g(ab^{-1}) db. \tag{1.71}$$

And recall Lemma 1.1.1 of that same section—a lemma which gave the properties of these convolution operators.

The following proposition is necessary for the study of spherical functions as well as analogues of Poisson summation for $\mathcal{P}_n / GL(n, \mathbb{Z})$. In order to state it, we need to define the **Helgason–Fourier transform** of a function $f \in C_c(\mathcal{P}_n / K)$, which is:

$$\widehat{f}(s) = \int_{Y \in \mathcal{P}_n} f(Y) \, \overline{p_s(Y)} \, d\mu, \quad \text{for } s \in \mathbb{C}^n. \tag{1.72}$$

This transform will be scrutinized as carefully as we can manage in the next section. In the special case under consideration (i.e., when the function f is K-invariant) this transform can be identified with the spherical transform whose inversion formula was obtained by Harish-Chandra. And in the context of the present discussion it can also be called the "Selberg transform." We have named the transform for Helgason since his lectures [275] demonstrate clearly that the transform really does behave like the usual Fourier or Mellin transform and can be used to solve some of the sorts of problems that Fourier transforms are traditionally used to solve in applied mathematics (e.g., those connected with the wave equation on a symmetric space). But the reader should be cautioned that this transform has a plethora of names in the literature.

Part (2) of the following proposition is **Selberg's basic lemma** that eigenfunctions of invariant differential operators are eigenfunctions of invariant integral operators. It is of central importance for the trace formula.

Proposition 1.2.4. *(1) The spherical function h corresponding to the eigenvalues $(\lambda_1, \ldots, \lambda_n) \in \mathbb{C}^n$, with*

$$\mathrm{Tr}((Y\partial/\partial Y)^i)h = \lambda_i h,$$

is unique. Here the invariant differential operators are from Theorem 1.1.2 of Section 1.1.5.

*(2) **Eigenfunctions of Invariant Differential Operators Are Eigenfunctions of Invariant Integral Operators.***

Let $f \in C^\infty(\mathcal{P}_n)$ be an eigenfunction of all the G-invariant differential operators $L \in D(\mathcal{P}_n)$; i.e., $Lf = \lambda_L f$, for some $\lambda_L \in \mathbb{C}$. Define

$$s \in \mathbb{C}^n \ \text{by} \ Lp_s = \lambda_L p_s,$$

where p_s denotes the power function (1.41). If $g \in C_c^\infty(K\backslash G/K)$; i.e., if g is K-bi-invariant and infinitely differentiable with compact support, and if we assume, in addition, that $g(x) = g(x^{-1})$, for all $x \in G$, then f is an eigenfunction of the convolution operator C_g in (1.71). More precisely,

$$C_g f = f * g = \widehat{g}(\overline{s})f,$$

with $\widehat{g}$ denoting the Helgason–Fourier transform (1.72). Conversely, suppose that $f \in C(\mathcal{P}_n)$ is an eigenfunction of all the convolution operators C_g in (1.71), with $g \in C_c^\infty(\mathcal{P}_n/K)$. Then f is also an eigenfunction of all the invariant differential operators.

Proof (Selberg [543, pp. 53–56]).

(1) Let h_i, $i = 1, 2$, be two spherical functions corresponding to the eigenvalues λ_L; i.e., $Lh_i = \lambda_L h_i$, $i = 1, 2$, for $L \in D(\mathcal{P}_n)$. Since h_i is a solution of an elliptic partial differential equation with analytic coefficients, by a theorem of Bernstein, h_i must be real analytic (see John [332, p. 142] or Garabedian [201, p. 164]). We want to show that all the terms in a Taylor expansion of $h_1 - h_2$ must be zero.

What is the Taylor expansion of a function f on a symmetric space like $\mathcal{P}_n$? It is best to view f as a function on $G = GL(n, \mathbb{R})$. The **Taylor expansion** of f is then

$$f(\exp Xg) = \sum_{n \geq 0} \frac{1}{n!}(\widetilde{X}^n f)(g),$$

where $(\widetilde{X}^n f)(g) = [d^n/ds^n f(\exp(sX)g)]_{s=0}$. Then $\widetilde{X}^n$ is a right-invariant differential operator on G (see Helgason [275, p. 16] and Chapter 2 of this volume).

Suppose $L \in D(\mathcal{P}_n)$. We can think of L as a differential operator on G commuting with right translation. Form a left $K = O(n)$-invariant differential operator $L^{\#}$ by taking the K-average of the transforms L^k of L under inner automorphism by $k \in K$. That is, let $i_k(x) = kxk^{-1}$, for $x \in G$. Then

$$L^k(f) = L(f \circ i_k) \circ i_k^{-1}, \quad L^{\#} = \int_{k \in K} L^k \, dk.$$

See Helgason [275, pp. 41–43] or [273, Chs. I and X], or [282] for more details on these constructions. The conclusion is that the differential operators in the Taylor series for f on the group G correspond to G-invariant differential operators on the symmetric space $K \backslash G$, with the same value at the identity.

It follows from our original hypothesis that all terms in the Taylor series for $h_1 - h_2$ must vanish at the identity. But we can translate everything by $g \in G$ to complete the proof.

(2) Here we imitate the proof of Lemma 3.7.3 in Vol. I. Define an operator M which averages functions over the compact group $K = O(n)$:

$$Mf(Y) = \int_{k \in K} f(Y[k]) dk.$$

Since f is assumed to be an eigenfunction of the G-invariant differential operators $L \in D(\mathcal{P}_n)$, it follows from part (1) that Mf is unique up to a constant. So we find that

$$Mf(Y) = f(I)h_s(Y),$$

where $h_s(Y)$ is the spherical function defined by (1.70); i.e.,

$$h_s(Y) = \int_{k \in K} p_s(Y[k]) \, dk.$$

Here we are using Exercise 1.1.11 of Section 1.2.1.

It follows that

$$M(f*g)(a) = (Mf*g)(a) = f(I)(h_s*g)(a).$$

Evaluate this at $a = I$ to find that

$$\frac{(f*g)(I)}{f(I)} = \frac{(h_s*g)(I)}{h_s(I)},$$

since $h_s(I) = 1$ and $Mf(I) = f(I)$. Therefore this quotient is independent of the chosen eigenfunction f for all the $L \in D(\mathcal{P}_n)$. Thus, in particular, we can replace f by the power function p_s and get the same result. This means that

$$\frac{f*g}{f}(I) = \frac{p_s*g}{p_s}(I).$$

Next note that if $Lf = \lambda f$ and if $a \in G$, then, defining f^a as in Lemma 1.1.1 of Section 1.1.4, we have $Lf^a = \lambda f^a$, by the G-invariance of L. It follows that

$$\frac{f*g}{f}(a) = \frac{f^a*g}{f^a}(I) = \frac{p_s*g}{p_s}(I) = \widehat{g}(\overline{s}).$$

This completes the proof of the first statement in part (2).

For the converse, look at $(f*g) = \lambda_g f$ and apply the G-invariant differential operator L to this equality to obtain:

$$\lambda_g Lf = L(f*g) = f*(Lg) = \lambda_{Lg}f.$$

This implies that

$$Lf = (\lambda_{Lg}/\lambda_g)f.$$

It is easy to see that we can force $\lambda_g \neq 0$ by taking g to run through a Dirac sequence at the identity. For then $f*g$ approaches $f = f*\delta$ and the eigenvalues λ_g must approach 1. ∎

Exercise 1.2.22. (a) Fill in the details in the proof of part (2) of Proposition 1.2.4. For example, what happens if $f(a) = 0$?

(b) What happens in part (2) of Proposition 1.2.4 if we do not assume that $g(x) = g(x^{-1})$ for all $x \in G$?

Now we can give some other characterizations of spherical functions.

Theorem 1.2.3 (Equivalent Definitions of Spherical Functions). *Here* $G = GL(n, \mathbb{R})$, $K = O(n)$.

(1) *Eigenfunctions of Convolution Operators.*

Let $\mathcal{A}$ denote the set of all $f : G \to \mathbb{C}$ which are continuous with compact support and K bi-invariant; i.e., $f(k_1 a k_2) = f(a)$, for all $k_i \in K$, $a \in G$. Then $\mathcal{A}$ is a commutative algebra under pointwise sum and convolution product. A function $h : G \to \mathbb{C}$ which is K bi-invariant with $h(I) = 1$ is a spherical function if and only if h is a common eigenfunction of all the convolution equations:

$$f * h = \lambda_f h, \quad \text{for all } f \in \mathcal{A}.$$

Here the eigenvalue is λ_f.

(2) **Homomorphisms of** $\mathcal{A}$ (Gelfand).

The spherical functions are the functions $h : G \to \mathbb{C}$ which are K bi-invariant and continuous, with $h(I) = 1$, such that the mapping

$$f \mapsto (f * h)(I)$$

defines an algebra homomorphism of $\mathcal{A}$ onto $\mathbb{C}$.

(3) **More Integral Equations—The Analogue of the Funk-Hecke Theorem** (Gelfand).

A function $h : G \to \mathbb{C}$ which is nonzero, continuous, and K bi-invariant, is spherical if and only if

$$\int_{v \in K} h(xvy)\, dv = h(x)h(y) \quad \text{for all } x, y \in G.$$

(4) **Harish-Chandra's Integral Formula** (Selberg [543, pp. 53–59]).

A spherical function must be of the form (1.70) and if we use the r-variables from formula (1.43) and Proposition 1.2.1 in Section 1.2.1, we have for $Y \in \mathcal{P}_n$

$$h_s(Y) = \int_{k \in K} p_s(Y[k])\, dk = \int_{k \in K} \varphi_r(Y[k])\, dk, \quad \text{if } \varphi_r(I[t]) = \prod_{j=1}^{n} t_{jj}^{2v_j}$$

where $2v_j = 2(s_j + \cdots + s_n) = 2r_j + j - (n+1)/2$, for $r \in \mathbb{C}^n$, and $t \in T_n$, the group of upper triangular $n \times n$ matrices with positive diagonal entries. Moreover $h_{s(r)} = h_{s(r')}$ if and only if $r' = (r_{\sigma(1)}, \ldots, r_{\sigma(n)})$ for some permutation σ of n elements.

(5) **Connection with Group Representations** (Gelfand and Naimark).
     *A spherical function h is called **positive definite** if for any $f : G \to \mathbb{C}$ which is continuous with compact support, the following inequality holds:*

$$\int_G \int_G h(a^{-1}b) \, \overline{f(a)} \, f(b) \, da \, db \geq 0.$$

Positive definite spherical functions h can be expressed in the form

$$h(a) = (x_0 | U_a x_0), \ a \in G,$$

*for some irreducible unitary representation U of G of class one. Here x_0 is a K-fixed vector in the Hilbert space X on which U acts and $(x|y)$ denotes the Hilbert space inner product of x and y in X. **Class one** means that such a K-fixed x_0 must exist in X.*

Proof.

(1) $\Leftrightarrow$ (h is spherical).

This follows from Proposition 1.2.4 and the following exercise.

Exercise 1.2.23. Show that $\mathcal{A}$ is a commutative algebra.
Hint. Imitate the proof of part (4) of Lemma 3.7.2 in Vol. I. This was essentially Exercise 1.1.26 of Section 1.1.4.

(1) $\Rightarrow$ (2).
Suppose that $(f*h) = \lambda_f h$ for all $f \in \mathcal{A}$. Then $(f*h)(I) = \lambda_f h(I) = \lambda_f$. Therefore

$$\lambda_{f*g} = ((f*g)*h)(I) = (f*(g*h))(I) = \lambda_f \lambda_g.$$

(2)$\Rightarrow$(3).

Suppose that h is as in (2); i.e., suppose that upon setting $\lambda_f = (f*h)(I)$, we have $\lambda_{f*g} = \lambda_f \lambda_g$, for all f, g in $\mathcal{A}$. We want to show that h satisfies the integral equation:

$$\int_K h(xvy) \, dv = h(x)h(y).$$

Now $\lambda_{f*g} = \lambda_f \lambda_g$ implies that

$$\int \int f(y^{-1})g(x^{-1})h(yx) \, dy \, dx = \int \int f(y^{-1})g(x^{-1})h(x)h(y) \, dx \, dy.$$

And the left-hand side of this equality may be rewritten as:

$$\int \int f(y^{-1})g(x^{-1}) \int_K h(yvx) \, dv \, dy \, dx.$$

It follows that

$$\int_K h(yvx) \, dv = h(x)h(y) \text{ almost everywhere}$$

and continuity completes the proof.

(3)$\Rightarrow$ (h is spherical).

To see that (3) implies that h is infinitely differentiable, suppose that g is in $C_c^\infty(G)$ with the property that $\int_G g(y)h(y)dy \neq 0$. By part (3) we have:

$$h(x) \int_G h(y)g(y) \, dy = \int_G g(y) \int_K h(xky) \, dk \, dy = \int_K \int_G g(k^{-1}x^{-1}u)h(u) \, du \, dk.$$

It follows that $h(x)$ is infinitely differentiable, since $g(x)$ is.

To see that indeed part (3) implies that h is a spherical function, we must show that h is an eigenfunction for the G-invariant differential operators $L \in D(\mathcal{P}_n)$. This follows from the following considerations:

$$(L_x h(x))h(y) = L_x \int_K h(xvy) \, dv = \int_K (Lh)(xvy) \, dv.$$

Set $x = I$ to obtain $(Lh)(I)h(y) = (Lh)(y)$. This completes the proof that h is a spherical function, since the integral formula satisfied by h clearly implies that $h(I) = 1$.

This completes the proof that (1)–(3) are all equivalent to the definition of spherical function.

(4)$\Leftrightarrow$ (h is spherical).

Only the $\Leftarrow$ needs some discussion. We know from Theorem 1.1.2 of § 1.1.5 that the algebra $D(\mathcal{P}_n)$ is the polynomial algebra over $\mathbb{C}$ generated by the algebraically independent operators $L_j = \text{Tr}((Y\partial/\partial Y)^j)$, for $j = 1, 2, \ldots, n$. Any spherical function h gives a homomorphism of the algebra $D(\mathcal{P}_n)$ into $\mathbb{C}$, defined by sending L in $D(\mathcal{P}_n)$ to the eigenvalue λ_L, with $Lh = \lambda_L h$.

We know from Theorem 1.2.1 of § 1.2.1 that $L_j h_r = \lambda_j(r)h_r$, $j = 1, 2, \ldots, n$, where $\lambda_j(r)$ is a symmetric polynomial of degree j in $r_1, \ldots, r_n$, such that the highest degree homogeneous term is $r_1^j + \cdots + r_n^j$. It follows that the λ_j form a basis for the symmetric polynomials in $r_1, \ldots, r_n$. Now suppose we are given a spherical function

h and thus a set of eigenvalues λ_j, with $L_j h = \lambda_j h$, $j = 1, \ldots, n$. The uniqueness of spherical functions (proved in part 1) of Proposition 1.2.4 implies that $h = h_r$.

Suppose next that $h_r = h_{r'}$, for $r, r' \in \mathbb{C}^n$. Then each polynomial $\lambda_i(r) = \lambda_i(r')$, for $i = 1, \ldots, n$. But then all the symmetric polynomials agree on r and r'. This implies that r' is obtained from r by permuting the entries r_j.

In this proof we used some facts about symmetric polynomials which are proved in Exercise 1.2.10 of Section 1.2.1.

(5) We omit the proof of part (5) of Theorem 1.2.3. Proofs can be found in Helgason [273, pp. 414–417] or Maurin [437, p. 233], for example. ∎

Harish-Chandra's result in the preceding theorem, generalizing the unique characterization of spherical functions for the Poincaré upper half plane, is very remarkable. For it is much harder to obtain uniqueness results for solutions of partial differential equations than for solutions of ordinary differential equations. One might ask whether the method of proof could be used to obtain an analogue for Bessel functions for $\mathcal{P}_n$. So far, this does not appear to be possible. For there is no obvious way to replace the operator M which averages functions over the compact group K with its analogue for the noncompact group $N(m, n-m)$ in (1.58) of Section 1.2.2. However, for Whittaker functions this might be possible. Indeed we will see in the next section that integrals over K (actually K/M) can be replaced by integrals over N (actually the opposite group $\overline{N}$ of lower triangular matrices with ones on the diagonal). In fact, the multiplicity one result of Shalika mentioned in Section 1.5.4 and proved in Goldfeld [230, p. 155] provides a uniqueness result for Whittaker functions.

In Vol. I, p. 194, we claimed again (sorry) that the central limit theorem for rotation invariant densities on the Poincaré upper half plane followed from the following asymptotic formula:

$$h_{1/2+ip,0} \begin{pmatrix} e^{(1/2)r} & 0 \\ 0 & e^{-(1/2)r} \end{pmatrix} = P_{-1/2+ip}(\cosh r) \sim J_0(pr), \quad \text{as } r \to 0, \tag{1.73}$$

$$\sim 1 - \tfrac{1}{4} r^2 p^2, \quad \text{as } r \to 0.$$

Here $P_v(z)$ is the Legendre function and $J_0(z)$ is the Bessel function. Actually for the central limit theorem we need to know second order terms exactly and thus we really need the following formulas, using the standard power series for the Gauss hypergeometric function:

$$P_v(z) = {}_2F_1\left(-v, \; v + 1; 1; \frac{1}{2}(1 - z)\right)$$

$$= \sum_{k \geq 1} \frac{(-v)_k (v+1)_k}{k!(1)_k} \left(\frac{1-z}{2}\right)^k, \quad |z - 1| < 2.$$

It follows that, as $r \to 0$,

$$P_{-(1/2)+ip}(\cosh r) \sim 1 + \frac{1}{2}\left|\frac{1}{2} - ip\right|^2 (1 - \cosh r) \sim 1 - \frac{1}{4}\left(\frac{1}{4} + p^2\right)r^2.$$

Thus when $n = 2$ the zonal spherical function has the following asymptotic expansion:

$$h_{(1/2)+ip}\begin{pmatrix} e^r & 0 \\ 0 & e^{-r} \end{pmatrix} \sim 1 - \left(\frac{1}{4} + p^2\right)r^2, \quad \text{as } r \to 0. \qquad (1.74)$$

This does not change the fact that the central limit theorem holds for identically distributed rotation-invariant sequences of random variables on H. In fact (1.74) is even "better" than (1.73) in the sense that the coefficient of r^2 is exactly the eigenvalue of the Laplacian corresponding to the spherical function; i.e.,

$$\Delta h_{1/2+ip}\begin{pmatrix} e^r & 0 \\ 0 & e^{-r} \end{pmatrix} = -\left(\frac{1}{4} + p^2\right)h_{1/2+ip}\begin{pmatrix} e^r & 0 \\ 0 & e^{-r} \end{pmatrix}.$$

The arguments of Vol. I, pp. 194–195, go through exactly as before.

Despite the problem noted above, it is still useful to generalize (1.73) to the space $\mathcal{P}_n$. We can do this for general n using Helgason [279, Lemma 4.3] and some expansions of James [328–330]. This will not give us a central limit theorem, however. For that, we must generalize (1.74). In this section, we shall do that only in the case of $\mathcal{P}_3$, using the Taylor expansion of the zonal spherical function. In Section 1.2.5 we will consider Richards' extension of this result to the general case.

Let us first discuss Helgason [279, Lemma 4.3] in our case. First we will need some preliminaries from the theory of symmetric spaces and Lie groups. We saw in Exercise 1.1.12 of Section 1.1.3 that any matrix g in $GL(n, \mathbb{R})$ can be decomposed (uniquely) into **Iwasawa coordinates**:

$$\left.\begin{array}{l} g = kan, \quad k \in K, \quad a \in A, \quad n \in N. \\ \text{Write } k = K(g), \ a = A(g), \ n = N(g). \end{array}\right\} \qquad (1.75)$$

Corresponding to (1.75), we have an Iwasawa decomposition of the Lie algebra $\mathfrak{g}$ of G which is the tangent space to G at the identity equipped with a Lie bracket coming from identification of tangent vectors with left G-invariant first order differential operators (vector fields) on G. In our case, we have:

$$\mathfrak{g} = \mathfrak{gl}(n, \mathbb{R}) = \mathbb{R}^{n \times n}, \quad \text{with Lie bracket } [X, Y] = XY - YX. \qquad (1.76)$$

See Chapter 2 for more information on Lie algebras.

The **Lie algebra Iwasawa decomposition** is:

$$
\left.
\begin{aligned}
&\mathfrak{g} = \mathfrak{k} \oplus \mathfrak{a} \oplus \mathfrak{n}, \quad \text{where} \\
&\mathfrak{k} = \mathfrak{o}(n) = \{X \in \mathfrak{g} \mid {}^tX = -X\}, \\
&\mathfrak{a} = \{X \in \mathfrak{g} \mid X \text{ diagonal}\}, \\
&\mathfrak{n} = \{X \in \mathfrak{g} \mid X \text{ is upper triangular}, 0 \text{ on the diagonal}\}.
\end{aligned}
\right\}
\tag{1.77}
$$

The **tangent space** to $\mathcal{P}_n$ at I can be identified with:

$$
\mathfrak{p} = \{X \in \mathbb{R}^{n \times n} \mid {}^tX = X\}.
\tag{1.78}
$$

We can clearly write

$$
\mathfrak{p} = \mathfrak{a} \oplus \mathfrak{q}, \quad \text{with } \mathfrak{a} \text{ as in (1.77)}
$$

and

$$
\mathfrak{q} = \{X \in \mathfrak{p} \mid X \text{ is } 0 \text{ on the diagonal}\}.
\tag{1.79}
$$

We will use the following notation for $X \in \mathfrak{p}$:

$$
X = H + Y, \qquad H \in \mathfrak{a}, \qquad Y \in \mathfrak{q}, \qquad H = a(X), \qquad Y = q(X).
\tag{1.80}
$$

Because $\mathcal{P}_n$ is a symmetric space coming from the Lie group G, we have a **Cartan involution** $\theta : \mathfrak{g} \to \mathfrak{g}$ given by $\theta(X) = -\,{}^tX$. If σ_I denotes the geodesic-reversing isometry of $\mathcal{P}_n$ at the identity, then $(d\sigma)_I = \theta|_{\mathfrak{p}}$. We can therefore write, for $X \in \mathfrak{p}$:

$$
q(X) = -(Z + \theta(Z)) + 2Z = Z + {}^tZ, \quad \text{for some } Z \in \mathfrak{n}.
\tag{1.81}
$$

Now, it is *not*, in general, true that for $X, Y \in \mathfrak{g}$, we have $\exp(X)\exp(Y) = \exp(X + Y)$. Instead there is an expansion called the **Campbell-Baker-Hausdorff formula.** Here we give only the first two terms:

$$
\exp(tX)\exp(tY) = \exp\left\{t(X + Y) + \left(\frac{1}{2}\right)t^2[X, Y] + O(t^3)\right\},
\tag{1.82}
$$

for $X, Y \in \mathfrak{g}$ and $t \in \mathbb{R}$ assumed to be small.

It follows that if $H \in \mathfrak{a}$, $k \in K$, $t \in \mathbb{R}$ (small),

$$
\exp(tH[k] + O(t^2)) = \exp(-t(Z + \theta(Z)))\exp(ta(H[k]))\exp(2tZ),
\tag{1.83}
$$

for some $Z \in \mathfrak{n}$. Note that in (1.83),

$$\exp(-t(Z + \theta(Z))) \in K, \quad \exp(ta(H[k])) \in A, \quad \exp(2tZ) \in N.$$

Thus, the following lemma has been proved.

Lemma 1.2.1 (Helgason [279, Lemma 4.3]). *Suppose that $H \in \mathfrak{a}$ and $k \in K = O(n)$, using the notation (1.77). Then we have the following asymptotic relation as t approaches 0:*

$$A(\exp(tH[k]) \sim \exp(ta(H[k])),$$

where we use the notation set up in (1.75) and (1.80).

The preceding lemma has as an immediate consequence an asymptotic formula for the spherical function (1.70). Suppose that for $H \in \mathfrak{a}$ we write

$$H = \begin{pmatrix} h_1 & & 0 \\ & \ddots & \\ 0 & & h_n \end{pmatrix}, \quad h_j \in \mathbb{R}.$$

Then we normalize the power function as in part (4) of Theorem 1.2.3, to obtain:

$$p_s(\exp H) = \prod_{j=1}^{n} \exp(h_j v_j), \tag{1.84}$$

with $v_j = s_j + \cdots + s_n = r_j + \frac{1}{2}j - \frac{1}{4}(n+1)$. Lemma 1.2.1 says that

$$p_s(\exp(tH[k])) \sim p_s(\exp(ta(H[k]))), \quad \text{as } t \to 0.$$

Now $a(H[k])$ is just the diagonal part of the symmetric matrix

$$H[k] = ({}^t k_i H k_j)_{1 \le i,j \le n}, \text{ for } k = (k_1 \cdots k_n) \in K. \tag{1.85}$$

Here k_j denotes the jth column of the orthogonal matrix k. So we find that

$$p_s(\exp(ta(H[k]))) = \prod_{j=1}^{n} \exp(tH[k_j] v_j) = \exp\left\{ t \sum_{j}^{n} H[k_j] v_j \right\}$$

$$= \exp\left\{ t \sum_{i,j=1}^{n} h_i v_j k_{ij}^2 \right\} = \exp\{t \operatorname{Tr}(H[k]V)\},$$

where V is the diagonal matrix with jth diagonal entry v_j. The following result has now been proved.

Theorem 1.2.4 (An Asymptotic Formula for the Spherical Function at the Identity). *If*

$$H = \begin{pmatrix} h_1 & & 0 \\ & \ddots & \\ 0 & & h_n \end{pmatrix} \in \mathfrak{a}, \quad s \in \mathbb{C}^n,$$

we have the following asymptotic formula for the spherical function in (1.70):

$$h_s(\exp(tH)) \sim \int\limits_{k \in K = O(n)} \exp\{t\mathrm{Tr}(H[k]V)\}\, dk, \quad \text{as } t \to 0,$$

where

$$V = \begin{pmatrix} v_1 & & 0 \\ & \ddots & \\ 0 & & v_n \end{pmatrix}, \quad v_j = s_j + \cdots + s_n = r_j + \frac{1}{4}(2j - n - 1).$$

Helgason [282, pp. 423–467] considers such functions as the integral appearing in Theorem 1.2.4. James [328–330] has extensively studied these functions with a view towards statistical applications (see also Farrell [173, 174]), Muirhead [468], and Takemura [585]). We need to review some of this work. In the notation of James, the integral of interest is:

$$_0F_0^{(n)}(X, Y) = \int\limits_{k \in K} \exp\left\{\mathrm{Tr}\left(XkYk^{-1}\right)\right\}\, dk, \quad \text{for } X, Y \in \mathfrak{p}. \tag{1.86}$$

James obtains an expansion of this integral in a series of zonal polynomials associated with partitions κ of k :

$$\kappa = (k_1, \ldots, k_n) \text{ of } k; \text{ i.e., } k = k_1 + \cdots + k_n,$$

with $k_1 \geq k_2 \geq \cdots \geq k_n \geq 0$, and $k_j \in \mathbb{Z}$. Note that we are allowing the parts k_j to vanish.

If $\kappa = (k_1, \ldots, k_n)$ and $\lambda = (l_1, \ldots, l_n)$ are two partitions of k and if $k_j > l_j$ for the first index j for which the parts are unequal, then we say $\kappa > \lambda$ and the monomial

$$x_1^{k_1} \cdots x_n^{k_n}$$

is of **higher weight** than

$$x_1^{l_1} \cdots x_n^{l_n}.$$

That is, we use the lexicographic order.

Now define the **zonal polynomial** $C_\kappa(Y)$, for partition κ, $Y \in \mathcal{P}_n$ with eigenvalues $a_1, \ldots, a_n$, to be a symmetric homogeneous polynomial of degree k in $a_1, \ldots, a_n$ with the following three properties:

(1) The term of highest weight is $d_\kappa a_1^{k_1} \cdots a_n^{k_n}$;
(2) C_κ is an eigenfunction of the Laplacian and

$$\Delta^\# C_\kappa = \lambda C_\kappa, \quad \text{where } \Delta^\# = \text{the part of } \left\{ \Delta + \frac{n-3}{2} \sum_j a_j \frac{\partial}{\partial a_j} \right\}$$

coming from the a_j variables; that is (cf. (1.36) in § 1.1.6),

$$\Delta^\# = \sum_j a_j^2 \frac{\partial^2}{\partial a_j^2} + \sum_{i \neq j} \left(\frac{a_i^2}{a_i - a_j} \right) \frac{\partial}{\partial a_i}; \tag{1.87}$$

(3) $(\mathrm{Tr} Y)^k = \sum_\kappa C_\kappa(Y)$, where the sum runs over all partitions of k into n parts some of which may vanish. Here we are using the well-known property of Euler's operator $\sum_i a_i \frac{\partial}{\partial a_i}$. See formula (1.89) below.

Lemma 1.2.2. *The eigenvalue λ in the preceding definition of C_κ is*

$$\lambda = \sum_i k_i(k_i - i) + k(n - 1).$$

Exercise 1.2.24. Prove Lemma 1.2.2.
Hint. See Muirhead [468, p. 229].

It is possible to express the zonal polynomials C_κ in terms of the following **monomial symmetric functions** corresponding to a partition $\kappa = (k_1, \ldots, k_p)$. If $Y = a[k]$, for $a \in A$, $k \in K$, then define M_κ by:

$$M_\kappa(Y) = \sum_\iota a_{i_1}^{k_1} a_{i_2}^{k_2} \cdots a_{i_p}^{k_p},$$

where the sum is over all choices $\iota = \{i_1, \ldots, i_p\}$ of p distinct integers in $\{1, 2, \ldots, n\}$. For example,

$$M_{(1)}(Y) = a_1 + \cdots + a_n = \mathrm{Tr}(Y),$$
$$M_{(2)}(Y) = a_1^2 + \cdots + a_n^2 = \mathrm{Tr}(Y^2),$$
$$M_{(1,1)}(Y) = a_1 a_2 + \cdots + a_1 a_n + a_2 a_3 + \cdots + a_{n-1} a_n.$$

Clearly

$$M_{(1,1)}(Y) = \frac{1}{2}\left\{M_{(1)}^2(Y) - M_{(2)}(Y)\right\}.$$

Now it is easy to compute the first few zonal polynomials.

Example 1.2.5. $k = 1$.

$$C_{(1)}(Y) = M_{(1)}(Y) = Tr(Y).$$

Example 1.2.6. $k = 2$.

$$C_{(2)}(Y) = M_{(2)}(Y) + (2/3)M_{(1,1)}(Y).$$
$$C_{(1,1)}(Y) = (4/3)M_{(1,1)}(Y).$$

Exercise 1.2.25. Fill in the details for the preceding examples.
Hint. See Muirhead [468, pp. 231–232].

Finally we get the expansion for $_0F_0^{(n)}(X, Y)$ obtained by James.

Proposition 1.2.5.

$$_0F_0^{(n)}(X, Y) = \int\limits_{k \in K} \exp\left\{Tr\left(XkYk^{-1}\right)\right\}\ dk$$

$$= \sum_{k \geq 0} \frac{1}{k!} \sum_{\kappa} \frac{C_\kappa(X)C_\kappa(Y)}{C_\kappa(I)},$$

where the second sum is over all partitions κ of k.

Proof (Following Muirhead [468, pp. 243–244, 258–260]).
    Clearly

$$\exp\left\{Tr\left(XkYk^{-1}\right)\right\} = \sum_{l \geq 0} \frac{1}{l!} \sum_{\lambda} C_\lambda\left(XkYk^{-1}\right).$$

One can complete the proof by showing that

$$\int\limits_{k \in K} C_\kappa\left(XkYk^{-1}\right)\ dk = \frac{C_\kappa(X)C_\kappa(Y)}{C_\kappa(I)}. \tag{1.88}$$

To see this, consider the left-hand side as a function of Y and call it $f(Y)$. Then $f(Y) = f(Y[k])$ for all $k \in K$. Thus $f(Y)$ depends only on the eigenvalues *of* Y, and moreover, it must be a homogeneous symmetric polynomial of degree k in these eigenvalues. Now use the G-invariance of Δ_Y and standard facts about the **Euler differential operator**

$$x_1 \partial/\partial x_1 + \cdots + x_n \partial/\partial x_n \tag{1.89}$$

acting on homogeneous functions (see Apostol [17, Vol. II, p. 287]) to see that if $\Delta^{\#}$ is as in (1.87)

$$\Delta^{\#} f(Y) = \lambda f(Y).$$

It follows that $f(Y) = \alpha C_\kappa(Y)$ for some scalar α. Set $Y = I$ and note that $f(I) = C_\kappa(X)$ to complete the proof.   ∎

Thus, in particular, if $H \in \mathfrak{a}$, $h_j \in \mathbb{R}$, $v_j \in \mathbb{C}$:

$$H = \begin{pmatrix} h_1 & & 0 \\ & \ddots & \\ 0 & & h_n \end{pmatrix}, \qquad V = \begin{pmatrix} v_1 & & 0 \\ & \ddots & \\ 0 & & v_n \end{pmatrix}, \tag{1.90}$$

we obtain the following expansion:

$$\int_{k \in K} \exp\left\{\mathrm{Tr}(HkVk^{-1})\right\} dk$$
$$= 1 + \frac{1}{n}\mathrm{Tr}(H)\mathrm{Tr}(V) + \frac{4}{3n(n-1)} \sum_{i<j} h_i h_j \sum_{i<j} v_i v_j$$
$$+ \frac{3}{2n(n+2)} \left(\sum_i h_i^2 + \frac{2}{3}\sum_{i<j} h_i h_j\right)\left(\sum_i v_i^2 + \frac{2}{3}\sum_{i<j} v_i v_j\right) \tag{1.91}$$
$$+ \text{higher order terms.}$$

Since Theorem 1.2.4 is only good to first order, we do not expect this expansion to hold for the spherical function $h_s(\exp H)$, though we do expect some similarity.

Next let us do the Taylor expansion of $h_s(\exp H)$ directly (using brute force) when $n = 3$. We will consider the general case in Section 1.2.5. Here H is as in (1.90). The differential operators appearing in the expansion will be evaluated at $H = 0$ and they can be identified with G-invariant differential operators $L \in D(\mathcal{P}_n)$ evaluated at I (see Helgason [273, 282] and the proof of part (1) of Proposition 1.2.4 above). So, making use of the symmetry of the function in the h_j, we obtain the following form of the spherical function expansion on $\mathcal{P}_3$, where $\omega(i) = i_1 + \cdots + i_p$:

$$
\left.
\begin{aligned}
h_s(\exp H) &= \sum_p \frac{1}{p!} \sum_{\substack{i=(i_1,\ldots,i_n)\in\mathbb{Z}^n \\ i_j\geq 0,\ \omega(i)=p}} \left[\frac{\partial^p h_s}{\partial h_1^{i_1}\cdots\partial h_n^{i_n}}\right]_{h_j=0} h_1^{i_1}\ldots h_n^{i_n} \\
&= 1 + \left\{\alpha_1\left(\sum_i r_i\right) + \beta_1\right\}\left(\sum_i h_i\right) \\
&\quad + \left\{\alpha_2\left(\sum_i r_i^2\right) + \beta_2\left(\sum_{i\neq j} r_i r_j\right) + \gamma_2\left(\sum_i r_i\right) + \delta_2\right\}\left(\sum_i h_i^2\right) \\
&\quad + p(r)\left(\sum_{i\neq j} h_i h_j\right) + \text{ higher order terms.}
\end{aligned}
\right\}
\tag{1.92}
$$

Here $\alpha_i, \beta_i, \gamma_i$ are constants and $p(r)$ is a symmetric polynomial of degree 2 in r and the r-variables are related to the s-variables by

$$
s_j + \cdots + s_n = r_j + \frac{1}{4}(2j - n - 1).
\tag{1.93}
$$

The **_first order terms_** in (1.92) are easy to find when $n = 3$. Here let $k = (k_1 k_2 k_3) \in K = O(3)$; i.e., k_j denotes the jth column *of* the 3×3 rotation matrix k. Then if $a = \exp H$,

$$
\begin{aligned}
p_s(a[k]) &= a[k_1]^{s_1}\,|a[k_1 k_2]|^{s_2}\,|a|^{s_3} \\
&= a[k_1]^{s_1}\left\{a[k_1]a[k_2] - ({}^t k_1 a k_2)^2\right\}^{s_2}|a|^{s_3},
\end{aligned}
$$

and

$$
a[k_j] = \sum_{i=1}^{3} a_i k_{ij}^2, \qquad {}^t k_i a k_j = \sum_{l=1}^{3} k_{li} a_l k_{lj}, \qquad |a| = a_1 a_2 a_3.
$$

Therefore, we have, setting

$$
w_i(a,k) = k_{i1}^2 a[k_2] + k_{i2}^2 a[k_1] - 2({}^t k_1 a k_2) k_{i1} k_{i2},
\tag{1.94}
$$

$$
\begin{aligned}
\frac{\partial}{\partial h_i} p_s(a[k]) &= s_1 a[k_1]^{s_1-1} k_{i1}^2 a_i |a[k_1 k_2]|^{s_2}|a|^{s_3} + s_3 a[k_1]^{s_1}|a[k_1 k_2]|^{s_2}|a|^{s_3} \\
&\quad + s_2 a[k_1]^{s_1}|a[k_1 k_2]|^{s_2-1}|a|^{s_3} a_i\, w_i(a,k).
\end{aligned}
\tag{1.95}
$$

If we set $h_1 = h_2 = h_3 = 0$, then $a[k] = I$ and we obtain:

$$
\left[\frac{\partial}{\partial h_i} p_s(a[k])\right]_{\substack{h_j=0 \\ j=1,2,3}} = s_1 k_{i1}^2 + s_2(k_{i1}^2 + k_{i2}^2) + s_3.
\tag{1.96}
$$

The representation theory of $O(3)$ helps to evaluate the integrals we need. In particular, it is known (see Section 2.1.5 of Vol. I and Broecker and tom Dieck [81]) that since k_{ij} is the entry of an irreducible representation of $O(3)$, we have the formula:

$$\int_{k\in K} k_{ij}^2 \, dk = 1/3. \tag{1.97}$$

You can check this using the **Euler angle decomposition** (see Vilenkin [635, pp. 106, 435–440], Wawrzyńczyk [657, pp. 287–291], or Volume I, p. 115):

$$
\left.
\begin{aligned}
&0 \le \alpha, \gamma \le 2\pi, \ 0 \le \beta \le \pi, \qquad dk = (8\pi^2)^{-1} \sin\beta \, d\alpha \, d\beta \, d\gamma, \\
&k = \begin{pmatrix} \cos\alpha & \sin\alpha & 0 \\ -\sin\alpha & \cos\alpha & 0 \\ 0 & 0 & 1 \end{pmatrix} \begin{pmatrix} 1 & 0 & 0 \\ 0 & \cos\beta & \sin\beta \\ 0 & -\sin\beta & \cos\beta \end{pmatrix} \begin{pmatrix} \cos\gamma & \sin\gamma & 0 \\ -\sin\gamma & \cos\gamma & 0 \\ 0 & 0 & 1 \end{pmatrix} \\
&= \begin{pmatrix} \cos\alpha\cos\gamma - \cos\beta\sin\alpha\sin\gamma & \cos\alpha\sin\gamma + \cos\beta\sin\alpha\cos\gamma & \sin\alpha\sin\beta \\ -\sin\alpha\cos\gamma - \cos\alpha\cos\beta\sin\gamma & -\sin\alpha\sin\gamma + \cos\alpha\cos\beta\cos\gamma & \cos\alpha\sin\beta \\ \sin\beta\sin\gamma & -\sin\beta\cos\gamma & \cos\beta \end{pmatrix}
\end{aligned}
\right\}
\tag{1.98}
$$

Thus, for example,

$$(2\pi)^2 \frac{1}{8\pi^2} \int_0^\pi \cos^2\beta \, \sin\beta \, d\beta = \frac{1}{3}.$$

One can use permutation matrices to see that all the integrals

$$\int_{k\in K} k_{ij}^2 \, dk$$

must be equal. We could also evaluate them using James' formula (1.91). It follows that

$$\int_{k\in K} \left[\frac{\partial}{\partial h_i} p_s(a[k]) \right]_{\substack{h_j = 0 \\ j = 1,2,3}} dk = \frac{1}{3}(s_1 + 2s_2 + 3s_3). \tag{1.99}$$

Recall that (as in (1.84))

$$s_3 = r_3 + \frac{1}{2} = v_3, \ s_2 = r_2 - r_3 - \frac{1}{2} = v_2 - v_3, \ s_1 = r_1 - r_2 - \frac{1}{2} = v_1 - v_2. \tag{1.100}$$

Therefore the right-hand side of (1.99) can be rewritten as

$$\frac{1}{3}(r_1 + r_2 + r_3) = \frac{1}{3}(v_1 + v_2 + v_3),$$

which is exactly the first order term in (1.91) for $n = 3$.

Let us now compute the second derivative (a sum of ten terms), defining $w_i(a, k)$ as in formula (1.94):

$$
\begin{aligned}
\frac{\partial^2}{\partial h_j \partial h_i} p_s(a[k]) = {} & s_1(s_1-1)\, a[k_1]^{s_1-2} k_{i1}^2 k_{j1}^2 a_i a_j\; |a[k_1 k_2]|^{s_2}\, |a|^{s_3} \\
& + \delta_{ij} s_1\, a[k_1]^{s_1-1} k_{i1}^2 a_i\; |a[k_1 k_2]|^{s_2} |a|^{s_3} \\
& + s_1 s_2\, a[k_1]^{s_1-1} k_{i1}^2\; |a[k_1 k_2]|^{s_2-1} a_i a_j\, w_j(a, k)\; |a|^{s_3} \\
& + s_2(s_2-1)\, a[k_1]^{s_1} |a[k_1 k_2]|^{s_2-2} a_i a_j\; |a|^{s_3}\, w_i(a, k)\, w_j(a, k) \\
& + s_1 s_2\, a[k_1]^{s_1-1} k_{j1}^2 |a[k_1 k_2]|^{s_2-1} a_i a_j\, w_i(a, k)\; |a|^{s_3} \\
& + s_1 s_3\, a[k_1]^{s_1-1} \left(k_{i1}^2 a_i + k_{j1}^2 a_j \right) |a[k_1 k_2]|^{s_2} |a|^{s_3} \\
& + s_2\, a[k_1]^{s_1} |a[k_1 k_2]|^{s_2-1} a_i a_j\, \left\{ k_{i1}^2 k_{j2}^2 + k_{i2}^2 k_{j1}^2 - 2k_{i1} k_{i2} k_{j1} k_{j2} \right\}\; |a|^{s_3} \\
& + s_2\, a[k_1]^{s_1} |a[k_1 k_2]|^{s_2-1} a_i\, \delta_{ij}\, w_i(a, k)\; |a|^{s_3} \\
& + s_2 s_3\, a[k_1]^{s_1} |a[k_1 k_2]|^{s_2-1} |a|^{s_3} \left\{ a_i\, w_i(a, k) + a_j\, w_j(a, k) \right\} \\
& + s_3^2\, a[k_1]^{s_1} |a[k_1 k_2]|^{s_2} |a|^{s_3}.
\end{aligned}
$$

If we set $h_1 = h_2 = h_3 = 0$, we obtain:

$$
\left.\begin{aligned}
&\left[\frac{\partial^2}{\partial h_j \partial h_i} p_s(a[k]) \right]_{\substack{h_k = 0 \\ k = 1, 2, 3}} \\[2mm]
&= s_1(s_1-1) k_{i1}^2 k_{j1}^2 + \delta_{ij} k_{i1}^2 s_1 + s_1 s_2 k_{i1}^2 (k_{j1}^2 + k_{j2}^2) + \delta_{ij} s_2 (k_{i1}^2 + k_{i2}^2) + s_1 s_2 k_{j1}^2 (k_{i1}^2 + k_{i2}^2) \\
&\quad + s_2 \{ k_{i1}^2 k_{j2}^2 + k_{i2}^2 k_{j1}^2 - 2k_{i1} k_{i2} k_{j1} k_{j2} \} + s_2(s_2-1)(k_{i1}^2 + k_{i2}^2)(k_{j1}^2 + k_{j2}^2) \\
&\quad + s_1 s_3 (k_{i1}^2 + k_{j1}^2) + s_2 s_3 (k_{i1}^2 + k_{i2}^2 + k_{j1}^2 + k_{j2}^2) + s_3^2.
\end{aligned}\right\}
$$

$$
\text{(1.101)}
$$

To integrate over K, we need some integral formulas:

$$
\int\limits_{k \in K} k_{ii}^4\, dk = \frac{1}{5}, \qquad \int\limits_{k \in K} k_{i1}^2 k_{i2}^2\, dk = \frac{1}{15}. \qquad (1.102)
$$

These formulas are easily proved using the Euler angle decomposition (1.98) or from formula (1.91).

It follows that when $i = j$ we have

$$
\int\limits_{k \in K} \left[\frac{\partial^2}{\partial h_i^2} p_s(a[k]) \right]_{\substack{h_j = 0 \\ j = 1, 2, 3}} dk = \tfrac{1}{5} s_1(s_1 - 1) + \tfrac{1}{3}(s_1 + 2s_3) + \tfrac{8}{15} s_1 s_2
$$

$$
+ \tfrac{8}{15} s_2(s_2 - 1) + \tfrac{2}{3} s_1 s_3 + \tfrac{4}{3} s_2 s_3 + s_3^2.
$$

Using (1.100), this is the same as:

$$\tfrac{3}{15}(v_1^2 + v_2^2 + v_3^2) + \tfrac{2}{15}(v_1 v_2 + v_1 v_3 + v_2 v_3) + \tfrac{2}{15}(v_1 - v_3)$$

$$= \tfrac{3}{15}(r_1^2 + r_2^2 + r_3^2) + \tfrac{2}{15}(r_1 r_2 + r_1 r_3 + r_2 r_3) - \tfrac{1}{15}.$$

We are not interested in the terms with $i \neq j$ since they will disappear in the central limit theorem. So our asymptotic expansion for the spherical function on $\mathcal{P}_3$ becomes:

$$\left.\begin{aligned}
h_s(\exp H) \sim 1 + \tfrac{1}{3}(r_1 + r_2 + r_3)(h_1 + h_2 + h_3) \\
+ \tfrac{1}{30}\{3(r_1^2 + r_2^2 + r_3^2) + 2(r_1 r_2 + r_1 r_3 + r_2 r_3) - 1\}(h_1^2 + h_2^2 + h_3^2) \\
+ p(r)(h_1 h_2 + h_1 h_3 + h_2 h_3) + \text{ higher order terms,} \quad \text{as } H \to 0.
\end{aligned}\right\}$$
(1.103)

Here $p(r)$ is a symmetric polynomial of degree 2 which we will not need to evaluate.

Note that the coefficient of $h_1^2 + h_2^2 + h_3^2$ is not the eigenvalue of Δ given in Exercise 1.2.12 of Section 1.2.1. This has interesting consequences for the central limit theorem. D. St. P. Richards [511] generalized (1.103) to $\mathcal{P}_n$, for all n. See Section 1.2.5.

Exercise 1.2.26. (a) Check the evaluations of integrals over $O(3)$ given in formulas (1.97) and (1.102). Find $p(r)$ in (1.103).
(b) What happens to formula (1.103) if $\mathrm{Tr}(H) = 0$?
(c) Obtain the analogue of formula (1.103) for $\mathcal{P}_2$ and $\mathcal{SP}_2$.

This completes our discussion of spherical functions for the present. In the next section we will consider another sort of asymptotic formula for spherical functions. In the preceding formula Y approached I, but in that of Section 1.3 the variable Y will approach the boundary of the symmetric space. It will be necessary to find such an expansion in order to obtain the inversion formula for the Helgason–Fourier transform on $\mathcal{P}_n$.

It is possible to prove a Weyl character formula (cf. Chapter 2 of Vol. I) in the framework of spherical functions (see Harish-Chandra [263] and Berezin [43]). Gelfand and Naimark [215] do the special case of the symmetric space $SL(n, \mathbb{C})/SU(n)$ (cf. Chapter 2 of this volume). Recall that the Weyl character formula expresses the characters of the irreducible representations of compact semisimple Lie groups as ratios of exponential polynomials on a maximal abelian Lie subalgebra.

The continuous homomorphisms from the algebra $\mathcal{A}$ (defined in Theorem 1.2.3) onto the complex numbers are maps $f \mapsto (f*h)(I)$, provided that h is a bounded spherical function (see Helgason [273, p. 410]). Helgason and Johnson [283] characterize the bounded spherical functions in terms of the s-variables. Flensted-Jensen discovered relations between spherical functions on real semisimple Lie groups (such as $SL(n, \mathbb{R})$) and the more elementary spherical functions on the corresponding complex groups (e.g., $SL(n, \mathbb{C})$) (see Helgason [282, pp. 489–490]). Healy [266] has made a study of relations between Fourier analysis on $SL(2, \mathbb{C})/SU(2)$ and that on the sphere. See also Chapter 2 of this volume.

1.2.4   The Wishart Distribution

Now we consider applications of some of the results of the preceding subsections. In particular, we will obtain a theorem of Wishart which is important both for multivariate statistics and for the study of some number theoretical results to be discussed in Section 1.4.

References for this subsection are Anderson [8], Farrell [173, 174], Herz [293], James [328–330], Morrison [464], Muirhead [468], and Press [493].

A random matrix $Y \in \mathcal{P}_p$ is said to have the (central) **Wishart distribution** $W(\Sigma, p, n)$ **with scale matrix Σ and n degrees of freedom, $p \leq n$,** if the joint distribution of the entries of Y has the density function:

$$f(Y) = c|Y|^{(n-p-1)/2}|\Sigma|^{-n/2} \exp\left(-\tfrac{1}{2}\mathrm{Tr}\left(\Sigma^{-1}Y\right)\right), \quad \Sigma \in \mathcal{P}_p, \left.\vphantom{\prod_{j=1}^p}\right\}$$
$$\text{with } c^{-1} = 2^{np/2}\pi^{p(p+1)/4}\left(\prod_{j=1}^p \Gamma\left(\tfrac{n+1-j}{2}\right)\right) = 2^{np/2}\Gamma_p(0, \ldots, 0, n/2). \left.\vphantom{\prod_{j=1}^p}\right\}$$

$$\text{(1.104)}$$

The application of this density comes from the following theorem. This result was first obtained by Fisher in 1915 for the case that $p = 2$ and by Wishart [670] in 1928 for general p.

Exercise 1.2.27. Show that in order for

$$\int_{\mathcal{P}_n} f(Y) \prod_{i \leq j} dy_{ij} = 1,$$

the constant c in formula (1.104) must be as stated.
Hint. Use the formula for $\Gamma_p(0, \ldots, 0, s)$ in Section 1.2.1.

Theorem 1.2.5 (Wishart [670]). *Let X_i be a random variable in $\mathbb{R}^p$, for $i = 1, 2, \ldots, n$, where $p \leq n$. Suppose that $X_1, \ldots, X_n$ are mutually independent and distributed according to $N(0, \Sigma)$; that is, normal with mean 0 and covariance matrix $\Sigma \in \mathcal{P}_p$ (as in Section 1.1.7). Let $X = (X_1, \ldots, X_n) \in \mathbb{R}^{p \times n}$ and $Y = X\,{}^t X$. Then, with probability one, Y is in $\mathcal{P}_p$ and is distributed according to the Wishart density $W(\Sigma, p, n)$ defined by (1.104).*

Proof. First one must show that $Y = X\,{}^t X$ is positive definite with probability one. We leave this to the reader in Exercise 1.2.28 below.

The joint density of $X_1, \ldots, X_n$ is (cf. Section 1.1.7):

$$d(X) = (2\pi)^{-np/2} |\Sigma|^{-n/2} \exp\left(-\frac{1}{2} \sum_{j=1}^{n} {}^t X_j \Sigma^{-1} X_j\right)$$

$$= (2\pi)^{-np/2} |\Sigma|^{-n/2} \exp\left(-\frac{1}{2} \mathrm{Tr}\left(\Sigma^{-1}\left(X\,{}^t X\right)\right)\right).$$

To complete the proof, we simply need to see how to change variables from X in $\mathbb{R}^{p \times n}$ to $V = X\,{}^t X$ in the closure of $\mathcal{P}_p$. First note that

$$\int_{X \in \mathbb{R}^{p \times n}} h\left(X\,{}^t X\right) \left|X\,{}^t X\right|^{-n/2} dX, \quad \text{with} \quad dX = \prod_{i,j} dx_{ij},$$

defines a $GL(p, \mathbb{R})$-invariant measure on functions $h : \mathcal{P}_p \to \mathbb{C}$. Such measures on $\mathcal{P}_p$ are unique up to a constant. Exercise 1.2.27 shows that the constant given in (1.104) is correct. This completes the proof of Theorem 1.2.5. ∎

The proof of Theorem 1.2.5 also proves **Wishart's integral formula** for $p \leq n$:

$$\int_{X \in \mathbb{R}^{p \times n}} h\left(X\,{}^t X\right) \left|X\,{}^t X\right|^{-n/2} dX = w_{p,n} \int_{\mathcal{P}_p} h(Y)\, d\mu_p(Y),$$

$$w_{p,n} = \prod_{j=n-p+1}^{n} \pi^{j/2} \Gamma(\tfrac{j}{2})^{-1} = \pi^{np/2}\, \Gamma_p\left(0, \ldots, 0, \tfrac{n}{2}\right)^{-1}. \tag{1.105}$$

Later (see Section 1.4) this formula will be very important to us in our study of matrix series.

Exercise 1.2.28. (a) Show that under the hypotheses of Theorem 1.2.5, $Y = X\,{}^t X$ is in $\mathcal{P}_p$ with probability one.
(b) Show that if $p > n$, $X \in \mathbb{R}^{p \times n}$, then $Y = X\,{}^t X$ is singular.
Hint. (See Muirhead [468, pp. 82–83].) Note that if the columns of X are linearly independent then the matrix Y is non-singular.

The Wishart distribution is a matrix analogue of the chi-square distribution (which is the case $p = 1$). Univariate statistics makes frequent use of tables of the chi-square distribution, which is, in fact, an incomplete gamma function.

If the random variables in Theorem 1.2.5 did not all have mean zero, then one would be dealing with the noncentral Wishart distribution, written in terms of the integral

$$\int_{k \in K} \exp(\mathrm{Tr}(XkY))\, dk,$$

which is a generalization of the J-Bessel function (see Herz [293], James [328–330], Farrell [173, 174], Muirhead [468, pp. 441–449]). The noncentral Wishart distribution was first studied by T.W. Anderson in 1946 in special cases. In 1955 Herz expressed the distribution in terms of the $_0F_1$ matrix argument hypergeometric function. In the early 1960s James and Constantine gave the zonal polynomial expansion for it.

The Wishart distribution is important, for example, in factor analysis, which seeks to explain correlation between a set of random variables in terms of a minimum number of factors. Such analysis is useful in many of the social sciences. There are many examples and references in Press [493, Chapter 10]. One example given by Press involves a seven-factor analysis of prices of 63 securities. One of the factors appeared to have an effect on all securities. The remaining six factors tended to group the stocks by industry.

It is possible to use Proposition 1.2.4 of Section 1.2.3 to evaluate some special integrals which arise in number theory and statistics (see Maass [426, Ch. 7] and Muirhead [468, Ch. 7]). We want to evaluate an integral which appears in Muirhead [468, Theorem 7.2.7, p. 248]:

$$I(B, r, s) = \int_{Y \in \mathcal{P}_n} \exp[-\text{Tr}(B^{-1}Y)] \, |B^{-1}Y|^r \, h_s(Y) \, d\mu_n(Y), \qquad (1.106)$$

for $B \in \mathcal{P}_n$, $r \in \mathbb{C}$, $s \in \mathbb{C}^n$. Here $h_s(Y)$ denotes the spherical function defined by (1.70) in the preceding subsection. We can use Proposition 1.2.4 of the preceding subsection to evaluate $I(B, r, s)$ as

$$I(B, r, x) = \widehat{f}(\overline{s}^*) h_s(B), \quad \text{with } f(Y) = |Y|^r \exp[-\text{Tr}(Y)], \qquad (1.107)$$

and $s^* = (s_{n-1}, \ldots, s_2, s_1, -(s_1 + \cdots + s_n))$ as in part (4), Proposition 1.2.1, Section 1.2.1. Here $\widehat{f}$ denotes the Helgason–Fourier transform defined in formula (1.72).

We should perhaps discuss the proof of (1.107), since Proposition 1.2.4, Section 1.2.3 considered only functions $g \in C_c^\infty(K \backslash G / K)$ such that g is invariant under inversion. Our function $f(Y)$ does not satisfy these hypotheses, setting $Y = I[x]$, $x \in G$, to make it a function on G. We can easily do away with the hypothesis that f have compact support. But to do away with the hypothesis that f be invariant under inversion on G is impossible. But note that our function f has the property that $f(I[x]) = f(I[^t x])$ for all $x \in G$. Therefore we have the following equalities:

$$(p_s * f)(I) = \int_G p_s(I[x]) f(I[x^{-1}]) \, dx = \int_G p_s(I[x]) f(I[{}^t x^{-1}]) \, dx$$

$$= \int_{\mathcal{P}_n} p_s(Y) f(Y^{-1}) \, d\mu_n(Y) = \int_{\mathcal{P}_n} p_s(Y^{-1}) f(Y) \, d\mu_n(Y)$$

$$= \int_{\mathcal{P}_n} p_s(Y^{-1}[\omega]) f(Y) \, d\mu_n(Y) = \int_{\mathcal{P}_n} p_{s^*}(Y) f(Y) \, d\mu_n(Y),$$

where ω and s^* are as in part (4) of Proposition 1.2.1, Section 1.2.1. This last integral is $\widehat{f}(\bar{s}^*)$, and our discussion of (1.107) is finished.

Now the Helgason–Fourier transform $\widehat{f}$ is:

$$\widehat{f}(s) = \int_{Y \in \mathcal{P}_n} \overline{p_s(Y)} \, |Y|^r \, \exp[-\text{Tr}(Y)] \, d\mu_n(Y) = \Gamma_n \left((0, \ldots, 0, r) + \bar{s} \right).$$

Thus we have proved that:

$$\int_{Y \in \mathcal{P}_n} \exp[-\text{Tr}(B^{-1} Y)] \, |B^{-1} Y|^r \, h_s(Y) \, d\mu_n(Y) = \Gamma_n \left(s^* + (0, \ldots, 0, r) \right) h_s(B),$$

$$(1.108)$$

where $B \in \mathcal{P}_n$, $r \in \mathbb{C}$, $s \in \mathbb{C}^n$. This says that a spherical function is reproduced upon taking expectations with respect to the Wishart distribution. Muirhead [468, p. 260] uses (1.108) to show that one obtains the matrix argument $_{p-1}F_q$ function as the matrix Laplace transform (as defined in Exercise 1.2.2 of Section 1.2.1) of the matrix $_pF_q$ function. Herz [293] had defined the matrix argument hypergeometric functions recursively by taking matrix Laplace transforms in this way. See James [329] for a nice summary of the facts about the matrix argument hypergeometric functions and their statistical applications. For example, James notes [329, p. 481] that the zonal polynomials C_K that we considered in the preceding section can be expressed as integrals over $O(n)$ of the characters of the general linear group. James defined the matrix argument hypergeometric functions as series of these zonal polynomials.

Exercise 1.2.29. Prove that the integral $I(B, r, s)$ defined in (1.106) really is the one considered by Muirhead [468]. To do this, you must show that given a partition κ of k, one can find a special choice $s(\kappa) \in \mathbb{C}^n$ so that

$$\alpha h_s(Y) = C_K(Y), \text{ for some constant } \alpha.$$

Here $C_K(Y)$ is the zonal polynomial which we defined in the preceding subsection (see the discussion before Proposition 1.2.5).

Hint. Clearly $h_s(Y)$ is a K-invariant eigenfunction of the Laplacian. If you choose $s \in \mathbb{C}^n$ to be a vector of integers, then $h_s(Y)$ is indeed a symmetric polynomial.

1.2.5   Richards' Extension of the Asymptotics of Spherical Functions for $\mathcal{P}_3$ to $\mathcal{P}_n$ for General n

It takes a new approach and much cleverness to generalize the asymptotic formula (1.103) from $n = 3$ to general n. This is necessary in order to generalize the central limit theorem from $n = 3$ to general n. Richards managed this in [511]. We follow his proof. It makes use of some integral formulas for the integral power function $p_s(X)$, for X a real, square, but not necessarily symmetric, matrix with nonnegative integral s— formulas proved by Kushner [380].

Theorem 1.2.6 (Richards' Asymptotic Formula for Spherical Functions). *As the $n \times n$ diagonal matrix $H \to 0$, we have the following asymptotic formula for the spherical function $h_s(\exp H)$:*

$$
h_s(\exp H) \sim 1 + \frac{1}{n}\left(\sum_{i=1}^{n} r_i\right)\left(\sum_{i=1}^{n} h_i\right)
$$

$$
+ \frac{1}{2n(n+2)}\left(3\sum_{i=1}^{n} r_i^2 + 2\sum_{1 \le i < j \le n} r_i r_j - \frac{n^3 - n}{24}\right)\left(\sum_{i=1}^{n} h_i^2\right)
$$

$$
+ P(r_1, \ldots, r_n)\left(\sum_{1 \le i < j \le n} h_i h_j\right) + \text{higher order terms.}
$$

Here h_i denotes the ith diagonal entry of H and the relation between r and s is

$$
\left. \begin{array}{l} s_j + \cdots + s_n = r_j + \frac{2j-n-1}{4}, \quad j = 1, \ldots, n, \\ r_j - r_{j+1} = s_j - \frac{1}{2}, \ j = 1, \ldots, n-1, \ s_n = r_n - \frac{n-1}{4} \end{array} \right\} \tag{1.109}
$$

as in Theorem 1.2.3. The function $P(r)$ is a polynomial we will not need to determine.

Proof (From [511]).

Step 1. Derivatives of Power Functions.

First it helps to use the logarithmic derivative of the power function, since it is a product. We find that if $a = \exp H$ and $\Delta_j(x) = j$th principal minor of the $n \times n$ matrix x :

$$
\frac{\partial}{\partial h_1} \log p_s(a[k]) = \frac{1}{p_s(a[k])} \frac{\partial}{\partial h_1} p_s(a[k]) = \sum_{j=1}^{n} s_j \frac{1}{\Delta_j(a[k])} \frac{\partial}{\partial h_1} \Delta_j(a[k]). \tag{1.110}
$$

Define $\Delta^{b_1,\dots,b_j}_{c_1,\dots,c_j}(x)$ to be the $j \times j$ subdeterminant of the matrix x obtained by taking the rows from b and the columns from c. Then apply the Cauchy–Binet formula to get

$$\Delta_j(a[k]) = \sum_{1 \le b_1 < \cdots < b_j \le n} \sum_{1 \le c_1 < \cdots < c_j \le n} \Delta^{1,\dots,j}_{b_1,\dots,b_j}({}^t k)\ \Delta^{b_1,\dots,b_j}_{c_1,\dots,c_j}(a)\ \Delta^{c_1,\dots,c_j}_{1,\dots,j}(k).$$

$$(1.111)$$

As a is diagonal, $\Delta^b_c = 0$ unless $b = c$. One finds that $\Delta^b_b = a_{b_1} \cdots a_{b_j}$. Thus

$$\Delta_j(a[k]) = \sum_{1 \le b_1 < \cdots < b_j \le n} \left(\Delta^{1,\dots,j}_{b_1,\dots,b_j}({}^t k) \right)^2 a_{b_1} \cdots a_{b_j}.$$

$$(1.112)$$

Since $a_{b_i} = e^{h b_i}$, we find that

$$\frac{\partial}{\partial h_1} a_{b_1} \cdots a_{b_j} = \begin{cases} 0, & \text{if } b_1 \ne 1, \\ a_{b_1} \cdots a_{b_j} & \text{if } b_1 = 1. \end{cases}$$

For $\ell = 1, 2$, and $j = 1, \dots, n$, define

$$f_j(k) = \left[\frac{\partial^\ell}{\partial h_1^\ell} \Delta_j(a[k]) \right]_{H=0} = \sum_{1 = b_1 < \cdots < b_j \le n} \left(\Delta^{1,\dots,j}_{b_1,\dots,b_j}({}^t k) \right)^2.$$

$$(1.113)$$

Note that $f_j(k)$ is independent of ℓ and $f_n(k) = 1$.

Step 2. Integration over $K = O(n)$.

For any function $\phi \in L^1(K)$, define

$$\mathcal{E}(\phi) = \int_K \phi(k)\, dk, \qquad dk = \text{Haar measure on K.}$$

Then we need two formulas, for $j = 1, \dots, n$:

$$\left. \begin{array}{l} \textbf{Formula 1.} \quad \mathcal{E}\left(f_j(k)\right) = \frac{j}{n}. \\[2mm] \textbf{Formula 2.} \ \mathcal{E}\left(f_1(k)f_j(k)\right) = \frac{j+2}{n(n+2)}. \end{array} \right\}$$

$$(1.114)$$

Step 3. Proof of Formula 1.

The invariance of Haar measure allows us to see that $\mathcal{E}\left(\left(\Delta^{1,\dots,j}_{b_1,\dots,b_j}({}^t k) \right)^2 \right)$ is independent of b. Then, using (1.112) with $a = I$,

$$\sum_{1=b_1<\cdots<b_j\leq n} \left(\Delta_{b_1,\ldots,b_j}^{1,\ldots,j}\left({}^t k\right)\right)^2 = 1.$$

Apply $\mathcal{E}$ to this equation and obtain

$$\mathcal{E}\left(\left(\Delta_{b_1,\ldots,b_j}^{1,\ldots,j}\left({}^t k\right)\right)^2\right) = \frac{1}{\binom{n}{j}}.$$

Substitute this into (1.113) to see that

$$\mathcal{E}\left(f_j(k)\right) = \sum_{1=b_1<\cdots<b_j\leq n} \mathcal{E}\left(\left(\Delta_{b_1,\ldots,b_j}^{1,\ldots,j}\left({}^t k\right)\right)^2\right) = \frac{\binom{n-1}{j-1}}{\binom{n}{j}} = \frac{j}{n}.$$

Step 4. Proof of Formula 2.

Our usual invariance argument shows that

$$\mathcal{E}\left(f_1(k)f_j(k)\right) = \sum_{1=b_1<\cdots<b_j\leq n} \mathcal{E}\left(\Delta_1\left({}^t k\right)^2 \Delta_{b_1,\ldots,b_j}^{1,\ldots,j}\left({}^t k\right)^2\right)$$

$$= \binom{n-1}{j-1}\mathcal{E}\left(\Delta_1\left(k\right)^2 \Delta_j\left(k\right)^2\right).$$

Using Kushner's formula in Proposition 1.2.6 that follows, we obtain formula 2.

Step 5. Computation of the Coefficients in the Asymptotic Expansion.

Write

$$\left.\begin{aligned} h_s\left(\exp H\right) &\sim 1 + P_1(r_1,\ldots,r_n)\left(\sum_{i=1}^n h_i\right) + P_2(r_1,\ldots,r_n)\left(\sum_{i=1}^n h_i^2\right) \\ &\quad + P_3(r_1,\ldots,r_n)\left(\sum_{1\leq i<\leq n} h_i h_j\right) + \text{higher order terms,} \\ \text{where}\quad P_1(r_1,\ldots,r_n) &= \alpha_{11}\sum_{i=1}^n r_i + \alpha_{12}, \\ P_2(r_1,\ldots,r_n) &= \alpha_{21}\sum_{i=1}^n r_i^2 + \alpha_{22}\sum_{1\leq i<j\leq n} r_i r_j + \alpha_{23}\sum_{i=1}^n r_i + \alpha_{24}. \end{aligned}\right\}$$

$$(1.115)$$

We do not need to determine the polynomial $P_3(r_1,\ldots,r_n)$.

Recall the relation between the r variables and the s variables which is stated in the theorem. Also note that formula (1.115) is symmetric in $h_1, \ldots, h_n$ so that we only need to find the coefficients of h_1 and h_1^2.

We need to rewrite the polynomials $P_j(r_1, \ldots, r_n)$ in terms of the s variables. We obtain

$$P_1(r_1, \ldots, r_n) = \alpha_{11} \sum_{i=1}^{n} r_i + \alpha_{12} = \sum_{i=1}^{n} \beta_{11}^{(i)} s_i + \beta_{12}$$

and

$$P_2(r_1, \ldots, r_n) = \alpha_{21} \sum_{i=1}^{n} r_i^2 + \alpha_{22} \sum_{1 \le i < j \le n} r_i r_j + \alpha_{23} \sum_{i=1}^{n} r_i + \alpha_{24}$$

$$= \sum_{i=1}^{n} \beta_{21}^{(i)} s_i^2 + \sum_{1 \le i < j \le n} \beta_{22}^{(i,j)} s_i s_j + \sum_{i=1}^{n} \beta_{23}^{(i)} s_i + \beta_{24}.$$

(a) Computation of α_{11} and α_{12}.

With effort we see that

$$\left[\frac{\partial}{\partial h_1} h_s \left(e^H \right) \right]_{H=0} = \sum_{j=1}^{n} s_j \mathcal{E} \left(f_j(k) \right) = \sum_{j=1}^{n} s_j \frac{j}{n}.$$

Then one can use Abel partial summation (see Apostol [17, Vol. I, p. 407]) to see that

$$\sum_{j=1}^{n} j s_j = \sum_{j=1}^{n} r_j.$$

This implies that $\alpha_{11} = \frac{1}{n}$ and $\alpha_{12} = 0$.

(b) Computation of α_{21}.

After some work we find that

$$\left[\frac{\partial^2}{\partial h_1^2} h_s \left(e^H \right) \right]_{H=0} = \sum_{j=1}^{n} s_j \left(\mathcal{E} \left(f_j(k) \right) - \mathcal{E} \left(f_j(k) \right)^2 \right) + \mathcal{E} \left(\sum_{j=1}^{n} s_j f_j(k) \right)^2 \doteq Q(s)$$

$$(1.116)$$

As the Taylor expansion we seek is symmetric in the r_j, we need to only compute the coefficient of r_1^2 which is the same as the coefficient of s_1^2. Thus $\alpha_{21} = \mathcal{E} \left(f_1(k)^2 \right) = \frac{3}{n(n+2)}$, using formula 2.

(c) Computation of α_{22}.

We have

$$2\alpha_{21} + \alpha_{22} = \beta_{22}^{(1,2)} = 2\mathcal{E} \left(f_1(k) f_2(k) \right).$$

This implies

$$\alpha_{22} = 2\mathcal{E}\left(f_1(k)f_2(k)\right) - 2\alpha_{21} = 2\mathcal{E}\left(f_1(k)f_2(k)\right) - 2\mathcal{E}\left(f_1(k)^2\right)$$
$$= 2\mathcal{E}\left(f_1(k)f_2(k) - f_1(k)^2\right),$$

which implies $\alpha_{22} = \frac{2}{n(n+2)}$.

(d)   Computation of α_{23}.

Using similar arguments to the preceding, we find that

$$\alpha_{23} = \beta_{23} - \alpha_{21}\frac{n-1}{2} + \alpha_{22}\frac{n-1}{4} \quad \text{and} \quad \beta_{23} = \mathcal{E}\left(f_1(k) - f_1(k)^2\right).$$

Plug in the previously computed coefficients $\alpha_{21} = \mathcal{E}\left(f_1(k)^2\right)$ and $\alpha_{22} = 2\mathcal{E}\left(f_1(k)f_2(k) - f_1(k)^2\right)$ to find that $\alpha_{23} = 0$.

(e)   Computation of α_{24}.

From (1.116) we see that $Q(0) = 0$ and

$$Q(s) = \left[\frac{\partial^2}{\partial h_1^2}h_s\left(e^H\right)\right]_{H=0} = \alpha_{21}\sum_{i=1}^{n}r_i^2 + \alpha_{22}\sum_{1\leq i<j\leq n}r_ir_j + \alpha_{24}.$$

If $s = 0$, then $r_j = \frac{n+1-2j}{4}$, for $j = 1, \ldots, n$. Evaluate everything at $s = 0$ to find after some effort that

$$\alpha_{24} = \frac{-(n-1)(n+1)}{24(n+2)}.$$

This completes the proof of the theorem.   ∎

Next we must prove the integral formulas needed for the preceding theorem. They are generalizations of (1.97). Kushner [380] proves a much more general result but we will just consider the formulas we need. First we extend the power function $p_n(X)$ to any square matrix $X \in \mathbb{R}^{k \times k}$ (not necessarily positive definite symmetric) if $n = (n_1, \ldots, n_k)$, where the n_i **are nonnegative integers**. If X_i denotes the upper left $i \times i$ part of X, then clearly the **integral power function**:

$$p_n(X) = \prod_{i=1}^{k}|X_i|^{n_i} \tag{1.117}$$

makes sense. We want to evaluate certain integrals of the form

$$\mathcal{E}(p_n) = \int_{O(k)} p_n(X)dX,$$

where dX is Haar measure on the orthogonal group. Here $\mathcal{E}$ stands for expectation.

We need to know that $p_n(X)$ is an eigenfunction of the differential operator $|X| \left| \frac{\partial}{\partial x_{ij}} \right|$. That is we need the following exercises.

Exercise 1.2.30. Compute the eigenvalue of the differential operator $D_1 = |X| \partial_X$, $\partial_X = \det \left(\frac{1}{2} (1 + \delta_{ij}) \frac{\partial}{\partial x_{ij}} \right)$ on the usual the power function $p_s(X)$ for $X \in \mathcal{P}_k$. The answer is

$$D_1 p_s(X) = \prod_{i=1}^{k} \left(r_i + \frac{k-1}{4} \right) p_s(X),$$

if, as usual,

$$s_j + \cdots + s_k = r_j + \tfrac{2j-k-1}{4}, \quad j = 1, \ldots, k,$$
$$r_j - r_{j+1} = s_j - \tfrac{1}{2}, \ j = 1, \ldots, k-1, \ s_k = r_k - \tfrac{k-1}{4}.$$

Hint. This is proved by Maass [424, p. 83.].
 Start with

$$J_s(X) = p_s(X) \Gamma_k(s) = \int_{Y \in \mathcal{P}_k} p_s(Y) e^{-Tr(YX^{-1})} d\mu_k,$$

where $d\mu_k = |Y|^{-(k+1)/2} \prod_{1 \le i \le j \le k} dy_{ij}$. Let $\sigma(Y) = Y^{-1}$. We know that $L^* = \overline{L^\sigma}$ and $p_s\left(Y^{-1}[\omega]\right) = p_{s^*}(Y)$, with ω and s^* defined by Proposition 1.2.1.
 For $\widehat{X} = X^{-1}$ and $s + 1 = (s_1, s_2, \ldots, s_{k-1}, s_k + 1)$

$$J_s(I) D_1 \left(\widehat{X}[\omega] \right) p_{s^*} \left(\widehat{X}[\omega] \right) = J_s(I) D_1 \left(\widehat{X} \right) p_s(X) = D_1 \left(\widehat{X} \right) J_s(X)$$

$$= \int_{Y \in \mathcal{P}_k} p_s(Y) D_1(\widehat{X}) e^{-Tr(Y\widehat{X})} d\mu_k$$

$$= (-1)^k \left| \widehat{X} \right| \int_{Y \in \mathcal{P}_k} |Y| p_s(Y) D_1(Y) e^{-Tr(Y\widehat{X})} d\mu_k$$

$$= (-1)^k J_{s+1}(I) p_s(X) = (-1)^k J_{s+1}(I) p_{s^*} \left(\widehat{X}[\omega] \right).$$

Then use formula (1.45)

$$\Gamma_k(s) = \Gamma_k(s) = \pi^{k(k-1)/4} \prod_{j=1}^{k} \Gamma \left(s_j + \cdots + s_k - \frac{j-1}{2} \right).$$

Exercise 1.2.31. Show that if $D_X = \left|\frac{\partial}{\partial x_{ij}}\right|$ and $p_n(X)$ denotes the power function extended to $X \in \mathbb{R}^{k \times k}$ when the powers are $n = (n_1, \ldots, n_k)$, with n_j nonnegative integers:

$$|X| \, D_X \, p_n(X) = \sigma_k\left(\frac{n}{2}\right) p_n(X),$$

where

$$\sigma_k(n) = \tau_k(t) = \prod_{i=1}^{k} (k - i + 2t_i), \qquad (1.118)$$

with $t_i - t_{i+1} = n_i$, for $i = 1, \ldots, k$.

Hint. This is Kushner's Theorem 2 (see [380, p. 697]). First one needs to connect the eigenvalue for $|X| \left|\frac{\partial}{\partial x_{ij}}\right|$ of the extended power function to that of $D_1 = |X| \, \partial_X$, $\partial_X = \det\left(\frac{1}{2}\left(1 + \delta_{ij}\right)\frac{\partial}{\partial x_{ij}}\right)$ on the usual power function with $X \in \mathcal{P}_k$ which we computed in the preceding exercise. Assume that the entries of n are *even* nonnegative integers. Note that

$$p_n(XT) = p_n(X)p_n(T) = p_{n/2}({}^tTT)p_n(X),$$

for upper triangular T.

Next we need to prove formula 2 in (1.114) above which means we need to prove

$$\mathcal{E}\left(\Delta_1 \, (k)^2 \, \Delta_r \, (k)^2\right) = \frac{r + 2}{k(k + 2)\binom{k-1}{r-1}}.$$

To do this, we follow Kushner [380]. For $n = (n_1, \ldots, n_k)$, with n_j nonnegative even integers he defines

$$g(X) = e^{\frac{1}{2}Tr({}^tXX)} \quad \text{and} \quad (2\pi)^{k^2/2} \, c_n = \int_{X \in \mathbb{R}^{k \times k}} p_n(X)g(X)dX, \qquad (1.119)$$

where dX is the usual Lebesgue measure on the space $\mathbb{R}^{k \times k}$.

Note that we are seeking to evaluate $\mathcal{E}(p_n)$, where $n = (2, 0, \ldots, 0, 2, 0, \ldots, 0)$, the second 2 being in the rth place for $1 < r \leq k$.

Exercise 1.2.32 (Kushner's Integral Formula). Suppose that dX denotes Haar measure on $O(k)$ and p_n denotes the integral power function with $n = (n_1, \ldots, n_k)$, such that n_j are nonnegative even integers. Show that, if c_n is as in (1.119), we have

$$\mathcal{E}(p_n) = \int\limits_{O(k)} p_n(X)dX = \frac{c_n (2\pi)^{k^2/2}}{g_k \Gamma_k \left(\frac{n+k}{2}\right) p_{\frac{n+k}{2}} (2I)}, \tag{1.120}$$

where $2^k g_k = Vol(O(k))$ using the $k^{-1}dk$ measure on $O(k)$ which was computed in Proposition 1.2.2. Note that if any of the n_j were odd, the integral would vanish.
Hint. Kushner proves this [380, p. 698]. First make the change of variables $X = HT$, where $H \in O(k)$ and T is an upper triangular $k \times k$ matrix with nonnegative diagonal entries. Then if dH is Haar measure on $O(k)$ and dT is the usual Lebesgue measure

$$dX = 2^k g_k \prod_{i=1}^{k} t_{ii}^{k-i} \, dH \, dT.$$

It follows that

$$c_n = (2\pi)^{-k^2/2} 2^k g_k \int\limits_{T_k} p_n(T)g(T) \prod_{i=1}^{k} t_{ii}^{k-i} dT \int\limits_{H \in O(k)} p_n(H)dH.$$

Now make the change of variables $V = {}^tTT$. Then

$$dV = \prod_{1 \le i \le j \le k} dv_{ij} = 2^k \prod_{i=1}^{k} t_{ii}^{k+1-i} dT.$$

This shows that

$$c_n = (2\pi)^{-k^2/2} g_k \int\limits_{V \in \mathcal{P}_k} p_{n/2}(V)e^{-\frac{1}{2}Tr(V)} |V|^{-\frac{1}{2}} \, dV \int\limits_{H \in O(k)} p_n(H) \, dH.$$

Proposition 1.2.6 (Special Case of Kushner's Integral Formula).

(1) If our k-vector of integers is $n = (2, 0, \ldots, 0, 2, 0, \ldots, 0)$, the second 2 being in the rth place for $1 < r \le k$, the constant c_n in (1.119) is

$$c_n = \prod_{i=1}^{k} (k - i + t_i - 1) = (r + 2)(r - 1)!.$$

(2) With n as in part (1), Kushner's integral formula (1.120) becomes

$$\mathcal{E}(p_n) = \frac{r + 2}{k(k + 2) \dbinom{k-1}{r-1}}.$$

Proof. (1) We make use of the differential operator $D_X = \left.\frac{\partial}{\partial x_{ij}}\right|$. For $n = (2, 0, \ldots, 0, 2, 0, \ldots, 0)$ and
$$\underset{r}{}$$

$$g(X) = e^{\frac{1}{2}Tr({}^tXX)},$$

write $X = \begin{pmatrix} S & T \\ U & V \end{pmatrix}$, with $S \in \mathbb{R}^{r \times r}$, and let $p_{n|_r}$ denote the power function on $S \in \mathbb{R}^{r \times r}$ defined by taking the first r parts of n as the powers. Define $c_{n|_r}$ to be the corresponding integral over $S \in \mathbb{R}^{r \times r}$. First note that

$$\int\limits_{T \in \mathbb{R}^{r \times (k-r)}} g(T)\, dT = (2\pi)^{r(k-r)/2}.$$

It follows that

$$(2\pi)^{k^2/2} c_n = \int\limits_{X \in \mathbb{R}^{k \times k}} p_n(X) g(X)\, dX$$

$$= \int\limits_{T \in \mathbb{R}^{r \times (k-r)}} g(T)\, dT \int\limits_{U \in \mathbb{R}^{(k-r)) \times r}} g(U)\, dU \int\limits_{V \in \mathbb{R}^{(k-r) \times (k-r)}} g(V)\, dV\, (2\pi)^{r^2/2} c_{n|_r} = c_{n|_r}.$$

So it suffices to work out the formula when $r = k$. In this case, using Exercise 1.2.31 and $D_X g = (-1)^k\, |X|\, g$,

$$(2\pi)^{k^2/2} c_n = \int\limits_{X \in \mathbb{R}^{k \times k}} p_n(X) g(X)\, dX = (-1)^k \int\limits_{X \in \mathbb{R}^{k \times k}} p_{n-1}(X)\, D_X g(X)\, dX$$

$$= \int\limits_{X \in \mathbb{R}^{k \times k}} D_X p_{n-1}(X)\, g(X)\, dX$$

$$= \tau_k \left(\frac{t-1}{2}\right) \int\limits_{X \in \mathbb{R}^{k \times k}} p_{n-2}(X)\, g(X)\, dX, \quad \text{with } \tau_k \text{ as in (1.118).}$$

Here since $r = k$, we make the same change of variables $X = \begin{pmatrix} s & T \\ U & V \end{pmatrix}$, now with $s \in \mathbb{R}$, and find that if τ_k is as in (1.118)

$$(2\pi)^{k^2/2} c_n = \tau_k \left(\frac{t-1}{2}\right) \int\limits_{X \in \mathbb{R}^{k \times k}} p_{n-2}(X)\, g(X)\, dX$$

$$= \tau_k \left(\frac{t-1}{2}\right) \int\limits_{s \in \mathbb{R}} s^2 e^{-s^2/2} ds \,(2\pi)^{(k^2-1)/2} = (2\pi)^{k^2/2} \tau_k \left(\frac{t-1}{2}\right).$$

Thus

$$c_n = \tau_k \left(\frac{t-1}{2}\right) = \prod_{i=1}^{k} (k - i + t_i - 1) = (k+2)(k-1)!$$

as $n_i = t_i - t_{i+1}, t_{k+1} = 0$, when $n = (2, 0, \ldots, 0, 2)$ which says $t_1 = 4, t_2 = 2, \ldots, t_k = 2$.

It follows that in the general case that $n = (2, 0, \ldots, 0, \underset{r}{2}, 0, \ldots, 0)$, we have $c_n = c_{n|r} = (r+2)(r-1)!$ and part (1) of the Proposition is proved.

(2) Now we seek to show that

$$\mathcal{E}\left(\Delta_1 (k)^2 \Delta_r (k)^2\right) = \frac{r+2}{k(k+2)\dbinom{k-1}{r-1}}.$$

It will help to know the duplication formula for the gamma function (see Lebedev [398, p. 4]):

$$2^{2z-1} \Gamma(z)\, \Gamma\left(z + \frac{1}{2}\right) = \sqrt{\pi}\, \Gamma(2z).$$

From part (1) and the preceding exercise,

$$\mathcal{E}(p_n) = \frac{c_n (2\pi)^{k^2/2}}{g_k \Gamma_k \left(\frac{n+k}{2}\right) p_{\frac{n+k}{2}}(2I)}$$

$$= \frac{(r+2)\,(r-1)!\,(2\pi)^{k^2/2}\, \Gamma\left(\frac{1}{2}\right) \Gamma\left(\frac{2}{2}\right) \cdots \Gamma\left(\frac{k}{2}\right)}{\pi^{(k^2+k)/4} \pi^{k(k-1)/4} \displaystyle\prod_{j=1}^{k} \left(\frac{t_j+k-j+1}{2}\right) p_{\frac{n+k}{2}}(2I)}.$$

Now

$$t_1 = 4, t_2 = \cdots = t_r = 2, t_{r+1} = \cdots = t_k = 0.$$

It follows with quite a bit of effort that

$$\mathcal{E}(p_n) = \frac{r+2}{(k+2)\, r \binom{k}{r}} = \frac{r+2}{(k+2)\, k \binom{k-1}{r-1}}.$$

■

This ends our discussion of matrix argument special functions.

1.3 Harmonic Analysis on $\mathcal{P}_n$ in Polar Coordinates

...die Schwierigkeit beginnt da, wo es sich darum handelt, aus diesem Labyrinth
von Formeln einen Ausweg zu finden....[1] Frobenius (quoted by Siegel [565, Vol. III,
p. 373]).

1.3.1 Properties of the Helgason–Fourier Transform on $\mathcal{P}_n$

The main goal of this section is the discussion of an inversion formula for the
Helgason–Fourier transform defined in formula (1.72) of Section 1.2.3 when the
function is K-invariant. The subject contains a labyrinth of formulas, similar to that
occurring in any higher rank symmetric space (as in the quote of Frobenius above,
which refers to the formulas for multidimensional theta functions). The discussion
is intended to provide a way through the labyrinth—a route which follows that
set out by Harish-Chandra and Helgason, particularly Helgason's Battelle lectures
[275]. In outlining the path, we will not provide all the details of the arguments. For
example, our discussion of Fourier inversion (Theorem 1.3.1) will use the analogue
of the asymptotics/functional equations principle from Section 3.2 of Vol. I. We
will not give a rigorous justification of the principle. That would require an analysis
similar to that given in discussions of the Paley–Wiener theorem. See Helgason
[282, pp. 55–56, 452–453] for a careful treatment of the argument that we are
omitting.

The Helgason–Fourier inversion formula for $\mathcal{P}_n$ (in part (1) of Theorem 1.3.1)
writes a smooth compactly supported function on $\mathcal{P}_n$ as a superposition of eigenfunctions of the generalized Laplacians in $D(\mathcal{P}_n)$. This provides the fundamentals
of harmonic analysis on $\mathcal{P}_n$ and an analogue of Theorem 3.2.3 in Volume I. This
result can also be viewed as an analogue of the Mellin inversion formula (see
Exercise 1.4.1 of Vol. I), which is the case $n = 1$.

---

[1] The difficulty begins when one must find a way out of this labyrinth of formulas.

A reader interested in delving further into the labyrinth of formulas associated with Fourier analysis on Lie groups and symmetric spaces might consult some of the references below. We discuss some of these other points of view briefly at the end of this section.

References for this section include: Bhanu-Murthy [48], Ehrenpreis and Mautner [155], Flensted-Jensen [181], Gangolli [196–198], Gelfand and Naimark [215], Gurarie [254], Harish-Chandra [263], Helgason [273–282], Rebecca Herb and Joe Wolf [288], Koornwinder [363], Jorgenson and Lang [333], Rosenberg [516], Varadarajan [624, 625], Michèle Vergne [629], Wallach [651, 653], Warner [655], and Wawrzyńczyk [657].

Define the **Helgason–Fourier Transform** of a (sufficiently nice) function f : $\mathcal{P}_n \to \mathbb{C}$, for $s \in \mathbb{C}^n$, $k \in K = O(n)$, by:

$$\mathcal{H}f(s,k) = \int\limits_{Y \in \mathcal{P}_n} f(Y)\,\overline{p_s(Y[k])}\,d\mu_n(Y). \tag{1.121}$$

Here $p_s(Y)$ denotes the power function in formula (1.41) of Section 1.2.1. The Helgason–Fourier transform can be viewed as a Mellin transform on $\mathcal{P}_n$.

Note that if we set $f(Y) = \exp\{-\text{Tr}(Y)\}$ in (1.121), we obtain

$$\mathcal{H}f(s,k) = \Gamma_n(\bar{s}),$$

with the gamma function as in (1.44) of Section 1.2.1.

If we consider the Helgason–Fourier transform $\mathcal{H}f(s,k)$ as a function of its second variable $k \in K = O(n)$, the function depends only on the coset $kM = \bar{k}$ in the boundary K/M of the symmetric space, where M is the group of all diagonal matrices with entries ± 1. See the discussion of the boundary and formulas (1.19) and (1.20) as well as Exercise 1.1.22 in Section 1.1.4. The inversion formula for the Helgason–Fourier transform will involve an integral over the boundary K/M and we shall assume that the measure $d\bar{k}$ on K/M is normalized to give

$$\int_{\bar{k} \in K/M} d\bar{k} = 1. \tag{1.122}$$

It is this inversion formula that gives us harmonic analysis on $\mathcal{P}_n$; i.e., the spectral resolution of the generalized Laplace operators on $\mathcal{P}_n$

Theorem 1.3.1 (Properties of the Helgason–Fourier Transform).

(1) Inversion Formula.

Suppose that $f : \mathcal{P}_n \to \mathbb{C}$ is infinitely differentiable with compact support. If $\mathcal{H}f(s,k)$ denotes the Helgason–Fourier transform defined by formula (1.121), then

$$f(Y) = \omega_n \int\limits_{\substack{s \in \mathbb{C}^n \\ \mathrm{Re}\, s = -\rho}} \int\limits_{\bar{k} \in K/M} \mathcal{H}f(s, k)\, p_s(Y[k])\, d\bar{k}\, |c_n(s)|^{-2}\, ds,$$

where $\rho = \left(\frac{1}{2}, \ldots, \frac{1}{2}, \frac{1}{4}(1-n)\right)$,

$$\omega_n = \prod_{j=1}^{n} \frac{\Gamma(j/2)}{j(2\pi i)\pi^{j/2}},$$

and $c_n(s)$ denotes the **Harish-Chandra c-function** given by:

$$c_n(s) = \prod_{1 \leq i \leq j \leq n-1} \frac{B\left(\frac{1}{2}, s_i + \cdots + s_j + \frac{1}{2}(j - i + 1)\right)}{B\left(\frac{1}{2}, \frac{1}{2}(j - i + 1)\right)}.$$

Here $B(x, y) = \Gamma(x)\Gamma(y)/\Gamma(x + y)$, the beta function.

(2) Convolution Property.

If either f or g is a K-invariant function on $\mathcal{P}_n$ satisfying the hypothesis of part (1), then defining convolution as in formula (1.24) of Section 1.1.4:

$$\mathcal{H}(f * g) = \mathcal{H}f \cdot \mathcal{H}g.$$

(3) G-invariant Differential Operators Changed to Multiplication by a Polynomial.

If $L \in D(\mathcal{P}_n)$ and f is as in (1), then

$$\mathcal{H}(Lf)(s, k) = \overline{\lambda_{L^*}(s)}\mathcal{H}f(s, k), \quad \text{where } Lp_s(Y) = \lambda_L(s)p_s(Y).$$

Here L^* denotes the adjoint of L (see Theorem 1.2.1 in Section 1.2.1). Note that the eigenvalue $\lambda_{L^*}(s)$ is a polynomial in s.

(4) Plancherel Theorem in the K-invariant Case.

Let $\alpha(s) = \omega_n |c_n(s)|^{-2}$. For f as in part (1) and K-invariant, if $\widehat{f}(s) = \mathcal{H}f(s, I)$, where I is the identity matrix, as in (1.72), we have

$$\int\limits_{\mathcal{P}_n} |f(Y)|^2\, d\mu_n(Y) = \int\limits_{\mathrm{Re}\, s = -\rho} \left|\widehat{f}(s)\right|^2 \alpha(s)\, ds.$$

Moreover, the Helgason–Fourier transform can be extended to an isometry between $L^2(\mathcal{P}_n/K, d\mu_n)$ and functions of s which are square integrable with respect to $\alpha(s)ds$ and invariant under permutations of the r-variables which are related to the s-variables as in (1.123) below (see also Proposition 1.2.1 of Section 1.2.1).

The discussion of the proof of Theorem 1.3.1 will extend through the next several sections. There are four main steps. The first step (in Section 1.3.2) is a reduction to the corresponding inversion formula for the determinant one surface $\mathcal{SP}_n$. The second step (also in Section 1.3.2) is due to Helgason (see [275, pp. 60–61]). Using arguments similar to those which we gave to prove Fourier inversion on $\mathbb{R}^n$ (see the proof of Theorem 1.2.1 in Vol. I), Helgason demonstrates that it suffices to show the inversion formula at any point Y in the symmetric space and for any Dirac family of functions $g(Y)$. Thus, in particular, it suffices to show the inversion formula for $O(n)$-invariant functions. The third step in the proof of the inversion formula (found in Section 1.3.3) is due to Harish-Chandra [263] and Bhanu-Murthy [48]. It proves the inversion formula in the $O(n)$-invariant case, when the Helgason–Fourier transform is often called the **spherical transform**. The idea is similar to that which we used in Section 3.2 of Vol. I to prove the Kontorovich-Lebedev and Mehler-Fock inversion formulas. Thus one reduces the computation of the spectral measure or Harish-Chandra c-function to the determination of the asymptotics and functional equations of spherical functions. As we mentioned above, we will not give a rigorous justification of this principle—only a heuristic argument. We have already found that the spherical functions on $\mathcal{P}_n$ satisfy $n!$ functional equations (see Theorem 1.2.3 of Section 1.2.3). The fourth step (also in Section 1.3.3) is the determination of the asymptotic behavior of these functions as the symmetric space variable approaches the boundary. To do this, we shall rewrite Harish-Chandra's integral formula for spherical functions, in part (4) of the theorem just cited, as an integral over $\overline{N}$, which is the group of lower triangular matrices with ones on the diagonal. This is analogous to the discussion indicated in Exercise 3.2.10 in Vol. I for the case of the Poincaré upper half plane.

The details of the asymptotics/functional equations argument require Helgason's version of the Paley–Wiener theorem for $\mathcal{P}_n$ (see Helgason [273, 274, 276–278, 281, 282], Gangolli [196]). We will not discuss this here or give the proof of part (4) of Theorem 1.3.1 which is the Fourier inversion formula for L^2 functions on $\mathcal{P}_n$. Nor shall we discuss the Helgason–Fourier transform on Harish-Chandra's Schwartz space for $\mathcal{P}_n$. See Helgason [282, p. 489] for an exercise on the subject. Or see Gangolli [198, pp. 78–82]. Another possible reference is Jorgenson and Lang [333].

Before beginning our discussion of Theorem 1.3.1, we need to make a few preliminary remarks and do some exercises.

Remarks. (1) Using the change of variables from formula (1.49) of Section 1.2.1, we have

$$p_s(I[t]) = \varphi_r(t) = \prod_{t=1}^{n} t_{ii}^{2r_i+i-(n+1)/2},$$

for $t \in T_n$, which is the group of upper triangular $n \times n$ real matrices with positive diagonal entries. Here then we have

$$2r_i + i - (n+1)/2 = 2(s_i + \cdots + s_n). \tag{1.123}$$

It follows that if we write Harish-Chandra's c-function in terms of the r-variables, we obtain

$$c_n(s) = \prod_{1 \leq i \leq j \leq n-1} \frac{B\left(\frac{1}{2}, r_i - r_{j+1}\right)}{B\left(\frac{1}{2}, \frac{1}{2}(j - i + 1)\right)}. \tag{1.124}$$

Furthermore, Re $s = -\rho = -\left(\frac{1}{2}, \ldots, \frac{1}{2}, \frac{1}{4}(1 - n)\right)$ implies that Re $r_i = 0$. Of course, we cannot allow the second argument of the beta function to be zero.

(2) In the case $n = 1$, Theorem 1.3.1 is just ordinary Mellin inversion (see Exercise 1.4.1 of Volume I).

(3) When $n = 2$, Theorem 1.3.1 is just ordinary Mellin inversion plus the Helgason inversion formula in Theorem 3.2.3 in Vol. I. To see this, suppose that $f : \mathcal{P}_2 \to \mathbb{C}$ is $O(2)$-invariant. Then the Helgason–Fourier transform of f is:

$$\mathcal{H}f(s, k) = \frac{\pi}{2} \int_{a \in A} f(a)\overline{h_s(a)}\gamma(a)\, da,$$

$$= \pi \int_{a \in A^+} f(a)\overline{h_s(a)}\gamma(a)\, da,$$

where $A^+ = \{a \in A \mid a_1 > a_2\}$ (a positive Weyl chamber),

$$da = \frac{da_1\, da_2}{a_1\, a_2}, \quad \gamma(a) = |a|^{-1/2}|a_1 - a_2|, \quad h_s(a) = \int_{k \in K} p_s(a[k])\, dk.$$

Here we have used the integral formula for polar coordinates (see formula (1.37) from Section 1.1.6).

Now change variables according to

$$a = \begin{pmatrix} a_1 & 0 \\ 0 & a_2 \end{pmatrix} = v^{1/2} \begin{pmatrix} u & 0 \\ 0 & u^{-1} \end{pmatrix}, \quad \text{writing } f(a) = f(v, u).$$

Then we see that

$$a_1 = u\sqrt{v}, \quad a_2 = \sqrt{v}/u,$$

and

$$\left| \frac{\partial(a_1, a_2)}{\partial(u, v)} \right| = \left| \begin{matrix} \sqrt{v} & -\sqrt{v}/u^2 \\ u/(2\sqrt{v}) & 1/(2u\sqrt{v}) \end{matrix} \right| = 1/u.$$

If we write $f(a) = f(v, u)$, then the Helgason–Fourier transform is:

$$\mathcal{H}f(s_1, s_2) = \pi \int_{u \geq 1} \int_{v>0} f(v, u) \, h_{\bar{s}}(v, u)(u - u^{-1}) \, \frac{du}{u} \frac{dv}{v}.$$

The spherical function is:

$$h_s(v, u) = \frac{v^{s_2}}{2\pi} \int_0^{2\pi} (a_1 \cos^2 \theta + a_2 \sin^2 \theta)^{s_1} \, d\theta$$

$$= v^{s_2 + s_1/2} P_{s_1}(\cosh(\log u)),$$

where P_s denotes the Legendre function from Section 3.2.3 of Vol. I.
To see the last formula, write

$$Y = \begin{pmatrix} a_1 & 0 \\ 0 & a_2 \end{pmatrix} \begin{bmatrix} \cos \theta & -\sin \theta \\ \sin \theta & \cos \theta \end{bmatrix} = \begin{pmatrix} y_1 & * \\ * & * \end{pmatrix}$$

and note that

$$y_1 = a_1 \cos^2 \theta + a_2 \sin^2 \theta = a_1 + (a_2 - a_1) \sin^2 \theta$$

$$= a_1 + \frac{1}{2}(a_2 - a_1)(1 - \cos(2\theta)) = \frac{1}{2}(a_1 + a_2) - \frac{1}{2}(a_2 - a_1) \cos(2\theta).$$

Thus

$$y_1 = v^{1/2} \left\{ \frac{u + u^{-1}}{2} - \frac{u - u^{-1}}{2} \cos(2\theta) \right\}.$$

So the inversion formula which we seek becomes, upon setting $x = \cosh \log u$ and $x_1 = \cosh \log u_1$:

$$f(v, u) = 2\pi\omega_2 \int_{\mathrm{Re}\, s = -\rho} \int_{\substack{v_1 > 0 \\ x_1 > 1}} f(v_1, u_1) v_1^{\overline{s_2 + \frac{s_1}{2}}} \overline{P_{s_1}(x_1)} \, dx_1 \, \frac{dv_1}{v_1} v^{s_2 + \frac{s_1}{2}} P_{s_1}(x) \, |c_2(s)|^{-2} \, ds,$$

for the constant ω_2 of part (1) in Theorem 1.3.1. By formula (3.28) in Section 3.2 of Volume I, we see that

$$\left| c_2 \left(-\frac{1}{2} + it \right) \right|^{-2} = \pi t \tanh(\pi t).$$

Thus we obtain the result of part (1) of Theorem 1.3.1 in the case $n = 2$ using ordinary Mellin inversion (Exercise 1.4.1 in Volume I) and the Mehler-Fock inversion formula (formulas (3.26) and (3.27) in Section 3.2 of Volume I). For this, we need:

Re $s_1 = -1/2$ and Re $s_2 = 1/4$, with $\omega_2 = (2\pi i)^{-2}(2\pi)^{-1} = -(2\pi)^{-3}$.

(4) Note that if we consider the spectral measure $|c_n(s)|^{-2}$ from part (1) of Theorem 1.3.1, as a function of $r = it$, in formula (1.123), we are looking at:

$$\prod_{1 \leq i \leq j \leq n-1} \left| B\left(\frac{1}{2}, r_i - r_{j+1}\right) \right|^{-2};$$

that is,

$$\prod_{1 \leq i \leq j \leq n-1} \pi \left| t_i - t_{j+1} \right| \tanh\left(\pi \left| t_i - t_{j+1} \right|\right). \tag{1.125}$$

Exercise 1.3.1 (Properties of the Helgason–Fourier Transform).

(a) Prove part (2) of Theorem 1.3.1 above. Show that the hypothesis of K-invariance is necessary.

(b) Prove part (3) of Theorem 1.3.1.
    Hint. (a) Use the power function identity (1.138) below. Note that the convolution property implies that $f * g = g * f$

(5) **The K-Invariant Case of Theorem 1.3.1.** Suppose the function $f(Y)$ in (1.121) is K-invariant; i.e., $f(Y[k]) = f(Y)$ for all $Y \in \mathcal{P}_n$ and $k \in K = O(n)$. Then the Helgason–Fourier transform (1.121) is really only a function of the s-variable and we will write

$$\mathcal{H}f(s, k) = \widehat{f}(s), \tag{1.126}$$

for K-invariant functions f, as in (1.72) of Section 1.2.3. Let $h_s(Y)$ denote the spherical function defined as in formula (1.70) of Section 1.2.3; i.e.,

$$h_s(Y) = \int_{\overline{k} \in K/M} p_s(Y[k])\, d\overline{k}, \tag{1.127}$$

with $d\overline{k}$ normalized as in (1.122). Then we see from formula (1.37) of Section 1.1.6 that

$$\widehat{f}(s) = b_n \int\limits_{a_i > a_{i+1}} f(a)\overline{h_s(a)}\, J(a) \prod da_j, \quad \text{where}$$

$$a = \begin{pmatrix} a_1 & & 0 \\ & \ddots & \\ 0 & & a_n \end{pmatrix}, \quad b_n = \pi^{(n^2+n)/4} \prod_{j=1}^{n} \frac{1}{\Gamma\left(\frac{j}{2}\right)}, \tag{1.128}$$

$$J(a) = \prod_{j=1}^{n} a_j^{-(n+1)/2} \prod_{1 \le i < j \le n} (a_i - a_j).$$

As we said earlier, the transform (1.128) is often called the "spherical transform" (see Helgason [282, p. 449]).

Exercise 1.3.2. Prove that the spherical function $h_s(Y)$ is bounded when $\operatorname{Re} s = -\rho$, for $\rho = \left(\frac{1}{2}, \ldots, \frac{1}{2}, \frac{1}{4}(1-n)\right)$. Note that in the r-variables of (1.123), we are just saying that the spherical function is bounded when the powers r_i are purely imaginary. Helgason and Johnson [283] obtain a more precise result (see Helgason [282, pp. 458–466]). In fact, Helgason shows [282] that the spherical functions are bounded on a tube domain where $\widehat{f}(s)$ is holomorphic for K-invariant f. Moreover Helgason [282, p. 480] shows, using the Riemann–Lebesgue lemma, that for $r = it$, the spherical function approaches 0 as $\|t\| \to \infty$. See Lebedev [398, p. 191] for the case $G = SL(2, \mathbb{R})$, when h_s is $P_{-1/2+it}$.
Hints. Another reference is Gangolli [198]. You can use the fact that a holomorphic function of $s \in \mathbb{C}^n$ which is bounded on a region must be bounded on the convex hull of that region. The exercise requires use of the functional equations of the spherical function. Recall (1.123) relating the r- and s-variables. The permutation of r_i and r_{i+1} corresponds in the s-variables to the map from s to s' given by:

$$\begin{aligned} s'_i &= -1 - s_i, \\ s'_{i\pm1} &= s_{i\pm1} + s_i - \tfrac{1}{2}, \\ s'_j &= s_j, \quad j \ne i, i \pm 1. \end{aligned}$$

So $\operatorname{Re} s_i = 0$ corresponds to $\operatorname{Re} s'_i = -1$. The convex hull contains $\operatorname{Re} s_i = -\frac{1}{2}$.

Exercise 1.3.3. Suppose that f is a K-invariant function on $\mathcal{P}_n$ of the sort considered in part (1) of Theorem 1.3.1; i.e., assume that $f(a) = 0$, for $a \in A$ such that the diagonal entries a_j satisfy

$$\Sigma \log a_j^2 > R^2.$$

This means that the geodesic distance $d(I, I[a])$ is greater than R. Prove that for every G-invariant differential operator $L \in D(\mathcal{P}_n)$ such that the eigenvalue polynomial $\lambda_L(s)$ does not vanish; i.e., $Lp_s = \lambda_L(s)p_s$, with $\lambda_L(s) \ne 0$, there is a positive constant C such that

$$|\widehat{f}(s)| \le C \exp(\|u\|R)|\lambda_L(s)|^{-1}.$$

Here

$$u \in \mathbb{C}^n, \quad u_j = \mathrm{Re}\,(s_j + \cdots + s_n), \quad \|u\| = \left(\Sigma u_j^2\right)^{1/2}.$$

Note that we can find operators L to make $\lambda_L(s)$ have arbitrarily high degree. This result is the converse of the Paley–Wiener theorem proved by Helgason [282, pp. 450–454].

Hint. Note that

$$|p_s(a)| \le \prod_{j=1}^n \exp\left(u_j \log a_j\right).$$

Use Iwasawa coordinates to write $\widehat{f}(s)$ as a composition of a Mellin transform over A and the **Harish transform**

$$F_f(I[a]) = \alpha(a)^{1/2} \int_{n \in N} f(I\,[an])\,dn \tag{1.129}$$

where $\alpha(a)$ is the Jacobian of Iwasawa coordinates as given in (1.38) of Section 1.1.6. If is not hard to see that $d(I, I[a]) > R$ implies that $F_f(I[a]) = 0$. For you need only note that

$$d(I, I[a]) \le d(I, I[an]) \quad \text{for all } a \in A,\ n \in N.$$

When we restrict everything to the determinant one surface, so that $|a| = 1$, we will write $\alpha(a)$ in formula (1.38) as $\alpha(a) = p_{2\rho}(I[a])$, where $\rho_j = \frac{1}{2}$, $j = 1, \ldots, n-1$. In this notation we have:

$$\int_{\substack{Y \in SP_n}} \overline{p_s(Y)} f(Y)\,dY = 2^n \int_{\substack{a \in A \\ |a|=1}} \overline{p_{s+\rho}(I[a])}\,F_f(a)\,da,$$

where F_f denotes the Harish transform defined by (1.129). Note that if, for $j = 1, \ldots, n-1$, $\mathrm{Re}\,s_j = -\frac{1}{2}$, as in the Fourier inversion formula, then $\mathrm{Re}(s + \rho)_j = 0$, $j = 1, \ldots, n-1$.

Exercise 1.3.4. Show that, with f as in Exercise 1.3.3, c_n and ρ as in Theorem 1.3.1, we have

$$\int_{\mathrm{Re}\,s=-\rho} \widehat{f}(s)\,|c_n(s)|^{-2}\,ds < \infty.$$

Hint. You need a bound on $|c_n(s)|^{-2}$ when $\mathrm{Re}\,s = -\rho$. For this, use formula (1.125).

1.3.2 Beginning of the Discussion of Part (1) of Theorem 1.3.1—Steps 1 and 2

Now we start to contemplate how one might prove something like part (1) of Theorem 1.3.1.

Step 1 (Pulling Out the Determinant). Write $Y = v^{1/n}W$ with $v > 0$ and $W \in \mathcal{SP}_n$, which is the determinant one surface in $\mathcal{P}_n$. Then by formula (1.21) of Section 1.1.4, we can normalize measures so that

$$d\mu_n(Y) = v^{-1}dv \, dW, \tag{1.130}$$

where dW is an $SL(n, \mathbb{R})$-invariant measure on $\mathcal{SP}_n$. It follows that the Helgason–Fourier transform can be rewritten as:

$$\mathcal{H}f(s, k) = \int\limits_{v>0} v^{\bar{r}} \int\limits_{W \in \mathcal{SP}_n} f(v^{1/n}W) \, \overline{p_s(W[k])} \, dW \, \frac{dv}{v}, \qquad \text{where} \quad r = \frac{1}{n}\sum_{j=1}^{n} js_j. \tag{1.131}$$

It suffices to assume that $f(v^{1/n}W) = f_1(v)f_2(W)$. Then

$$\mathcal{H}f(s, k) = Mf_1(\bar{r})\mathcal{H}^0 f_2(s, k), \tag{1.132}$$

where Mf_1 is the ordinary Mellin transform of f_1 (see Section 1.4, Vol. I) and $\mathcal{H}^0 f_2$ denotes **the Helgason–Fourier transform on the determinant one surface**:

$$\mathcal{H}^0 f_2(s, k) = \int\limits_{W \in \mathcal{SP}_n} f_2(W) \, \overline{p_s(W[k])} \, dW. \tag{1.133}$$

It follows from the ordinary Mellin inversion formula that the Harish-Chandra c-function does not depend on the variable s_n. Moreover, we need $\mathrm{Re}\, r = 0$ in the final inversion formula, and thus using formula (1.131):

$$\mathrm{Re}\, s_n = -\frac{1}{n}\sum_{j=1}^{n-1} \mathrm{Re}(js_j).$$

Since we shall show that we need $\mathrm{Re}\, s_j = -\frac{1}{2}$, for $j = 1, \ldots, n-1$, it follows that we need:

$$\mathrm{Re}\, s_n = (n-1)/4. \tag{1.134}$$

For the rest of proof of part (1) of Theorem 1.3.1 we will replace $\mathcal{P}_n$ by $\mathcal{SP}_n$ and G will denote $SL(n, \mathbb{R})$.

Step 2 (Helgason's Reduction to the Case That f is K-Invariant). We base the following analysis on Helgason [275, pp. 60–61]. The idea is to imitate the proof of the Euclidean Fourier inversion formula (Theorem 1.2.1 of Volume I). Let $\mathcal{T}$ denote the **inverse transform** defined for nice functions:

$$F : \mathbb{C}^{n-1} \times (K/M) \to \mathbb{C}$$

and for $Y \in \mathcal{SP}_n$ by:

$$\mathcal{T}F(Y) = \int\limits_{\text{Re}\,s=-\rho} \int\limits_{\bar{k}\in K/M} F(s,\bar{k})\, p_s(Y[k])\, |c_n(s)|^{-2}\, d\bar{k}\, ds. \tag{1.135}$$

Lemma 1.3.1 (Two Properties of the Helgason–Fourier Transform and Its Inverse Transform).

(1) Let $\mathcal{T}$ be defined by (1.135) and choose $\rho = (\frac{1}{2}, \ldots, \frac{1}{2}) \in \mathbb{C}^{n-1}$. For $x \in G$ and $f : \mathcal{SP}_n \to \mathbb{C}$, define

$$f^x(W) = f(W[x]), \text{ for } W \in \mathcal{SP}_n.$$

Then $\mathcal{T}\mathcal{H}^0$ commutes with the action of G; i.e.,

$$\mathcal{T}\mathcal{H}^0(f^x) = (\mathcal{T}\mathcal{H}^0 f)^x, \text{ for all } x \in G = SL(n,\mathbb{R}).$$

(2) If f and g are infinitely differentiable functions with compact support on $\mathcal{SP}_n$, then

$$\int\limits_{W\in\mathcal{SP}_n} f(W)\, \overline{\mathcal{T}\mathcal{H}^0 g(W)}\, dW = \int\limits_{W\in\mathcal{SP}_n} (\mathcal{T}\mathcal{H}^0 f)(W)\, \overline{g(W)}\, dW;$$

i.e., $\mathcal{T}\mathcal{H}^0$ is self-adjoint.

Proof. (1) First we need to show that $\mathcal{T}\mathcal{H}^0$ makes sense for functions f which are smooth and compactly supported. This follows from the estimate in Exercise 1.3.3 of the preceding section.

The main fact needed to prove part (1) is a certain identity satisfied by the power function $p_s(Y)$. This identity (1.138) provides a kind of substitute for the following property of exponentials:

$$\exp(x+y) = \exp(x)\exp(y), \quad \text{for } x,y \in \mathbb{R}.$$

To describe this identity, we need to recall the Iwasawa decomposition (see Exercise 1.1.20 of Section 1.1.4) of $x \in G = GL(n,\mathbb{R})$ into

$$x = K(x)A(x)N(x),$$

with $K(x) \in K = O(n)$, $A(x) \in A$, the group of positive diagonal matrices in G, and $N(x) \in N$, the (nilpotent or unipotent) group of upper triangular elements of G with ones on the diagonal. An element $x \in G$ acts on the boundary element $\bar{k} = kM$, with M defined to be the group of diagonal matrices lying in K, according to the formula:

$$x(\bar{k}) = K(xk)M \tag{1.136}$$

(see formula (1.20) of Section 1.1.4). Here $K(xk)$ denotes the K-part of xk in the Iwasawa decomposition.

It will help to use the following **notation for the power function**:

$$p_s(Y[k]) = p_s(Y, kM) = p_s(Y, \bar{k}). \tag{1.137}$$

Then we have the **power function identity**:

$$p_s(Y, \bar{k}) = p_s(Y[x], x^{-1}(\bar{k}))p_s(I[x^{-1}], \bar{k}), \tag{1.138}$$

for all $x \in G$, $Y \in SP_n$, $\bar{k} \in K/M$. To prove formula (1.138), let $x^{-1}k = k_1 a_1 n_1$, with $k_1 \in K$, $a_1 \in A$, $n_1 \in N$. Clearly

$$p_s(Y[k]) = p_s(Y[xx^{-1}k]) = p_s(Y[xk_1])p_s(I[a_1]),$$

which gives the desired identity.

To prove part (1) of Lemma 1.3.1, note that for $x \in G$:

$$\mathcal{T}\mathcal{H}^0(f^x)(Y) = \int\int\int f(W[x]) \, p_{\bar{s}}(W[k]) \, p_s(Y[k]) \, |c_n(s)|^{-2} \, dW \, d\bar{k} \, ds$$

$$= \int\int\int f(V) \, p_{\bar{s}}(V[x^{-1}k]) \, p_s(Y[k]) \, |c_n(s)|^{-2} \, dV \, d\bar{k} \, ds$$

$$= \int\int\int f(V) \, p_{\bar{s}}(V, x^{-1}(\bar{k})) \, p_{\bar{s}}(I[x^{-1}], \bar{k}) \, p_s(Y[k]) \, |c_n(s)|^{-2} \, dV \, d\bar{k} \, ds$$

$$= \int\int\int f(V) \, p_{\bar{s}}(V, \bar{k}) \, p_{\bar{s}}(I[x^{-1}], x(\bar{k})) \, p_s(Y, x(\bar{k})) \, \alpha^{-1}(A(xk)) \, |c_n(s)|^{-2}$$
$$\times dV \, d\bar{k} \, ds.$$

The integrals are over $\operatorname{Re} s = -\rho$, $\bar{k} \in K/M$, and $W \in SP_n$. We have first changed variables via $V = W[x]$, then used (1.138), and finally replaced $\bar{k}$ by $x(\bar{k})$, using Exercise 1.1.22 of Section 1.1.4 to give us the Jacobian of this change of variables, which is:

$$\alpha(A(xk)) = p_{2\rho}(I[x], \bar{k}) = p_{-2\rho}(I[x^{-1}], x(\bar{k})) = \prod_{i=1}^{n} a_i^{n-2i+1} \tag{1.139}$$

with $\rho \in \mathbb{C}^{n-1}$ given by formula (1.141) below.

Now choose

$$s = -\rho + i\lambda \quad \text{so that } \bar{s} + 2\rho = -s. \tag{1.140}$$

It follows that

$\mathcal{TH}^0(f^x)(Y)$

$$= \int \int \int f(V)\, p_{\bar{s}}(V, \bar{k})\, p_{\bar{s}+2\rho}(I[x^{-1}], x(\bar{k}))\, p_s(Y, x(\bar{k}))\, |c_n(s)|^{-2}\, dV\, d\bar{k}\, ds$$

$$= \int \int \int f(V)\, p_{\bar{s}}(V, \bar{k})\, p_s(Y[x], \bar{k})\, |c_n(s)|^{-2}\, dV\, d\bar{k}\, ds$$

$$= (\mathcal{TH}^0 f)^x(Y).$$

Here we used formulas (1.138)–(1.140) to see that

$$p_{\bar{s}+2\rho}(I[x^{-1}], x(\bar{k}))\, p_s(Y, x(\bar{k})) = p_s(Y[x], \bar{k}).$$

Now we want to compute ρ in formula (1.139). Exercise 1.1.20 of Section 1.1.4 shows that if

$$a = \begin{pmatrix} a_1 & & 0 \\ & \ddots & \\ 0 & & a_n \end{pmatrix} \quad \text{with } |a| = 1,$$

then

$$\alpha(a) = \prod_{j=1}^{n} a_j^{n-2j+1} = (a_1 \cdots a_{n-1})^{n-1} \prod_{j=1}^{n-1} a_j^{n-2j+1} = \prod_{j=1}^{n-1} a_j^{2(n-j)}.$$

Thus if $|a| = 1$ we find that

$$\alpha(a) = \prod_{j=1}^{n-1} a_j^{2(n-j)} = \prod_{j=1}^{n-1} (a_1 \cdots a_j)^{4\rho_j} = \prod_{j=1}^{n-1} a_j^{4(\rho_j + \cdots + \rho_{n-1})}.$$

It follows that $4\rho_j = 2(n - j - n + (j + 1))$ and therefore

$$\rho_j = \frac{1}{2}, \qquad (j = 1, \ldots, n - 1). \qquad (1.141)$$

(2) To prove part (2) of Lemma 1.3.1, note that, assuming we can change the order of integration, we have

$$\int_{S\mathcal{P}_n} f(W)\overline{TF(W)}\, dW = \int_{\mathrm{Re}\, s=-\rho} \int_{\bar{k}\in K/M} \mathcal{H}^0 f(s,k)\, \overline{F(s,\bar{k})}\, |c_n(s)|^{-2}\, d\bar{k}\, ds.$$

It follows that

$$\int_{S\mathcal{P}_n} f(W)\, \overline{T\mathcal{H}^0 g(W)}\, dW = \int_{\mathrm{Re}\, s=-\rho} \int_{\bar{k}\in K/M} \mathcal{H}^0 f(s,k)\, \overline{\mathcal{H}^0 g(s,\bar{k})}\, |c_n(s)|^{-2}\, d\bar{k}\, ds,$$

and therefore the operator $T\mathcal{H}^0$ is self-adjoint, completing the proof of Lemma 1.3.1. ∎

Exercise 1.3.5. Show that the interchange of integration orders is legal in the proof of part (2) of Lemma 1.3.1.

Exercise 1.3.6. Give an example of a Dirac sequence of functions $f_m : S\mathcal{P}_n \to \mathbb{C}$ such that f_m approaches δ_I, as $m \to \infty$.

In order to finish our discussion of Step 2 in the proof of part (1) of Theorem 1.3.1, note that it suffices to prove the inversion formula for any Dirac sequence of functions $g_m : S\mathcal{P}_n \to \mathbb{C}$ approaching δ_I, the Dirac delta function at the identity matrix in $S\mathcal{P}_n$. One can always take such a Dirac sequence to be K-invariant. To see that the general inversion formula follows from that for such a Dirac sequence, one proceeds as in the analogous argument for $\mathbb{R}^n$ (which can be found in the proof of Theorem 1.2.1 in Vol. I). For part (2) of Lemma 1.3.2 implies that if $T\mathcal{H}^0 g_m = g_m$, then

$$\int f\, g_m = \int (T\mathcal{H}^0 f)\, g_m,$$

assuming the functions g_m are real-valued. Now the left-hand side of the above equality will approach $f(I)$, while the right-hand side will approach $T\mathcal{H}^0 f(I)$, as m goes to infinity. Moreover, part (1) of Lemma 1.3.1 implies that it suffices to prove that $f(Y) = T\mathcal{H}^0 f(Y)$ at the point $Y = I$, or any other fixed point of $S\mathcal{P}_n$.

Remarks. (1) Of course it is easy to believe that it should not really matter what point in G/K is chosen at which to try to prove the inversion formula for the

Helgason–Fourier transform, since the result is an eigenfunction expansion for differential operators that commute with the action of G on the symmetric space G/K.

(2) We will use our results from Step 2 to push the support of the function in the inversion formula out to infinity. By this we mean out towards the boundary of the symmetric space where asymptotic expansions and ordinary Mellin inversion take over.

(3) The "lines" of integration in the inversion formula (part 1) of Theorem 1.3.1 are $\operatorname{Re} s_j = -\frac{1}{2}$, for $j = 1, \ldots, n-1$ and $\operatorname{Re} s_n = (n-1)/4$. If we change variables according to formula (1.123), then these lines correspond to $\operatorname{Re} r_j = 0$, $j = 1, \ldots, n$. These lines are certainly fixed by the $n!$ permutations of the variables r_j which represent the functional equations of the spherical functions (see part 4 of Theorem 1.2.3 in Section 1.2.3).

Note that when $\operatorname{Re} r_j = 0$ the eigenvalues of the Laplacian on $\mathcal{P}_n$ are negative, since the eigenvalues are:

$$r_1^2 + \cdots + r_n^2 + \frac{n - n^3}{48}$$

(see Exercise 1.2.12 of Section 1.2.1).

The Harish-Chandra c-function in part (1) of Theorem 1.3.1 is quickly seen to have poles along the domain of integration for the inverse transform; i.e., when $r_i = r_{j+1}$. See formula (1.124) of the preceding section.

All of these phenomena occurred in the inversion formula for the Helgason–Fourier transform on the Poincaré upper half plane (see Section 3.2 of Volume I).

1.3.3   End of the Discussion of Part (1) of Theorem 1.3.1—Steps 3 and 4

Step 3 (Asymptotics and Functional Equations). This part of the proof of part (1) of Theorem 1.3.1 was obtained by Harish-Chandra [263, Vol. II, pp. 409–539], who gave the general theory for G a semisimple real Lie group, and by Bhanu-Murthy [48], who made Harish-Chandra's results explicit when $G = SL(n, \mathbb{R})$. Gindikin and Karpelevic [220] computed the spectral measure explicitly for general G. Helgason [282, pp. 425–466] gives a very detailed treatment of this last step in our discussion. We will stick to a heuristic version of the argument. See also Varadarajan [625].

Suppose that $f(Y) = f(Y[k])$ for all Y in $\mathcal{P}_n$ and $k \in K = O(n)$. The Helgason–Fourier transform of f is obtained in formula (1.128) in section 1.3.1. If we pull out the determinant in formula (1.128), as in Step 1 of the proof, then we must change variables via

$$a = \begin{pmatrix} a_1 & & 0 \\ & \ddots & \\ 0 & & a_n \end{pmatrix} = v^{1/n} \begin{pmatrix} u_1 & & 0 \\ & \ddots & \\ 0 & & u_n \end{pmatrix}, \text{ with } v = |a| \text{ and } u_1 \cdots u_n = 1.$$

Then formula (1.128) becomes:

$$\widehat{f}(s) = b_n \int_{v>0} \int_{u_i > u_{i+1}} f(v, u) v^{r-1} \prod_{i<j}(u_i - u_j) \, \overline{h_s(u)} \, dv \, du, \tag{1.142}$$

where $f(a) = f(v, u)$,

$$du = \prod_{i=1}^{n-1} u_i^{-1} du, \tag{1.143}$$

r is defined by formula (1.131), and b_n is defined by (1.128).

According to Step 1, we need to show that if we are given a function f on $\mathcal{SP}_n$ which is $O(n)$-invariant and if u and a denote positive diagonal matrices of determinant one, then:

$$f(u) = (2\pi i)\omega_n b_n \int_{\mathrm{Re}\, s = -\rho} \int_{a_i > a_{i+1}} f(a) \prod_{i<j}(a_i - a_j) \, \overline{h_s(a)} \, da \, h_s(u) \, |c_n(s)|^{-2} \, ds, \tag{1.144}$$

with da as in formula (1.143), b_n as in (1.128), and ω_n as in part (1) of Theorem 1.3.1.

Suppose now that a_i/a_{i-1} is near zero for all $i = 2, \ldots, n$. Then

$$\prod_{i<j}(a_i - a_j) = \prod_{i=1}^{n-1} a_i^{n-i} \prod_{i<j}(1 - a_j/a_i),$$

which approaches

$$\prod_{i=1}^{n-1} a_i^{n-i}, \text{ as } a_i/a_{i-1} \to 0, \; i = 2, \ldots, n. \tag{1.145}$$

Assume next that the function $f(a)$ is supported on positive diagonal matrices a with a_i/a_{i-1} near 0, for all $i = 2, \ldots, n$. Then the inversion formula of Harish-Chandra can be obtained from an asymptotics/functional equations principle similar to that seen in Section 3.2 of Vol. I, if we can show that the spherical functions have **asymptotic expansions** of the form:

$$h_s(a) \sim c_n(s)p_s(a), \text{ if } a_j/a_{j-1} \to 0. \tag{1.146}$$

Here the argument a of the spherical function is a diagonal determinant one matrix with jth diagonal entry a_j and the parameter $s \in \mathbb{C}^{n-1}$ is fixed with $\operatorname{Re} s_j$ sufficiently large. In fact, $\operatorname{Re} s_j$ must be so large that s_j is outside of the domain of integration in the inversion transform. For such a parameter s, the term on the right in (1.146) should be replaced by a sum of $n!$ terms coming from the functional equations of the spherical functions. This is proved by Harish-Chandra [263, Vol. II, pp. 409–539]. See also Helgason [282] and Varadarajan [625].

According to formulas (1.144)–(1.146), if $f(Y)$, $Y \in \mathcal{SP}_n$, is K-invariant and supported on $a \in A$ with a_j/a_{j-1} near 0 for all $j = 2, \ldots, n$, the **inversion formula of part (1) of Theorem 1.3.1 would look approximately like**:

$$f(u) = n!(2\pi i)\omega_n b_n \int\limits_{\substack{s \in \mathbb{C}^{n-1} \\ \operatorname{Re} s = -\rho}} \int\limits_{\substack{a \in A \\ \frac{a_i}{a_{i-1}} \text{ near } 0}} f(a) \prod_{i=1}^{n-1} a_i^{n-i} \, \overline{p_s(a)} \, da \, p_s(u) \, ds,$$

(1.147)

for $u \in A$ with u_j/u_{j-1} near 0, $j = 2, \ldots, n$. The spectral measure in part (1) of Theorem 1.3.1 is thus seen to be chosen to cancel out the term $\overline{c_n(s)}c_n(s)$ coming from the asymptotic formula (1.146) for the spherical function. The extra $n!$ comes from the $n!$ functional equations of the spherical function, which replaces the right-hand side of (1.146) by a sum of similar terms, where the sum is over the group of permutations acting on the r-variables for the power functions. Here one needs to note the orthogonality of the different exponentials that are summed over the permutation group.

Clearly (1.147) is the same as

$$f(u) = n!(2\pi i)\omega_n b_n \int\limits_{\substack{s \in \mathbb{C}^{n-1} \\ \operatorname{Re} s = -\rho}} \int\limits_{\substack{a \in A \\ \frac{a_i}{a_{i-1}} \text{ near } 0}} f(a) \prod_{i=1}^{n-1} a_i^{\overline{s_i + \cdots + s_{n-1}} + n - i - 1} \, da_i \prod_{i=1}^{n-1} u_i^{s_i + \cdots + s_{n-1}} \, ds.$$

Then ordinary Mellin inversion and formula (1.128) imply that we must choose ω_n as in part (1) of Theorem 1.3.1. For if $\operatorname{Re} s_j = -\frac{1}{2}$, we find that the exponent of a_j is $e_j - 1$ while that of u_j is $-e_j$, which is just what is required for ordinary Mellin inversion. To give a rigorous justification of this argument would require us to delve into the proof of the Paley–Wiener theorem on $\mathcal{P}_n$. We shall not do this here but see Helgason [282, pp. 450–454]. The miracle is that one needs only the main term in the asymptotic expansion of the spherical function. This is very similar to the standard miracle of Paley–Wiener theory, as well as the result of Lemma 3.7.1 of Volume I. And this miracle seems believable, recalling the orthogonality of the distinct power functions.

Anyway, we are accordingly reduced to the computation of the Harish-Chandra c-function in formula (1.146).

Step 4 (Computation of Harish-Chandra's c-Function). We must compute the coefficient in the main term of the asymptotic formula (1.146) for the spherical function. This computation is most easily accomplished by changing variables from K/M to $\overline{N}$ in Harish-Chandra's integral formula (1.127) for the spherical function. Here $\overline{N}$ is the (nilpotent or unipotent) subgroup of $G = SL(n, \mathbb{R})$ consisting of all lower triangular matrices with one's on the diagonal. Harish-Chandra [263, Vol. II, p. 455] shows how to produce this change of variables as a consequence of the Bruhat decomposition to be discussed later. And Bhanu-Murthy [48] carried out the computation of the c-function explicitly for $G = SL(n, \mathbb{R})$. See also Helgason [282, pp. 198, 434–448]

Lemma 1.3.2. *If $P = MAN$, then P is the minimal parabolic subgroup of $G = SL(n, \mathbb{R})$ consisting of all upper triangular matrices and $P\overline{N}$ is an open subset of G with lower dimensional complement. Thus we can realize the boundary $B = K/M$ as $\overline{N}$ as far as integration is concerned, obtaining the integral formula for the change of variables from the boundary B to $\overline{N}$:*

$$\int_B f(b) \, db = \kappa \int_{\overline{N}} f(\overline{n}(M)) \, p_{-2\rho}(I[\overline{n}]) \, d\overline{n}, \text{ where } \kappa^{-1} = \int_{\overline{N}} p_{-2\rho}(I[\overline{n}]) \, d\overline{n}.$$

Here $\overline{n}(M)$ denotes the result of letting $\overline{n}$ act on the coset M in $B = K/M$. If $K(\overline{n}) = k_1$ is the K-component in the Iwasawa decomposition of $\overline{n}$, then

$$\overline{n} = k_1 a_1 n_1, \ k_1 \in K, \ a_1 \in A, \ n_1 \in N,$$

and

$$\overline{n}(M) = K(\overline{n})M = k_1 M.$$

The measure $d\overline{n}$ is defined by:

$$d\overline{n} = \prod_{i>j} dx_{ij}, \ \text{if} \ \overline{n} = \begin{pmatrix} 1 & & & 0 \\ & 1 & & \\ & & \ddots & \\ x_{ij} & & & 1 \end{pmatrix}.$$

Proof (Cf. Helgason [282, p. 198]). To see the first statement of the lemma, multiply the matrices below.

$$
\begin{pmatrix}
v_{11} & v_{12} & \cdots & v_{1,n-1} & v_{1n} \\
0 & v_{22} & \cdots & v_{2,n-1} & v_{2n} \\
\vdots & \vdots & \ddots & \vdots & \vdots \\
0 & 0 & \cdots & v_{n-1,n-1} & v_{n-1,n} \\
0 & 0 & \cdots & 0 & v_{nn}
\end{pmatrix}
\begin{pmatrix}
1 & 0 & \cdots 0 & 0 \\
u_{21} & 1 & \cdots 0 & 0 \\
\vdots & \vdots & \ddots \vdots & \vdots \\
u_{n-1,1} & u_{n-1,2} & \cdots 1 & 0 \\
u_{n1} & u_{n2} & \cdots u_{n,n-1} & 1
\end{pmatrix}
$$

$$
=
\begin{pmatrix}
v_{11} + v_{12}u_{21} + \cdots + v_{1n}u_{n1} & \cdots & v_{1,n-1} + v_{1n}u_{n,n-1} & v_{1n} \\
v_{22}u_{21} + \cdots + v_{2n}u_{n1} & \cdots & v_{2,n-1} + v_{2n}u_{n,n-1} & v_{2n} \\
\vdots & \ddots & \vdots & \vdots \\
v_{n-1,n-1}u_{n-1,1} + v_{n-1,n}u_{n1} & \cdots & v_{n-1,n-1} + v_{n-1,n}u_{n,n-1} & v_{n-1,n} \\
v_{nn}u_{n1} & \cdots & v_{nn}u_{n,n-1} & v_{nn}
\end{pmatrix}.
$$

This is an element of $G = SL(n, \mathbb{R})$ such that the lower right $k \times k$ corner matrix is non-singular for all $k = 1, 2, \ldots, n$. Thus $P\overline{N}$ is indeed an open subset of G with lower dimensional complement.

As an **Exercise**, prove these last two statements. **Hint.** We can multiply block matrices as follows:

$$
(0 \ I)
\begin{pmatrix} F & H \\ 0 & G \end{pmatrix}
\begin{pmatrix} A & 0 \\ C & B \end{pmatrix}
\begin{pmatrix} 0 \\ I \end{pmatrix}
= GB.
$$

Thus we can define a mapping

$$
\psi : \overline{N} \to K/M
$$

$$
\overline{n} \mapsto \overline{n}(M) = K(\overline{n})M.
$$

And we can use this mapping to identify $\overline{N}$ and the boundary $B = K/M \cong G/P$, where $P = MAN$ (recalling formula (1.20) of Section 1.1.4).

Next we seek the Jacobian $J(\overline{n})$ of the mapping ψ; i.e.,

$$
\int_{B=K/M} f(b) \, db = \int_{\overline{N}} f(n(M)) \, J(\overline{n}) \, d\overline{n}. \tag{1.148}
$$

Now the integral formula for the action of G on B (see Exercise 1.1.22 of Section 1.1.4 and formula (1.139) above) gives:

$$
\int_{b=\overline{k}\in K/M} f(\overline{n}_0(b))\alpha^{-1}(A(\overline{n}_0 k)) \, d\overline{k} = \int_{b\in K/M} f(b) \, db \tag{1.149}
$$

with $\alpha(A(\overline{n}_0 k)) = p_{2\rho}(I[\overline{n}_0 k])$, for $\rho \in \mathbb{C}^{n-1}$ as in (1.141). Combining (1.148) and (1.149) yields:

$$\int\limits_{\bar{n}\in\bar{N}} f(\bar{n}_0\bar{n}(M))\,\alpha^{-1}(A(\bar{n}_0 K(\bar{n})))\,J(\bar{n})\,d\bar{n} = \int\limits_{\bar{n}\in\bar{N}} f(\bar{n}_0\bar{n}(M))\,J(\bar{n}_0\bar{n})\,d\bar{n},$$

$$J(\bar{n}_0\bar{n}) = \alpha^{-1}(A(\bar{n}_0 K(\bar{n})))J(\bar{n}).$$

Set $\bar{n} = I$ to obtain

$$J(\bar{n}_0) = p_{-2\rho}(I[\bar{n}_0])J(I).$$

The constant $J(I)$ is determined by demanding that the total volume of the boundary be one. This completes the proof of Lemma 1.3.2.   ∎

If we apply Lemma 1.3.2 to the integral formula (1.127) for the spherical function, we find the **second integral formula for the spherical function:**

$$h_s(I[a]) = \kappa \int\limits_{\overline{N}} p_s(I[a], \bar{n}(M))\, p_{-2\rho}(I[\bar{n}])\, d\bar{n} \qquad (1.150)$$

with the constant κ given in Lemma 1.3.2.

To discover the asymptotics of (1.150), we need the **second power function identity:**

$$p_s(I[a], \bar{n}(M)) = p_{2s}(a)p_s(I[\bar{n}^a])p_{-s}(I[\bar{n}]), \text{ with } \bar{n}^a = a\bar{n}a^{-1}. \qquad (1.151)$$

To prove this, write $\bar{n} = k_1 a_1 n_1$, with $k_1 \in K$, $a_1 \in A$, $n_1 \in N$. Then

$$\bar{n}^a = ak_1 a_1 n_1 a^{-1} = ak_1 a_1 a^{-1}(an_1 a^{-1}).$$

Since $(an_1 a^{-1}) \in N$, it follows from the definition of the power function that

$$p_s(I[\bar{n}^a]) = p_s(I[a], \bar{n}(M))\, p_s(I[\bar{n}])\, p_s(I[a^{-1}]).$$

This implies (1.151).

Combining (1.150) and (1.151) gives the **third integral formula for the spherical function:**

$$h_s(I[a]) = \kappa\, p_{2s}(a) \int\limits_{\overline{N}} p_s(I[\bar{n}^a])\, p_{-s}(I[\bar{n}])\, p_{-2\rho}(I[\bar{n}])\, d\bar{n}, \qquad (1.152)$$

with κ as in Lemma 1.3.2.

For Re s_j all sufficiently large, we can let a_j/a_{j-1} approach zero inside the integral in (1.152). If

$$\bar{n} = \begin{pmatrix} 1 & & 0 \\ & \ddots & \\ x_{ij} & & 1 \end{pmatrix} \quad \text{and} \quad \bar{n}^a = \begin{pmatrix} 1 & & 0 \\ & \ddots & \\ y_{ij} & & 1 \end{pmatrix},$$

then $y_{ij} = a_i x_{ij} a_j^{-1}$, for $i > j$. It follows that if a_i/a_{i-1} approaches zero, then y_{ij} approaches zero if $i > j$.

Thus for Re s_j sufficiently large, $j = 1, \ldots, n-1$, as $a_j/a_{j-1} \to 0, j = 2, \ldots, n$, we see that formula (1.152) approaches

$$\kappa \, p_{2s}(a) \int_{\bar{N}} p_{-(s+2\rho)}(I[\bar{n}]) \, d\bar{n}.$$

This means that the **Harish-Chandra c-function** is:

$$c_n(s) = \kappa \int_{\bar{N}} p_{-(s+2\rho)}(I[\bar{n}]) \, d\bar{n}. \tag{1.153}$$

Now we want to use mathematical induction to evaluate $c_n(s)$. First, note that when $n = 2$, we have $\alpha(a) = a_1^2$ and $\rho = \frac{1}{2}$. Thus we find that $c_2(s)$ is evaluated as follows in terms of beta functions (see Lebedev [398, p. 13]) $B(x, y) = \Gamma(x)\Gamma(y)/\Gamma(x+y)$;

$$c_2(s) = \kappa \int_{x \in \mathbb{R}} (1+x^2)^{-(s+1)} \, dx = \kappa B\left(\frac{1}{2}, s + \frac{1}{2}\right) = \frac{B\left(\frac{1}{2}, s + \frac{1}{2}\right)}{B\left(\frac{1}{2}, \frac{1}{2}\right)}.$$

In the general case we define

$$b_n(s) = \int_{\bar{N}} p_{-s}(I[\bar{n}]) \, d\bar{n}. \tag{1.154}$$

This is a special case of the Whittaker function in Section 1.2.2. Clearly $c_n(s) = b_n(s + 2\rho)/b_n(2\rho)$. We need to relate b_n with b_{n-1}. To do this, write the element $\bar{n}$ in $\bar{N}$ as:

$$\bar{n} = \begin{pmatrix} \bar{m} & 0 \\ {}^t x & 1 \end{pmatrix},$$

for $x \in \mathbb{R}^{n-1}$, $\bar{m}$ in the group $\bar{N}$ for $SL(n-1, \mathbb{R})$, a group that we shall denote $\bar{N}_{n-1}$. Then

$$p_s(I[\bar{n}]) = p_s({}^t\bar{m}\,\bar{m} + x\,{}^t x),$$

and it follows that:

$$b_n(s) = \int\limits_{\overline{m} \in N_{n-1}} \int\limits_{x \in \mathbb{R}^{n-1}} p_{-s}\left({}^t\overline{m}\,\overline{m} + x\,{}^tx\right) dx\, d\overline{m}.$$

Now write ${}^t\overline{m}\,\overline{m} = I[t]$, for t upper triangular with positive diagonal. Make the change of variables $x = {}^ttu$ in the last formula for $b_n(s)$. This gives:

$$b_n(s) = b_{n-1}(s_1, \ldots, s_{n-2})b_n'(s_1, \ldots, s_{n-1}),$$

where

$$b_n'(s) = \int\limits_{u \in \mathbb{R}^{n-1}} p_{-s}(I + u\,{}^tu)\, du. \tag{1.155}$$

To evaluate this last integral, write $u = (v, w)$, with $v \in \mathbb{R}^{n-2}$ and $w \in \mathbb{R}$. Also define $s = (r, s_{n-1})$, with $r \in \mathbb{C}^{n-2}$. Note that the determinant of $I + u\,{}^tu$ is:

$$\left|I + u\,{}^tu\right| = 1 + {}^tuu = 1 + u_1^2 + \cdots + u_{n-1}^2.$$

This is an easy consequence of the spectral theorem, since $u\,{}^tu$ is an $(n-1) \times (n-1)$ matrix with only one nonzero eigenvalue. And that eigenvalue is the square of the norm of the vector u. By these remarks

$$b_n'(s) = \int\limits_{v \in \mathbb{R}^{n-2}} p_{-r}\left(I + v\,{}^tv\right) \int\limits_{w \in \mathbb{R}} \left(\left|I + v\,{}^tv\right| + w^2\right)^{-s_{n-1}} dw\, dv.$$

Next change variables via $w = \left|I + v\,{}^tv\right|^{1/2} y$. Thus we obtain:

$$b_n'(s) = b_{n-1}'\left(s_1, \ldots, s_{n-3}, s_{n-2} + s_{n-1} - \frac{1}{2}\right) B\left(\frac{1}{2}, s_{n-1} - \frac{1}{2}\right)$$

$$= \prod_{j=1}^{n-1} B\left(\frac{1}{2}, s_j + s_{j+1} + \cdots + s_{n-1} - \frac{n-j}{2}\right).$$

It follows from this that

$$b_n(s) = \prod_{1 \leq i \leq j \leq n-1} B\left(\frac{1}{2}, s_i + s_{i+1} + \cdots + s_j - \frac{j-i+1}{2}\right),$$

which quickly leads to the formula in part (1) of Theorem 1.3.1. Hopefully we have given the reader enough insight into the proof of Theorem 1.3.1 to find the theorem believable. Note that $\mathrm{Re}\ s_j > -\frac{1}{2}$, for $j = 1, \ldots, n-1$, is required for the absolute convergence of the integral in the beta functions. Thus we find that the lines of integration in Theorem 1.3.1 are outside the region of absolute convergence and care must be taken because of this.

This completes our discussion of part (1) of Theorem 1.3.1. As we said earlier, we will not prove part (4). And we relegated parts (2) and (3) of Theorem 1.3.1 to Exercise 1.3.1.

Remark. The function b_n' defined by formula (1.155) is a k-Bessel function of singular argument (see formula (1.60) in Section 1.2.2). The function $b_n(s)$ in (1.154) is a Whittaker function of singular argument (see the end of Section 1.2.2). One can also consider these functions to be analogues of the beta function (cf. Gindikin [218]).

Finally we give a table of Helgason–Fourier transforms on $\mathcal{P}_n$ (see Table 1.1). Sadly it is quite short.

Exercise 1.3.7. Check Table 1.1.

The Helgason–Fourier transform of a non-K-invariant function involves a variable $k \in K$. But one can use the Fourier inversion formula for K itself (see Chapter 2 of Vol. I) to replace functions on K with functions of $\pi \in \widehat{K}$, which is the set of equivalence classes of irreducible unitary representations of K. Thus we could replace $\mathcal{H}f(s, k)$, $k \in K$, with a matrix-valued transform

$$\mathcal{H}f(s, \pi), \ \pi \in \widehat{K}.$$

Would that make us happier?

Table 1.1 A short table of Helgason transforms

$f(Y)$	$\widehat{f}(\bar{s}) = \mathcal{H}f(\bar{s}, I) = \displaystyle\int_{\mathcal{P}_n} f(Y)p_s(Y)d\mu_n(Y)$	
$\exp\left[-\mathrm{Tr}\left(X^{-1}Y\right)\right], \ X \in \mathcal{P}_n$	$p_s(X)\Gamma_n(s),$	§ 1.2.1, Exercise 1.2.4
$\exp\left[-\mathrm{Tr}\left(VY + WY^{-1}\right)\right]$	$K_n(s \mid V, W),$	§ 1.2.2, Formula (1.61)
$\exp\left[-\mathrm{Tr}\left(XY\right)\right] p_{r-s-(n+1)/2}(I + Y)$	$\Psi_n(s, r; X),$	§ 1.2.2, Formula (1.65)
$K_n(r \mid I, Y)$	$\Gamma_n(s)\,\Gamma_n(s + r),$	§1.2.2, Exercise 1.2.18

1.3.4   Applications—Richards' Central Limit Theorem
for K-Invariant Functions on $\mathcal{P}_n$

Analysis on $\mathcal{P}_n$ has seen many applications in multivariate statistics, as we have already noted. See Sections 1.1.7 and 1.2.4. But it is only recently that the Helgason–Fourier transform has appeared in such applications. The paper of Haff et al. [256] gives a method of minimax estimation of Wishart mixtures, with applications to a study of the stochastic volatility of stocks. Their proofs rely on the properties of the Helgason–Fourier transform.

There are also many applications in random matrix theory, but you might wish for more. We save this discussion for the next section.

The main result of this section is a generalization of the central limit theorem for rotation-invariant random variables on $\mathcal{P}_n$ which was obtained by Richards using his Theorem 1.2.6. In the old edition of this volume, we only did the case $n = 3$. Our discussion of the central limit theorem will use methods related to those of Cramér [126], Dym and McKean [149], and Feller [178], as in Section 1.2.4 of Volume I. The proof is analogous to that of the central limit theorem proved in Section 3.2.7 of Volume I using harmonic analysis on the Poincaré upper half plane. We will find that there is a significant difference between the situation for $GL(n, \mathbb{R})$ and that for $SL(2, \mathbb{R})$. The limiting density is not the same as the fundamental solution of the heat equation (see Exercise 1.3.8 below). See also Bougerol [72] and Graczyk [240]. Other references are the volumes edited by H. Heyer appearing in *Springer Lecture Notes in Math.* titled *Probability Measures on Groups* (e.g., [295]).

Our methods are special to the case of limit theorems for groups on which one can do harmonic analysis. Methods based on martingales, semigroups, stochastic difference, or differential equations can produce more general results. But the methods of harmonic analysis can give more detailed information. There are many papers on central limit theorems for Lie groups as we mentioned already in Vol. I.

Exercise 1.3.8 (The Heat Equation on $\mathcal{P}_n$).   Suppose that f is a K-invariant function on $\mathcal{P}_n$ which is continuous with compact support. We seek a solution $u(Y, t)$, $Y \in \mathcal{P}_n$, $t > 0$, to the heat equation:

$$\begin{cases} u_t = \Delta u, \ \Delta = \operatorname{Tr}((Y\partial/\partial Y)^2) = \text{Laplacian on } \mathcal{P}_n, \\ u(Y, 0) = f(Y). \end{cases}$$

Show that the solution is $u(Y, t) = G_t * f$, with convolution as in formula (1.156) below, where the fundamental solution G_t, also known as the **heat kernel** or **normal density**, is given by:

$$G_t(Y) = \omega_n \int\limits_{\text{Re}\,s = -\rho} \exp[\lambda_2(s)t]\, h_s(Y)\, |c_n(s)|^{-2}\, ds.$$

Here $\Delta p_s = \lambda_2(s)p_s$, with $\lambda_2^n(r) = r_1^2 + \cdots + r_n^2 + (n - n^3)/48$. The relation between the r and s variables is as stated in Theorem 1.2.6 (see Exercise 1.1.12 of

Section 1.2.1). The constants ω_n, ρ as well as $c_n(s)$ are defined in Theorem 1.3.1. Here $h_s(Y)$ denotes the spherical function in formula (1.127).

Hints. (See Gangolli [197, pp. 108 ff].) Imitate the method used to solve the heat equation on the non-Euclidean upper half plane in Section 3.2 of Vol. I. Note that G_t approaches the Dirac delta distribution at the identity, as t approaches 0 from above, by the same argument that proved the analogous result in Vol. I. Moreover G_t cannot have compact support, since it is a solution of a parabolic partial differential equation and is thus an analytic function of Y. Therefore we really need an extension of Theorem 1.3.1 to Schwartz functions on $\mathcal{P}_n$ in order to do this exercise rigorously. Another reference for the heat kernel is Jorgenson and Lang [333, p. 365].

Remarks. Ólafsson and Schlichtkrull [479] consider the holomorphic extension of $G_t * f$, for f an L^2 function on a general symmetric space such as $\mathcal{SP}_n$. The extension is to the **complex crown of the symmetric space**—a space introduced by Akhiezer and Gindikin in 1990. The crown of G/K is a G-invariant subdomain of the complexified symmetric space G^c/K^c (obtained for a group like $G = SL(n, \mathbb{R})$ by replacing $\mathbb{R}$ by $\mathbb{C}$). We will say a little more about this later.

It would also be interesting and useful to consider other partial differential equations on $\mathcal{P}_n$. For example, Helgason [275] and [282, pp. 342–343] investigates the wave equation on symmetric spaces and extensions of Huyghen's principle.

We consider **random variables** Y in $\mathcal{P}_n$ with **density** $f(Y)$ in $L^1(\mathcal{P}_n, d\mu_n), f \geq 0$, where $d\mu_n$ denotes the G-invariant measure on $\mathcal{P}_n$. Then if S is a measurable subset of $\mathcal{P}_n$, the **probability** that the random variable Y, with density f, is in S is

$$P(Y \in S) = \int_{Y \in S} f(Y) \, d\mu_n(Y) = \int_{I[g] \in S} f(I[g]) \, dg,$$

where dg denotes Haar measure on $G = GL(n, \mathbb{R})$. There are many possible analogues of the mean and the standard deviation as is the case for $\mathbb{R}^n$. We will return to this subject below (see (1.157) and (1.158)).

Here **we consider only K-invariant** random variables Y; i.e., we will *always* assume that the density function $f = f_Y$ satisfies:

$$f(Y[k]) = f(Y), \text{ for all } Y \in \mathcal{P}_n \text{ and } k \in K = O(n).$$

It will often be helpful to identify such a function f on $\mathcal{P}_n$ with a **K-bi-invariant** function on the group G via

$$f(I[g]) = f(g), \text{ for all } g \in G.$$

The **composition** $Y_1 \circ Y_2$ of two K-invariant random variables Y_1 and Y_2 on $\mathcal{P}_n$ is defined to be that coming from multiplication of the corresponding group elements. If Y_j has density function f_j, $j = 1, 2$, then $Y_1 \circ Y_2$ has density the **convolution** $f_1 * f_2$, assuming that Y_1 and Y_2 are independent:

$$(f_1 * f_2)(x) = \int_G f_1(y) \, f_2(y^{-1}x) \, dy. \qquad (1.156)$$

Note the difference with equation (1.24) of Section 1.1.4. This does not matter for K-invariant functions since then convolution is commutative.

To see that $Y_1 \circ Y_2$ has density $f_1 * f_2$, note that

$$
\begin{aligned}
P(Y_1 \circ Y_2 \in S) &= \int\!\!\int_{g_1 g_2 \in S} f_1(g_1) f_2(g_2) \, dg_1 \, dg_2 \\
&= \int_{g_1 \in G} f_1(g_1) \int_{w \in S} f_2(g_1^{-1}w) \, dw \, dg_1 \\
&= \int_{w \in S} (f_1 * f_2)(w) \, dw.
\end{aligned}
$$

If we now seek to imitate our discussion from Vol. I, Section 3.2.7, we find that the result for $\mathcal{P}_n$ is somewhat different from that for the Poincaré upper half plane.

Let $\{Y_\nu\}_{\nu \geq 1}$ be a sequence of independent K-invariant random variables on $\mathcal{P}_n$, each having the same density function $f(Y)$. We will assume the vanishing of the means with respect to the $h_j = \log a_j$, where a_j denotes the jth eigenvalue of Y, and we will also assume that the covariance matrix with respect to the h_j is the identity; i.e., we assume via the change to polar coordinates that the following integral formulas hold:

$$c_n \int_{H \in \mathfrak{a}} h_j f(\exp H) \, J(\exp H) \, dh = 0, \; j = 1, \ldots, n; \qquad (1.157)$$

$$c_n \int_{H \in \mathfrak{a}} h_i h_j f(\exp H) \, J(\exp H) \, dh = \delta_{ij}, \; 1 \leq i, j \leq n. \qquad (1.158)$$

Here

$$\mathfrak{a} = \left\{ H = \begin{pmatrix} h_1 & \cdots & 0 \\ \vdots & \ddots & \vdots \\ 0 & \cdots & h_n \end{pmatrix} \, \middle| \, h_j \in \mathbb{R} \right\}, \quad dh = \prod_{j=1}^n dh_j.$$

Note that $\mathfrak{a}$ is the tangent space to A at the identity (cf. (1.77) in Section 1.2.3). Formulas (1.38), (1.40) in Section 1.1.6 give the constant c_n and the Jacobian $J(a) = \gamma(a)/a$, for $a \in A$.

Consider the composition $S_\nu = Y_1 \circ \cdots \circ Y_\nu$ which was defined in the paragraph preceding (1.156). We will normalize S_ν as follows. Let $h_j = \log a_j$, where a_j

denotes the jth eigenvalue of Y. Then normalize by replacing h_j by $v^{-1/2}h_j$. Call the resulting random variable $S_v^{\#}$. The **characteristic function** of the normalized random variable is:

$$\varphi_{S_v^{\#}}(s) = \left\{ c_n \int_{H \in \mathfrak{a}} f(\exp H)\, h_s(\exp(v^{-1/2}H))\, J(\exp H)\, dH \right\}^v . \tag{1.159}$$

Here we have used the convolution property of the Helgason–Fourier transform from Theorem 1.3.1.

Now use Richards' Theorem 1.2.6 to see that, as v approaches infinity, the term inside the braces in (1.159) is asymptotic to:

$$c_n \left\{ \int_{H \in \mathfrak{a}} f(\exp H)J(\exp H)\, dH + \int_{H \in \mathfrak{a}} \left(\sum_{j=1}^{n} h_j \right) f(\exp H)J(\exp H)\, dH \frac{1}{n\sqrt{v}} \left(\sum_{j=1}^{n} r_j \right) \right.$$

$$+ \int_{H \in \mathfrak{a}} \left(\sum_{j=1}^{n} h_j^2 \right) f(\exp H)J(\exp H)\, dH$$

$$\times \frac{1}{2n(n+2)v} \left(3 \sum_{j=1}^{n} r_j^2 + 2 \sum_{1 \le i < j \le n} r_i r_j - \frac{n^3 - n}{24} \right)$$

$$\left. + \int_{H \in \mathfrak{a}} \left(\sum_{1 \le i < j \le n} h_i h_j \right) f(\exp H)J(\exp H) \frac{P(r)}{v}\, dH \right\} .$$

Formulas (1.157) and (1.158) imply that as v approaches infinity:

$$\varphi_{S_v^{\#}}(s) \sim \left\{ 1 + \frac{1}{v} \frac{n}{2n(n+2)} \left(3 \left(\sum_{i=1}^{n} r_i^2 \right) + 2 \left(\sum_{1 \le i < j \le n} r_i r_j \right) - \frac{n^3 - n}{24} \right) \right\}^v$$

$$\sim \exp \left\{ \frac{1}{2(n+2)} \left(3 \left(\sum_{i=1}^{n} r_i^2 \right) + 2 \left(\sum_{1 \le i < j \le n} r_i r_j \right) - \frac{n^3 - n}{24} \right) \right\} .$$

Here $s \in \mathbb{C}^n$ is the function of $r \in \mathbb{C}^n$ specified in Theorem 1.2.6 and Theorem 1.3.2 below.

Recalling the formula for the eigenvalue of the Laplacian in Exercise 1.3.8 above, we see that the limit characteristic function does not appear to be related in a simple way to

$$\exp\left\{t\left(\sum_{i=1}^{n} r_i^2 - \frac{n^3 - n}{48}\right)\right\},$$

the Fourier transform of the fundamental solution G_t of the heat equation.
By the convolution theorem we see that the limit density is

$$\exp\left(\frac{n^3 - n}{96(n+2)}\right) G_{3/(2(n+2))} * F_{1/(n+2)}(Y),$$

where F_t is the function on $\mathcal{P}_n$ whose Helgason–Fourier transform is

$$\exp\left\{t\left(\sum_{1\le i<j\le n} r_i r_j\right)\right\}.$$

Here G_t is the fundamental solution of the heat equation from Exercise 1.3.8 above.

Theorem 1.3.2 (Richards' Central Limit Theorem for $\mathcal{P}_n$). *Suppose that $\{Y_m\}_{m\ge1}$ is a sequence of independent, $O(n)$-invariant random variables in $\mathcal{P}_n$, each having the same density function $f(Y)$. And suppose that the density satisfies (1.157) and (1.158). Let $S_m = Y_1 \circ \cdots \circ Y_m$ be normalized as in (1.159). The normalized variable has density function $f_m^\#$. Then for measurable sets S in $\mathcal{P}_n$ we have, as $m \to \infty$:*

$$\int_S f_m^\#(Y)\, d\mu(Y) \sim \exp\left(\frac{n^3 - n}{96(n+2)}\right) \int_S G_{3/(2(n+2))} * F_{1/(n+2)}(Y)\, d\mu(Y).$$

Here G_t is the fundamental solution of the heat equation from Exercise 1.3.8 and its Helgason–Fourier transform is

$$\widehat{G}_t(s(r)) = \exp\left\{t\left(\sum_{i=1}^{n} r_i^2 - \frac{n^3 - n}{48}\right)\right\}$$

while F_t has Helgason–Fourier transform:

$$\widehat{F}_t(s(r)) = \exp\left\{t\left(\sum_{1\le i<j\le n} r_i r_j\right)\right\}.$$

Note that we have reparametrized the Helgason–Fourier transform using the change of variables from s-variables to r-variables as in Theorem 1.2.6

$$s_j + \cdots + s_n = r_j + \frac{2j-n-1}{4}, \quad j = 1, \ldots, n,$$
$$r_j - r_{j+1} = s_j - \frac{1}{2}, \quad j = 1, \ldots, n-1, \quad s_n = r_n - \frac{n-1}{4}.$$

Proof. As in the case $n = 2$, for the Poincaré upper half plane, we need only to argue that the limiting behavior of densities mirrors that of their Fourier transforms. To see this, recall the inversion and Plancherel formulas from Theorem 1.3.1. Let β be an infinitely differentiable function with compact support on $\mathcal{P}_n$. Let $d\sigma(r)$ denote the spectral measure in part (1) of Theorem 1.3.1, using the r-variables from (1.123) rather than the s-variables of Theorem 1.3.1. Then, by Theorem 1.2.6 and the dominated convergence theorem, we have:

$$\lim_{m \to \infty} \int_{\mathcal{P}_n} f_m^\#(Y)\, \beta(Y)\, d\mu(Y) = \lim_{m \to \infty} \int_{\mathrm{Re}\, r=0} \widehat{f_m^\#}(s(r))\, \widehat{\beta}(s(r))\, d\sigma(r)$$

$$= \int_{\mathrm{Re}\, r=0} \exp\left\{ \frac{1}{2(n+2)} \left(3\left(\sum_{i=1}^{n} r_i^2 \right) \right.\right.$$

$$\left.\left. +2\left(\sum_{1 \le i < j \le n} r_i r_j \right) - \frac{n^3 - n}{24} \right)\right\} \widehat{\beta}(s(r))\, d\sigma(r)$$

$$= \exp\left(\frac{n^3 - n}{96\,(n+2)} \right) \int_{\mathcal{P}_n} (G_{3/(2(n+2))} * F_{1/(n+2)})(Y)\, \beta(Y)\, d\mu.$$

Here we are using the fact (proved by Helgason [274, p. 458] that spherical functions are bounded on the lines of integration for the inverse transform and thus if f is in $L^1(\mathcal{P}_n, d\mu)$, then $\widehat{f}(s(r))$ is bounded for $\mathrm{Re}\, r = 0$. Next let β approximate the indicator function of a set S in $\mathcal{P}_n$ (i.e., the function that is 1 on S and 0 off S) to complete the proof. ∎

Another reference for central limit theorems on Lie groups is the volume edited by Cohen et al. [115]. It would be useful to compare our results here with the limit theorem of Oseledec discussed by several authors in this volume. See also Furstenberg [193].

Exercise 1.3.9. What is the central limit theorem for $SO(n)$-invariant functions on $\mathcal{SP}_n$?

1.3.5   Quantum Chaos and Random Matrix Theory

Physicists have long studied spectra of Schrödinger operators and random matrices thanks to the implications for quantum mechanics. This is often found under the hashtag quantum chaos. Number theorists and geometers have similarly studied spectra of Laplacians on Riemannian manifolds. Sarnak has termed this "arithmetic quantum chaos" when the manifold in question is a quotient of a symmetric space with an arithmetic group. Equivalently one is investigating the zeros of the Selberg zeta function. Parallels with the statistics of the zeros of Riemann's zeta function have been known for some time. Here we only give a brief taste of a large subject. Some references are: Gernot Akemann et al. [3], Greg Anderson, Alice Guionnet, and Ofer Zeitouni [7], Oriol Bohigas and Marie-Joya Giannoni [57], Folkmar Bornemann [71], Barry Cipra [112, pp. 2–17], Brian Conrey [121], Alan Edelman and N. Raj Rao [151], Todd Kemp [347], Madan Lal Mehta [441], Stephen J. Miller and Ramin Takloo-Bighash [448], Andrew Odlyzko [477], Michael Rubinstein [518], Peter Sarnak [527], Craig A. Tracy and Harold Widom [615], Eugene Wigner [667], and my surveys in [610] and [611, Chapter 5]. Many of these references and lots more are available on line. Just Google random matrices.

In the 1950s Wigner (see [667]) modeled Schrödinger eigenvalues with the eigenvalues of large real symmetric $n \times n$ matrices whose entries are independent Gaussian random variables. He found that the histogram of such eigenvalues looks like a semi-circle (or, more precisely, a semi-ellipse). This has been named the **Wigner semi-circle distribution** (aka the Sato-Tate distribution in number theory). For example, he considered the eigenvalues of 197 "random" real symmetric 20×20 matrices. It is easy now to do an analogous experiment to that of Wigner using Matlab. See the upper histogram in Figure 1.1. However, Wigner notes on p. 5 of [667]: "What is distressing about this distribution is that it shows no similarity to the observed distribution in spectra."

So physicists have devoted more attention to histograms of level spacings rather than levels. This means that you arrange the energy levels (eigenvalues) E_i in decreasing order:

$$E_1 \geq E_2 \geq \cdots \geq E_n.$$

Assume that the eigenvalues are normalized so that the mean of the level spacings $(E_i - E_{i+1})$ is 1. Then one can ask for the shape of the histogram of the normalized level spacings. There are (see Sarnak [527]) two main sorts of answers to this question: **Poisson level spacings**, meaning e^{-x}, and **GOE spacings** (see Mehta [441]) which is more complicated to describe exactly but looks roughly like $\frac{\pi}{2} x e^{-\frac{\pi x^2}{4}}$ (the **Wigner surmise**). The spacings in the lower histogram of Figure 1.1 do roughly approximate this function which vanishes at the origin—unlike e^{-x}. This is interpreted as **level repulsion**, meaning that the eigenvalues find each other repulsive.

In 1957 Wigner (see [667]) gave an argument for the surmise that the level spacing histogram for levels having the same values of all quantum numbers is given by $\frac{\pi}{2} x e^{-\frac{\pi x^2}{4}}$ if the mean spacing is 1. In 1960 Gaudin and Mehta found the correct distribution function which is surprisingly close to Wigner's conjecture but different. There are many experimental studies comparing GOE prediction and nuclear data. See Bohigas [57], Bohigas and Giannoni [58], and Bohigas et al. [59].

Wigner's argument for the Wigner surmise from [667] is rather simple. He first derives the Wigner surmise for the Poisson level (or eigenvalue) spacing density as follows. If the location of the eigenvalues is independent, the probability v that

Fig. 1.1 Histograms for eigenvalues of a random normal symmetric real 1001×1001 matrix. The *upper histogram* shows the eigenvalues and the *lower one* shows the normalized spacings of the eigenvalues

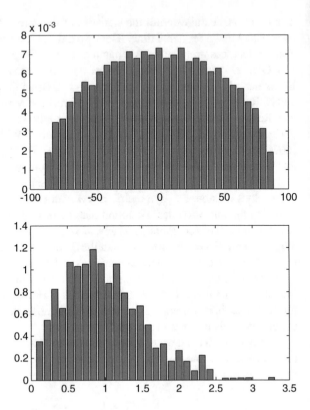

there is no eigenvalue within distance S of the given eigenvalue, for small value of h, satisfies

$$v(S + h) = v(S) - v(S)hc \tag{1.160}$$

where c is the probability of an eigenvalue in an interval of length 1. This leads to the ODE

$$\frac{dv}{dS} = -cv \quad \text{with solution} \quad v = e^{-cS}.$$

Then the spacing density is proportional to $\frac{dv}{dS} = e^{-cS}$. This is the spacing for events distributed with the Poisson distribution. If we wish to find the spacing for random matrices from the GUE or GOE distribution (and more generally) then instead of formula (1.160), we start with

$$v(S + h) = v(S) - v(S)hcS.$$

This leads to the differential equation

$$\frac{dv}{dS} = -cvS \text{ with solution } v = e^{-cS^2/2}.$$

Then if the mean spacing is 1, we find that the spacing density is $\frac{dv}{dS} = \frac{\pi}{2}Se^{-\frac{\pi S^2}{4}}$.

Andrew Odlyzko (see [477]) has investigated the level spacing distribution for the nontrivial zeros of the Riemann zeta function. He considers only zeros which are high up on the $\operatorname{Re} s = \frac{1}{2}$ line. Assume the Riemann hypothesis and look at the zeros ordered by imaginary part

$$\left\{ \gamma_n \;\middle|\; \zeta\left(\frac{1}{2} + i\gamma_n\right) = 0, \; \gamma_n > 0 \right\}.$$

To normalize the level spacings, replace γ_n by $\widetilde{\gamma}_n = \frac{1}{2\pi}\gamma_n \log \gamma_n$, since we want the mean spacing to be one. Here one needs to know that the number of γ_n such that $\gamma_n \leq T$ is asymptotic to $\frac{1}{2\pi}T \log T$ as $T \longrightarrow \infty$. Odlyzko's experimental results show that the spacings $(\gamma_{n+1} - \gamma_n)$, for large n, look like spacings of the Gaussian unitary ensemble (GUE); i.e., the eigenvalue distribution of a random complex Hermitian matrix (which differs slightly from the GOE spacing). See Figure 1.2, which Odlyzko just emailed to me in August 2015. It shows Odlyzko's results for the spacing of high zeros of the Riemann zeta function and GUE eigenvalue spacing. Another reference for this subject is Barry Cipra's expository article [112].

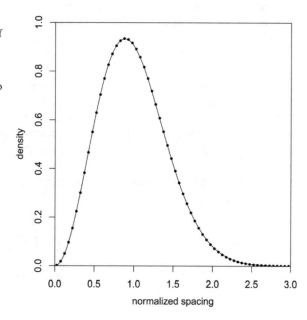

Fig. 1.2 Odlyzko's comparison of the spacings of the zeros of Riemann's zeta function and the GUE eigenvalue spacings curve. The fit is good for the close to one billion zeros near zero number $10^{23} + 17,368,588,794$

Katz and Sarnak and others (see [341, 342, p. 23]) have investigated many zeta and L-functions of number theory and have found that "the distribution of the high zeroes of any L-function follow the universal GUE Laws, while the distribution of the low-lying zeroes of certain families follow the laws dictated by symmetries associated with the family. The function field analogs of these phenomena can be established...." More precisely (see [342, p. 11]) they show that "the zeta functions of almost all curves C [over a finite field $\mathbb{F}_q$] satisfy the Montgomery-Odlyzko law [GUE] as q and g [the genus] go to infinity." For the details, see Katz and Sarnak [341].

It is quite surprising that the spacings of eigenvalues of the Poincaré Laplacian detect arithmeticity of Γ for compact Riemannian manifolds $\Gamma \backslash H$, $H=$ the Poincaré upper half plane or equivalently the unit disc. This is shown in Figure 1.3 from C. Schmit [536]. The top of the figure shows the fundamental domain of an arithmetic group in the unit disc. C. Schmit [536] found 1500 eigenvalues for the Dirichlet problem of the Poincaré Laplacian on the triangle OLM with angles $\pi/8, \pi/2, \pi/3$. The histogram of level spacings for this problem is the lower right part of Figure 1.3. Schmit also considered the Dirichlet problem for a non-arithmetic triangle with angles $\pi/8, \pi/2, 67\pi/200$. He found that the level spacing histogram for this non-arithmetic triangle is given in the lower left of Figure 1.3. Schmit concludes: "The spectrum of the tessellating [arithmetic] triangle exhibits neither level repulsion nor spectral rigidity and there are strong evidences that asymptotically the spectrum is of Poisson type, although the billiard is known to be a strongly chaotic system. The spectrum of the non-tessellating [non-arithmetic] triangle, whose classical properties are not known but which is probably a chaotic system too, exhibits the essential features of a generic chaotic system, namely the level repulsion and the spectral rigidity of GOE, as already observed in other chaotic systems."

Let's summarize a bit of the theory that ultimately derives the exact GUE and GOE level spacing. We follow Madan Lal Mehta [441] mostly. Other references are listed at the beginning of this section. One of the recent ones is Greg Anderson, Alice Guionnet, and Ofer Zeitouni [7]. We will mostly restrict ourselves to the GUE—the **Gaussian unitary ensemble**. An $N \times N$ random Hermitian matrix H is distributed according to the GUE distribution if the joint probability density is proportional to $\exp\left(-Tr(H^2)\right)$. A real symmetric random $N \times N$ matrix X is distributed according to the GOE distribution if the joint probability density is proportional to $\exp\left(-\frac{1}{2}Tr(X^2)\right)$. Set $\beta = 1$ for the GOE distribution and $\beta = 2$ for the GUE distribution. Then one can show that the joint probability density of the eigenvalues $x_1, \ldots, x_N$ is

$$P_{N\beta}(x_1, \ldots, x_N) = \text{constant } |\Delta(x)|^\beta \exp\left(\frac{-1}{\beta} \sum_{i=1}^n x_i^2\right),$$
$$\text{where } \Delta(x) = \prod_{1 \le i < j \le N} (x_i - x_j).$$

(1.161)

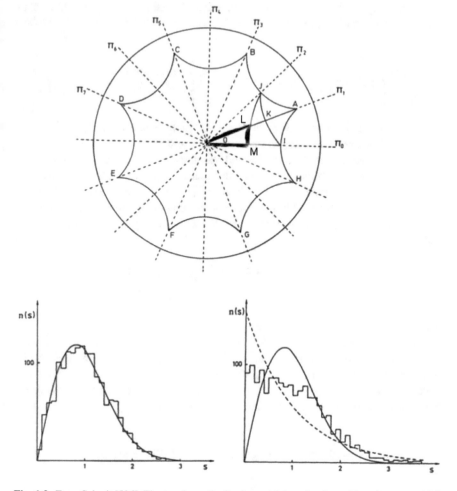

Fig. 1.3 From Schmit [536]. The *top* shows the fundamental domain of an arithmetic group which tessellates the Poincaré disk. The *lower right* shows the level spacing histogram for the Dirichlet problem on the triangle OLM using the 1500 eigenvalues computed by Schmit using the method of collocation. The *lower left* is the analogous histogram for a non-arithmetic triangle. The *solid line* is the GOE distribution and the *dashed one* is Poisson (e^{-x})

Note that $\Delta(x)$ is the **Vandermonde determinant**

$$\Delta(x) = \det\left(x_i^{j-1}\right)_{1 \le i,j \le n} = \prod_{i<j}(x_j - x_i).$$

See Mehta [441] or Anderson et al. [7] for a number of proofs of (1.161). At least in the GOE case, the density in formula (1.161) looks similar to that in the polar decomposition used to evaluate the gamma integral on $\mathcal{P}_n$. The constant of proportionality can be evaluated in various ways. A method used by Anderson et al. [7] involves the **Selberg integral formula** which says that for all positive numbers a, b, c

$$\frac{1}{n!} \int_0^1 \cdots \int_0^1 |\Delta(x)|^{2c} \prod_{i=1}^n x_i^{a-1} (1 - x_i)^{b-1} \, dx_i = \prod_{j=0}^{n-1} \frac{\Gamma(a + jc)\Gamma\,(b + jc)\,\Gamma\,((j + 1)\,c)}{\Gamma(a + b + (n + j - 1)\,c)\Gamma\,(c)}.$$

(1.162)

A method used by Kemp [347] involves the representation of $P_{N\beta}(x_1, \ldots, x_N)$ obtained below involving a kernel made up of Hermite functions.

The results on eigenvalue density and spacing of Gaudin and Mehta plus the connection with the Painlevé ordinary differential equations are to be found in Mehta [441] or Anderson et al. [7, Chapter 3]. Let us state them for the GUE case. The **Wigner semi-circle law** (known to number theorists as the Sato-Tate law) says that in the limit as the matrix size N goes to infinity, the eigenvalue density approaches

$$\sigma_N(x) = \begin{cases} \frac{1}{\pi} \left(2N - x^2\right)^{\frac{1}{2}}, & \text{if } |x| < \sqrt{2N} \\ 0, & \text{if } |x| > \sqrt{2N}. \end{cases}$$

We discuss this result later in this section.

The main results as stated in Anderson et al. [7, Chapter 3] are the following two theorems. The first really writes the limiting probabilities as a Fredholm determinant. This was defined by Fredholm in 1903 in his famous paper on integral equations. Fredholm sought solutions u to the Fredholm equation involving the **Fredholm integral operator** with kernel $K(x, y)$ given by

$$(I - zL_K)u = f, \quad \text{where } L_K u(x) = \int_a^b K(x, y)u(y)dy.$$

He introduced the **Fredholm determinant**

$$d(z) = \sum_{k=0}^{\infty} \frac{z^n}{n!} \int_a^b \cdots \int_a^b \det\left(K\left(t_i, t_j\right)\right)_{1 \le i, j \le n} \, dt_1 \cdots dt_n.$$

Fredholm showed that the Fredholm equation is uniquely solvable if and only if $d(z) \neq 0$. Hilbert transformed this theory to the theory of compact operators and really eliminated the determinants. However the subject has now shown its usefulness. Bornemann [71] has given a means to compute Fredholm determinants beginning with a short Matlab program. This enables one to avoid thinking about computing solutions of the nonlinear ordinary differential equation satisfied by the

Painlevé 5th transcendent in Theorem 1.3.4. More information on integral operators can be found in the paper of Bornemann [71] and the book by Hochstadt [302], for example.

The Fredholm determinant can be viewed as an infinite determinant of the operator L_K in the case under consideration. The requirements to be able to do this are summarized by Bornemann [71]. One wants to use the formula (hopefully converging for small z)

$$\log \det (I - zL_K) = \operatorname{Tr} \log (I - zL_K) = \operatorname{Tr} \left(-\sum_{n\geq 1} \frac{1}{n} (zL_K)^n \right) = -\sum_{n\geq 1} \frac{1}{n} \operatorname{Tr} (zL_K)^n .$$

The problem in general is that the operator L_K might not have a finite trace. The kernels K_N associated with $P_{N\beta}(x_1,\ldots,x_N)$ will be what is called a **self-reproducing kernel**, meaning that $L_K^j = L_K^{j-1}$, for $j \geq 2$. They are also **separable (or degenerate) kernels**; i.e.,

$$K(x, y) = \sum_{j=1}^{n} \alpha_j(x)\beta_j(y).$$

This really means that the operator L_K is essentially a finite matrix operator for the image of L_K is then spanned by the β_j. See Courant and Hilbert [125]. Moreover our kernels K_N are **symmetric** since $\beta_j = \alpha_j$. We will be interested in the limit of the K_N as $N \to \infty$. This will lead to the following theorem as we will explain in a little more detail later. One would like to write the Fredholm determinant as an infinite product over the eigenvalues of L_K. Hochstadt [302] derives such formulas in his last chapter.

Theorem 1.3.3 (Gaudin-Mehta). *Let X be a random matrix with GUE probability law P_{n2} and eigenvalues $\lambda_1 \leq \lambda_2 \leq \cdots \leq \lambda_n$. For any compact set $A \subset \mathbb{R}$*

$$\lim_{n \to \infty} P_{n2} \left(\sqrt{n}\lambda_1, \ldots, \sqrt{n}\lambda_n \notin A \right)$$

$$= 1 + \sum_{k=1}^{\infty} \frac{(-1)^k}{k!} \int_A \cdots \int_A \det \left(K_{\text{sinc}}(x_i, x_j)_{1 \leq i,j \leq k} \right) dx_1 \cdots dx_k,$$

where

$$K_{\text{sinc}}(x, y) = \frac{1}{\pi} \operatorname{sinc}(x - y) = \begin{cases} \frac{1}{\pi} \frac{\sin(x-y)}{x-y}, & \text{if } x \neq y, \\ \frac{1}{\pi}, & \text{if } x = y. \end{cases}$$

Using this result one can show that one can connect the level spacings with a solution of the nonlinear Painlevé V ordinary differential equation.

Theorem 1.3.4 (Jimbo-Miwa-Môri-Sato). *Let X be a random matrix with GUE probability law P_{n2} and eigenvalues $\lambda_1 \leq \lambda_2 \leq \cdots \leq \lambda_n$. Then*

$$\lim_{n \to \infty} P_{n2}\left(\sqrt{n}\lambda_1, \ldots, \sqrt{n}\lambda_n \notin (-t/2, t/2)\right) = 1 - F(t),$$

where for $t \geq 0$

$$1 - F(t) = \exp\left(\int_0^t \frac{\sigma(x)}{x} dx\right)$$

with σ the solution of the Painlevé V ordinary differential equation

$$\left(t\sigma''\right)^2 + 4(t\sigma' - \sigma)\left(t\sigma' - \sigma + \left(\sigma'\right)^2\right) = 0$$

such that

$$\sigma = -\frac{t}{\pi} - \left(\frac{t}{\pi}\right)^2 - \left(\frac{t}{\pi}\right)^3 + O\left(t^4\right) \text{ as } t \downarrow 0.$$

Let us give a brief summary of Mehta's discussion of Theorem 1.3.3 stated above (see Mehta [441]). First one needs to recall the Hermite polynomials from Volume I, Chapter 1. The jth **Hermite polynomial** is (using what Wikipedia calls the "physics normalization")

$$H_j(x) = e^{x^2}\left(-\frac{d}{dx}\right)^j e^{-x^2} = j! \sum_{i=0}^{\lfloor j/2 \rfloor} \frac{(-1)^i (2x)^{j-2i}}{i!(j-2i)!}. \tag{1.163}$$

Thus, for example,

$$H_0(x) = 1, H_1(x) = 2x, H_2(x) = 4x^2 - 2, H_3(x) = 8x^3 - 12x.$$

Define the **Hermite functions** φ_j by

$$\varphi_j(x) = \left(2^j j! \sqrt{\pi}\right)^{-1/2} \exp\left(-x^2/2\right) H_j(x). \tag{1.164}$$

Physicists and Mathematica calls the φ_j functions the **quantum mechanical oscillator wave functions**. Figure 1.4 shows graphs of $\varphi_j(x)$ for a small value of j.

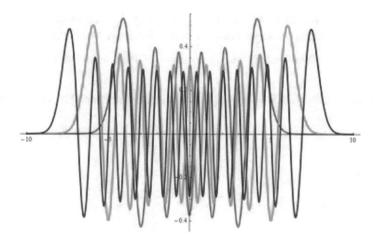

Fig. 1.4 Graphs of Hermite functions φ_j, for $j = 10, 20, 30$

One has the **orthogonality relations**:

$$\int_{-\infty}^{\infty} \varphi_i \varphi_j = \delta_{ij} = 1, \ \text{if} \ i = j \ \text{and} \ = 0 \ \text{otherwise.}$$

It follows from properties of determinants that the Vandermonde determinant $\Delta(x)$ can be expressed as a determinant of Hermite functions; in particular, it satisfies

$$\exp\left(-\frac{1}{2}\sum_{i=1}^{N} x_i^2\right) \Delta(x) = \text{constant det}(M), \ \text{if} \ M_{ij} = \varphi_{i-1}\left(x_j\right), 1 \le i, j \le N.$$

$$(1.165)$$

One sees that

$$P_{N2}(x_1, \ldots, x_N) = \frac{1}{N!} \det\left({}^t MM\right) = \frac{1}{N!} \det\left(\left(K_N\left(x_i, x_j\right)\right)_{1 \le i, j \le N}\right), \qquad (1.166)$$

where $K_N(x, y)$ denotes the symmetric and separable kernel made up of Hermite functions φ_j :

$$K_N(x, y) = \sum_{k=0}^{N-1} \varphi_k(x)\varphi_k(y). \qquad (1.167)$$

To see that the normalization constant in formula (1.166) is correct, one can argue as in Kemp [347] using properties of the kernel $K_N(x, y)$. Recall that in the theory of Fredholm integral operators, separable kernels really are finite matrix operators.

Moreover it is not hard to see that the kernels $K_N(x, y)$ are self-reproducing kernels, using the orthogonality relations of the Hermite functions. See the exercise below.

Exercise 1.3.10 (The Density of Eigenvalues for Matrices with GUE Distribution).

(a) Prove formula (1.165) using properties of determinants and Hermite polynomials.

(b) Show that for $K(x, y) = K_N(x, y)$ as in formula (1.167), the Fredholm integral operator

$$L_K u(x) = \int_{-\infty}^{\infty} K(x, y) u(y) dy$$

satisfies $(L_K)^j = (L_K)^{j-1}$, for $j = 2, 3, \ldots..$

(c) Then show that the normalization in formula (1.166) is correct using the orthogonality relations of the Hermite functions.

We want to integrate over $N - n$ variables $x_{n+1}, \ldots, x_N$. Thanks to the orthogonality properties of the Hermite functions this leads to

$$\int_{\mathbb{R}^{N-n}} \cdots \int P_{N2}(x_1, \ldots, x_N) dx_{n+1} \cdots dx_N = R_n = \det\left((K_N(x_i, x_j))_{1 \le i, j \le n}\right). \quad (1.168)$$

Using various values of n one obtains various densities of interest. When $n = 1$ we get the eigenvalue (or level) density which can be seen (as in Mehta [441, Appendix A.9]) to approach the semi-circle distribution as $N \to \infty$. When $n = 2$, one has the 2-point correlation which can be seen to approach $\left(\frac{\sin(\pi r)}{\pi r}\right)^2$ as $N \to \infty$ (see [441, Appendix A.10]). When $n = 0$ one has the eigenvalue (or level) spacing density. We are most interested in this last entity, thanks to Odlyzko's experiments with the spacings of high zeros of the Riemann zeta function.

But first let's look at the case $n = 1$ and its approach to the Wigner semi-circle distribution. We have

$$R_1 = R_1(x) = K_N(x, x) = \sum_{k=0}^{N-1} \varphi_k(x)^2 = N\varphi_N(x)^2 - \sqrt{N(N+1)}\varphi_{N-1}(x)\varphi_{N+1}(x),$$

$$(1.169)$$

using facts about the Hermite functions φ_j (see Erdélyi et al. [170]). Then one needs to understand the asymptotics as $N \to \infty$. We content ourselves with letting Mathematica show us what is happening. Figure 1.5 shows the results of computing $\left(\frac{\pi}{\sqrt{2N}}\right) R_1\left(\sqrt{2N}x\right)$, for $N = 5, 25, 45$. If we let N get much bigger than this, however, Mathematica seems to go crazy making giant oscillations that cannot be

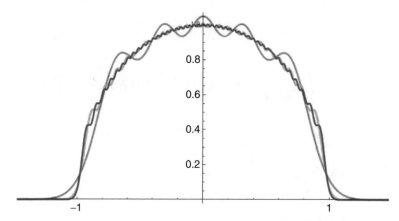

Fig. 1.5 Graphs of $\left(\frac{\pi}{\sqrt{2N}}\right) R_1\left(\sqrt{2N}x\right)$, for $N = 5, 25, 45$ in *red, green,* and *blue*

right. In any case, the figure shows that the graphs are looking more like the Wigner semi-circle. But the oscillations are indeed increasing.

Now let us try to understand the level spacing density, which is the case $n = 0$ in formula (1.168). We need an identity of Gram from linear algebra. We take our proof from Conrey [121]. The identity says for any interval S and integrable functions φ_j and ψ_j on S we have **Gram's formula**:

$$\frac{1}{N!} \int_{S^N} \det\left(\varphi_j\left(\theta_k\right)\right) \det\left(\psi_j\left(\theta_k\right)\right) d\theta_1 \cdots d\theta_N = \det\left(\int_S \varphi_j(\theta)\psi_k(\theta) \, d\theta\right).$$

$$(1.170)$$

Edelman and Rao [151] note that this is a continuous version of the Cauchy–Binet formula. Conrey [121] proves formula (1.170) by brute force from the definition of the $N \times N$ determinant as a sum over the symmetric group S_N. This says that the left-hand side of the formula is

$$\int_{S^N} \sum_{\sigma \in S_N} sgn(\sigma) \prod_{j=1}^{N} \varphi_j\left(\theta_{\sigma j}\right) \sum_{\tau \in S_N} sgn(\tau) \prod_{k=1}^{N} \psi_k\left(\theta_{\tau k}\right) d\theta_1 \cdots d\theta_N$$

$$\underset{\tau \to \sigma\tau}{=} \int_{S^N} \sum_{\sigma \in S_N} \sum_{\tau \in S_N} sgn(\tau) \prod_{j=1}^{N} \varphi_j\left(\theta_{\sigma j}\right) \prod_{k=1}^{N} \psi_k\left(\theta_{\sigma\tau k}\right) d\theta_1 \cdots d\theta_N$$

$$\underset{k \to \tau^{-1}k}{=} \int_{S^N} \sum_{\sigma \in S_N} \sum_{\tau \in S_N} sgn(\tau) \prod_{j=1}^{N} \prod_{k=1}^{N} \varphi_j\left(\theta_{\sigma j}\right) \psi_{\tau^{-1}k}\left(\theta_{\sigma k}\right) d\theta_1 \cdots d\theta_N$$

$$= \int_{S^N} \sum_{\sigma \in S_N} \sum_{\tau \in S_N} sgn(\tau) \prod_{j=1}^{N} \varphi_j \left(\theta_{\sigma j}\right) \psi_{\tau^{-1} j} \left(\theta_{\sigma j}\right) d\theta_1 \cdots d\theta_N$$

$$\underset{\tau \to \tau^{-1}}{=} \int_{S^N} \sum_{\sigma \in S_N} \sum_{\tau \in S_N} sgn(\tau) \prod_{j=1}^{N} \varphi_j \left(\theta_{\sigma j}\right) \psi_{\tau j} \left(\theta_{\sigma j}\right) d\theta_1 \cdots d\theta_N$$

$$= \sum_{\sigma \in S_N} \sum_{\tau \in S_N} sgn(\tau) \prod_{j=1}^{N} \int_S \varphi_j (\theta) \psi_{\tau j} (\theta) d\theta$$

$$= N! \sum_{\tau \in S_N} sgn(\tau) \prod_{j=1}^{N} \int_S \varphi_j (\theta) \psi_{\tau j} (\theta) d\theta,$$

which is the right-hand side of the formula.

Define $A_2(\theta)$ =the probability that the interval $(-\theta, \theta)$ contains no level $x_1, \ldots, x_N$. Then by Gram's formula (1.170):

$$A_2(\theta) = \int_{|x_1| \geq \theta} \cdots \int_{|x_N| \geq \theta} P_{N2}(x_1, \ldots x_N) dx_1 \cdots dx_N$$

$$= \frac{1}{N!} \int_{|x_1| \geq \theta} \cdots \int_{|x_N| \geq \theta} \left(\det \left((\varphi_{i-1}(x_j))_{1 \leq i,j \leq n}\right)\right)^2 dx_1 \cdots dx_N = \det G,$$

where
$$G_{ij} = \int_{|x| \geq \theta} \varphi_{i-1}(x) \varphi_{j-1}(x) dx = \delta_{ij} - \int_{-\theta}^{\theta} \varphi_{i-1}(x) \varphi_{j-1}(x) dx.$$

Since $K_N(x, y)$ is a symmetric separable kernel, if $\lambda_0, \ldots, \lambda_{N-1}$ are the eigenvalues of the corresponding Fredholm integral operator on $[-\theta, \theta]$ one has

$$A_2(\theta) = \prod_{i=0}^{N-1} (1 - \lambda_i).$$

To take the limit as $N \to \infty$, we need to renormalize K_N and write

$$Q_N(\xi, \eta) = \frac{\pi t}{\sqrt{2N}} K_N(x, y), \quad \text{with} \quad \pi t = \theta \sqrt{2N}, \; \pi t \xi = x\sqrt{2N}, \; \pi t \eta = y\sqrt{2N}.$$

Mehta shows (see [441, Appendix A.10]) that (with $\mathrm{sinc}(x) = \sin(x)/x$)

$$\lim_{N \to \infty} Q_N(\xi, \eta) = \frac{\sin (\pi t (\xi - \eta))}{\pi (\xi - \eta)} = t \, \mathrm{sinc} (\pi t (\xi - \eta)).$$

One has (see Erdélyi et al. [170]) for $x \neq y$,

$$K_N(x, y) = \sum_{k=0}^{N-1} \varphi_k(x)\varphi_k(y) = \sqrt{\frac{N}{2}} \left(\frac{\varphi_N(x)\varphi_{N-1}(y) - \varphi_N(y)\varphi_{N-1}(x)}{x - y} \right).$$

$$(1.171)$$

Using (1.171) we can compare 3D plots of the surface

$$z = K_N(x, y) \quad \text{and} \quad z = \text{sinc}(x - y)$$

using Mathematica and small values of N. See Figures 1.6 and 1.7. There does appear to be a similarity even for $N = 5$.

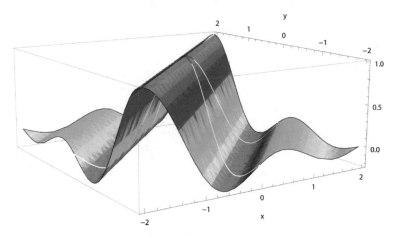

Fig. 1.6 A 3D plot of $z = \text{sinc}(x - y)$ in Mathematica

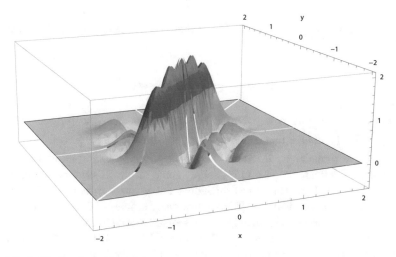

Fig. 1.7 A 3D plot of $z = K_5(x, y)$

The sinc kernel is a familiar Fredholm kernel from the theory of Fourier transforms and uncertainty in Volume I, Chapter 1. The eigenfunctions are prolate spheroidal wave functions. They have been well studied. This permits the computation of the limiting density.

Mehta [441] replaces the eigenvalue problems with some involving sin and cos:

$$\mu_{2j} f_{2j}(x) = 2 \int_0^1 \cos(\pi xyt) f_{2j}(y) dy$$

$$\mu_{2j+1} f_{2j+1}(x) = 2i \int_0^1 \sin(\pi xyt) f_{2j+1}(y) dy.$$

Write for $s = 2t$

$$E_2(0, s) = \lim_{N \to \infty} A_2 \left(\frac{\theta \sqrt{2N}}{\pi} \right) = \prod_{i=0}^{\infty} \left(1 - \frac{t}{2} |\mu_i|^2 \right).$$

Using these facts, Mehta (see [441, Appendix A.13]) finds that

$$E_2(0, s) = 1 - s + \frac{\pi^2}{36} s^4 - \frac{\pi^4}{675} s^6 + O\left(s^8\right).$$

Then the density

$$p_2(0, s) = \frac{d^2}{ds^2} E_2(0, s) = \frac{\pi^2}{3} s^2 - \frac{2\pi^4}{45} s^4 + O\left(s^6\right).$$

The corresponding result for the GOE random matrices (again in [441, Appendix A.13]) is

$$p_1(0, s) = \frac{\pi^2}{6} s - \frac{\pi^4}{60} s^3 + O\left(s^4\right).$$

If you compare this with the Taylor expansion for the Wigner surmise $\frac{\pi}{2} s \exp\left(\frac{-\pi x^2}{4}\right)$, you see that the lead terms bear some resemblance.

A few last remarks are in order. The reader might be interested in a few more references. Edelman and Rao [151] survey some results in random matrix theory for the many kinds of ensembles, including the Wishart ensemble. *The Oxford Handbook of Random Matrices* [3] includes many articles with applications of the subject; e.g., to finance. Freeman Dyson has an interesting introduction in which he notes that the times between buses in Cuernavaca, Mexico have been found to agree with GUE spacings according to some physicists from the Czech Republic.

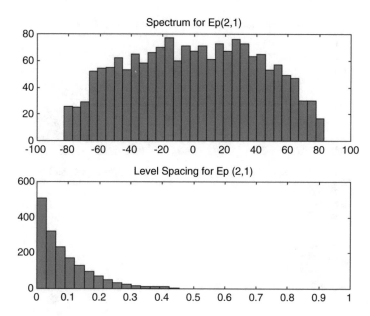

Fig. 1.8 The *top histogram* is the spectrum (without multiplicity) of the adjacency matrix of the finite Euclidean plane graph $X(\mathbb{F}_p^2, S)$, where $p = 1723$ and S consists of solutions to the congruence $x_1^2 + x_2^2 \equiv 1 \pmod{p}$, as defined in Volume I, p. 91. The *bottom* is the unnormalized eigenvalue spacing histogram for the same graph

We and many others have performed experiments on spacings of eigenvalues of adjacency matrices of graphs as well as zeros of Ihara zeta functions of graphs. Some of the results are considered in [610] and [611]. For example consider Figure 1.8 in which the top histogram is the spectrum (without multiplicity) of the adjacency matrix of the finite Euclidean plane graph $X(\mathbb{F}_p^2, S)$, where $p = 1723$ and S consists of solutions to the congruence $x_1^2 + x_2^2 \equiv 1 \pmod{p}$, as defined in Volume I, p. 91. The bottom is the unnormalized eigenvalue spacing histogram for the same graph. Figure 1.9 shows histograms related to spectra (without multiplicity) of the finite upper half plane graph $X_{353}(3,3)$ defined in Volume I, p. 223. In our program to compute Soto-Andrade sums we needed to know that a generator of the multiplicative group of $\mathbb{F}_{353}(\sqrt{3})$ is $1 + 5\sqrt{3}$. The top histogram is for the spectrum and the lower one is for the unnormalized eigenvalue spacings. The lower histograms in Figures 1.8 and 1.9 do indeed appear to be Poisson (i.e., e^{-x}). This is in accordance with the arithmetic used in constructing the graphs. If instead one creates the analogous histograms for random regular graphs, one finds that the eigenvalue spacing histogram looks more like the GOE density. This is seen in Figure 1.10 from Derek Newland's Ph.D. thesis [473]. In Figure 1.10, the upper part compares the histogram of the eigenvalues of the adjacency matrix for a random regular graph of degree 52 having 2000 vertices with the semi-circle density. The lower part of Figure 1.10 compares the histogram of normalized eigenvalue spacings for the same graph with the Wigner surmise for the GOE density.

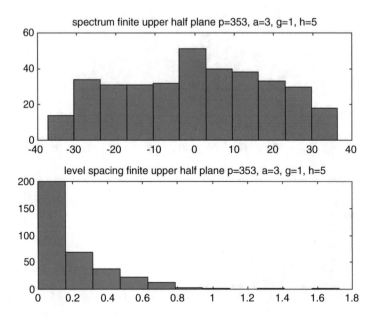

Fig. 1.9 Histograms related to spectra (without multiplicity) of the finite upper half plane graph $X_{353}(3,3)$ defined in Volume I, p. 223. The *top histogram* is for the spectrum and the *lower one* is for the unnormalized eigenvalue spacings

One can also study the statistics of eigenvalues of non-Hermitian matrices. For a random $n \times n$ matrix, one expects to see that the eigenvalues are dense in a circle of radius $\sqrt{n}$. This is the Girko circle law. We discuss the subject with some experiments from graph theory in [611, Chapter 26]. See also Tao and Vu [592].

1.3.6 Other Directions in the Labyrinth

Our discussion of inversion for the Helgason–Fourier transform on $\mathcal{P}_n$ mainly followed the path of Helgason [275]. At this point, the reader might like to travel some other paths. Varadarajan's introduction to the collected works of Harish-Chandra [263] gives a good historical introduction to the representation-theoretic road to harmonic analysis, as it was traveled by Harish-Chandra and others. Jorgenson and Lang [333] give a more recent treatment of the subject. See Rebecca Herb and Paul Sally [287] for a survey of Plancherel formulas on real and p-adic groups.

One part of the route involves **orbital integrals**:

$$\int_{G/T} f(xtx^{-1}) \, dx, \text{ for } t \text{ in a maximal abelian subgroup } T \text{ of } G.$$

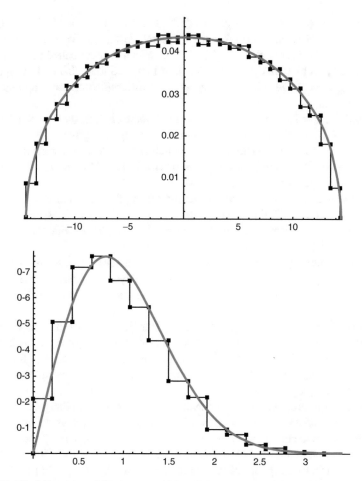

Fig. 1.10 The histograms of the spectrum of the adjacency matrix (*top*) and eigenvalue spacings (*bottom*) for a random regular graph with degree 53 and 2000 vertices created by Mathematica from the thesis of Derek Newland [473]

Weyl [666] already made great use of these integrals in his development of the theory of representations of compact Lie groups. See also Broecker and tom Dieck [81] or Helgason [282]. Gelfand and Graev [213] use such an approach for complex groups such as $SL(n, \mathbb{C})$. In particular they utilize formulas for the residues of certain integrals of M. Riesz type defined by:

$$R(s) = \int_{\substack{x \in \mathbb{R}^m \\ Q[x] \geq 0}} f(x)\, Q[x]^s\, dx, \text{ for } s \in \mathbb{C},$$

where Q is a symmetric matrix in $\mathbb{R}^{m \times m}$ and $f : \mathbb{R}^m \to \mathbb{C}$ is sufficiently differentiable. The residue formulas involve certain differential operators. This leads to a version of Fourier analysis of $f : G \to \mathbb{C}$ which is often called the Plancherel formula, for groups G like $SL(n, \mathbb{C})$, $U(n)$, at first, and then also real groups. One writes $f(e)$, $e = $ the identity of the group, as a differential operator applied to an orbital integral.

Michèle Vergne [629] provides a view of harmonic analysis on $G = SL(2, \mathbb{R})$ and other groups, which is close to that of Kirillov for nilpotent groups. Again the main direction is given by the **orbit method** which is used to classify representations according to orbits of the Adjoint action of G on the dual of its Lie algebra. For matrix groups, the Adjoint is conjugation.

Ehrenpreis and Mautner [155] give an interesting discussion of Fourier analysis on $SL(2, \mathbb{R})$ and $SL(2, \mathbb{R})/\Gamma$ from the point of view of "classical" analysis (after Laurent Schwartz), including a Riemann–Lebesgue lemma and a readable discussion of the Schwartz space.

Flensted-Jensen [181] gives some relations between analysis on symmetric spaces like $GL(n, \mathbb{R})/O(n)$ and $GL(n, \mathbb{C})/U(n)$. Analysis is much easier on the latter space. Healy [266] studies relations between harmonic analysis on $GL(2, \mathbb{C})/U(2)$ and that on $SU(2)$.

Rebecca Herb and Joe Wolf [288] note that Harish-Chandra's Plancherel formula was not proved for all real connected semisimple Lie groups; e.g., the universal covers of groups like $SL(2, \mathbb{R})$. They obtain the Plancherel formula for all real semisimple groups using different methods from Harish-Chandra. Once more, orbital integrals play a key role.

Orbital integrals are also of fundamental importance in the Selberg trace formula. See Section 1.5.5 as well as Section 3.7 of Volume I. Another reference is the conference volume edited by Hejhal et al. [272] which contains many papers on that subject; e.g., that of James Arthur, Rebecca Herb, and Paul Sally. Orbital integrals are also intrinsic to the theory of the Radon transform. See Helgason [280].

When $G = SL(2, \mathbb{R})$, for example, the Plancherel formula involves a series as well as an integral (see Lang [388]). Why doesn't this happen for $G/K \cong H$? Or for $\mathcal{P}_n$? Equivalently, one wonders why there are no square-integrable eigenfunctions of the G-invariant differential operators on $\mathcal{P}_n$? One answer to this question comes from thinking about discrete subgroups Γ of $GL(n, \mathbb{R})$ or $SL(n, \mathbb{R})$. If there is a nonzero function f in $L^2(\mathcal{P}_n)$, such that $Lf = \lambda f$ for all L in $D(\mathcal{P}_n)$, it follows that $f \in L^2(\mathcal{P}_n/\Gamma)$ for *all* discrete subgroups Γ of $GL(n, \mathbb{R})$. This is absurd.

Furstenberg [194] defines a **boundary** M for a Lie group G to be a compact space such that there is a continuous G action $(g, x) \mapsto gx$ taking $G \times M$ into M such that the **group action** has the following three properties:

(1) **associative** : $(g_1 g_2)x = g_1(g_2 x)$;

(2) **transitive:**   for each $x, y \in M$, there is a $g \in G$ so that $gx = y$;

(3) for each probability measure π on M, $\exists\, g_n \in G$
    such that $g_n \pi$ converges to a point measure on M.

A **maximal boundary** $B(G)$ has the property that, for any boundary M of G, there is a map from $B(G)$ to M preserving the G actions. Furstenberg [194] proves that a maximal boundary for $G = SL(n, \mathbb{R})$ is the boundary G/MAN appearing in Theorem 1.3.1. Here MAN consists of the upper triangular matrices in G and we can identify G/MAN with K/M by the Iwasawa decomposition of G. Furstenberg's result is actually more general and he goes on to show that Poisson's integral formula for bounded harmonic functions can be generalized using the maximal boundary. We will discuss this further in Chapter 2.

Exercise 1.3.11 (Boundaries of $G = SL(n,\mathbb{R})$).

(a) Show that if $G = SL(n, \mathbb{R})$ and MAN is the group of upper triangular matrices of determinant one, then the maximal boundary G/MAN can be identified with the **flag manifold** F_n of $(n-1)$-tuples $(V_1, V_2, \ldots, V_{n-1})$ where V_i denotes an i-dimensional vector subspace of $\mathbb{R}^n$ and $V_1 \subset V_2 \subset \cdots \subset V_{n-1}$. The action of G on F_n is the obvious one defined via $gV_i = \{gx \mid x \in V_i\}$.

(b) Define $G_{i,n-1}$ to be the **Grassmann variety** of i-dimensional subspaces of $\mathbb{R}^n$. Show that the mapping $(V_1, \ldots, V_{n-1}) \mapsto V_i$ sends F_n onto $G_{i,n-1}$ and preserves the G-actions. Thus $G_{i,n-1}$ is also a boundary of G. In particular, the projective space $G_{1,n-1} = \mathbb{P}^{n-1}$ is a boundary of G.

(c) Let $P(i, n-i)$ denote the parabolic subgroup of G consisting of matrices with block form

$$\begin{pmatrix} A & B \\ 0 & C \end{pmatrix}, \ A \in GL(i, \mathbb{R}), \ C \in GL(n-i, \mathbb{R}).$$

Show that we can identify the Grassmann variety $G_{i,n-1}$ of part (b) with $G/P(i, n-i)$.

As we noted earlier, holomorphic extensions of eigenfunctions of the invariant differential operators on a symmetric space to the crown domain of the symmetric space have been studied and applied to various topics such as the heat operator, estimating Maass cusp forms and Helgason's conjecture on eigenfunctions of the invariant differential operators being reconstructible from their hyperfunction boundary values. See the papers of Gindikin [219], Krötz and Opdam [372], and Ólafsson and Schlichtkrull [479].

1.4 Fundamental Domains for $\mathcal{P}_n/GL(n,\mathbb{Z})$

Seit meiner ersten Studienzeit war mir Minkowski der beste und zuverlässigste Freunde, der an mir hing mit der ganzen ihm eigenen Tiefe und Treue. Unsere Wissenschaft, die uns das liebste war, hatte uns zusammengeführt; sie erschien uns wie ein blühender Garten; in diesem Garten gibt es geebnete Wege, auf denen man mühelos geniesst, indem man sich umschaut, zumal an der Seite eines

Gleichempfindenden. Gern suchten wir aber auch verborgene Pfade auf und entdeckten manche neue, uns schön dünkende Aussischt, und wenn der eine dem andern sie zeigte und wir sie gemeinsam bewunderten, war unsere Freude vollkommen.[3]

From Hilbert's speech in memory of Minkowski (see Minkowski [453, Vol. I, XXX]).

1.4.1   Introduction

In this section we study the action of the modular group $GL(n, \mathbb{Z})$ consisting of $n \times n$ matrices with integer entries and determinant ± 1 on the space $\mathcal{P}_n$ of positive matrices. A **fundamental domain** D for $\mathcal{P}_n/GL(n, \mathbb{Z})$ is a subset of $\mathcal{P}_n$ which behaves like the quotient space $\mathcal{P}_n/GL(n, \mathbb{Z})$, at least up to boundary identifications. The fundamental domains for $\mathcal{P}_n/GL(n, \mathbb{Z})$ are much more difficult to visualize than those for $SL(2, \mathbb{Z}) \backslash H$ which were considered in Section 3.3 of Volume I, since $\mathcal{P}_n$ is a subset of $(n(n + 1)/2)$-dimensional Euclidean space. Thus the smallest dimension for a picture of such a fundamental domain (for $n \geq 3$) would be six. If we consider only the determinant one surface $S\mathcal{P}_n/GL(n, \mathbb{Z})$, this reduces the dimension by one, making our picture five-dimensional. We will include some pictures of projections of points in a fundamental domain for $S\mathcal{P}_3/GL(3, \mathbb{Z})$ in Section 1.4.3 as well as a "movie" of the region obtained by projecting onto the three x-variables as the two y-variables dance around near $(1, 1)$. See Figure 1.26.

Much of this section is due to Minkowski, who was the first to describe a fundamental domain for $GL(n, \mathbb{Z})$ (see Section 1.4.2). We will discuss another fundamental domain—that of Grenier [241] (also Hermite, Korkine and Zolotareff) in Section 1.4.3. The latter domain has the advantage of looking more like the one for $SL(2, \mathbb{Z})$ which we used in Section 3.3 of Volume I. There are indeed many unusual flowers in these higher dimensional gardens. The names of those who have cultivated these flowers include: Gauss, Hermite, Minkowski, Voronoi, Siegel, Weyl, Weil, Satake, Baily, Borel, Serre, Harish-Chandra, Mostow, Tamagawa, Mumford, Delone, Korkine and Zolotareff, Ryskov, ...

The reader may be wondering why one would want to wander about in these higher dimensional gardens. As we mentioned in Section 1.1.1, our main motivation is the desire to study some **relatives of Riemann's zeta function**. We will see that we can generalize Riemann's method of analytic continuation of the Riemann zeta function—a method used in Theorem 1.4.1 in Volume I. This method involves

---

[3] Since my first days as a student, Minkowski, with his typical depth and faith, was my best and most reliable friend. Our beloved science had brought us together; it seemed to us like a blooming garden; in this garden there were smooth (well-tended) paths that one enjoyed effortlessly while looking around, especially at the side of someone with the same feelings. But we also liked to seek out the hidden paths and discovered several new views which were beautiful in our opinion and when one of us showed them to the other and we both admired them, our joy was complete.

taking a Mellin transform of a theta function. If $X \in \mathcal{P}_n$, and $Y \in \mathcal{P}_m$, for $1 \leq m \leq n$, we define the **theta function** by:

$$\theta(Y,X) = \sum_{A \in \mathbb{Z}^{n \times m}} \exp\{-\pi \, \text{Tr}(X[A]Y)\}. \tag{1.172}$$

This theta function is related to a zeta function generalizing Epstein's zeta function from Section 1.4 of Volume I as well as the zeta function introduced in formula (1.4) of Section 1.1.1. The zeta function in question is called **Koecher's zeta function** because it was first studied by Koecher [359] and it is defined for $1 \leq m \leq n$ by:

$$Z_{m,n-m}(X,s) = \sum_{\substack{A \in \mathbb{Z}^{n \times m}/GL(m,\mathbb{Z}) \\ \text{rank } A = m}} |X[A]|^{-s}, \quad \text{if} \quad \text{Re } s > \frac{n}{2}. \tag{1.173}$$

Here the sum is over $n \times m$ integral matrices A of rank m running through a complete set of representatives for the equivalence relation

$$A \sim B \text{ iff } A = BU \text{ for some } U \in GL(m,\mathbb{Z}).$$

Note that if $m = 1$, the theta function (1.172) is just that considered in Exercise 1.4.6 in Section 1.4.2 of Volume I, and in this case, Koecher's zeta function reduces to the Epstein zeta function defined in Section 1.4.2 of Volume I. When $n = m$, Koecher's zeta function is the function in formula (1.4) of Section 1.1.1 of this Volume and we will prove that—in this case—it is a product of Riemann zeta functions (see Lemma 1.4.7 below)

$$Z_{n,0}(X,s) = |X|^{-s} \prod_{j=0}^{n-1} \zeta(2s-j). \tag{1.174}$$

In fact $Z_{n,0}(I,s)$ is the analogue of the Dedekind zeta function (considered in Section 1.4 of Volume I) for the simple algebra of all $n \times n$ rational matrices.

In order to imitate the proof of the analytic continuation of the Epstein zeta function given in Section 1.4 of Volume I, we need to Mellin transform the theta function (1.172). The Mellin transform used here is not a transform over all Y in $\mathcal{P}_m$, but instead over $\mathcal{P}_m/GL(m,\mathbb{Z})$. This is necessary because $\theta(Y[U],X) = \theta(Y,X)$ for all $U \in GL(m,\mathbb{Z})$. Explicitly, the **Mellin transform** is:

$$\int_{\mathcal{P}_m/GL(m,\mathbb{Z})} |Y|^s \, \theta_m(Y,X) \, d\mu_m(Y) = 2\pi^{-ms} \, \Gamma_m(0,\ldots,0,s) \, Z_{m,n-m}(X,s).$$

$$\tag{1.175}$$

Here Γ_m denotes the gamma function defined by (1.44) in Section 1.2.1 and θ_m denotes the partial sum of (1.172) over all $A \in \mathbb{Z}^{n \times m}$ such that the rank of A is m. Here we always assume that $1 \leq m \leq n$.

This kind of example motivates the search for analogues of **Hecke's correspondence** (see Section 3.6 of Volume I) which would relate Siegel modular forms such as the theta function in (1.172) with Dirichlet series of several variables. One needs more variables than the one complex variable s appearing in (1.175) in order to invert the matrix analogue of the Mellin transform on the fundamental domain for $GL(2, \mathbb{Z})$. This inversion was used by Kaori Imai (Ota) [317] in the case of cuspidal Siegel modular forms for $Sp(2, \mathbb{Z})$ to generalize Hecke's correspondence. Her results say that there is a dictionary which translates between the languages:

$$\text{Siegel modular forms for } Sp(2, \mathbb{Z}) \Leftrightarrow \begin{array}{l} \text{Dirichlet series ``twisted'' by Maass wave} \\ \text{forms for } GL(2, \mathbb{Z}) \text{ with functional equations.} \end{array}$$

The $\Longrightarrow$ can be found in Maass [426, Section 16] for Siegel modular forms for $Sp(n, \mathbb{Z})$ with n arbitrary, in fact. The converse correspondence $\Longleftarrow$ is proved for cusp forms by Kaori Imai (Ota) [317] using the Roelcke–Selberg–Mellin inversion formula on $\mathcal{P}_2/GL(2, \mathbb{Z})$. See also Chapter 2, as well as Maass [419] and Roelcke [512]. Weissauer [663] extended the converse result to congruence subgroups of $Sp(n, \mathbb{Z})$ for all n.

The main goal of this chapter is to present some of the ideas necessary for harmonic analysis on $\mathcal{P}_n/GL(n, \mathbb{Z})$, from the same point of view that worked in the preceding section for $\mathcal{P}_n$ itself. The theory is still not in its final form, however. But it is this goal that motivates our detailed study of the fundamental domain.

Jacquet et al. [325] have shown that the adelic version of the Hecke converse theorem for $GL(n)$ does not require "twists" by Maass forms for $GL(m)$, $m \leq n-1$, but only those for $GL(m)$, $m \leq n-2$. Such a converse theorem can be used to show, for example, that zeta and L-functions for totally real cubic number fields correspond to cusp forms for the adelized version of $GL(3)$, i.e., cusp forms for congruence subgroups of $GL(3)$. Thus when one sees L-functions with the right gamma factors in their functional equations, one expects to find corresponding cusp forms for $GL(n)$. But, in general, one must also have functional equations for L-functions "twisted" by Maass forms for $GL(m)$, $m \leq n-2$. Making use of the Rankin–Selberg convolution, which leads to L-functions with an Euler product that indicates the presence of a Maass form for $GL(3)$, Gelbart and Jacquet [210] obtain a lifting of Maass forms from $GL(2)$ to $GL(3)$. See Goldfeld [230].

The aforementioned results are part of a vast program of Langlands and many coworkers which gives a theory of L-functions attached to adelic irreducible automorphic representations of reductive groups over global fields such as $\mathbb{Q}$ and $\mathbb{F}_p(x)$. This theory is surveyed by Borel in Borel and Casselman [66, Vol. II, pp. 27–61] and by Gelbart [209]. Newer references are the 2002 Park City Conference Proceedings edited by Sarnak and Shahidi [529], Bernstein and Gelbart [47], Booker [60], Frenkel [187], Goldfeld [230], and Goldfeld and Hundley [232]. Langlands has attached L-functions to an automorphic representation of the adelic $GL(n)$ by defining an Euler product over primes p. Langlands made a conjecture about his L-functions which generalizes the Artin reciprocity law in the theory of abelian extensions of number fields. This conjecture of Langlands would imply the

Artin conjecture that the Artin L-functions are entire (excluding cases which are obviously not entire; e.g., when the character is trivial), since the Langlands L-function is entire for any nontrivial cuspidal representation of $GL(n)$. Attempts to prove the Artin conjecture this way have indeed made progress in the case of degree 2 representations of the Galois group of the extension (see Bernstein and Gelbart (Eds.) [47], Goldfeld [230], Langlands [394], Sarnak and Shahidi (Eds.) [529], Tunnell [620, 621]). This progress involves the "twisted" Selberg trace formula and "base change." There are several L-functions websites that provide a wealth of information.

At first sight, the Langlands L-function defined by an Euler product sounds rather different from an L-function defined by a Dirichlet series or a Mellin transform over a fundamental domain for $GL(n,\mathbb{Z})$. However, as we saw in Sections 3.6.4 and 3.6.5 of Volume I, Hecke L-functions can be defined in either way, if the corresponding automorphic form is an eigenform for all the Hecke operators. We will find that an analogous result holds for $GL(n,\mathbb{Z})$ in Section 1.5 which follows. Thus we will study L-functions using Mellin transforms over $\mathcal{P}_n/GL(n,\mathbb{Z})$; and these L-functions will indeed have Euler products when the corresponding Maass form is an eigenfunction of all the Hecke operators for $GL(n,\mathbb{Z})$. Of course, these Mellin transforms can also be used to study the Eisenstein series generalizing Koecher's zeta function (1.173). Such Eisenstein series need not have Euler products, except in certain special cases, such as that of (1.174), where the Euler product comes from that for the Riemann zeta function. Bump [83] and Goldfeld [230] provide more connections between the adelic point of view and the Dirichlet series point of view. See; in particular, Section 1.5.4 for more information on L-functions for $GL(n,\mathbb{Z})$.

If you are not interested in these L-functions for $GL(n,\mathbb{Z})$, there are still lots of reasons to study fundamental domains for $\mathcal{P}_n/GL(n,\mathbb{Z})$. We listed some of these at the beginning of Section 1.1.1. Let's go into more detail here.

The embedding used by Hecke to relate zeta functions of algebraic number fields with Epstein zeta functions (see Theorem 1.4.2 of Volume I) leads one to suspect that explicit fundamental domains for $\mathcal{P}_n/GL(n,\mathbb{Z})$ should lead to **explicit algorithms for the computation of class numbers and units of number fields**. This was indeed the case for imaginary quadratic fields (see Section 3.3.3 of Volume I). The units in a number field are connected with a certain fundamental domain in a Euclidean space (see the proof of Theorem 1.4.2 in Section 1.4.3 of Volume I). The units and class number also influence the fundamental domains for $SL(2,\mathfrak{O}_K)$, $\mathfrak{O}_K = $ the ring of integers of a number field K—groups to be considered in the next chapter. In many ways, $\mathcal{P}_n/GL(n,\mathbb{Z})$ is the prototype for all fundamental domains.

Another related issue is that of the closed geodesics in $SL(2,\mathbb{Z})\backslash H$ corresponding to hyperbolic elements of $SL(2,\mathbb{Z})$. Such a geodesic corresponds to an element z in a real quadratic number field—z being fixed by the hyperbolic matrix γ. Here γ in $SL(2,\mathbb{Z})$ is called **hyperbolic** if the eigenvalues of γ are distinct, real, and different from 1 or -1. If

$$\sigma \begin{pmatrix} \varepsilon & 0 \\ 0 & 1/\varepsilon \end{pmatrix} \sigma^{-1} = \gamma, \text{ for } \sigma \in SL(2, \mathbb{R}),$$

then the geodesic fixed by γ is the image of the positive y-axis under σ. Moreover the eigenvalue ε of γ is a unit in a real quadratic field and the columns of σ are eigenvectors of γ. The periodic continued fraction expansions of these quadratic numbers z come from the translations and inversions needed to map the half circle connecting z and its conjugate into the fundamental domain for $SL(2, \mathbb{Z})$ (see Exercise 3.7.20 of Volume I). You might wonder how z and its conjugate z' over $\mathbb{Q}$ relate to ε and its conjugate ε^{-1}. It is not hard to see that if

$$\gamma = \begin{pmatrix} a & b \\ c & d \end{pmatrix}, \text{ and } \gamma z = (az + b)/(cz + d) = z,$$

then

$$\{\varepsilon, \varepsilon^{-1}\} = \{cz + d, cz' + d\}.$$

This happens because $\gamma z = z$, $\gamma z' = z'$, and

$$\gamma \begin{pmatrix} z \\ 1 \end{pmatrix} = \begin{pmatrix} az + b \\ cz + d \end{pmatrix},$$

$$\gamma \begin{pmatrix} z & z' \\ 1 & 1 \end{pmatrix} = \begin{pmatrix} z & z' \\ 1 & 1 \end{pmatrix} \begin{pmatrix} cz + d & 0 \\ 0 & cz' + d \end{pmatrix}.$$

One wonders whether $GL(n, \mathbb{Z})$-analogues of the preceding remarks would lead to periodic algorithms for the approximation of elements of a totally real number field of degree n. Here **totally real field** K means that all the conjugate fields of K over $\mathbb{Q}$ are real. There is a long history of the search for a generalization of the theorem that a real number is quadratic if and only if its continued fraction expansion is periodic. Minkowski [453, Vol. I, pp. 357–371]) gives an algorithm which is periodic in some cases. There are many other algorithms generalizing continued fractions, but none seems to be completely satisfactory.

There is a generalization to $GL(n, \mathbb{Z})$ of the relation between units in real quadratic fields and closed geodesics in $SL(2, \mathbb{Z}) \backslash H$ (see Dorothy Wallace [643] for related results). A **hyperbolic element** γ in $GL(n, \mathbb{Z})$ is one with distinct real eigenvalues none of which are equal to ± 1. Thus γ has eigenvalues which are units in a totally real number field of degree n. If for $\sigma \in GL(n, \mathbb{R})$, we have $\sigma \gamma \sigma^{-1}$ is diagonal with jth diagonal entry ε_j, then the following totally geodesic submanifold is fixed by γ:

$$\bigcup_{a \in \mathbb{R}^n} G_a[\sigma],$$

where

$$G_a = \left\{ \left. \begin{pmatrix} e^{a_1 t} & \cdots & 0 \\ \vdots & \ddots & \vdots \\ 0 & \cdots & e^{a_n t} \end{pmatrix} \right| t \in \mathbb{R} \right\}.$$

Other references for continued fraction type algorithms are Brentjes [79], and Ferguson and Forcade [179]. There are many applications of higher dimensional continued fraction algorithms in coding and elsewhere (see Lagarias and Odlyzko [383]). Related references are: Ash et al. [30], Barrucand et al. [37], Cusick and Schoenfeld [128], Delone and Faddeev [133], Hirzebruch [296], and Williams and Broere [669].

It is also of interest to number theorists that the Euclidean volume of the subset of matrices Y in Minkowski's fundamental domain for $\mathcal{P}_n/GL(n, \mathbb{Z})$ such that $|Y| \leq 1$ involves a product of **Riemann zeta functions at odd as well as even positive integer arguments** $n > 1$ (see Theorem 1.4.4). For recall from Exercise 3.5.7 of Volume I that Euler found a nice formula for values of zeta at positive even integers, but no one has managed a similar result for odd integers. Siegel used formulas (1.174) and (1.175) above to prove Minkowski's formula for this volume (see Siegel [565, Vol. I, pp. 459–468 and Vol. III, pp. 328–333]). Weil [662, Vol. I, p. 561] notes:

Siegel était arrivé à Princeton en 1940; pendant tout mon séjour aux États-Unis, je l'avais vu souvent. Depuis longtemps, avec juste raison, il attachait une grande importance au calcul du volume des domaines fondamentaux pour les sous-groupes arithmétiques des groupes simples; il avail consacré à ce sujet, inauguré autrefois par Minkowski, plusieurs mémoires importants. À ce propos il s'était vivement intéressé à la formule générale de Gauss-Bonnet, d'où pouvait résulter, du moins pour les sous-groupes à quotient compact, une détermination topologique des volumes en question. Je crois même me souvenir qu'il avait cru un jour tirer de là des conclusions au sujet de valeurs de $\zeta(n)$ pour n impair > 1, et s'était donné quelque mal pour les vérifier numériquement, avant de s'apercevoir qu'il s'agissait d'un cas où la courbure de Gauss-Bonnet est nulle.[4]

To bring up a different and quite old question from number theory, define the representation numbers $A_Y(m)$ for **the number of representations of an integer** m **as m $=$ Y[a] for a positive definite quadratic form** Y in $\mathcal{P}_n$ with integer coefficients and an integral vector $a \in \mathbb{Z}^n$. We discussed some of this at the end

---

[4]Siegel arrived at Princeton in 1940; during my entire stay in the United States, I saw him often. For a long time, rightly, he attached a great importance to the calculation of the volume of the fundamental domain for arithmetic subgroups of simple groups; he had devoted several important papers to this subject which had been begun long before by Minkowski. In this regard he was keenly interested in the general Gauss–Bonnet formula, from which could result a topological characterization of the volume in question, at least for subgroups with compact quotient. I even believe that I remember that he once thought that he had derived conclusions from that on the subject of the values of $\zeta(n)$ for n odd > 1, and had taken some trouble to verify this numerically, before realizing that it was a question of a case where the Gauss–Bonnet curvature is zero.

of Section 3.4 in Volume I. Gauss treated the cases $n = 2, 3$. The case that $Y = I_n$ is the $n \times n$ identity matrix has received special attention. For example, in 1829, Jacobi proved that

$$A_{I_4}(n) = 8 \sum_{0 < d \mid n} d, \quad \text{if } n \text{ is odd.}$$

One can view the left-hand side of the equality as the Fourier coefficient of a theta function of weight 2 and the right-hand side as the Fourier coefficient of an Eisenstein series of weight 2.

In 1883 when Minkowski was 17, he and Smith split a prize for proofs of Eisenstein's formula for the mass of a genus of quadratic forms (see Minkowski [453, Vol. I, pp. 157–202] or Hancock [260]). Siegel developed a vast extension of these results in the 1930s (see Siegel [565, Vol. I, pp. 326–405, 410–443, 469–548; Vol. II, pp. 1–7, 20–40] and Milnor and Husemoller [451]). The general result can be viewed as an identity between Siegel modular forms. See also Freitag [185, pp. 285–297]. There is a brief exposition of Siegel's work and related developments in Cassels [99, pp. 374–388].

These studies of quadratic forms require a knowledge of the fundamental domain for $\mathcal{P}_n / GL(n, \mathbb{Z})$ since the usual fundamental domain for the Siegel modular group cannot be understood without first understanding a fundamental domain for $GL(n, \mathbb{Z})$, as we shall see in Chapter 2 of this volume.

In the 1960s Tamagawa, Weil, Ono, and Kneser obtained an adelic version of Siegel's results on quadratic forms. Some references are the article of Kneser in Cassels and Fröhlich [101, pp. 250–265], the articles of Mars in Borel and Mostow [68, pp. 133–142], and Weil [662, Vol. III pp. 1–157].

Fundamental domains for groups like $GL(n, \mathbb{Z})$ are not just of interest to number theorists. They also provide food for thought to those interested in **geometry and topology**. Ash et al. [30] have obtained smooth compactifications of such fundamental domains. This would allow one to use the Riemann–Roch theorem and other methods from geometry to compute dimensions of spaces of modular forms. These smooth compactifications are obtained explicitly using ideas of Minkowski and Voronoi, as well as the theory of toroidal embeddings. References include: Baily and Borel [33], Borel and Ji [67], Borel and Serre [69], Chai [103], Goresky [236], Mostow and Tamagawa [467], Namikawa [472], Satake [531, 533], and Yamazaki [673]. References related to the computation of cohomology of arithmetic groups are: Ash [28], Ash et al. [29], Borel [65], Borel and Serre [69], Borel and Wallach [70], Schwermer [539, 540], Serre [548], and Soulé [571, 572].

There are many places in **physics** where automorphic forms for $GL(n, \mathbb{Z})$ and $Sp(n, \mathbb{Z})$ have popped up. Of course, it should not be surprising to find that abelian integrals and thus Riemann theta functions such as (1.172) above should bear solutions to partial differential equations as their fruit. For example, classical theta functions such as those discussed in Section 3.4 of Volume I appear in the solutions by Euler, Lagrange, and Poisson of two special cases of the problem of describing the motion of a solid body rotating about a fixed point. The third known case of this problem was solved by Sofya Kovalevskaya [aka Sonya Kovalevsky] [368] using Siegel modular forms (Riemann theta functions). She was awarded the Prix Bordin for this work in 1888. Evidently no less a mathematician than Picard told Kovalevskaya in 1886 that he was skeptical that theta functions for $Sp(n, \mathbb{Z})$ "can be useful in the integration of certain differential equations" (see Dubrovin et al. [143]). But 90 years later, in the paper of Dubrovin et al. [143] theta functions are used to solve the Korteweg–deVries partial differential equation arising in the theory of solitons. For related papers and some short articles on Kovalevskaya's life see the volume edited by Linda Keen [345]. The books of Cooke [124], Ann Hibner Koblitz [357], and Pelageya Kochina [358] give more detailed discussions.

The theta functions for $Sp(n, \mathbb{Z})$ are also intrinsic to Siegel's work on quadratic forms mentioned above. This work has recently been connected with quantum mechanics via the Segal-Shale-Weil representation. References are Gérard Lion and Michèle Vergne [406], Shale [551], Wallach [653], and Weil [662, Vol. 3, pp. 1–157]. See also the book by Mumford [471].

Finding **densest lattice packings of spheres** in $\mathbb{R}^n$ is a part of Hilbert's eighteenth problem (see Cassels [99], Davenport [129], Milnor [449], Rogers [514], Siegel [561, 562], Sloane [568, 569]), and Thompson [614]). A **lattice** L [5]in $\mathbb{R}^n$ is a subgroup of the additive group of $\mathbb{R}^n$ of the form:

$$L = \mathbb{Z}v_1 \oplus \mathbb{Z}v_2 \oplus \cdots \oplus \mathbb{Z}v_n, \tag{1.176}$$

where the vectors $v_1, v_2, \ldots, v_n$ form a vector space basis of $\mathbb{R}^n$. It can be shown that this is equivalent to saying that L is a discrete subgroup of $\mathbb{R}^n$ such that $\mathbb{R}^n/L$ is compact; i.e., a discrete cocompact subgroup of $\mathbb{R}^n$. For a proof of this last remark see Siegel [562, pp. 9–12]. The problem of finding the densest lattice packings of spheres in $\mathbb{R}^n$ is that of finding a lattice L such that if nonoverlapping open spheres of equal radii are centered at each point of L, the largest possible volume is filled up. This sphere packing problem goes back to a book review that Gauss wrote in 1831.

There is an **identification between lattices** L (modulo rotation) in $\mathbb{R}^n$ **and positive matrices** Y in a fundamental domain for $\mathcal{P}_n/GL(n, \mathbb{Z})$ which is made as follows. Suppose we are given a lattice L as in formula (1.176). Define the positive matrix $Y(L)$ in $\mathcal{P}_n$ by:

$$Y(L) = I[v], \text{ for } v = (v_1 v_2 \cdots v_n) \in \mathbb{R}^n. \tag{1.177}$$

Since the lattice L remains the same upon change of $\mathbb{Z}$-basis, which amounts to replacing v by $v\gamma$, for some $\gamma \in GL(n, \mathbb{Z})$, we must consider $Y(L)$ as an equivalence class in $\mathcal{P}_n/GL(n, \mathbb{Z})$.

Using the identification (1.177), the problem of finding the lattice L giving the densest packing of spheres of equal radius r with centers at points in L turns out to be equivalent to the problem of choosing Y in $\mathcal{P}_n$ to maximize $m_Y |Y|^{-1/n}$, where m_Y denotes the **minimum over the integer lattice**:

---

[5]Some people use the word lattice for discrete subgroups of non-abelian Lie groups. We will not.

$$m_Y = \min \{Y[a] \mid a \in \mathbb{Z}^n - 0\}. \qquad (1.178)$$

To see this, note first that if $a \in \mathbb{Z}^n$, and we set

$$w = a_1 v_1 + \cdots + a_n v_n,$$

then $Y[a] = {}^t w w$ is the square of the distance from the lattice point $w \in L$ to the origin. Next note that the spheres must not intersect, which means that one should take them to have radius equal to one-half the minimum distance of any lattice point from the origin. This means that the radius r must be chosen to be $\frac{1}{2}(m_Y)^{1/2}$.

The density of space occupied by spheres of radius r centered at points in the lattice L is:

$$d_L = \lim_{X \to \infty} \frac{v_n(r) \cdot \#(L \cap (\text{cube of volume } X))}{X}, \qquad (1.179)$$

with $v_n(r)$ being the volume of the sphere of radius r in $\mathbb{R}^n$. Now the number of points of L in a cube of volume X is easily seen to be asymptotic to $|Y|^{-1/2} X$ as X approaches infinity. Therefore

$$d_L = r^n v_n(1) |Y|^{-1/2} = \left(\frac{m_Y}{4}\right)^{n/2} v_n(1) |Y|^{-1/2}, \qquad (1.180)$$

and

$$v_n(1) = \frac{\pi^{n/2}}{\Gamma(1 + n/2)}. \qquad (1.181)$$

Note that the density d_L is unchanged if we multiply the $\mathbb{Z}$-basis of L by a constant c (or equivalently if we multiply the corresponding matrix $Y(L)$ by c^2) for then r is multiplied by c and $|Y|^{-1/2}$ is multiplied by $1/c$. The fact that the density d_L must be less than or equal to one gives the **Minkowski upper bound for the minimum m_Y**:

$$m_Y \le c_n |Y|^{1/n}, \text{ with } c_n = \frac{4}{\pi} \Gamma(1 + n/2)^{2/n} \sim \frac{2n}{\pi e}, \text{ as } n \to \infty. \qquad (1.182)$$

The asymptotic behavior of c_n comes from Stirling's asymptotic formula for the gamma function (see Lebedev [398]).

Exercise 1.4.1. Show that any packing (whether the centers form a lattice or not) of spheres of equal radii in $\mathbb{R}^n$ such that no further spheres can be added without overlap has density $\ge 2^{-n}$.
Hint. Spheres of radius $2r$ completely cover $\mathbb{R}^n$.

We will give another (and more detailed) view of (1.182) throughout the next part of this section. Blichfeldt showed in 1914 that the constant c_n can be halved.

This is equivalent to showing that the density d_L cannot exceed about $2^{-.5n}$, for large n. Kabatiansky and Levenshtein [337] have shown that for large n the density cannot exceed about $2^{-.599n}$. This leads to an upper bound on m_Y, $Y \in \mathcal{SP}_n$, of about

$$\frac{n}{\pi e} 2^{-.198n}, \quad \text{for large } n.$$

In Section 1.4.4 we will consider Minkowski's result that there exist Y in $\mathcal{P}_n$ such that the minimum satisfies

$$m_Y > \frac{n}{2\pi e} |Y|^{1/n}, \quad \text{for large } n$$

(see Corollary 1.4.2 below). We will also consider the Minkowski-Hlawka theorem saying that there are lattice packings L of $\mathbb{R}^n$ such that $d_L \geq \zeta(n)2^{-n+1}$, where $\zeta(n)$ is Riemann's zeta function. This result does not, however, give a construction for these lattices. This lower bound on d_L has been improved by K.M. Ball [34] to $(n-1)\zeta(n)2^{-n+1}$. Stephanie Vance [622] has improved the lower bound further when n is divisible by 4.

See Henry Cohn's website for some interesting talks and papers on sphere packing and applications. Some work on finding dense lattice packings explicitly is surveyed by Sloane [568]. For example, Barnes and Sloane have constructed lattice packings in dimensions up to 100,000 with density roughly $2^{-1.25n}$. But Sloane notes that Minkowski's theorem guarantees that there exist packings that are 10^{4000} times denser. See also Rush [519] and Rush and Sloane [520].

It may surprise the reader to learn that the **Kepler conjecture** to the effect that the densest lattice packing in $\mathbb{R}^3$ actually gives the densest not necessarily lattice centered packing of spheres in $\mathbb{R}^3$ required such a complex computer proof that the dozen referees of the paper for the *Annals of Math.* said they were just 99 % convinced and had run out of energy to consider the matter further. In 2014 Hales announced completion of the Flyspeck project giving a formal proof of the Kepler conjecture, which it claims is more trustworthy than peer reviewed proofs. A **formal proof** is a mathematical proof that has been checked by computer. See the Flyspeck website for more information. The Wikipedia page on the Kepler conjecture gives a short summary of the story.

In 1983 Sigrist [567] gave a short survey on sphere packing. Sigrist noted the following quotes on the Kepler problem:

H.S.M. Coxeter: "It is conceivable that some irregular packing might be still denser."

C.A. Rogers: "Many mathematicians believe, and all physicists know, that the density cannot exceed $\pi/\sqrt{18}$."

Other references on sphere packing are Berger [45, Ch. 10], Conway and Sloane [122], J.H. Conway et al. [123], Hales and Ferguson [257].

See Figures 1.11 and 1.12 for the densest lattice packings in the plane and 3-space. It is known that the densest lattice packing in the plane also gives the densest packing, lattice or not. As we've said, this is known for 3-space, provided that you accept the computer proof and you acknowledge that there are equally dense non-lattice packings in 3-space. In fact, Berger [45, Ch. 10, p. 631] shows how to find an infinite number of non-lattice packings of 3-space with density equal to the face-centered cubic packing. He says: "We point out that a large number of metals crystallize in the lattice A3 (or D3), or cubic face centered lattice, also called cubic dense: Al, Ni, Cu, Ag, Au, while others crystallize in a hexagonal dense packing: Mn, Ca, Sr, Ti. But there are some that crystallize in still other forms that aren't always of maximal density."

Henry Cohn and Noam Elkies [117, 118] give upper bounds on sphere packing densities which are the best known for dimensions 4–36. Poisson summation and linear programming bounds for error-correcting codes are integral components of the proofs.

Exercise 1.4.2. (a) Prove formula (1.181) for the volume of the unit sphere in $\mathbb{R}^n$.
(b) Fill in the details in the rest of the discussion of formula (1.182) above.

Hints. (a) Note that

$$\int_{\mathbb{R}^n} \exp\left(-\,^t xx\right)\, dx = \pi^{n/2}.$$

Suppose now that w_n denotes the surface area of the unit sphere in $\mathbb{R}^n$; i.e.,

$$w_n = \text{surface area}\{x \in \mathbb{R}^n \mid \,^t xx = 1\}.$$

Use polar coordinates on the preceding integral to show that

$$\pi^{n/2} = \frac{w_n}{2} \Gamma\left(\frac{n}{2}\right).$$

On the other hand, polar coordinates can be used to show that

$$v_n(1) = w_n/n. \tag{1.183}$$

Formula (1.183) says that the volume of the unit sphere in $\mathbb{R}^n$ gets much smaller than the surface area as n goes to infinity. In fact, both w_n and $v_n(1)$ approach zero—a fact that we will use later in this section (see Corollary 1.4.2 in Section 1.4.4). Hamming [258, Ch. 9, 10] gives an interesting paradox related to these facts as well as applications to information theory. We'll consider this paradox in Section 1.4.4.

Denote by L_n, a lattice giving the densest lattice packing of spheres of equal radii in $\mathbb{R}^n$. For $n \le 5$, the lattice L_n was determined by Korkine and Zolotareff. For $n \le 8$, L_n was found by Blichfeldt. The lattice L_2 is often called the regular hexagonal lattice because the **Voronoi polyhedron**, which is the set of points in $\mathbb{R}^2$

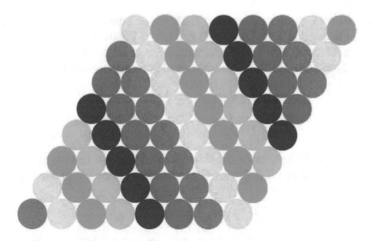

Fig. 1.11 Part of the densest lattice packing of circles of equal radii in the plane

Fig. 1.12 Part of the face-centered cubic lattice packing in 3-space

lying as close to the origin as any lattice point, is a regular hexagon (see Figure 1.11).
In 1831 Gauss proved that the lattice L_2 has $\mathbb{Z}$-basis $v_1 = \left(1, \sqrt{3}\right)$, $v_2 = (2,0)$
and thus corresponds to the positive matrix

$$2\begin{pmatrix} 2 & 1 \\ 1 & 2 \end{pmatrix}, \quad d_L = \frac{\pi}{2\sqrt{3}} \cong .9068.$$

The lattice L_3 is the face centered cubic lattice pictured in the crystallography discussion in Section 1.4 of Volume I as well as Figure 1.12. This lattice occurs in crystals of gold, silver, and aluminum, for example. It has $\mathbb{Z}$-basis $v_1 = (1, 1, 0)$, $v_2 = (1, 0, 1)$, $v_3 = (0, 1, 1)$ and thus corresponds to the positive matrix:

$$\begin{pmatrix} 2 & 1 & 1 \\ 1 & 2 & 1 \\ 1 & 1 & 2 \end{pmatrix}, \qquad d_L = \frac{\pi}{\sqrt{18}} \cong .7404.$$

The Voronoi polyhedron for L_3 is a rhombic dodecahedron (a solid bounded by 12 rhombuses). The lattices L_n, for $n \leq 8$, can be shown to correspond to root systems for the simple Lie groups

$$A_2, \ A_3, \ D_4, \ D_5, \ E_6, \ E_7, \ E_8.$$

See Chapter 2 for a discussion of root systems and see Milnor [449, p. 502] or Milnor and Husemoller [451] for a description of how to get Y out of the Dynkin diagram for the Lie group. See Thompson [614, Appendix] for a table of densest known sphere packings at the time of publication.

We know from Section 3.4 of Volume I that there are connections between sphere packings and **coding theory** (see Sloane [568, 569]). Shannon found that the problem of finding densest sphere packings in spherical space has applications to information theory (see Van der Waerden [639]). One can consider various non-Euclidean analogues of the sphere-packing problem (see Fejes Tóth [177]). Moreover, work on codes led to the discovery of the Leech lattice in $\mathbb{R}^{24}$ which gives rise to many of the densest known lattice packings as well as some new simple groups. Thompson [614] provides a survey of the connection. Elkies [163] gives a survey of lattices, codes, and connections with other parts of mathematics. Henry Cohn and Abhinav Kumar [119] show that the densest lattice packing in 24 dimensions comes from the Leech lattice.

Dyson [150] discusses some of the stories of the interplay between dense lattice packings, codes, and simple groups in an article about unfashionable mathematics. We quote:

> Roughly speaking, unfashionable mathematics consists of those parts of mathematics which were declared by the mandarins of Bourbaki not to be mathematics. A number of very beautiful mathematical discoveries fall into this category. To be mathematics according to Bourbaki, an idea should be general, abstract, coherent, and connected by clear logical relationships with the rest of mathematics. Excluded from mathematics are particular facts, concrete objects which just happen to exist for no identifiable reason, things which a mathematician would call accidental or sporadic. Unfashionable mathematics is mainly concerned with things of accidental beauty, special functions, particular number fields, exceptional algebras, sporadic finite groups. It is among these unorganized and undisciplined parts of mathematics that I would advise you to look for the next revolution in physics.

Analysis on the fundamental domain $\mathcal{P}_n / GL(n, \mathbb{Z})$ can also be applied to the problem of finding the best lattice of points in $\mathbb{R}^n$ to use for **numerical integration** (see Ryskov [522]).

Explicit fundamental domains for $\mathcal{P}_n/GL(n, \mathbb{Z})$ are pertinent to the problem of finding subgroups of $GL(n, \mathbb{Z})$ fixing some Y in $\mathcal{P}_n$, a problem which is of interest in **crystallography** and was thus solved long ago for $n \leq 3$. The cases $n = 4$ and 5 have been solved by Dade and Ryskov (see Ryskov [522] for the references).

There are also applications of reduction theory in **cryptography** (in the design of algorithms to break codes). Related applications to integer programming have also appeared. Here one needs the reduction of lattices rather than quadratic forms. The HKZ algorithm of Hermite, Korkine, and Zolotareff (which we call Grenier reduction) is used as well as the LLL algorithm of Lenstra, Lenstra, and Lovász. See Joux and Stern [336] or Lagarias et al. [382]. The LLL algorithm was discovered in the 1980s. Immediately thereafter Shamir used it to break the Merkle–Hellman public-key cryptographic code (based on the knapsack problem) in polynomial time.

Sarnak and Strömbergsson [530] prove that if L_n is a lattice yielding the densest lattice packing of n-space for $n = 4, 8$ and 24 and $s > 0$, the Epstein zeta function $Z(Y_{L_n}, s)$ has a strict local minimum at $L = L_n$, where $Y_{L_n} \in SP_n$ has ij coordinate ${}^t v_i v_j$ if the lattice L has $\mathbb{Z}$-basis $v_1, \ldots, v_n$ and the torus $\mathbb{R}^n/L$ is assumed to have volume 1 so that Y_L has determinant 1. They also show that in these cases the lattice L_n will have minimum **height** defined as a multiple of the derivative of the Epstein zeta function at 0. They define a lattice L_n to be **universal** (i.e., universally extremal for the Epstein zeta function) if

$$Z(Y, s) \geq Z(Y_{L_n}, s) \quad \text{for all } s > 0 \text{ and all } Y \in SP_n.$$

They conjecture that in dimensions $n = 4, 8, 24$ the lattice L_n is universal. They prove that for $n = 3$ the height of L_3 is minimal. Use is made of the incomplete gamma expansion of Epstein's zeta function as well as the Grenier fundamental domain for $SP_n/GL(n, \mathbb{Z})$ from § 1.4.3. Gruber and Lekkerkerker [251] provide more information on minima of Epstein zetas. They note [251, p. 531]: "On the one hand British mathematicians investigated the minimum of the Epstein zeta-function primarily for its number-geometric interest .... On the other hand the same problem was studied by the Russian school of geometry of numbers because of its implications for numerical integration."

This completes our list of reasons for beginning the study of the fundamental domain $\mathcal{P}_n/GL(n, \mathbb{Z})$. The main references for this section are: Borel [63, 65], Borel's article in Borel and Mostow [68, pp. 20–25], Cassels [99, 100], Davenport [129], Delone and Ryskov [134], Freitag [185], Grenier [241, 242], Gordon et al. [237], Gruber and Lekkerker [251], Hancock [260], Humphreys [308], O.-H. Keller [346], Maass [426], Minkowski [453], Raghunathan [496], Rogers [514], Ryskov [522], Ryskov and Baranovskii [523], Schwarzenberger [538], Séminaire Cartan [547], Siegel [561, 562, 566], Van der Waerden [638, 639], Weil [662, Vol. I, pp. 339–358], [658], and Weyl [666, Vol. III, pp. 719–757, Vol. IV, pp. 46–96]. Some of the earlier references are: Gauss [205, Vol. I, p. 188], Hermite [291, Vol. I, pp. 94–164], Korkine and Zolotareff [365, 366], Lagrange [384, Vol. III, pp. 693–758], Seeber [541], and Voronoi [636, 637].

1.4.2   Minkowski's Fundamental Domain

Before describing Minkowski's fundamental domain for $\mathcal{P}_n/GL(n, \mathbb{Z})$, we need to retrace Minkowski's steps and consider his most fundamental results in the geometry of numbers. These results have immediate applications in the very foundations of algebraic number theory. In general, they are useful when one wants to know whether some inequality has a solution in integers. Here we are interested in the size of the minimum of a quadratic form Y over the integer lattice; i.e., in the size of m_Y defined by (1.178) for $Y \in \mathcal{P}_n$. We have already given one approach to this problem, which led to the inequality (1.182). Now let us consider another approach. Define the **ellipsoid** in $\mathbb{R}^n$ **associated with the positive matrix** $Y \in \mathcal{P}_n$:

$$S_Y(t) = \{x \in \mathbb{R}^n \mid Y[x] < t\}, \quad \text{for } t > 0. \tag{1.184}$$

This is a **convex** set; i.e., if $x, y \in S_Y(t)$ and $a \in [0, 1]$, then

$$ax + (1 - a)y \in S_Y(t).$$

Exercise 1.4.3.   (a) Show that $S_Y(t)$ defined in (1.184) is a convex set. Show also that its closure is compact. Why do we call it an ellipsoid?

(b) Show that the volume of $S_Y(t)$ is $|Y|^{-\frac{1}{2}} v_n \left(\sqrt{t}\right)$, where $v_n (r)$ is the volume of the sphere of radius r, obtained using formula (1.181).

Minkowski used the fundamental facts in the Lemma below to see that

$$S_Y(t) \cap \mathbb{Z}^n \neq \{0\}, \quad \text{if Vol} (S_Y(t)) > 2^n.$$

This means that

$$t > \frac{4}{\pi} \Gamma \left(1 + \frac{n}{2}\right)^{2/n} |Y|^{1/n}$$

implies that there exists an $a \in \mathbb{Z}^n - 0$ such that $Y[a] < t$. The inequality (1.182) for m_Y follows from this result.

Lemma 1.4.1 (Minkowski's Fundamental Lemma in the Geometry of Numbers).

(1) Suppose that S is a Lebesgue measurable set in $\mathbb{R}^n$ with $\text{Vol}(S) > 1$. Then there are two points x, y in S such that $0 \neq x - y \in \mathbb{Z}^n$.

(2) Let S be a Lebesgue measurable subset of $\mathbb{R}^n$ which is convex and symmetric with respect to the origin (i.e., $x \in S$ implies $-x \in S$). If, in addition, $\text{Vol}(S) > 2^n$, then $S \cap \mathbb{Z}^n \neq \{0\}$.

Proof. (1) (From Weil [662, p. 36].) One has the following integral formula (as a special case of formula (1.7) from Section 1.1.1):

$$\int_{\mathbb{R}^n} f(x)dx = \int_{[0,1]^n} \sum_{a \in \mathbb{Z}^n} f(x + a) \, dx.$$

Let f be the indicator function of S; i.e., $f(x) = 1$ if $x \in S$ and 0 otherwise. If the conclusion of part (1) of the lemma were false, the inner sum over $\mathbb{Z}^n$ on the right-hand side of this integral formula would be less than or equal to one for all x in $[0, 1]^n$. This gives a contradiction to the hypothesis that $\text{Vol}(S) > 1$.

Another way to see part (1) is to translate S to the unit cube $[0, 1]^n$ by elements of $\mathbb{Z}^n$. If there were no overlap among these translates, the volume of S would be less than one. See Figure 1.13.

(2) The proof of this part of the lemma is illustrated in Figure 1.14. More explicitly, let us define S_0 to be the set $\frac{1}{2}S$ consisting of vectors of the form $\frac{1}{2}x$, for $x \in S$. Then $\text{Vol}(S_0) > 1$ by hypothesis and thus, by part (1) of this lemma, there are points x, y in S_0 such that $x - y \in \mathbb{Z}^n - 0$. It follows that $\frac{1}{2}(x - y)$ lies in S_0 by the convexity and symmetry of S. So $x - y$ lies in S. This completes the proof of Lemma 1.4.1.

Minkowski's fundamental lemma leads quickly to the finiteness of the class number of an algebraic number field K as well as to Dirichlet's unit theorem giving the structure of the group of units in the ring of integers of K (see Section 1.4 of Volume I and the references mentioned there). A lower bound on the absolute value of the discriminant of K is also a consequence.

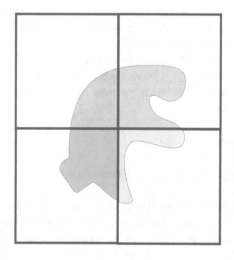

Fig. 1.13 Picture proof of part 1 of Minkowski's fundamental lemma in two dimensions. Each square is a unit square. The four parts of the big square are translated to the square on the lower right by integral translations. There are overlaps. Otherwise $\text{Vol}(S) < 1$

Fig. 1.14 Picture proof of
part 2 of Minkowski's
fundamental lemma in two
dimensions

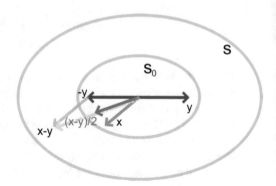

Next we need another lemma.

Lemma 1.4.2 (Vectors That Can Be First Columns of Elements of). $GL(n,\mathbb{Z})$.
A vector $a = {}^t(a_1,\ldots,a_n) \in \mathbb{Z}^n$ *can be made the first column of a matrix in*
$GL(n,\mathbb{Z})$ *if and only if the greatest common divisor* $\gcd(a_1,\ldots,a_n)$ *equals one.*

Proof.

$\Rightarrow$

This direction is clear upon expanding the determinant of A by its first column a.

$\Leftarrow$

Suppose that $a \in \mathbb{Z}^n$ has $\gcd(a_1,\ldots,a_n) = 1$. We need to obtain a matrix U in
$GL(n,\mathbb{Z})$ such that $Ua = {}^t(1,0,\ldots,0)$. For then a is the first column of U^{-1}.

Our goal can be attained by multiplying our column vector a on the left by
combinations of matrices giving rise to the elementary row operations on the vector.
These operations are essential for elementary divisor theory which is basic for ideal
theory in number fields and for the fundamental theorem of abelian groups.

These **elementary row operations** are:

(i)   changing the order of rows,
(ii) multiplying any row by $+1$ or -1,
(iii) adding an integral multiple of any row to any other row.

Operations (i) and (ii) allow us to assume that a_1 is positive and the smallest
nonzero entry of the vector a. We can use operation (iii) to replace any nonzero
entry a_j, for $j > 1$, by its remainder upon division by a_1. In this way, we can cause
all the entries $a_2,\ldots,a_n$ to lie in the interval $[0,a_1)$. Either all the a_j are zero for
$j \geq 2$, or there is a smallest positive entry $a_j, j \geq 2$. Use operation (i) to put this a_j in
the first row and continue in this way. The final result is a vector ${}^t(m,0,\ldots,0)$, with
$m > 0$. Thus $a = m\left(U^{-1}\right){}^t(1,0,\ldots,0)$. Since the greatest common divisor of the
entries in the vector a is one, it follows that $m = 1$ and the proof of Lemma 1.4.2 is
complete.

The preceding lemma has even been established in an analogous situation in which $\mathbb{Z}$ is replaced by $F[X_1, \ldots, X_n]$ for fields F, proving a conjecture of Serre (see Bass [40]).

Our main goal is to produce a **fundamental domain** for $\mathcal{P}_n/GL(n, \mathbb{Z})$ and study its properties. That is, we want to find a (nice) subset D of $\mathcal{P}_n$ having the following two properties:

(i) $\mathcal{P}_n = \bigcup_{A \in GL(n,\mathbb{Z})} D[A]$;

(ii) $Y, W \in D$, with $W = Y[A]$ for some $A \in GL(n, \mathbb{Z})$, with $A \neq \pm I$ implies that Y must lie in the boundary ∂D of D.

Given $Y \in \mathcal{P}_n$, we will also want a procedure for finding a matrix $A \in GL(n, \mathbb{Z})$ such that $Y[A]$ lies in the fundamental domain D. Such a procedure is called a **reduction algorithm**. We will also call the fundamental domain a **set of reduced matrices**.

In this section we will discuss Minkowski's fundamental domain which is a convex, closed subset of $\mathcal{P}_n$, bounded by a finite set of hyperplanes through the origin. The domain was found by Minkowski who was motivated by the work of mathematicians such as Lagrange, Gauss, Seeber, and Hermite. **Minkowski's fundamental domain** $\mathcal{M}_n$ is defined as follows:

$$\mathcal{M}_n = \left\{ Y = (y_{ij}) \in \mathcal{P}_n \,\middle|\, \begin{array}{l} Y[a] \geq y_{kk}, \text{ if } a \in \mathbb{Z}^n, \ \gcd(a_k, \ldots, a_n) = 1 \\ y_{k,k+1} \geq 0, \text{ for all } k \end{array} \right\}. \tag{1.185}$$

The domain in (1.185) appears to be bounded by an infinite number of hyperplanes. We will show in Theorem 1.4.1 that a finite number of inequalities actually suffices to give the region. We say $Y \in \mathcal{M}_n$ is **Minkowski-reduced**.

By Lemma 1.4.2, it is easily seen that $\mathcal{M}_n$ has the alternative definition:

$$\mathcal{M}_n = \left\{ Y \in \mathcal{P}_n \,\middle|\, \begin{array}{l} Y[a] \geq y_{kk} \text{ if } (e_1 e_2 \cdots e_{k-1} a * \cdots *) \in GL(n, \mathbb{Z}) \\ y_{k,k+1} \geq 0, \text{ for all } k \end{array} \right\}. \tag{1.186}$$

Here e_k denotes the kth element in the standard basis for $\mathbb{R}^n$; i.e., e_k is a vector with all its entries 0 except for the kth entry which is 1. To see (1.186), note that Lemma 1.4.2 implies that the condition on the vector a in (1.186) is equivalent to asking that the greatest common divisor $\gcd(a_k, \ldots, a_n) = 1$.

Exercise 1.4.4. (a) Prove that

$$\mathcal{M}_2 = \{Y \in \mathcal{P}_2 \mid 0 \leq 2y_{12} \leq y_{11} \leq y_{22}\}. \tag{1.187}$$

(b) Note that restricting to the determinant one surface

$$\mathcal{SM}_2 = \{W \in \mathcal{M}_2 \mid |W| = 1\}.$$

does not change the form of the inequalities in (1.187). We have a mapping which identifies the Poincaré upper half plane H and $\mathcal{SP}_2$ given by

$$z \longmapsto W_z = \begin{pmatrix} y^{-1} & 0 \\ 0 & y \end{pmatrix} \begin{bmatrix} 1 & -x \\ 0 & 1 \end{bmatrix} \in \mathcal{SP}_2, \quad \text{for } z = x + iy \in H$$

as we saw in Volume I, Exercise 3.1.9. Show that $W_z \in \mathcal{SM}_2$ is equivalent to the inclusion of z in the left half of the fundamental domain for $SL(2, \mathbb{Z}) \backslash H$ pictured in Figure 3.14 of Section 3.3 of Volume I or Figure 1.16 of this Volume. It is not the whole fundamental domain because $GL(2, \mathbb{Z})/SL(2, \mathbb{Z})$ has order 2 and the nontrivial coset comes from the matrix $\begin{pmatrix} -1 & 0 \\ 0 & 1 \end{pmatrix}$.

Hints. To see that $\mathcal{M}_2$ is actually given by (1.187), you must first show that if e_k, $k = 1, 2$, denotes the standard basis of $\mathbb{R}^2$, as usual, then

$$y_{11} \geq y_{22} \quad \text{comes from} \quad Y[e_2] = y_{22} \geq y_{11}$$

since the greatest common divisor of the entries of e_2 is 1. Similarly

$$y_{11} \geq 2y_{12} \quad \text{comes from} \quad Y\begin{bmatrix} 1 \\ -1 \end{bmatrix} = y_{11} - 2y_{12} + y_{22} \geq y_{22}.$$

To see that it suffices to use only the inequalities in (1.186) coming from the vectors $a = e_2$ and $'(-1, 1)$, you need to use the fact that for any vector $a \in \mathbb{Z}^n$ we have for Y in the set defined by (1.187):

$$Y[a] \geq y_{11} \left(a_1^2 - |a_1 a_2| + a_2^2 \right) = y_{11} \left\{ (|a_1| - |a_2|)^2 + |a_1 a_2| \right\}.$$

This is greater or equal to y_{11} if $\gcd(a_1, a_2) = 1$. And if $a_2 = \pm 1$, we see that if Y satisfies the inequalities in (1.187):

$$Y[a] = y_{11} a_1^2 + 2y_{12} a_1 a_2 + y_{22} a_2^2 \geq y_{22}.$$

Exercise 1.4.5 (Successive Minima). Show that $Y \in \mathcal{M}_2$ is equivalent to saying that Y satisfies the following inequalities:

$$y_{12} \geq 0;$$

$$Y\begin{bmatrix} 1 \\ 0 \end{bmatrix} = y_{11} = m_Y = \min \left\{ Y[a] \mid a \in \mathbb{Z}^2 - 0 \right\} = N_1;$$

$$Y\begin{bmatrix} 0 \\ 1 \end{bmatrix} = y_{22} = \min \left\{ Y[a] \;\middle|\; a = \begin{pmatrix} a_1 \\ a_2 \end{pmatrix} \in \mathbb{Z}^2, \; \begin{vmatrix} 1 & a_1 \\ 0 & a_2 \end{vmatrix} = a_2 \neq 0 \right\} = N_2.$$

We call N_1 and N_2 the **first and second successive minima** of Y respectively.

Similar ideas to those of Exercise 1.4.5 work for $GL(n,\mathbb{Z})$ with $n = 3$, as was found by Seeber [541] in 1831 and Gauss [205, Vol. II, p. 188]. That is, $Y \in \mathcal{M}_3$ is equivalent to requiring that y_{kk} be the minimum of the values $Y[a]$, $a \in \mathbb{Z}^3$ such that a is linearly independent of the standard basis vectors $e_1, \ldots, e_{k-1}$, for $k = 1, 2, 3$ as well as requiring $y_{k,k+1} \geq 0$, as usual.

For $n \geq 5$ reduction by successive minima is not possible because successive minima do not necessarily occur at vectors a_k which give matrices $(a_1 a_2 \cdots a_n)$ in $GL(n,\mathbb{Z})$.

We note here that Minkowski was probably inspired not only by the work of Gauss and Seeber, but also by that of Hermite [291, Vol. I, pp. 94–164]. However, with Minkowski, the theory progressed by a quantum leap. Minkowski [453, Vol. II, pp. 51–100] proved all of the following theorem as well as part of the Minkowski-Hlawka Theorem in Section 1.4.4. His proofs have been rewritten by many eminent mathematicians. See the references mentioned earlier for alternative treatments. The following theorem gives an explicit finite list of inequalities for Minkowski-reduced matrices when $n = 2, 3, 4$.

Theorem 1.4.1 (The Minkowski Fundamental Domain). *Let $\mathcal{M}_n$ denote the Minkowski fundamental domain defined by formula (1.185). This domain has the following properties.*

(1) For any Y in $\mathcal{P}_n$, there exists a matrix A in $GL(n,\mathbb{Z})$ such that $Y[A]$ lies in the Minkowski domain $\mathcal{M}_n$.

(2) Only a finite number of inequalities are necessary in the definition of $\mathcal{M}_n$. Thus $\mathcal{M}_n$ is a convex cone through the origin bounded by a finite number of hyperplanes.

(3) If Y and $Y[A]$ both lie in the Minkowski domain $\mathcal{M}_n$, and A is an element of $GL(n,\mathbb{Z})$ distinct from $\pm I$, then Y must lie on the boundary $\partial\mathcal{M}_n$. Moreover, $\mathcal{M}_n$ is bounded by a finitely number of images $\mathcal{M}_n[A]$, for A in $GL(n,\mathbb{Z})$. That is $\mathcal{M}_n \cap (\mathcal{M}_n[A]) \neq \emptyset$, for only finitely many $A \in GL(n,\mathbb{Z})$.

(4) When $n = 2, 3, 4$, we have

$$\mathcal{M}_n = \left\{ Y \in \mathcal{P}_n \mid y_{k,k+1} \geq 0; \ Y[a] \geq y_{kk}, \ \text{if } a_k = 1 \text{ and } a_j = 0 \text{ or } \pm 1, \text{ for all } k \right\}.$$

The fundamental domains $\mathcal{M}_5$ and $\mathcal{M}_6$ were determined explicitly by Minkowski [453, Vol. I, pp. 145–148, 154, 218] and $\mathcal{M}_7$ was found by Tammela [590]. These domains require more inequalities.

Proof. (1) Suppose $Y \in \mathcal{P}_n$ is given. We need to find some $A \in GL(n,\mathbb{Z})$ such that $Y[A] \in \mathcal{M}_n$. To find A we locate the columns $a^{(j)}$ of $A = (a^{(1)} \cdots a^{(n)})$ as follows. First choose $a^{(1)}$ so that

$$Y\left[a^{(1)}\right] = \min\left\{Y[a] \mid a \in \mathbb{Z}^n - 0\right\}.$$

Such a vector $a^{(1)} \in \mathbb{Z}^n - 0$ must exist because if c is the smallest eigenvalue of Y, we have the inequality:

$$cI[x] \leq Y[x], \text{ for all } x \in \mathbb{R}^n.$$

This implies the finiteness of the set of $a \in \mathbb{Z}^n$ such that $Y[a]$ is less than any given bound.

Now if $a^{(1)}$ minimizes $Y[a^{(1)}]$, it follows that the greatest common divisor of the entries of $a^{(1)}$ must be one. By Lemma 1.4.2, then $a^{(1)}$ is the first column of a matrix in $GL(n, \mathbb{Z})$. So there exists a vector $b \in \mathbb{Z}^n$ such that the $n \times 2$ matrix $(a^{(1)} \, b)$ is the first two columns of a matrix in $GL(n, \mathbb{Z})$. This means that there exists some $a^{(2)}$ in $\mathbb{Z}^n$ such that $(a^{(1)} \, a^{(2)})$ can be completed to a matrix in $GL(n, \mathbb{Z})$ and $Y\left[a^{(2)}\right]$ is minimal.

Continue inductively to obtain the matrix $A = \left(a^{(1)} a^{(2)} \cdots a^{(n)}\right)$ in $GL(n, \mathbb{Z})$. The column $a^{(k)}$ in $\mathbb{Z}^n$ is defined by requiring that the $n \times k$ matrix $(a^{(1)} \cdots a^{(k)})$ can be completed to a matrix in $GL(n, \mathbb{Z})$ and that

$$Y\left[a^{(k)}\right] = \min \left\{ Y[b] \mid \left(a^{(1)} \cdots a^{(k-1)} \, b \, * \cdots *\right) \in GL(n, \mathbb{Z}) \right\}.$$

And lastly we require that

$${}^t a^{(k-1)} Y a^{(k)} \geq 0, \quad \text{for all } k = 2, \ldots, n.$$

This last requirement is possible, since we can always multiply $a^{(k)}$ by -1, if necessary.

Now we must show that if $A = (a^{(1)} \cdots a^{(n)})$ is constructed as above, then $W = Y[A]$ must lie in the Minkowski domain $\mathcal{M}_n$. To see this, note that (1.186) is an equally good definition of the Minkowski domain. That is, W is Minkowski-reduced (assuming $w_{k,k+1} \geq 0$) if

$$W[c] \geq w_{kk} \text{ when } (e_1 \cdots e_{k-1} \, c * \cdots *) \in GL(n, \mathbb{Z}).$$

Here, as usual, e_j denotes the standard basis vector in $\mathbb{R}^n$, with all entries 0 but the jth which is 1. To complete the proof that $W \in \mathcal{M}_n$, observe that

$$A\left(e_1 \cdots e_{k-1} \, c * \cdots *\right) = (a^{(1)} \cdots a^{(k-1)} \, d * \cdots *),$$

if $Ac = d$. And

$$W[c] = Y[d] \geq Y[a^{(k)}] = w_{kk}, \tag{1.188}$$

by the construction of A. This concludes the proof that $W \in \mathcal{M}_n$. Conversely, note that if W is in $\mathcal{M}_n$, then we also have the inequality (1.188).

(4) The case $n = 2$ was Exercise 1.4.4. Suppose now that $n = 2, 3$, or 4 and set

$$\mathcal{M}_n^* = \left\{ Y \in \mathcal{P}_n \mid y_{k-1,k} \geq 0, \, Y[a] \geq y_{kk}, \text{ if } a_k = 1, a_j = 0 \text{ or } \pm 1, \right.$$

$$\left. \text{for } j \neq k, \, 1 \leq k \leq n \right\}.$$

Clearly $\mathcal{M}_n \subset \mathcal{M}_n^*$. To show the reverse inclusion, we need to prove that $Y \in \mathcal{M}_n^*$ and $m \in \mathbb{Z}^n$ with $\gcd(m_k, \ldots, m_n) = 1$ imply that $Y[m] \geq y_{kk}$. We can assume by induction and changing signs that all the m_j are positive.

Among the numbers $m_1, \ldots, m_n$, let m_t be the last occurrence of the minimum. Define vectors u and w in $\mathbb{R}^n$ by:

$$u_i = \begin{cases} m_t, & \text{when } i \neq t, \\ 0, & i = t; \end{cases}$$

and $w_i = 1$, for all i. Then we have the following equality (proved in the following exercise):

$$Y[m] - Y[m - u] = m_t^2 \, (Y[w] - y_{tt}) + 2 \sum_{i \neq t} (m_i - m_t) m_t \sum_{j \neq t} y_{ij}. \qquad (1.189)$$

Now $Y \in \mathcal{M}_n^*$ implies that $Y[w] \geq y_{tt}$, and if $n \leq 4$, $Y \in \mathcal{M}_n^*$ implies that we have the following (again proved in the following exercise)

$$\sum_{j \neq t} y_{ij} \geq 0, \quad \text{if } i \neq t. \qquad (1.190)$$

It follows from (1.189) and (1.190) that $Y[m] \geq Y[m - u]$. Since $m - u$ has smaller entries than m, the proof of part (4) of Theorem 1.4.1 is completed by induction on the norm of m.

(2) and (3). These proofs will appear later in this section after some preliminaries.

Exercise 1.4.6. (a) Prove formula (1.189).
(b) Prove formula (1.190) if $n \leq 4$. Show that the inequality fails if $n \geq 5$.

The following proposition gives some inequalities for Minkowski-reduced matrices.

Proposition 1.4.1 (Other Properties of Minkowski's Fundamental Domain).

(a) *If $Y \in \mathcal{M}_n$, then the entries of Y satisfy the following inequalities:*

$$y_{11} \leq y_{22} \leq \cdots \leq y_{nn}$$

and

$$|y_{ij}| \leq y_{ii}/2, \quad \text{if } 1 \leq i < j \leq n.$$

(b) *If $Y \in \mathcal{M}_n$, then Y satisfies the inequality:*

$$k_n y_{11} \cdots y_{nn} \leq |Y| \leq y_{11} \cdots y_{nn},$$

where k_n is a positive constant (depending only on n and not on Y). The right-hand inequality actually holds for any matrix Y in $\mathcal{P}_n$.

(c) If $Y \in \mathcal{M}_n$, let Y_0 denote the diagonal matrix formed by taking the diagonal out of Y :

$$Y_0 = \begin{pmatrix} y_{11} & & 0 \\ & \ddots & \\ 0 & & y_{nn} \end{pmatrix}.$$

Then there is a positive constant γ_n such that

$$\gamma_n^{-1} Y_0[x] \leq Y[x] \leq \gamma_n Y_0[x], \quad \text{for all } x \in \mathbb{R}^n.$$

The constant γ_n depends only on n and not on Y or x.

(d) Suppose that $Y \in \mathcal{P}_n$ has Iwasawa decomposition $Y = D[T]$ given by

$$D = \begin{pmatrix} d_1 & \cdots & 0 \\ \vdots & \ddots & \vdots \\ 0 & \cdots & d_n \end{pmatrix} \quad \text{and} \quad T = \begin{pmatrix} 1 & \cdots & t_{ij} \\ \vdots & \ddots & \vdots \\ 0 & \cdots & 1 \end{pmatrix}, \quad \text{with} \quad d_j > 0.$$

Then

$$d_i/d_{i+1} \leq \kappa_n \quad \text{and} \quad |t_{ij}| \leq \kappa_n,$$

for a positive constant κ_n depending only on n and not on Y.

Proof. (a) Let e_i denote the standard basis vectors of $\mathbb{R}^n$, $i = 1,\ldots,n$. From formula (1.185) defining $\mathcal{M}_n$, we have

$$y_{ii} = Y[e_i] \leq Y[e_{i+1}] = y_{i+1,i+1}$$

and

$$y_{ii} \pm 2y_{ij} + y_{jj} = Y[e_i \pm e_j] \geq Y[e_j] = y_{jj}, \quad \text{if } 1 \leq i < j \leq n.$$

The inequalities in part (a) are an easy consequence.

(b) The inequality $|Y| \leq y_{11} \cdots y_{nn}$, for $Y \in \mathcal{P}_n$, is easily proved by induction on n. To do so, use the partial Iwasawa decomposition:

$$Y = \begin{pmatrix} V & 0 \\ 0 & w \end{pmatrix} \begin{bmatrix} I & q \\ 0 & 1 \end{bmatrix}, \quad \text{for } V \in \mathcal{P}_{n-1}, w > 0, q \in \mathbb{R}^{n-1}.$$

Then $w = y_{nn} - V[q]$ which implies that $w \leq y_{nn}$. It follows that

$$|Y| = |V|\, w \leq |V|\, y_{nn}.$$

Note also that V is the upper left-hand corner of the matrix Y to complete the proof of the right-hand inequality in (b).

We will have to work harder to prove the left-hand inequality in (b) (cf. Freitag [185], Siegel [562, pp. 44–46], Maass [426, pp. 124–127], and Minkowski [453, Vol. II, pp. 63–67]). We shall follow an approach of Van der Waerden [638]. Let $N_1, \ldots, N_n$ denote the n **successive minima** of $Y \in \mathcal{P}_n$; i.e.,

$$
\begin{aligned}
&N_1 = Y[a_1] = m_Y = \min\{Y[a] \mid a \in \mathbb{Z}^n - 0\}, \\
&\text{and, given} \quad N_i = Y[a_i], \quad \text{for} \quad i = 1, \ldots, k-1, \\
&N_k = \min\{Y[a] \mid a \in \mathbb{Z}^n; \ a_1, \ldots, a_{k-1}, a \ \text{linearly independent}\}.
\end{aligned}
\tag{1.191}
$$

It is clear that $Y \in \mathcal{M}_n$ implies that $N_k \leq y_{kk}$, for $k = 1, \ldots, n$ and $N_1 = y_{11}$. We need two inequalities for $Y \in \mathcal{M}_n$:

$$
\left.
\begin{aligned}
&N_1 N_2 \cdots N_n \leq c_n^n\, |Y|, \quad \text{with } c_n = \tfrac{4}{\pi}\Gamma\left(1 + \tfrac{n}{2}\right)^{2/n} \text{ as in (1.182)}, \\
&y_{kk} \leq \delta_k N_k, \quad \text{where } \delta_k = 1, \text{ for } k \leq 4 \quad \text{and} \quad \delta_k = \left(\tfrac{5}{4}\right)^{k-4}, \ k \geq 4.
\end{aligned}
\right\}
\tag{1.192}
$$

These inequalities will be proved in Lemma 1.4.3. They imply that the constant k_n in part (b) is given by the following expression:

$$
k_n =
\begin{cases}
(\tfrac{\pi}{4})^n \Gamma(1 + \tfrac{n}{2})^{-2}, & \text{if } n \leq 4, \\[2mm]
(\tfrac{\pi}{4})^n (\tfrac{4}{5})^{p_n} \Gamma(1 + \tfrac{n}{2})^{-2}, & \text{if } n \geq 4,
\end{cases}
\tag{1.193}
$$

with

$$
p_n = \frac{(n-3)(n-4)}{2}.
$$

This completes the proof of (b).

(c) The inequality to be proved says that the eigenvalues of $W = Y\left[Y_0^{-1/2}\right]$ are bounded above and below by constants independent of Y. Call these eigenvalues $\rho_1, \ldots, \rho_n$. Since

$$
\rho_1 + \cdots + \rho_n = \mathrm{Tr}(W) = n,
$$

it is clear that $\rho_j < n$, for all j.

Using part (b), one has

$$\rho_1 \cdots \rho_n = |W| = |Y| / (y_{11} \cdots y_{nn}) \geq k_n > 0.$$

It follows that $\rho_j > k_n n^{1-n}$, for all j. This completes the proof of part (c).

(d) First we need to recall Exercise 1.1.13, which gave formulas for d_i and t_{ij} in terms of the entries of the matrix Y:

$$d_i = \frac{|Y_i|}{|Y_{i-1}|} \quad \text{and} \quad t_{ij} = \frac{Y(1, \ldots, i-1, i \mid 1, \ldots, i-1, j)}{|Y_i|}.$$

Here Y_i denotes the upper left-hand $i \times i$ corner of the matrix Y and $Y(*|*)$ stands for the subdeterminant of Y obtained by using the indicated rows and columns of Y.

Now to prove part (d), use part (b) to see that since $Y_i \in \mathcal{M}_i$, as is demonstrated in Exercise 1.4.7, we have the following inequalities:

$$\frac{d_i}{d_{i+1}} = \frac{|Y_i|\,|Y_i|}{|Y_{i-1}|\,|Y_{i+1}|} \leq \frac{(y_{11} \cdots y_{ii})^2}{k_{i-1}k_{i+1}(y_{11} \cdots y_{i-1,i-1})^2\, y_{ii}\, y_{i+1,i+1}} \leq \frac{1}{k_{i-1}k_{i+1}}$$

and

$$|t_{ij}| = \frac{|Y(1, \ldots, i-1, i \mid 1, \ldots, i-1, j)|}{|Y_i|} \leq \frac{i!\, y_{11} \cdots y_{ii}}{2k_i\, y_{11} \cdots y_{ii}} = \frac{i!}{2k_i}.$$

This completes the proof of Proposition 1.4.1.   ■

Exercise 1.4.7. Prove that if Y lies in the Minkowski domain $\mathcal{M}_n$ and Y_i denotes the upper left-hand $i \times i$ corner of Y, then Y_i lies in the Minkowski domain $\mathcal{M}_i$.

Before returning to the proof of Theorem 1.4.1, we need to prove the lemma which was used in the proof of part (b) of Proposition 1.4.1. The results in the Lemma should be compared with those in Gruber and Lekkerkerker [251], Lagarias, Lenstra and Schnorr [382], and Siegel [566].

Lemma 1.4.3 (More Inequalities for Minkowski-Reduced Matrices).

(i) *Minkowski's Inequality for the Product of the Successive Minima.* Let the successive minima N_k of $Y \in \mathcal{M}_n$ be defined by (1.191). Then

$$N_1 N_2 \cdots N_n \leq c_n^n\, |Y|,$$

where

$$c_n = \frac{4}{\pi} \Gamma \left(1 + \frac{n}{2}\right)^{2/n},$$

as defined in (1.182).

(ii) Mahler's Inequality Relating the Diagonal Entries of a Minkowski-Reduced Matrix and the Successive Minima. *If $Y \in \mathcal{M}_n$ then*

$$y_{kk} \leq \delta_k N_k, \qquad k = 1, 2, \ldots, n,$$

where $\delta_k = 1$ for $k \leq 4$ and $\delta_k = (5/4)^{k-4}$, for $k \geq 4$.

Proof. (i) (Cf. Minkowski [453, Section 51] and Van der Waerden [639].) Suppose that

$$N_k = Y[a_k] \quad \text{with} \quad A = (a_1 \cdots a_n) \in \mathbb{Z}^{n \times n} \text{ having rank } n.$$

And let $Y[A] = D[T]$ be the Iwasawa decomposition with:

$$D = \begin{pmatrix} d_1 & \cdots & 0 \\ \vdots & \ddots & \vdots \\ 0 & \cdots & d_n \end{pmatrix}, \ d_i > 0, \text{ and } T = \begin{pmatrix} 1 & \cdots & t_{ij} \\ \vdots & \ddots & \vdots \\ 0 & \cdots & 1 \end{pmatrix}. \tag{1.194}$$

Form the matrix

$$Y^{\#} = \begin{pmatrix} d_1/N_1 & \cdots & 0 \\ \vdots & \ddots & \vdots \\ 0 & \cdots & d_n/N_n \end{pmatrix} [T].$$

Part (i) will follow from formula (1.182) if we can show that, if $m_{Y^{\#}}$ is the minimum of $Y^{\#}[a]$ over $a \in \mathbb{Z}^n - 0$, then $m_{Y^{\#}}$ is greater than or equal to one.
    To see this, let

$$T = {}^t(\xi_1 \cdots \xi_n) \quad \text{with the } \xi_j \text{ being column vectors in } \mathbb{R}^n. \tag{1.195}$$

It follows that if x is a column vector in $\mathbb{Z}^n - 0$, then

$$Y^{\#}[x] = \frac{d_1}{N_1} \left({}^t\xi_1 x \right)^2 + \cdots + \frac{d_n}{N_n} \left({}^t\xi_n x \right)^2.$$

Let ${}^t\xi_k x$ be the last of the ${}^t\xi_j x$'s that is not zero. Then because

$${}^t\xi_k = {}^t(0, \ldots, 0, \underset{k}{1}, t_{k,k+1}, \ldots, t_{kn}),$$

we know that x must be linearly independent of $e_1, \ldots, e_{k-1}$. It follows that

$$Y^{\#}[x] \geq \frac{1}{N_k} \left(d_1 \left({}^t\xi_1 x \right)^2 + \cdots + d_n \left({}^t\xi_n x \right)^2 \right) = \frac{1}{N_k} Y[x] \geq 1,$$

by the definition of N_k.

(ii) See Mahler [432], Weyl [666, Vol. III, pp. 719–757], Remak [503], and Van der Waerden [639]. Clearly $\delta_1 = 1$. We will show that

$$\delta_k = \max\left(1, \frac{1}{4}\delta_1 + \cdots + \frac{1}{4}\delta_{k-1} + \frac{1}{4}\right), \tag{1.196}$$

which leads quickly to the desired formula for δ_k. This is Remak's formula for the constants involved.

Thus we want to use mathematical induction to show that $y_{kk} \leq \delta_k N_k$, with δ_k given by formula (1.196). Suppose that $Y[a] = N_k$ with

$$a = \begin{pmatrix} a_1 \\ ma_2 \end{pmatrix}, \qquad m \in \mathbb{Z}^+, \ a_1 \in \mathbb{Z}^{k-1}, \qquad a_2 \in \mathbb{Z}^{n-k+1}, \qquad \gcd(a_2) = 1.$$

Write

$$Y = \begin{pmatrix} Y_1 & 0 \\ 0 & W \end{pmatrix} \begin{bmatrix} I_{k-1} & B \\ 0 & I_{n-k+1} \end{bmatrix}.$$

Since $m = 1$ implies that $N_k \geq y_{kk}$, we may assume that $m \geq 2$. Then

$$N_k = Y[a] = Y_1[a_1 + Bma_2] + W[ma_2] \geq 4W[a_2],$$

which implies $W[a_2] \leq \frac{N_k}{4}$. Therefore

$$y_{kk} \leq Y\begin{bmatrix} c \\ a_2 \end{bmatrix} = Y_1[c + Ba_2] + W[a_2] \leq Y_1[c + Ba_2] + \frac{N_k}{4}. \tag{1.197}$$

Write $Y = D[T]$ with D, T as in (1.194) and (1.195). Then we can choose $c \in \mathbb{Z}^{k-1}$ so that

$$Y_1[c + Ba_2] \leq \frac{1}{4}(d_1 + \cdots + d_{k-1}). \tag{1.198}$$

To see this, let $x = c + Ba_2$ and note that, using (1.195), $Y_1[x]$ equals

$$Y_1[x] = \sum_{j=1}^{k-1} d_j \left({}^t\xi_j x_j\right)^2, \tag{1.199}$$

for

$${}^t\xi_j x_j = x_j + t_{j,j+1}x_{j+1} + \cdots + t_{j,k-1}x_{k-1}.$$

We can choose the entries of $c \in \mathbb{Z}^{k-1}$ so that all the terms ${}^t\xi_j x_j$ in the formula (1.199) are forced to lie in the interval $[-\frac{1}{2}, +\frac{1}{2}]$. Thus we obtain (1.198). It follows from the inequalities (1.197) and (1.198) that:

$$y_{kk} \leq Y \begin{bmatrix} c \\ a_2 \end{bmatrix} \leq \frac{1}{4}(d_1 + \cdots + d_{k-1}) + \frac{N_k}{4}$$

$$\leq \frac{1}{4}(y_{11} + \cdots + y_{k-1,k-1}) + \frac{N_k}{4}$$

$$\leq \frac{1}{4}(\delta_1 N_1 + \cdots + \delta_{k-1} N_{k-1} + N_k),$$

by induction, since $d_i \leq y_{ii}$ (using the analogous formula to (1.199) for $Y[x]$ with $x = e_i$). This completes the proof of (ii).

∎

We continue the intermission in the proof of Theorem 1.4.1 with a discussion of the Iwasawa coordinates (1.194) and certain domains related directly to these coordinates. In the next subsection we will make even greater use of these coordinates in obtaining a fundamental domain.

We define a **Siegel set** of matrices to be given by the following expression, assuming that u and v are positive numbers and $Y = D[T]$ as in (1.194):

$$S_{v,u} = \left\{ Y = D[T] \in \mathcal{P}_n \,\middle|\, \frac{d_i}{d_{i+1}} \leq v, \; |t_{ij}| \leq u, \; \text{all } i,j \right\}. \tag{1.200}$$

Siegel [562, p. 49] makes great use of these sets. The name "Siegel set" appears for example in Borel's article in Borel and Mostow [68, pp. 20–25], Borel [65], and Goldfeld [230]. Borel defines a **fundamental set** to be a subset S of $\mathcal{P}_n$ possessing the following two properties:

(a) $\displaystyle\bigcup_{A \in GL(n,\mathbb{Z})} S[A] = \mathcal{P}_n$, and

(b) $S \cap S[A] \neq \emptyset$, for only finitely many $A \in GL(n, \mathbb{Z})$.

One can show that the Siegel set $S_{v,u}$ is a fundamental set when $v = \frac{4}{3}$ and $u = \frac{1}{2}$ (see Exercise 1.4.8 below and Borel [65, p. 34] who uses the Bruhat decomposition to prove property (b)). But note that, for example, when $n = 2$, if one restricts to the determinant one surface, a Siegel set corresponds to a rectangle in the upper half plane and is thus **not a fundamental domain**.

Exercise 1.4.8. (a) Show that $S_{v,u}$ defined by (1.200) satisfies property (a) of a fundamental set if $v \geq \frac{4}{3}$ and $u \geq \frac{1}{2}$.

(b) Use part (a) to show that the Euclidean volume of

$$\{Y \in S_{v,u} \mid |Y| \leq 1\}$$

is finite.

Hint. See Borel [65, p. 14] or Raghunathan [496, pp. 160–161].

Part (d) of Proposition 1.4.1 shows that $\mathcal{M}_n \subset \mathcal{S}_{v,u}$ for some large constants v and u. In order to obtain a reverse inclusion, we need to replace $\mathcal{M}_n$ by a related set whose definition does not involve the inequalities $y_{k,k+1} \geq 0$. To do this, again motivated by Proposition 1.4.1, we define a set also considered by Siegel [562, p. 49]:

$$\mathcal{R}_a = \left\{ Y \in \mathcal{P}_n \,\middle|\, \frac{y_{kk}}{y_{k+1,k+1}} < a, \ |y_{ij}| < ay_{ii}, \ y_{11} \cdots y_{nn} < a\,|Y|, \ \text{all } i,j \right\}.$$
(1.201)

The proof of part (d) of Proposition 1.4.1 shows that given $a > 0$, there are positive numbers v and u such that $\mathcal{R}_a \subset \mathcal{S}_{v,u}$.

One can also show that given v and u, there is a positive number a depending on v and u such that $\mathcal{S}_{v,u} \subset \mathcal{R}_a$. To see this, suppose that $Y \in \mathcal{S}_{v,u}$, with Iwasawa coordinates given by (1.194); that is:

$$y_{kj} = d_k t_{kj} + \sum_{h=1}^{k-1} d_h t_{hk} t_{hj}, \qquad \text{for } 1 \leq k \leq j \leq n.$$
(1.202)

We shall assume that $v > 1$. Clearly:

$$|y_{kj}| \leq \{ u + v^{k-1} u^2 (k-1) \} \, d_k \leq \eta y_{kk}, \qquad \text{if } \ \eta = u + v^{n-1} u^2 (n-1),$$
(1.203)

since $d_k \leq y_{kk}$ (as is seen by setting $k = j$ in (1.202)). Also

$$y_{kk} \leq \eta v \, y_{k+1,k+1} \qquad \text{and} \quad y_{11} \cdots y_{nn} \leq \eta^n |Y|.$$

This completes the proof that $\mathcal{S}_{v,u} \subset \mathcal{R}_a$, for some $a = a(v,u)$.

Exercise 1.4.9 (Some Finiteness Results).

(a) Show that there are only finitely many integral matrices of determinant d in Minkowski's fundamental domain $\mathcal{M}_n$.

(b) Let $S \in \mathcal{R}_a$ as defined in (1.201) and suppose that S has the block matrix decomposition

$$S = \begin{pmatrix} S_1 & S_{12} \\ {}^t S_{12} & S_2 \end{pmatrix}.$$

Show that the matrix $S_1^{-1} S_{12}$ has all of its elements bounded in absolute value by a constant depending only on n and a.

(c) Show that $\mathcal{M}_n \subset \mathcal{R}_a$ for some value of a. Can one prove that $\mathcal{R}_a$ is contained in a finite union of images of $\mathcal{M}_n$ for sufficiently small a?

Hints. (a) See Freitag [185, p. 36]. Use the boundedness of the product of the diagonal entries in such an integral matrix.

(b) See Siegel [562, p. 51]. Use the Iwasawa decomposition:

$$S = D[T] = \begin{pmatrix} D_1 & 0 \\ 0 & D_2 \end{pmatrix} \begin{bmatrix} T_1 & T_{12} \\ 0 & T_2 \end{bmatrix},$$

and write the matrix of interest in terms of the D_i, T_i, T_{12}.

Now finally we turn again to the proof of Theorem 1.4.1.

Proof of Part (2) of Theorem 1.4.1. (See Minkowski [453, Vol. II, pp. 67–68] or Hancock [260, Vol. II, pp. 787–788].)

We know from (1.202) and (1.203) that if $Y \in \mathcal{M}_n$ and Y has the Iwasawa decomposition $Y = D[T]$ given in (1.194), then

$$y_{ii} \geq d_i \geq \lambda_i y_{ii}, \qquad (1.204)$$

with λ_i positive and independent of Y. Moreover the t_{ij} are bounded. As in (1.195) we set $T = {}^t(\xi_1, \ldots, \xi_n)$. Then, if $a \in \mathbb{Z}^n$, define

$$\zeta_j(a) = {}^t\xi_j a = a_j + t_{j,j+1}a_{j+1} + \cdots + t_{jn}a_n, \qquad (1.205)$$

and note that

$$Y[a] = \sum_{j=1}^n d_j \, \zeta_j^2(a).$$

It is easily seen that a *finite* set of $a \in \mathbb{Z}^n$ (independent of $Y \in \mathcal{M}_n$) satisfy the following inequalities (see Exercise 1.4.10) with $\zeta_j = \zeta_j(a)$ in (1.205):

$$\lambda_1 \zeta_1^2 < \frac{1}{4}, \quad \lambda_2 \zeta_2^2 < \frac{2}{4}, \quad \ldots, \quad \lambda_n \zeta_n^2 < \frac{n}{4}. \qquad (1.206)$$

Exercise 1.4.10. (a) Prove that the $a \in \mathbb{Z}^n$ satisfying (1.206) form a finite set independent of the particular $Y \in \mathcal{M}_n$.

(b) Compute the λ_i in formula (1.204) as a function of v and u such that $\mathcal{M}_n \subset \mathcal{S}_{v,u}$ for the Siegel set defined in (1.200). Show that we can take λ_i to be greater than or equal to λ_{i+1} for all $i = 1, 2, \ldots, n-1$.

Hint. (a) The triangular nature of the transformation $\zeta_j(a)$ in (1.205) allows one to bound the entries of a recursively.

Now we prove that to put Y in $\mathcal{M}_n$ we need only the inequalities $Y[a] \geq y_{kk}$ with $a \in \mathbb{Z}^n$ determined by the inequalities (1.206) and assuming that a has the property that $\gcd(a_k, \ldots, a_n) = 1$. This list of vectors a includes $a = e_j$ and $a = e_i \pm e_j$.

We know that these inequalities are all that are needed to obtain formula (1.204) which is:

$$y_{hh} \geq d_h \geq \lambda_n y_{hh}.$$

Suppose that $b \in \mathbb{Z}^n$ with $\gcd(b_j, \ldots, b_n) = 1$. In order that b not satisfy (1.206), there must exist an index $h \leq n$ such that

$$\lambda_h \, \zeta_h^2(b) \geq h/4.$$

Let $h \leq n$ denote the largest such index. So b satisfies the inequalities in (1.206) except for those corresponding to the indices $j = 1, \ldots, h$. Then, since $d_h \geq \lambda_h y_{hh}$, we have

$$\left. \begin{aligned} Y[b] &\geq d_h \zeta_h^2(b) + \cdots + d_n \zeta_n^2(b) \\ &\geq y_{hh}\tfrac{h}{4} + d_{h+1}\zeta_{h+1}^2(b) + \cdots + d_n \zeta_n^2(b). \end{aligned} \right\} \tag{1.207}$$

Here if $h = n$, the last $n - h$ terms will be nonexistent, of course.

Now form

$$b_1^*, \ldots, b_h^* \quad \text{such that} \quad \left|\zeta_j(b^*)\right| \leq 1/2,$$

by subtracting integers from $b_1, \ldots, b_h$. And set

$$b_{h+1}^* = b_{h+1}, \ldots, b_n^* = b_n.$$

Now $\gcd(b_j^*, \ldots, b_n^*) = 1$. Therefore, making use of (1.207), we have:

$$\begin{aligned} Y[b^*] &\leq \frac{1}{4}(d_1 + \cdots + d_h) + d_{h+1}\zeta_{h+1}^2(b) + \cdots + d_n \zeta_n^2(b) \\ &\leq \frac{1}{4}(y_{11} + \cdots + y_{hh}) + d_{h+1}\zeta_{h+1}^2(b) + \cdots + d_n \zeta_n^2(b) \\ &\leq \frac{1}{4}h y_{hh} + d_{h+1}\zeta_{h+1}^2(b) + \cdots + d_n \zeta_n^2(b) \leq Y[b]. \end{aligned}$$

Finally

$$Y[b] \geq Y[b^*] \geq y_{jj}$$

since b^* satisfies all the inequalities in (1.206). This completes the proof of part (2) of Theorem 1.4.1, as it is now clear that $\mathcal{M}_n$ is a convex cone through the origin which is bounded by a finite number of hyperplanes through the origin. ∎

Exercise 1.4.11. In the definition (1.185) of $\mathcal{M}_n$, call those inequalities "boundary inequalities" which occur with equality. These boundary inequalities imply all the other inequalities. Prove this.

Hint. Suppose that $Y_0 \in \mathcal{M}_n$ and Y_0 satisfies only strict inequalities. Let $Y_1 \notin \mathcal{M}_n$ satisfy all the boundary inequalities. Consider the line segment joining Y_0 and Y_1; i.e., the points $Y_t = (1-t)Y_0 + tY_1$, for $0 \le t \le 1$. Let

$$u = \text{l.u.b.}\,\{t \in [0,1] \mid y_t \in \mathcal{M}_n\}.$$

Show that Y_u must satisfy a boundary inequality and then deduce a contradiction, using the linearity of the inequalities in the Y's. Note that we use the finiteness of the number of inequalities needed to define $\mathcal{M}_n$ here, but see Weyl [666, Vol. III, p. 743] for a modification and a reformulation of Minkowski's proof of part (2) of Theorem 1.4.1.

Proof of Part (3) of Theorem 1.4.1. It is easy to see that $Y, W = Y[A] \in \mathcal{M}_n$, for $A \in GL(n,\mathbb{Z})$, $A \ne \pm I$, implies that Y lies in the boundary of $\mathcal{M}_n$. For suppose that A is diagonal, with diagonal entries $a_1, \ldots, a_n$. Then there must be a first sign change, say between a_k and a_{k+1}. It follows that $w_{k,k+1} = -y_{k,k+1}$. Then $w_{k,k+1} \ge 0$ and $y_{k,k+1} \ge 0$ imply that $y_{k,k+1} = 0$ and thus Y must lie on the boundary of $\mathcal{M}_n$.

If A is not diagonal, there is a first column of A, say a_k, such that $a_k \ne \pm e_k$, for e_k the standard kth basis vector for $\mathbb{R}^n$. Then the kth column of A^{-1} has the same property. Call it b_k. Suppose that

$$a_k = {}^t(\alpha_1, \ldots, \alpha_n) \quad \text{and} \quad b_k = {}^t(\beta_1, \ldots, \beta_n).$$

Then

$$\gcd(\alpha_k, \ldots, \alpha_n) = 1 = \gcd(\beta_k, \ldots, \beta_n).$$

So it follows that

$$y_{kk} \le Y[a_k] = w_{kk} \quad \text{and} \quad w_{kk} \le W[b_k] = y_{kk}.$$

Thus we must have $y_{kk} = Y[a_k]$ and Y lies on the boundary of $\mathcal{M}_n$.

To complete the proof of part (3) of Theorem 1.4.1, we need to show that Y and $Y[A]$ both lie in $\mathcal{M}_n$ for at most a finite number of $A \in GL(n,\mathbb{Z})$. Again we shall follow Minkowski [453, Vol. II, p. 70] or Hancock [260, Vol. II, pp. 790–794].

Suppose that Y and $W = Y[A]$ both lie in $\mathcal{M}_n$ for $A \in GL(n,\mathbb{Z})$. Write $A = (a_1 \cdots a_n)$ with $a_j \in \mathbb{Z}^n$. Then $Y[a_j] = w_{jj}$. Suppose that a_{hk} is the last nonzero entry of a_k. Then using (1.204) and Exercise 1.4.10 (b), we find that:

$$w_{kk} \ge y_{hh} \ge d_h \ge \lambda_h y_{hh} \ge \lambda_n y_{hh}.$$

Claim. For each j, we have $w_{jj} \ge \lambda_n y_{jj}$ and $y_{jj} \ge \lambda_n w_{jj}$.

Proof of Claim. Suppose that $w_{jj} < \lambda_n y_{jj}$ for some j. Then

$$w_{11} \le w_{22} \le \cdots \le w_{jj} < \lambda_n y_{jj} \le \lambda_n y_{j+1,j+1} \le \cdots \le \lambda_n y_{nn}.$$

So no quantity $a_{hk}(h = j, j + 1, \ldots, n; \ k = 1, \ldots, j)$ can be the last not equal to zero. This means that the numbers $a_{hk}, \ h = j, \ldots, n; \ k = 1, \ldots, j$, must all be zero which implies that the determinant of A is zero, a contradiction, proving the claim.

If $w_{kk} \geq \lambda_n y_{k+1,k+1}$, for $k = 1, \ldots, n-1$, then we can easily bound the elements of a_k. For then we have:

$$y_{11} \geq \lambda_n w_{11} \geq \lambda_n^2 y_{22} \geq \lambda_n^3 w_{22} \geq \cdots$$
$$\geq \lambda_n^{2k-1} w_{kk} \geq \lambda_n^{2k} y_{k+1,k+1} \geq \cdots .$$

It follows that $w_{kk}/y_{11} \leq \lambda_n^{1-2k}$. We know that there is a positive constant c such that

$$c y_{11} \, {}^t a_k a_k \leq c Y_0[a_k] \leq Y[a_k] = w_{kk}.$$

from part (c) of Proposition 1.4.1. Thus a_k has bounded norm.

Next suppose that j is the largest index such that

$$w_{j-1,j-1} < \lambda_n y_{jj}.$$

The considerations in the proof of the claim show that

$$A = \begin{pmatrix} A_1 & A_{12} \\ 0 & A_2 \end{pmatrix} \quad \text{with} \quad A_1 \in GL(j-1, \mathbb{Z}).$$

Induction says that A_1 has bounded entries, since

$$Y[A] = \begin{pmatrix} Y_1 & * \\ * & * \end{pmatrix} \begin{bmatrix} A_1 & A_{12} \\ 0 & A_2 \end{bmatrix} = \begin{pmatrix} Y_1[A_1] & * \\ * & * \end{pmatrix}.$$

The hypothesis on j implies that if $j < n$, then

$$w_{kk} \geq \lambda_n y_{k+1,k+1}, \quad \text{for} \ k = j, \ldots, n-1.$$

Thus we have for $k = j, \ldots, n$, the inequality:

$$y_{jj} \geq \lambda_n^{2k-2j} y_{kk} \geq \lambda_n^{2k-2j+1} w_{kk}.$$

It follows that if $k \geq j$ and $Y = D[T]$ with D and T as usual in the Iwasawa decomposition (1.194), then if $\zeta_i(x)$ is defined by (1.205), for $x \in \mathbb{R}^n$, we have the following inequality (making use of (1.204)):

$$w_{kk} = Y[a_k] = \sum_{i=1}^{n} d_i \zeta_i^2(a) \geq \lambda_n y_{jj} \left(\zeta_j^2(a_k) + \cdots + \zeta_n^2(a_k) \right).$$

Therefore

$$\lambda_n^{-(2k-2j+2)} \geq \zeta_j^2(a_k) + \cdots + \zeta_n^2(a_k), \quad \text{for } k = j, \ldots, n,$$

which implies that the matrix A_2 has bounded entries. To see this, argue as in part (a) of Exercise 1.4.10.

It only remains to show that the matrix A_{12} has bounded entries. Since $W \in \mathcal{M}_n$, we know that for $k = j, \ldots, n$, $x_i \in \mathbb{Z}$, we have

$$W[x] \geq w_{kk}, \quad \text{if } x = {}^t(x_1, \ldots, x_{j-1}, \delta_{kj}, \ldots, \delta_{kn}).$$

Here δ_{ki} denotes the Kronecker delta; i.e., it is 1 if $k = i$ and 0 otherwise. Our inequality means that

$$Y[Ax] \geq Y[a_k] \quad \text{with} \quad Ax = {}^t(x_1^*, \ldots, x_{j-1}^*, a_{jk}, \ldots, a_{nk}) = a^*,$$

for an arbitrary vector $(x_1^*, \ldots, x_{j-1}^*) \in \mathbb{Z}^{j-1}$. We can choose this arbitrary vector to insure that $|\zeta_i(a^*)| \leq \frac{1}{2}$, for $i = 1, \ldots, j-1$. This implies that

$$d_1 \zeta_1^2(a^*) + \cdots + d_n \zeta_n^2(a^*) \geq d_1 \zeta_1^2(a_k) + \cdots + d_n \zeta_n^2(a_k)$$

and

$$\frac{1}{4}(d_1 + \cdots + d_{j-1}) \geq d_1 \zeta_1^2(a_k) + \cdots + d_{j-1} \zeta_{j-1}^2(a_k).$$

Thus $\zeta_i^2(a_k)$ is bounded for $i = 1, \ldots, j-1$, which implies that the entries of A_{12} are bounded, completing the proof of part (3) of Theorem 1.4.1, and thus the entire proof of Theorem 1.4.1—at last. ∎

Thus we have finished the discussion of the basic properties of the Minkowski fundamental domain $\mathcal{M}_n$ for $GL(n, \mathbb{Z})$. We tried to follow Minkowski's own reasoning in many places, despite the fact that we found it rather tortuous. In the next section we consider another sort of fundamental domain for the $GL(n, \mathbb{Z})$. The discussion will be much simpler. And in the last section we will look at the formula for the Euclidean volume of the subset of $\mathcal{M}_n$ consisting of all matrices of determinant less than or equal to one. Before this, let's consider a few miscellaneous geometric questions related to $\mathcal{M}_n$.

Exercise 1.4.12 (Edge Forms). Minkowski defined an **edge form** Q in $\mathcal{M}_n$ to be a reduced form such that $Q = Y + W$ with $Y, W \in \mathcal{M}_n$ implies that Y is a positive scalar multiple of W. Show that an edge form must be on an edge of $\mathcal{M}_n$. Define an equivalence relation between edge forms Y, W by saying that Y is equivalent to W if there is a positive real number c such that $Y = cW$. Show that there are only finitely many equivalence classes of edge forms. Show finally that every Y in $\mathcal{M}_n$ has an expression as a linear combination of edge forms with nonnegative scalars.

Hint. See Minkowski [453, Vol. II, p. 69] or Hancock [260, Vol. II, pp. 790–791].

Exercise 1.4.13 (The Determinant One Surface).

(i) Suppose that Y and W lie in the determinant one surface $\mathcal{SP}_n$ and $Y \neq W$. If $t \in (0,1)$, show that $tY + (1-t)W$ is in $\mathcal{P}_n$ with determinant greater than one.
(ii) Show that the determinant one surface in $\mathcal{M}_n$ is everywhere convex as seen from the origin; i.e., the surface lies on the side of the tangent plane away from the origin.

Hint. See Minkowski [453, Vol. II, p. 73] or Hancock [260, Vol. II, pp. 794–797]. Use the fact that there is a $k \in SL(n, \mathbb{R})$ such that both $Y[k]$ and $W[k]$ are diagonal. You will also need the inequality below which holds for positive real numbers $a_1, \ldots, a_n,\ b_1, \ldots, b_n$:

$$\left(\prod_{i=1}^{n} (a_i + b_i) \right)^{1/n} \geq \left(\prod_{i=1}^{n} a_i \right)^{1/n} + \left(\prod_{i=1}^{n} b_i \right)^{1/n},$$

with equality only if $a_i/b_i = a_{i+1}/b_{i+1}$, for all $i = 1, \ldots, n-1$.

Extreme forms $Y \in \mathcal{P}_n$ are defined to be forms for which the first minimum $m_Y |Y|^{-1/n}$ takes on a local maximum value. Recall that the problem of finding the densest lattice packings of spheres in Euclidean space seeks to find the Y with global maximum value of $m_Y |Y|^{-1/n}$. Korkine and Zolotareff [366] found these extreme forms in low dimensions ($n = 4, 5$). They used a reduction theory which makes use of the Iwasawa decomposition (see the next subsection and Ryskov and Baranovskii [523]). Minkowski [453, Vol. II, p. 76] shows that an extreme form of $\mathcal{M}_n$ must be an edge form. The proof uses Exercise 1.4.13 on the convexity of the determinant one surface in the fundamental domain. Minkowski finds representatives of the classes of positive edge forms with $m_Y = 2$ for $n = 2, 3, 4$ [453, p. 79] and claims that they are all extreme. Elsewhere [453, p. 218] he considers the cases $n = 5$ and 6. The inequivalent positive edge forms with $m_Y = 2$, for $n = 2, 3, 4$, are:

$$\begin{pmatrix} 2 & 1 \\ 1 & 2 \end{pmatrix} \quad \begin{pmatrix} 2 & 1 & 1 \\ 1 & 2 & 1 \\ 1 & 1 & 2 \end{pmatrix} \quad \begin{pmatrix} 2 & 1 & 1 & 1 \\ 1 & 2 & 1 & 1 \\ 1 & 1 & 2 & 1 \\ 1 & 1 & 1 & 2 \end{pmatrix} \quad \begin{pmatrix} 2 & 0 & 0 & 1 \\ 0 & 2 & 0 & 1 \\ 0 & 0 & 2 & 1 \\ 1 & 1 & 1 & 2 \end{pmatrix}.$$

Korkine and Zolotareff [366] also proved the following results.

(i) If f_n denotes the number of distinct vectors $a \in \mathbb{Z}^n - 0$ (modulo $\pm I$) such that $Y[a] = m_Y$ for an extreme form Y, then

$$n(n+1)/2 \leq f_n \leq (3^n - 1)/2.$$

(ii) The representations of the minimum m_Y of an extreme form Y determine Y completely, up to multiplication by a positive scalar.

Voronoi [636] gives an algorithm for finding all extreme forms and worked out the cases $n = 3, 4, 5$. Barnes [36] carried out the computation in the case $n = 6$. Voronoi's methods have also turned out to be useful in resolving the singularities of compactifications of the fundamental domain (see Ash et al. [30, pp. 145–150] and Namikawa [472, pp. 85–112]). Important in this work is the **Voronoi map** which is defined to be

$$V : \mathbb{R}^n \to \text{Closure}(\mathcal{P}_n)$$

$$x \mapsto V(x) = x\,{}^t x.$$

Note that $V(x)$ is positive semi-definite, since $V(x)[a] = ({}^t x a)^2 \geq 0$ for any $a \in \mathbb{R}^n$. The **Voronoi points** are the points x in $\mathbb{Z}^n$ with relatively prime coordinates. The **Voronoi cell** or **polyhedron** $\Pi(n)$ is the closure of the convex hull of the set of $V(x)$ such that x is a Voronoi point. See Ryskov and Baranovskii [523, pp. 40 ff.] for more details on Voronoi's theory.

Exercise 1.4.14 (Hermite–Mahler). Show that a subset of Minkowski's fundamental domain has compact closure if m_Y (the first minimum defined by (1.191)) is bounded from below and the determinant $|Y|$ is bounded from above. In the language of lattices, this says that the minimum separation of the lattice points is bounded from below and the volume of the fundamental parallelepiped is bounded from above.
Hint. Use parts (a) and (b) of Proposition 1.4.1.

The theory of smooth compactifications involves finding $GL(n, \mathbb{Z})$-invariant cone decompositions of the closure of $\mathcal{P}_n$. For $n \leq 3$, this comes from the following exercise.

Exercise 1.4.15. Define the **fundamental cone** by

$$C_0 = \left\{ X \in \mathbb{R}^{n \times n} \,\middle|\, {}^t X = X, \; x_{ij} \leq 0, \; \text{all } i \neq j, \; \sum_{i,j=1}^{n} x_{ij} \geq 0 \right\}.$$

Show that the closure of $\mathcal{P}_n$ is a union of images $C_0[A]$ over A in $GL(n, \mathbb{Z})$, when $n \leq 3$. According to Mumford: "This illustrates the interesting fact that only in 4 space or higher do lattice packing problems and related geometry of numbers problems get interesting" (see Ash et al. [30, p. 146]).

Exercise 1.4.16. (a) Use Theorem 1.4.1 to show that $GL(n, \mathbb{Z})$ is finitely generated. In fact, show that the $A \in GL(n, \mathbb{Z})$ such that $\mathcal{M}_n \cap (\mathcal{M}_n[A]) \neq \emptyset$ generate $GL(n, \mathbb{Z})$.

(b) Prove that $GL(n, \mathbb{Z})$ is generated by the matrices which give rise to the elementary row and column operations; i.e., by the matrices of the following three forms:

  (i) diagonal with ± 1 as the entries,
  (ii) permutation matrices,
  (iii) upper triangular matrices with ones on the diagonal and all elements above the diagonal equal to 0 except one which is equal to 1.

Minkowski [453, Vol. II, p. 95] showed that if $H(d)$ is the number of integral $Y \in \mathcal{P}_n$ of determinant d and inequivalent modulo $GL(n, \mathbb{Z})$, then

$$\lim_{x \to \infty} \left(x^{-(n+1)/2} \sum_{d=1}^{X} H(d) \right) = \text{Euclidean Volume} \{ Y \in \mathcal{M}_n \mid |Y| \leq 1 \}.$$

Note that $H(d)$ is finite by Exercise 1.4.9.

1.4.3   Grenier's Fundamental Domain

We begin with Figure 1.15. What do you see? The fact that I see bears should not prevent you from seeing other sorts of creatures. I explain this sort of figure later in the section as coming from a projection of Hecke points in Grenier's fundamental domain.

Here we consider a fundamental domain and reduction algorithm discussed by Grenier [241, 242] and pictured for $SL(3, \mathbb{Z})$ in Gordon et al. [237]. Grenier's

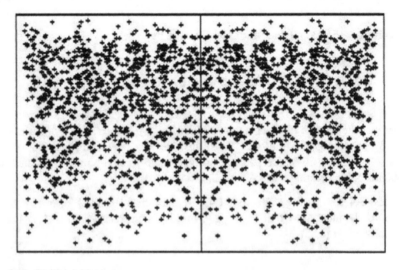

Fig. 1.15 $GL(3)$ ink blot test

method is analogous to that used by Siegel for $Sp(n,\mathbb{Z})$ in Siegel [565, Vol. II, pp. 105–113, 298–317] (see also Maass [426] and Gottschling [238]). As we have noted earlier, we should perhaps call the fundamental domain the HKZ domain for Hermite, Korkine, and Zolotareff. A more recent reference is Jorgenson and Lang [334]. Papers using the domain include Brenner and Sinton in the volume dedicated to Serge Lang [233, pp. 69–109], S. Catto et al. [102], Goldfarb [229], Sarnak and Strömbergsson [530], W. Schmidt [535], Eric Stade and Dorothy Wallace [574].

We need to consider another fundamental domain than Minkowski's for the reasons that follow. Recall that Minkowski's fundamental domain for $GL(3,\mathbb{Z})$ is:

$$\mathcal{M}_3 = \left\{ Y = (y_{ij}) \in \mathcal{P}_3 \;\middle|\; \begin{array}{l} y_{11} \leq y_{22} \leq y_{33}, \; 0 \leq y_{12} \leq \tfrac{1}{2}y_{11} \\ 0 \leq y_{23} \leq \tfrac{1}{2}y_{22}, \; |y_{13}| \leq \tfrac{1}{2}y_{11} \\ Y[e] \geq y_{33}, \; e = \pm e_1 \pm e_2 \pm e_3 \end{array} \right\}.$$

Here the e_j are the standard basis vectors for $\mathbb{R}^3$. Suppose Y has the Iwasawa decomposition:

$$
\begin{aligned}
Y &= \begin{pmatrix} y_1 & 0 & 0 \\ 0 & y_2 & 0 \\ 0 & 0 & y_3 \end{pmatrix} \begin{bmatrix} 1 & x_1 & x_2 \\ 0 & 1 & x_3 \\ 0 & 0 & 1 \end{bmatrix} \\
&= \begin{pmatrix} y_1 & y_1 x_1 & y_1 x_2 \\ y_1 x_1 & y_1 x_1^2 + y_2 & y_1 x_1 x_2 + y_2 x_3 \\ y_1 x_2 & y_1 x_1 x_2 + y_2 x_3 & y_1 x_2^2 + y_2 x_3^2 + y_3 \end{pmatrix}.
\end{aligned}
\tag{1.208}
$$

Thus if Y lies in Minkowski's fundamental domain, we find that:

$$0 \leq x_1 \leq \frac{1}{2}, \quad |x_2| \leq \frac{1}{2}, \quad 0 \leq y_1 x_1 x_2 + y_2 x_3 \leq \frac{1}{2}\left(y_1 x_1^2 + y_2\right).$$

If follows that:

$$-\left(\frac{1}{4}\right)\frac{y_1}{y_2} \leq x_3 \leq \frac{1}{2} + \left(\frac{3}{8}\right)\frac{y_1}{y_2}.
\tag{1.209}$$

This means that Minkowski's fundamental domain has only an **approximate box shape** as Y approaches the boundary; e.g., as $y_1/y_2 \to 0$. We prefer an **exact box shape**; that is, we prefer to see the inequality $0 \leq x_3 \leq \frac{1}{2}$, particularly when computing the integrals of Eisenstein series over truncated fundamental domains (a necessary prelude to generalizing the Selberg trace formula to $SL(3,\mathbb{Z})$).

Grenier [241, 242] describes a fundamental domain for $GL(n,\mathbb{Z})$ which makes essential use of the Iwasawa coordinates. Moreover, Grenier's domain has an exact box shape at the boundary. And Grenier gives a reduction algorithm to move Y in $\mathcal{P}_n$ into this fundamental domain via a "highest point method." See also Korkine and Zolotareff [365] for this fundamental domain using Iwasawa coordinates and successive minima. The method goes back to Hermite. We will (perhaps evilly) stick to our earlier nomenclature.

N.V. Novikova (aka N.V. Zaharova) [475] found exact boundaries of the domain for $n \leq 8$. R.A. Pendavingh and S.H.M. Van Zwam say in [483]: "It is unfortunate that the proofs were omitted from her [Novikova's] paper, as it appears to be a significant challenge to determine these irredundant sets. We were only able to verify her claims for $n \leq 4$. For $n \in \{5, 6\}$ we find sufficient sets that are slightly larger, and for larger n the sets we compute are much smaller. ...." Here we restrict ourselves to $n = 3$ and will not spend time worrying whether our domains are the same as Novikova's. Instead we attempt to visualize the five-dimensional domain using Hecke points. See Figures 1.18, 1.19, 1.20, 1.21, 1.22, 1.23, 1.24, and 1.25. Other references for fundamental domains are: Ryskov [521], Ryskov and Baranovskii [523], Siegel [565, Vol. II, pp. 105–113, 298–317], and B.A. Venkov [628]. We will say a little more about some of these other methods at the end of the section.

Before defining Grenier's fundamental domain, we need to fix a set of partial Iwasawa coordinates for $Y \in \mathcal{P}_n$ given by

$$Y = \begin{pmatrix} v & 0 \\ 0 & W \end{pmatrix} \begin{bmatrix} 1 & {}^t x \\ 0 & I_{n-1} \end{bmatrix}, \quad \text{for } v > 0, \ W \in \mathcal{P}_{n-1}, \ x \in \mathbb{R}^{n-1}. \quad (1.210)$$

Next let us consider how the action of a matrix in $GL(n, \mathbb{R})$ affects the v-coordinate. Suppose

$$M = \begin{pmatrix} a & {}^t b \\ c & D \end{pmatrix}, \quad \text{for } a \in \mathbb{Z}, \quad b, c \in \mathbb{Z}^{n-1}, \ D \in \mathbb{Z}^{(n-1)\times(n-1)}.$$

If $Y[M] = Y^*$ has partial Iwasawa coordinates v^*, W^*, and x^*, then we find that

$$v^* = v[a + {}^t xc] + W[c].$$

We want to think of the coordinate v as the reciprocal of the height of Y. This agrees with the idea of height in the case $n = 2$ which was used in Section 3.4 of Volume I to put a point into the fundamental domain for $SL(2, \mathbb{Z})$ by the highest point method. Thus we want Y to have coordinate v such that

$$v \leq v[a + {}^t xc] + W[c]$$

for any a, c forming the first column of a matrix in $GL(n, \mathbb{Z})$. It is thus natural to make the following definition.

Grenier's fundamental domain for $GL(n, \mathbb{Z})$ is defined to be the set $\mathcal{F}_n$ of Y in $\mathcal{P}_n$ satisfying the following three inequalities:

(1) $v \leq v[a + {}^t xc] + W[c]$, for $a \in \mathbb{Z}$, $c \in \mathbb{Z}^{n-1} - 0$, and

$$M = \begin{pmatrix} a & {}^t b \\ c & D \end{pmatrix} \in GL(n, \mathbb{Z});$$

(2) $W \in \mathcal{F}_{n-1}$, the fundamental domain for $GL(n-1, \mathbb{Z})$;
(3) $0 \le x_1 \le \frac{1}{2}$, $|x_i| \le \frac{1}{2}$, for $i = 2, \ldots, n-1$.

Note that for $n = 2$, this fundamental domain is just the same as Minkowski's. But for $n \ge 3$, it differs. In particular, it puts the x-variables into a "box" shape (see Exercise 1.4.19 below). The fundamental domain $\mathcal{F}_n$ was considered by Hermite also Korkine and Zolotareff (see Cassels [100, p. 259]). But Cassels notes that it doesn't appear that anyone (before Grenier) managed to show that the domain is a more reasonable one to use than Minkowski's (especially for those wishing to do a trace formula for $GL(3, \mathbb{Z})$), since it is defined by inequalities more similar to those that worked in the case $n = 2$ than those giving Minkowski's domain. However, as we noted earlier, the cryptography community has been successfully using the reduction theory attached this domain for lattices rather than quadratic forms.

We take the discussion of the following results from Grenier [241, 242].

Theorem 1.4.2. $\mathcal{F}_n$ is a fundamental domain for $GL(n, \mathbb{Z})$.

Before proving this result we need a lemma.

Lemma 1.4.4. *If Y is in Grenier's fundamental domain $\mathcal{F}_n$ and v, W are as in (1.210) with v the upper left-hand entry of Y (and the **inverse height** of Y), then if w_{jj} is the jth diagonal entry in W, we have the inequality:*

$$w_{jj} \ge \frac{3v}{4}.$$

Proof. If Y is in $\mathcal{F}_n$, then we know that $v \le v[a + {}^txc] + W[c]$ for any $a \in \mathbb{Z}, c \in \mathbb{Z}^{n-1}$ that can be made the first column of a matrix in $GL(n, \mathbb{Z})$. Take $a = 0$ and $c = e_j$, the jth element of the standard basis of $\mathbb{R}^{n-1}$. Then

$$v \le vx_j^2 + w_{jj} \le \frac{1}{4}v + w_{jj}.$$

The result follows.

Proof (of Theorem 1.4.2). We need to show two properties of $\mathcal{F}_n$:

(1) For any $Y \in \mathcal{P}_n$, there is a matrix $M \in GL(n, \mathbb{Z})$ such that $Y[M] \in \mathcal{F}_n$;
(2) For Y and $Y[M]$ both in $\mathcal{F}_n$ with $M \ne \pm I$, Y must be on the boundary of $\mathcal{F}_n$

We proceed by induction on n. The case $n = 2$ is already done. So assume that $\mathcal{F}_{n-1}$ is a fundamental domain for $GL(n-1, \mathbb{Z})$.

(1) To prove property (1), suppose $Y \in \mathcal{P}_n$ is given with partial Iwasawa decomposition (1.210). If $M \in GL(n, \mathbb{Z})$ has first column given, as usual, by ${}^t(ac)$, for $a \in \mathbb{Z}, c \in \mathbb{Z}^{n-1}$, then $Y[M]$ has its upper left corner $v^* = v[a + {}^txc] + W[c]$. It is easily seen that there are only finitely many a and c forming the first column of a matrix M in $GL(n, \mathbb{Z})$ such that v^* stays less than any given bound. Thus we may choose a and c so that v^* is minimal.

Then locate D^* in $GL(n-1,\mathbb{Z})$ so that $W[D^*] \in \mathcal{F}_{n-1}$ by the induction hypothesis. This means that we want to set

$$M_1 = M \begin{pmatrix} 1 & 0 \\ 0 & D^* \end{pmatrix} \in GL(n,\mathbb{Z}),$$

in order to say that $Y[M_1]$ satisfies the first two prerequisites for being a member of $\mathcal{F}_n$. Next we can put the x-coordinates of $Y[M_1]$ in the desired intervals by acting on it by:

$$N = \begin{pmatrix} \pm 1 & {}^t b^* \\ 0 & I \end{pmatrix}.$$

Thus $Y[M_1 N] \in \mathcal{F}_n$ and we're done with the proof of (1).

(2) To prove property (2), we use induction again. The case $n = 2$ is done. Now assume that the case of $\mathcal{F}_{n-1}$ is proved. And suppose that both Y and $Y^* = Y[M]$ lie in $\mathcal{F}_n$ for $M \in GL(n,\mathbb{Z})$. Then Y^* has partial Iwasawa decomposition

$$Y^* = \begin{pmatrix} v^* & 0 \\ 0 & W^* \end{pmatrix} \begin{bmatrix} 1 & {}^t x^* \\ 0 & I_{n-1} \end{bmatrix}, \quad \text{for } v^* > 0,\ W^* \in \mathcal{P}_{n-1},\ x^* \in \mathbb{R}^{n-1}.$$

We know that in terms of the inverse heights, since $Y^*[M^{-1}] = Y$:

$$v \le v\left[a + {}^t xc\right] + W[c] = v^* \quad \text{and similarly } v^* \le v.$$

Thus we see that $v = v^* = v\left[a + {}^t xc\right] + W[c]$. If $c \ne 0$, this equality puts Y on the boundary of $\mathcal{F}_n$ because it is close to something outside of $\mathcal{F}_n$. If $c = 0$, the equality means that $v = va^2$ and thus $a = \pm 1$ and

$$M = \begin{pmatrix} \pm 1 & {}^t b \\ 0 & D \end{pmatrix} \quad \text{and} \quad D \in GL(n-1,\mathbb{Z}).$$

Then $W^* = W[D]$ and W both lie in $\mathcal{F}_{n-1}$ and the induction hypothesis implies that W lies on the boundary of $\mathcal{F}_{n-1}$ and thus Y is on the boundary of $\mathcal{F}_n$ unless $D = \pm I$. In the latter case, by looking at the effect of M on the x's, we see that:

$$x \text{ and } \pm x \pm b \in \left[-\frac{1}{2}, \frac{1}{2}\right]^{n-1}, \quad \text{with } x_1,\ \pm x_1 \pm b_1 \ge 0.$$

Since $b \in \mathbb{Z}^{n-1}$, we see that either:

$$x_i = \pm\frac{1}{2} \quad \text{and} \quad b_i = \pm 1, \quad i = 2,\ldots,n-1; \quad x_1 = \frac{1}{2}, \quad b_1 = 1$$

or

$$x_i \neq \pm\frac{1}{2} \quad \text{and} \quad b_i = 0, \quad i = 1, \ldots, n - 1.$$

If all the $b_i = 0$ and if $M \neq \pm I$, it follows that $x_1 = 0$ and Y must lie on the boundary of $\mathcal{F}_n$.

Theorem 1.4.3 (Grenier). *$\mathcal{F}_n$ has a finite number of boundary inequalities, which can be explicitly given for small values of n.*

Proof. Once more we use induction on n. We already know the case $n = 2$ and now we assume the result for $\mathcal{F}_{n-1}$. We need to show that only a finite number of inequalities of the form

$$v \leq v\,[a + \,^t xc] + W[c], \qquad \begin{cases} \forall \; a, c \text{ forming the 1st column} \\ \quad \text{of a matrix in } GL(n, \mathbb{Z}), \end{cases}$$

are necessary to place Y in $\mathcal{F}_n$. Using the partial Iwasawa coordinate decomposition (1.210), we know that $W \in \mathcal{F}_{n-1}$ and so we can write

$$W = \begin{pmatrix} v' & 0 \\ 0 & W' \end{pmatrix} \begin{bmatrix} 1 & \,^t x' \\ 0 & I_{n-2} \end{bmatrix}, \quad v' > 0, \quad W' \in \mathcal{F}_{n-1}, \quad x' \in \left[-\frac{1}{2}, \frac{1}{2} \right]^{n-1}.$$

By the induction hypothesis there are only finitely many inequalities:

$$v' \leq v'\left[a' + \,^t x'c'\right] + W'[c']$$

which must be considered. Why?
    And clearly

$$W[c] = v'\left[c_1 + \,^t x'c'\right] + W'\left[c'\right] \quad \text{with} \quad \,^t c' = (c_2, \ldots, c_n).$$

So we now have a finite number of vectors c. Given the bounds on the x_i, this leads to a finite number of a.
    To describe what is happening more explicitly, continue the partial Iwasawa decompositions until you reach the full Iwasawa decomposition and obtain:

$$v\,[a + \,^t xc] + W[c] = v\,(a + \,^t xc)^2 + v'\,(c_1 + \,^t x'c')^2 + \cdots$$
$$+ v^{(n-2)}\left(c_{n-2} + x^{(n-2)}c_{n-1}\right)^2 + v^{(n-1)}c_{n-1}^2.$$

By repeated application of Lemma 1.4.4, we know that

$$v^{(i)} \geq \kappa^i v, \quad \text{for } \kappa = \frac{3}{4}.$$

Thus

$$v\left[a + {}^txc\right] + W[c]$$

$$\geq v \left\{ \left(a + {}^txc\right)^2 + \kappa \left(c_1 + {}^tx'c'\right)^2 + \cdots + \kappa^{n-2} \left(c_{n-2} + x^{(n-2)}c_{n-1}\right)^2 + \kappa^{n-1}c_{n-1}^2 \right\}.$$

This means that we need to only consider the a and c such that

$$\left(a + {}^txc\right)^2 + \kappa \left(c_1 + {}^tx'c'\right)^2 + \cdots + \kappa^{n-1}c_{n-1}^2 \leq 1.$$

Since the x_j's are bounded by $\frac{1}{2}$ in absolute value, this bounds the a's and c's.

One can go on to determine the exact list of inequalities for small values of n. See Grenier [241, 242] for the cases $n \leq 5$. See Novikova [475] for $n \leq 8$, also Pendavingh and Van Zwam [483].

Exercise 1.4.17. Show that on the determinant one surface, Grenier's fundamental domain $\mathcal{SF}_3$ for $SL(3, \mathbb{Z})$ is the set of $Y \in \mathcal{SP}_3$ with partial Iwasawa coordinates given by

$$W = \begin{pmatrix} w & 0 \\ 0 & 1/w \end{pmatrix} \begin{bmatrix} 1 & x_3 \\ 0 & 1 \end{bmatrix},$$

$$Y = Y(v, w, x) = \begin{pmatrix} v & 0 \\ 0 & v^{-1/2}W \end{pmatrix} \begin{bmatrix} 1 & (x_1, x_2) \\ 0 & I_2 \end{bmatrix}, \quad x = (x_1, x_2, x_3),$$

and satisfying the following inequalities:

$$0 \leq x_1 \leq \frac{1}{2}, \quad |x_2| \leq \frac{1}{2}, \quad 0 \leq x_3 \leq \frac{1}{2},$$

$$1 \leq w^{-2} + x_3^2,$$

$$v \leq v(a + {}^txc)^2 + v^{-1/2}W[c] \quad \text{for} \begin{cases} a = 0, & {}^tc = (1, 0), \ (0, 1), \ (1, -1), \\ a = 1, & {}^tc = (-1, 1). \end{cases}$$

Hint. See Grenier [241, 242] or Gordon et al. [237]. Note that $GL(3, \mathbb{Z})/SL(3, \mathbb{Z})$ has order 2 and the nontrivial coset is represented by $-I$ which does nothing to an element $Y \in \mathcal{SP}_3$, so the fundamental domain for $SL(3, \mathbb{Z})$ is the same as that for $GL(3, \mathbb{Z})$.

Setting $t^{-2} = v^{-3/2}w$, the **explicit inequalities for** $\mathcal{SF}_3$ **are:**

$$
\left.
\begin{aligned}
&\text{(i)} \quad 1 \le (1 - x_1 + x_2)^2 + t^{-2}\{(1 - x_3)^2 + w^{-2}\} \\
&\text{(ii)} \quad 1 \le (x_1 - x_2)^2 + t^{-2}\{(1 - x_3)^2 + w^{-2}\} \\
&\text{(iii)} \quad 1 \le x_1^2 + t^{-2} \\
&\text{(iv)} \quad 1 \le x_2^2 + t^{-2}(x_3^2 + w^{-2}) \\
&\text{(v)} \quad 1 \le x_3^2 + w^{-2} \\
&\text{(vi)} \quad 0 \le x_1 \le \tfrac{1}{2} \\
&\text{(vii)} \quad 0 \le x_3 \le \tfrac{1}{2} \\
&\text{(viii)} \quad -\tfrac{1}{2} \le x_2 \le \tfrac{1}{2}.
\end{aligned}
\right\}
\tag{1.211}
$$

Note that inequalities (iii) (v), (vi), and (vii) say that $x_3 + iw^{-1}$ and $x_1 + it^{-1}$ lie in the fundamental domain for $GL(2, \mathbb{Z})\backslash H$ which is half that for $SL(2, \mathbb{Z})\backslash H$, given in Exercise 3.3.1 of Volume I.

Exercise 1.4.18. Suppose $Y \in \mathcal{P}_3$. Write the full Iwasawa decomposition $Y = D[T]$, where D is positive diagonal having entries d_i on the diagonal, T upper triangular with 1's on the diagonal and entries t_{ij} above the diagonal. Show that, using the notation of the preceding exercise, $t^2 = d_1/d_2$ and $w^2 = d_2/d_3$.

Exercise 1.4.19. (a) Show that $w \le 1$ and $t \le 1$ for $t^{-2} = v^{-3/2}w$, plus inequalities (vi)–(viii) from (1.211) imply that the point $Y = Y(v, w, x)$ with Iwasawa coordinates given in Exercise 1.4.17 must lie in the fundamental domain $\mathcal{SF}_3$. That is, in this case, we do not need all of the inequalities (1.211). This shows that Grenier's fundamental domain does have an exact box shape at infinity, unlike Minkowski's fundamental domain (cf. (1.209)). Note that, in general,

$$Y \in \mathcal{SF}_3 \text{ implies that } w \text{ and } t \text{ are both } \le 2/\sqrt{3}.$$

(b) In the case of $GL(3, \mathbb{Z})$, compare the Grenier fundamental domain with the Siegel set $\mathcal{S}_{v, u}$ defined in formula (1.200) of Section 1.4.2. In particular, show that for the 3×3 case we have:

$$\mathcal{SF}_3 \subset \mathcal{S}_{4/3, 1/2} \cap \mathcal{SP}_3$$

and

$$\mathcal{S}_{1,1/2} \cap \mathcal{SP}_3 \cap \{Y = Y(v, w, x) \mid x_1, x_3 \ge 0\} \subset \mathcal{SF}_3.$$

Here $Y(v, w, x)$ is defined in Exercise 1.4.17. This is a similar situation to that which occurred for $GL(2, \mathbb{Z})$.

Boundary identifications for the fundamental domain $\mathcal{SF}_3$ come from completing the $^t(a\ c)$ in Exercise 1.4.17 to matrices in $SL(3, \mathbb{Z})$, a process which can be carried out as follows:

$$T1 = \begin{pmatrix} 1 & 1 & 0 \\ 0 & 1 & 0 \\ 0 & 0 & 1 \end{pmatrix}, \quad T2 = \begin{pmatrix} 1 & 0 & 1 \\ 0 & 1 & 0 \\ 0 & 0 & 1 \end{pmatrix}, \quad T3 = \begin{pmatrix} 1 & 0 & 0 \\ 0 & 1 & 1 \\ 0 & 0 & 1 \end{pmatrix},$$

$$S1 = \begin{pmatrix} 0 & 0 & 1 \\ 1 & 0 & 0 \\ 0 & 1 & 0 \end{pmatrix}, \quad S2 = \begin{pmatrix} 0 & 1 & 0 \\ 0 & 0 & 1 \\ 1 & 0 & 0 \end{pmatrix}, \quad S3 = \begin{pmatrix} 0 & 1 & 0 \\ 1 & 0 & -1 \\ -1 & 0 & 0 \end{pmatrix},$$

$$S4 = \begin{pmatrix} 1 & 0 & 0 \\ -1 & 1 & 0 \\ 1 & 0 & 1 \end{pmatrix}, \quad S5 = \begin{pmatrix} 1 & 0 & 0 \\ 0 & 0 & 1 \\ 0 & -1 & 0 \end{pmatrix},$$

$$U1 = \begin{pmatrix} -1 & 0 & 0 \\ 0 & -1 & 0 \\ 0 & 0 & 1 \end{pmatrix}, \quad U2 = \begin{pmatrix} 1 & 0 & 0 \\ 0 & -1 & 0 \\ 0 & 0 & -1 \end{pmatrix}.$$

$$(1.212)$$

Note. This gives more than enough generators for $SL(3, \mathbb{Z}) / \{\pm I\}$, but we do not appear to be able to get rid of any of the inequalities in (1.211) (cf. Exercise 1.4.16).

Grenier's reduction algorithm is a "highest point method" where the height of Y is $1/v$, for $v =$ the entry y_{11}. Korkine and Zolotareff [365] view it as a method of successive minima. The algorithm goes as follows.

Grenier's Reduction Algorithm

Here we use the matrices defined by (1.212) and the coordinates from Exercise 1.4.17.

Step I. Set $S0 = I$ and pick j to minimize the v-coordinate of $Y[Sj]$, for $j = 0, 1, 2, 3, 4$. Then replace Y by $Y[Sj]$.

Step II. Let $W(Y)$ denote the element of $\mathcal{SP}_2$ defined by the equations in Exercise 1.4.17. Put $W(Y)[\delta]$ in $\mathcal{SF}_2$ using $\delta \in GL(2, \mathbb{Z})$. That is, we make w and x_3 satisfy inequalities (v) and (vii) in (1.211). Replace Y by $Y[\gamma]$ for

$$\gamma = \begin{pmatrix} \pm 1 & 0 \\ 0 & \delta \end{pmatrix} \in SL(3, \mathbb{Z}).$$

Here $\gamma = S5$, $U1$, or $(T3)^n$ for some $n \in \mathbb{Z}$.

Step III. Translate the x_1, x_2-coordinates of Y in Exercise 1.4.17 by applying $\gamma = (Tj)^p$ to Y, for $p = \lfloor \frac{1}{2} - x_j \rfloor$, $j = 1, 2$. Here $\lfloor x \rfloor$ denotes the floor of x; i.e., the greatest integer $\leq x$.

Step IV. Make $x_1 \geq 0$ by replacing Y by $Y[U2]$, if necessary.

Keep performing Steps I–IV until the process stabilizes.

Historical Note. Jeff Stopple suggested that we use this last test; i.e., see whether the process has repeated itself to stop the program. This idea is useful since it allows us not to test all the inequalities at each step, as some might be tempted to do. On the other hand one might worry that the program would get into an infinite loop. This does not happen if one is careful in writing the code. However, one must be rather cautious because there can be great loss of precision due to subtraction and division. Thus we found that we had to use double precision when performing the algorithm in BASIC on the UCSD VAX in the 1980s.

Note that in obtaining the matrices Sj, $j = 1, 2, 3, 4, 5$, we completed the matrices ${}^t(a \ c)$ from Exercise 1.4.17 to 3×3 matrices in $SL(3, \mathbb{Z})$. This can be done in a number of ways—each differing by matrices of the form

$$\begin{pmatrix} 1 & {}^t q \\ 0 & R \end{pmatrix}$$

with $q \in \mathbb{Z}^2$ and $R \in \mathbb{Z}^{2 \times 2}$. The choice of Sj will affect the reduction algorithm, but not the final result that the algorithm does send a point into the fundamental domain.

Exercise 1.4.20. Write a program to carry out Grenier's reduction algorithm.

Let us now consider the results of some computer experiments we did with Dan Gordon and Doug Grenier which were published in Gordon et al. [237]. Our aim was understanding what $\mathcal{SF}_3$ looks like.

First recall that we saw in Figure 3.39 from Section 3.5, Vol. I, and in Exercise 3.6.19, Vol. I, that one can use Hecke operators to help us visualize the fundamental domain for $SL(2, \mathbb{Z})$ in the Poincaré upper half plane. Now we would like to do something similar to visualize Grenier's fundamental domain $\mathcal{SF}_3$. Before attempting that, let us look at Figure 1.16 which is a picture of the standard fundamental domain for $SL(2, \mathbb{Z})$. Then look at Figure 1.17 which consists of images of the point $z = 0.4i$ under matrices from the Hecke operator for the prime $p = 7919$, this time making the transformation $z = x + iy \longmapsto x + i/y$. Figure 1.17 was constructed using Mathematica on a PC with a similar program to that used to create Figure 3.39 in Volume I except that we applied the map $z = x + iy \longmapsto x + i/y$ rather than the map $z \longmapsto -1/z$.

Since $\mathcal{SF}_3$ is five-dimensional, we take the easy way out and look at graphs of two coordinates from the five coordinates $(t, w, x1, x2, x3)$ from Exercise 1.4.17. So there are ten possible graphs. The most interesting is that of (w, v) showing the shape of the **cuspidal region**, where v or w approaches 0.

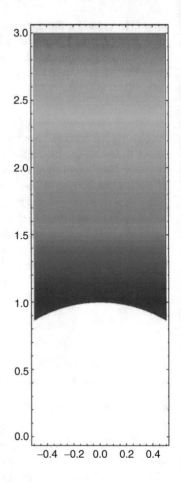

Fig. 1.16 The standard fundamental domain $\mathcal{SF}_2$ for $SL(2, \mathbb{Z})$

We quickly see that from (1.211), as in Exercise 1.4.13, that

$$t, w \leq 2(3^{-1/2}) \cong 1.154701. \qquad (1.213)$$

Therefore

$$v \leq 4/3 \cong 1.333333.$$

Hecke operators for $\Gamma_3 = GL(3, \mathbb{Z})$ will be discussed in Section 1.5. Here we need to only consider the simplest aspects of the theory. For $\Gamma_3 = GL(3, \mathbb{Z})$, $f : \mathcal{SP}_3/\Gamma_3 \to \mathbb{C}$ and $m \in \mathbb{Z}^+$, define the **Hecke operator** T_m by:

$$T_m f(Y) = \sum_{A \in M_m/\Gamma_3} f(Y[A]^0) \qquad (1.214)$$

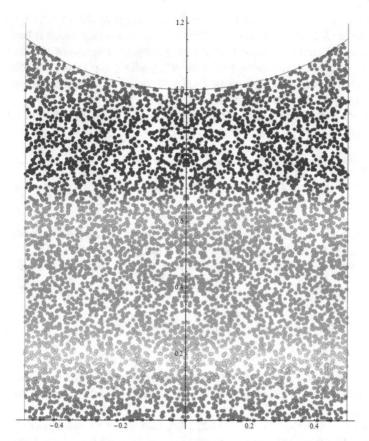

Fig. 1.17 Images of Hecke points $(z_0 + j)/p$, $0 \leq j \leq p - 1$, in the fundamental domain for $SL(2, \mathbb{Z})$, with $z_0 = 0.4i$ and $p = 7919$. The fundamental domain is transformed by the map $z = x + iy \longmapsto x + i/y$

where

$$M_m = \left\{ A \in \mathbb{Z}^{3 \times 3} \;\middle|\; |A| = m \right\}$$

and

$$Y^0 = |Y|^{-1/3} \, Y \in \mathcal{SP}_3.$$

It is easily seen (as in Lemma 1.4.6 which follows) that one can take representatives of M_m/Γ_3 of the form

$$\begin{pmatrix} d_1 & d_{12} & d_{13} \\ 0 & d_2 & d_{23} \\ 0 & 0 & d_3 \end{pmatrix}, \quad d_i > 0, \quad \prod_{i=1}^{3} d_i = m, \quad 0 \leq d_{ij} < d_i. \tag{1.215}$$

Maass [414] studied Hecke operators for the Siegel modular group $Sp(n, \mathbb{Z})$ in 1951. We are imitating his version of the theory. It is a theory which is basic to the study of automorphic forms on higher rank symmetric spaces G/K and it connects with many questions in representation theory, p-adic group theory, combinatorics, and number theory. Applications of Hecke operators to numerical integration on spheres are given by Lubotzky et al. [409, 410].

It is not hard to see that the Hecke operators for $SL(3, \mathbb{Z})$ have the following properties:

(i) $T_n T_m = T_{mn}$,   if $\gcd(m, n) = 1$,

(ii) $\displaystyle\sum_{r \geq 0} T_{p^r} X^r = (I - T_p X + [(T_p)^2 - T_{p^2}]X^2 - p^3 X^3)^{-1}$,   for $p =$ prime

It follows that L-functions associated with eigenforms f of the Hecke operators must have Euler products. We will discuss all these things in Section 1.5.

Here we graph points for the Hecke operator T_p, $p =$ prime. We use only the matrices

$$M(p; a, b) = p^{-1/3} \begin{pmatrix} p\ a\ b \\ 0\ 1\ 0 \\ 0\ 0\ 1 \end{pmatrix}, \quad 0 \leq a, b \leq p - 1. \tag{1.216}$$

The other matrices in T_p do not appear to be necessary. *Moreover, we will restrict b so that $b \equiv 5a + 163$ (mod p).* This will restrict the number of points to p rather than p^2 points.

Figures 1.18, 1.19, 1.20, 1.21, 1.22, 1.23, 1.24, and 1.25 show plots of pairs of Iwasawa-type coordinates of Γ_3-images of Hecke points in the fundamental domain $\mathcal{SF}_3$ or in the union of fundamental domains obtained by letting the x-coordinates run between $-\frac{1}{2}$ and $+\frac{1}{2}$. This allows us to produce more pleasing symmetric pictures than those in Gordon et al. [237]. By "Hecke points" we mean points of the form

$$Y_0[M(p; a, b)], \quad 0 \leq a, b \leq p - 1, \tag{1.217}$$

for $M(p; a, b)$ as in (1.216), $p = 3001$, and fixed Y_0 with its Iwasawa coordinates, as in Exercise 1.4.17, given by:

$$(v, w, x_1, x_2, x_3) = (.7815, .6534, .2123, .0786, .3312). \tag{1.218}$$

In Figure 1.18, the graph shows t versus w, where $t = v^{3/4} w^{-1/2}$. Note that t is the square root of the quotient of the first two diagonal entries in the Iwasawa decomposition of Y. The coordinate w is the square root of the quotient of the last two diagonal entries in the Iwasawa decomposition of Y. And we have seen in (1.211) that t and w play a more similar role than v and w, the variables

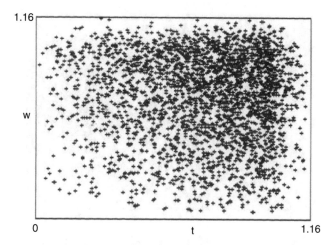

Fig. 1.18 The (t, w) coordinates of $SL(3, \mathbb{Z})$-images of Hecke points $Y_0[M(p; a, b)]$ in the fundamental domain $\mathcal{SF}_3$ for $SL(3, \mathbb{Z})$ using the notation (1.216), with Y_0 defined in (1.217), (1.218), and $p = 3001$. Here $b \equiv 5a + 163 (\mathrm{mod}\, p)$

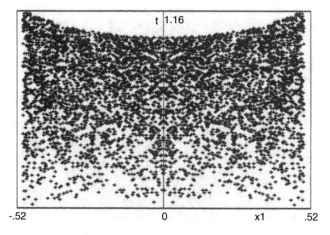

Fig. 1.19 The (x_1, t) coordinates of $SL(3, \mathbb{Z})$-images of Hecke points $Y_0[M(p; a, b)]$ in the union of the projection of the fundamental domain $\mathcal{SF}_3$ for $SL(3, \mathbb{Z})$ and its mirror image under the reflection across the t-axis. We use the notation (1.216), with Y_0 defined in (1.217), (1.218), and $p = 3001$. Here $b \equiv 5a + 163 (\mathrm{mod}\, p)$

we graphed in Gordon et al. [237]. Moreover, if $Y = D[T]$ is the Iwasawa decomposition of Exercise 1.4.18, then t^2 and w^2 are the simple roots d_1/d_2 and d_2/d_3, respectively. Roots will be discussed later. See Chapter 2.

Figures 1.19 and 1.20 show plots of the coordinates (x_1, t) and (x_3, w) of Hecke points for the prime $p = 3001$. These are the variables in copies of the fundamental domain of $SL(2, \mathbb{Z})$ in the Poincaré upper half plane. The figures *do* give a good approximation to Figure 1.17, as expected.

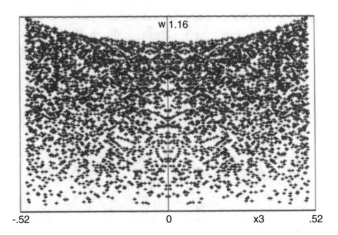

Fig. 1.20 The (x_3, w) coordinates of $SL(3, \mathbb{Z})$-images of Hecke points $Y_0[M(p; a, b)]$ in the union of the projection of the fundamental domain $\mathcal{SF}_3$ for $SL(3, \mathbb{Z})$ and its mirror image under reflection across the w-axis. We use the notation (1.216), with Y_0 defined in (1.217), (1.218), and $p = 3001$. Here $b \equiv 5a + 163 \pmod{p}$

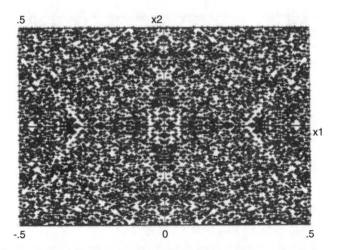

Fig. 1.21 The (x_1, x_2) coordinates of $SL(3, \mathbb{Z})$-images of Hecke points $Y_0[M(p; a, b)]$ in the union of the projection of the fundamental domain $\mathcal{SF}_3$ for $SL(3, \mathbb{Z})$ and its mirror image under reflection across the x_2-axis. We use the notation (1.216), with Y_0 defined in (1.217), (1.218), and $p = 3001$. Here $b \equiv 5a + 163 \pmod{p}$

Figures 1.21 and 1.22 give plots of $(x1, x2)$ and $(x1, x3)$, respectively. The plots look like randomly placed points in $[-\frac{1}{2}, \frac{1}{2}]^2$.

Figures 1.23, 1.24, and 1.25 are plots of (x_2, t), (x_3, t), and (x_2, w). The result should be compared with Figures 1.19 and 1.20. If we do so, we see that the top curves of Figures 1.23, 1.24, and 1.25 cannot be those of Figures 1.19 and 1.20. The variables in Figures 1.23, 1.24, and 1.25 are less closely related.

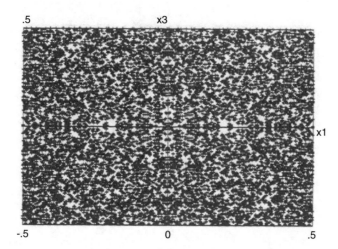

Fig. 1.22 The (x_1, x_3) coordinates of $SL(3, \mathbb{Z})$-images of Hecke points $Y_0[M(p; a, b)]$ in the union of the projection of the fundamental domain $\mathcal{SF}_3$ for $SL(3, \mathbb{Z})$ and its mirror images under reflections across the x_1 and x_3 axes. We use the notation (1.216), with Y_0 defined in (1.217), (1.218), and $p = 3001$. Here $b \equiv 5a + 163 \pmod{p}$

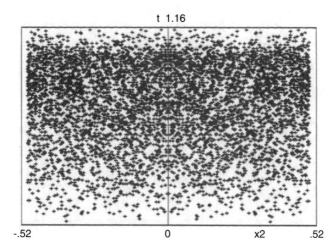

Fig. 1.23 The (x_2, t) coordinates of $SL(3, \mathbb{Z})$-images of Hecke points $Y_0[M(p; a, b)]$ in the fundamental domain $\mathcal{SF}_3$ for $SL(3, \mathbb{Z})$ using the notation (1.216), with Y_0 defined in (1.217), (1.218), and $p = 3001$. Here $b \equiv 5a + 163 \pmod{p}$

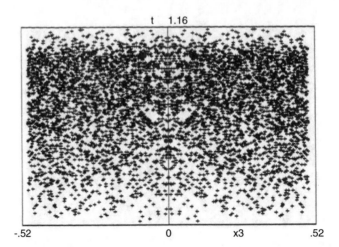

Fig. 1.24 The (x_3, t) coordinates of $SL(3, \mathbb{Z})$-images of Hecke points $Y_0[M(p; a, b)]$ in the union of the projection of the fundamental domain $\mathcal{SF}_3$ for $SL(3, \mathbb{Z})$ and its mirror image under reflection across the t-axis. We use the notation (1.216), with Y_0 defined in (1.217), (1.218), and $p = 3001$. Here $b \equiv 5a + 163 (\mathrm{mod}\, p)$

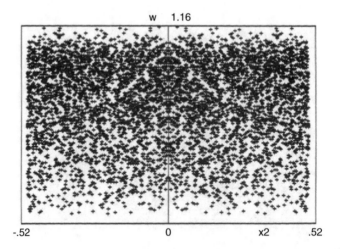

Fig. 1.25 The (x_2, w) coordinates of $SL(3, \mathbb{Z})$-images of Hecke points $Y_0[M(p; a, b)]$ in the fundamental domain $\mathcal{SF}_3$ for $SL(3, \mathbb{Z})$ using the notation (1.216), with Y_0 defined in (1.217), (1.218), and $p = 3001$. Here $b \equiv 5a + 163 (\mathrm{mod}\, p)$

These figures should also be compared with those in Gordon et al. [237] where, for example, the points were plotted as points. Here we are plotting small cross marks.

One might complain that our graphs still do not give a real five-dimensional feeling for the fundamental domain. We hope to make "$\mathcal{SF}_3$ THE MOVIE" some day, making use of motion and color. This would be a non-Euclidean analogue of Banchoff's movie of a rotating four-dimensional cube. For you may view our region $\mathcal{SF}_3$ as a five-dimensional non-Euclidean crystal. It would also be nice to produce a figure representing the tessellation of the five-dimensional space $\mathcal{SP}_3$ corresponding to $SL(3, \mathbb{Z})$ images of $\mathcal{SF}_3$. These would be five-dimensional analogues of pictures that inspired the artist M.C. Escher. Such graphs could be obtained by plotting images of boundary curves under matrices generated by Si, Ti, Ui appearing in Grenier's reduction algorithm.

Figures 1.18, 1.19, 1.20, 1.21, 1.22, 1.23, 1.24, and 1.25 were obtained in the late 1980s with my equipment at the time—a Fujitsu printer using an Atari 1040ST with Snapshot and Degas Elite to process the points computed on the UCSD VAX computer. If I were a good person I would update these figures. But I have to say that the old Atari programs were much easier to use that what I am using now on my PC. I could write a Mathematica program to do this but it would take more time than I have to translate all those goto statements and loops into proper Mathematica. It would be a good exercise though.

Using Mathematica we have managed to produce a movie of the 3D coordinates (x_1, x_2, x_3) for a matrix in $\mathcal{SF}_3$ as the y-values move around at the "bottom" of their heights. Here $y_1 = w^{-1}$ and $y_2 = t^{-1}$ (or vice versa). The main part of the Mathematica Program for the movie described at the end of the last paragraph is:

```
haro[t_, w_] := haro[t, w] =
    RegionPlot3D[
        (1 - x1 + x2)^2 + t^2*((1 - x3)^2 + w^2) >= 1
          && (x1 - x2)^2 + t^2*((1 - x3)^2 + w^2) >= 1
          && x1^2 + t^2 >= 1
          && x2^2 + t^2*(x3^2 + w^2) >= 1
          && x3^2 + w^2 >= 1,
        {x1, 0, .5}, {x2, -.5, .5}, {x3, 0, .5},
      Mesh->None,PlotStyle->Directive[Opacity[0.5],Pink,
        Specularity[White, 20]], PlotPoints -> 60]
```

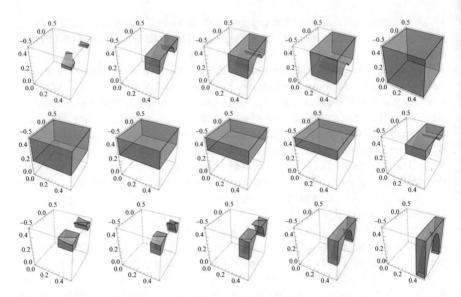

Fig. 1.26 We use the Mathematica command haro[y1,y2] repeatedly with (y_1, y_2) running through the following table of values with $y_1 = t^{-1}$, $y_2 = w^{-1}$ from (1.211), the *rows* corresponding to the *rows* in the figure, likewise the *columns*.

$$(.92, .92) \; (.94, .94) \; (.96, .96) \; (.98, .98) \; (1, 1)$$
$$(1, .98) \quad (1, .96) \quad (1, .94) \quad (1, .92) \quad (.96, .92)$$
$$(.94, .92) \; (.92, .94) \; (.92, .96) \; (.92, .98) \; (.92, 1)$$

Since a book cannot at present contain a movie, we have put various frames of the movie in Figure 1.26. It shows a bridge being built out of nothing, then turning into a solid block, shrinking, then growing again. There is an actual movie on my website.

There are various ways of understanding why the Hecke points should be dense in $\mathcal{SF}_3$. One could imitate an argument of Zagier using Eisenstein series (see page 314 of Volume I for the $SL(2, \mathbb{Z})$-version of the argument) to show that the image of a horocycle C_Y becomes dense in $\mathcal{SF}_3$ as Y approaches the boundary of $\mathcal{SP}_3$. Here, by a **horocycle** we mean the set:

$$C_Y = \{Y[n] \mid n \text{ is upper triangular with 1 on the diagonal}\}.$$

Patrick Chiu [105] proves the following result. Suppose

$$S(p) = \left\{ \begin{pmatrix} p & a & b \\ 0 & 1 & c \\ 0 & 0 & 1 \end{pmatrix} \middle| \begin{array}{l} 0 \leq a, b, c \leq p-1, \text{ and either} \\ a = b = c = 0, \text{ or } b = c = 0, \\ \text{or } a = 0 \end{array} \right\}.$$

If Ω is a bounded set in the fundamental domain for $SL(3, \mathbb{Z})$, then there is a number N_Ω such that for all primes $p > N_\Omega$, the Hecke points $S(p)$ cover Ω with covering radius $\varepsilon = Ap^{-1/10}$, where A is an absolute constant. Covering radius ε means any point is in a ball centered at a point of $S(p)$ of radius ε. Chiu uses the spectral resolution of the differential operators on the fundamental domain, known results on the Ramanujan conjecture for estimates of Hecke eigenvalues for Maass cusp forms for $SL(3, \mathbb{Z})$. We discuss the Ramanujan conjecture in §1.5.4.

Chiu's result is also related to standard results in ergodic theory for connected noncompact simple Lie groups G with finite center (e.g., $G = SL(3, \mathbb{R})$) saying that if H is a closed noncompact subgroup of G and Γ is an (irreducible) discrete subgroup of G (e.g., $\Gamma = SL(3, \mathbb{Z})$) then H acts ergodically on G/Γ. Here we are closest to looking at an equally spaced finite set of points in

$$H = \left\{ \begin{pmatrix} 1 & x & y \\ 0 & 1 & 0 \\ 0 & 0 & 1 \end{pmatrix} \,\middle|\, x, y \in \mathbb{R} \right\}.$$

For we are looking at points for T_p acting on a fixed $Y_0 \in \mathcal{SP}_3$ via

$$Y_0 \left[\begin{pmatrix} p & 0 & 0 \\ 0 & 1 & 0 \\ 0 & 0 & 1 \end{pmatrix} \begin{pmatrix} 1 & a/p & b/p \\ 0 & 1 & 0 \\ 0 & 0 & 1 \end{pmatrix} \right],$$

with $0 \leq a, b \leq p - 1$. For the ergodic theory result, see Zimmer [676, p. 19 ff.]. Burger and Sarnak [91, Thm. 5.2] suggested that Hecke point equidistribution results should follow from Marina Ratner's measure classification theorem [499].

Ultimately one would hope to be able to use the points $M(p; a, b)$ from (1.216) to generalize the computations of Maass wave forms for $SL(2, \mathbb{Z})$ by Stark [577] given as Table 3.10 of Section 3.5 of Volume I. This would require programs for the computation of matrix argument K-Bessel or Whittaker functions. See §1.5.4 for a discussion of recent computations of Maass forms for $GL(3)$.

More general Hecke point equidistribution results are to be found in Laurent Clozel, Hee Oh, and Emmanuel Ullmo [113], Alex Eskin and Hee Oh [171], and Hee Oh [478]. In particular, Eskin and Oh find a short ergodic theory proof of the equidistribution of Hecke points for Lie groups G such as $SL(n, \mathbb{R})$ and subgroups Γ such as $SL(n, \mathbb{Z})$. To state their result, we need some notation. We say that two subgroups A and B of G are **commensurable** if $A \cap B$ has finite index in both A and B. Define the **commensurator group** of Γ by

$$\mathrm{Comm}\,(\Gamma) = \left\{ g \in G \,\middle|\, \Gamma \ \& \ g\Gamma g^{-1} \text{ are commensurable with each other} \right\}.$$

Note that Margulis [433, Chapter IX, Thm. 6.5] has proved that $[\mathrm{Comm}\,(\Gamma) : \Gamma] = \infty$ characterizes **arithmetic** Γ. Define the degree of $a \in \mathrm{Comm}\,(\Gamma)$ by

$$\deg(a) = \#(\Gamma \backslash \Gamma a \Gamma) = \left[\Gamma : \Gamma \cap \left(a^{-1}\Gamma a \right) \right].$$

Then the theorem of Eskin and Oh [171] says that for any bounded continuous function f on $\Gamma \backslash G$ and for any $x \in \Gamma \backslash G$ and any sequence $\{a_i\}$ of elements of Comm(Γ) such that

$$\lim_{i \to \infty} \deg(a_i) = \infty,$$

we have

$$\frac{1}{\deg(a_i)} \sum_{\gamma \in \Gamma \backslash \Gamma a_i \Gamma} f(\gamma x) \to \int_{\Gamma \backslash G} f(g) \, d\bar{g}, \quad \text{as } i \to \infty.$$

At this point, there are various natural questions.

Questions.

(1) Exercise 1.4.16 gives generators of $GL(n, \mathbb{Z})$. What happened to Poincaré's generators and relations theorems in this context (cf. Exercise 3.3.1 in Volume I)?
(2) Is there some way of visualizing the tessellation of $\mathcal{SP}_n$ produced by writing

$$\mathcal{SP}_n = \bigcup_{\gamma \in \Gamma_n} \mathcal{SF}_n[\gamma]?$$

Perhaps we should take a hint from topology and look at retracts (cf. Ash [28]).
(3) In the classical case of $SL(2, \mathbb{Z})$, the reduction algorithm for putting a point $z \in H$ into the standard fundamental domain, using a sequence of translations and flips, is the same as the algorithm for finding a continued fraction expansion of a real number. Thus Grenier's algorithm for putting a matrix $Y \in \mathcal{SP}_n$ into $\mathcal{SF}_n$ by some combination of matrices from those listed after Exercise 1.4.19 gives an analogue of a continued fraction algorithm. This should be compared with the continued fraction algorithms of Ferguson and Forcade [179] and other work mentioned in the introduction to Section 1.1.
(4) There is an analogue for $GL(n, \mathbb{Z})$ of the method of perpendicular bisectors which writes the fundamental domain D for a discrete subgroup Γ of $GL(n, \mathbb{R})$ as follows, for a point $W \in \mathcal{P}_n$ such that $W \neq W[\gamma]$ if $\gamma \in \Gamma$ and $\gamma \neq \pm I$:

$$D = \{Y \in \mathcal{P}_n \mid d(Y, W) \leq d(Y, W[\gamma]), \text{ for all } \gamma \in \Gamma\}.$$

Here d denotes the distance obtained from the Riemannian structure. See Siegel [565, Vol. II, pp. 298–301]. Why can (and should) we choose the point W as stated?

Siegel [565, p. 310] notes: "The application of the general method ... [given above] would lead to a rather complicated shape of the frontier of [boundary of the fundamental domain] F." However, the method does lead to the standard fundamental domain for $SL(2, \mathbb{Z})$ if the point W in the Poincaré upper half plane is chosen to be $2i$, for example.

B.A. Venkov [628] considers a related domain defined for a fixed $H \in \mathcal{P}_n$ by:

$$\{Y \in \mathcal{P}_n \mid \text{Tr}(YH) \leq \text{Tr}(YH[\gamma]) \text{ for all } \gamma \in \Gamma\}.$$

When the point H is such that $H = H[\gamma]$ implies $\gamma = \pm I$, then this Venkov domain is a fundamental domain for $GL(n, \mathbb{Z}) = \Gamma$. See also Ryskov [521].

The question here is to compare all these domains with those of Minkowski and Grenier.

(5) One should relate our fundamental domain to that which would be obtained if one replaced $\Gamma = GL(n, \mathbb{Z})$ by integral matrices with arbitrary nonzero determinant. The question concerns the relationship between Minkowski reduction and reduction by successive minima. Or one could consider replacing Γ by $\Gamma \cap (A\Gamma A^{-1})$, where A is some integral matrix of positive determinant d. This has something to do with Hecke operators.

(6) One should consider compactifications of the fundamental domain. See Borel and Ji [67], Goresky [236], Jorgenson and Lang [334].

(7) What geodesics of $\mathcal{P}_n$, if any, induce dense geodesics on the fundamental domain for $\mathcal{P}_n/GL(n, \mathbb{Z})$?

Recall that in 1835 Jacobi showed that a line with irrational slope in $\mathbb{R}^2$ induces a densely wound line in the torus $\mathbb{R}^2/\mathbb{Z}^2$. Weyl [666] further developed the theory in 1916. Artin [26, pp. 499–504] showed in 1924 that almost all geodesics in the Poincaré upper half plane will induce densely wound lines in the standard fundamental domain for $SL(2, \mathbb{Z})$. See Figure 1.27. In fact, geodesics are typically dense in the unit tangent bundle for the $SL(2, \mathbb{Z})$ case. However, the geodesic flow is not ergodic on the unit tangent bundle in higher rank (cf. Mautner [438, pp. 419–421]). This still leaves open the question of density in the fundamental domain for the higher rank case. Perhaps it is more sensible to look for $(n-1)$-dimensional totally geodesic submanifolds. See Zimmer [676, especially pp. 18–19].

One can ask analogous questions about "horocycles" (i.e., conjugates of the group N of upper triangular matrices with ones on the diagonal). This sort of question was already considered in connection with our pictures of images of Hecke points in the fundamental domain $\mathcal{SF}_3$.

The next exercise gives the finiteness of the volume of $\mathcal{SF}_n$. In the next section we will obtain an exact formula for the volume.

Exercise 1.4.21 (Finiteness of the Volume of the Fundamental Domain in the Determinant One Surface).

(a) Show that, if $W \in \mathcal{SP}_n$ and

$$W = \begin{pmatrix} u & 0 \\ 0 & u^{-1/(n-1)}V \end{pmatrix} \begin{bmatrix} 1 & q \\ 0 & I_{n-1} \end{bmatrix}, \qquad \text{for } u > 0,\ V \in \mathcal{SP}_{n-1},\ q \in \mathbb{R}^{n-1},$$

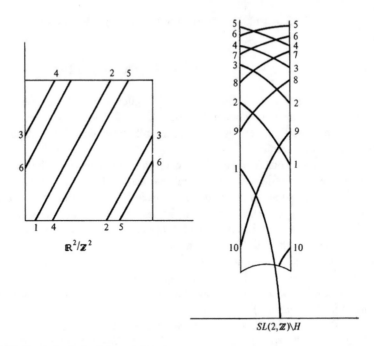

Fig. 1.27 Geodesics in fundamental domains

then

$$dW = u^{-1+n/2} \, du \, dV \, dq.$$

See Exercise 1.1.32 in §1.1.6.

(b) Use part (a) to show that

$$\mathrm{Vol}(\mathcal{SF}_n) \leq \text{ constant } \mathrm{Vol}(\mathcal{SF}_{n-1}).$$

(c) Conclude that the volume of $SO(n)\backslash SL(n, \mathbb{R})/SL(n, \mathbb{Z})$ is finite.

1.4.4   Integration over Fundamental Domains

Next we turn to integral formulas on fundamental domains $\mathcal{D}$ for $\mathcal{P}_n/GL(n, \mathbb{Z})$. Here we will often take $\mathcal{D}$ to be the Minkowski domain $\mathcal{M}_n$. Define the determinant one surface in the fundamental domain to be $\mathcal{SD} = \mathcal{D} \cap \mathcal{SP}_n$. Recall Exercise 1.1.23 of Section 1.1.4 which gave **the relation between the $GL(n,\mathbb{R})$-invariant measure $d\mu_n$ on $\mathcal{P}_n$ and an $SL(n,\mathbb{R})$-invariant measure dW on the determinant one surface $\mathcal{SP}_n$:**

$$Y = t^{1/n}W, \quad Y \in \mathcal{P}_n, \quad t = |Y| > 0, \quad W \in \mathcal{SP}_n, \quad d\mu_n(Y) = t^{-1}dt\, dW,$$
$$(1.219)$$

where the ordinary Euclidean volume element dY on $\mathcal{P}_n$ is related to the invariant measure $d\mu_n(Y)$ by:

$$d\mu_n(Y) = |Y|^{-(n+1)/2}dY. \tag{1.220}$$

Exercise 1.4.22. Compute the Jacobian of the change of variables $Y = t^{1/n}W$, from $Y \in \mathcal{P}_n$ to $t > 0$ and $W \in \mathcal{SP}_n$, using all but one of the entries of $W = Y^0$ above or on the diagonal (e.g., leave out w_{nn}).
Answer.

$$|dY/d(t, W)| = (nt)^{-1}t^{(n+1)/2}.$$

It follows from the preceding Exercise that formula (1.219) normalizes measures "wrong" by throwing away the factor of $1/n$. This does not really matter.

If we set $G = SL(n, \mathbb{R})$, $\Gamma = SL(n, \mathbb{Z})$, the quotient space G/Γ has a G-invariant volume element $d\bar{g}$, which is unique up to a positive constant multiple (see Lang [388] or Weil [659, pp. 42–45]). Therefore we can normalize $d\bar{g}$ to obtain:

$$\int_{\mathcal{SP}_n/SL(n, \mathbb{Z})} f(W)\, dW = \int_{G/\Gamma} f\left({}^t g g\right)\, dg. \tag{1.221}$$

Our first goal is to compute the volume of the fundamental domain in the determinant one surface. We know that this volume is finite by Exercise 1.4.21 of the preceding section.

Lemma 1.4.5. *The Euclidean volume of the set of matrices in $\mathcal{M}_n$ having determinant less than or equal to one is related to the $SL(n, \mathbb{R})$-invariant volume of $\mathcal{SM}_n$ (obtained using the measure dW in (1.219)) as follows:*

$$\textit{Euclidean } \mathrm{Vol}\{Y \in \mathcal{M}_n \mid |Y| \leq 1\} = 2\mathrm{Vol}(\mathcal{SM}_n)/(n+1).$$

Proof. By formulas (1.219) and (1.220), we have:

$$\int_{\substack{|Y|\leq 1 \\ Y\in\mathcal{M}_n}} dY = \mathrm{Vol}(\mathcal{SM}_n) \int_{t=0}^{1} t^{(n-1)/2}\, dt .$$

This clearly gives the stated formula. ∎

Our plan is to determine $\mathrm{Vol}(\mathcal{SM}_n)$ using an inductive procedure which derives from work of Minkowski [453, Vol. II, pp. 80–94], Siegel [565, Vol. III, pp. 39–46], and Weil [662, Vol. I, pp. 339–358]. Weil writes [662, Vol. I, p. 561] that he was able to use his simplification of Siegel's work on this subject to calculate the Tamagawa number, which gives the adelic formulation of Siegel's main theorem on quadratic forms. The determination of $\mathrm{Vol}(\mathcal{SM}_n)$ is closely related to the following proposition.

Proposition 1.4.2 (Siegel's Integral Formula in the Geometry of Numbers). *Let $G = SL(n, \mathbb{R})$, $\Gamma = SL(n, \mathbb{Z})$, and $f : \mathbb{R}^n \to \mathbb{C}$ be an integrable function. Then we have the following equalities:*

$$\frac{1}{\mathrm{Vol}(G/\Gamma)} \int_{\overline{g} \in G/\Gamma} \sum_{a \in \mathbb{Z}^n - 0} f(ga)\, d\overline{g} = \int_{\mathbb{R}^n} f(x)\, dx,$$

$$\frac{\zeta(n)}{\mathrm{Vol}(G/\Gamma)} \int_{\overline{g} \in G/\Gamma} \sum_{\substack{a \in \mathbb{Z}^n - 0 \\ \gcd(a)=1}} f(ga)\, d\overline{g} = \int_{\mathbb{R}^n} f(x)\, dx.$$

Here dx denotes Lebesgue measure on $\mathbb{R}^n$, $d\overline{g}$ is a G-invariant measure on G/Γ, the vectors $a \in \mathbb{Z}^n - 0$ are column vectors, and ga denotes the column vector that results from multiplying a by the $n \times n$ matrix g on the left.

Proof (Weil [662, Vol. I, pp. 339–358]). The main idea is to use the following integration formula which holds for a unimodular locally compact topological group G with closed unimodular subgroup G_1. Here "unimodular" means that right and left Haar measures coincide. The integral formula in question is:

$$\int_{G/G_1} \left(\int_{G_1} f(gg_1)\, dg_1 \right) d\overline{g} = c \int_G f(g)\, dg, \tag{1.222}$$

where c is a positive constant, dg and dg_1 are Haar measures on G and G_1, respectively, $d\overline{g}$ is a G-invariant measure on G/G_1. References for this result are Helgason [273, 282], Lang [388], and Weil [659, p. 45]. Formula (1.222) can be extended to non-unimodular G and G_1—provided that the modular functions of G and G_1 are equal on G_1. The modular function δ is defined in formula (2.37) in Chapter 2. It relates right and left Haar measures.

Two applications of formula (1.222) are required to prove Siegel's integral formula. There are, in fact, two quotients in Siegel's integral formula. The obvious quotient is G/Γ and the other is $\mathbb{R}^n - 0 \cong G/H$, where H is the subgroup:

$$H = \{g \in G \mid ge_1 = e_1\}, \tag{1.223}$$

and e_1 is the standard unit basis vector in $\mathbb{R}^n$; i.e., $e_1 = {}^t(1, 0, \ldots, 0)$. Note that the elements of H have the form:

$$\begin{pmatrix} 1 & * \\ 0 & * \end{pmatrix}.$$

The mapping that identifies G/H with $\mathbb{R}^n - 0$ is:

$$G/H \to \mathbb{R}^n - 0$$

$$gH \mapsto ge_1 = \text{the first column of } g.$$

Now let $\gamma = H \cap \Gamma$. Suppose that $f : G/H \to \mathbb{C}$ satisfies the hypotheses of the proposition. Then

$$c \int_{G/H} f(x)\, dx = \int_{G/\Gamma} \left(\int_{\Gamma/\gamma} f(gy)\, d\bar{y} \right) d\bar{g}, \tag{1.224}$$

for some positive constant c (independent of f). To see this, note that

$$\int_{G/H} \left(\int_{H/\gamma} f(gh)\, d\bar{h} \right) d\bar{g} = c_1 \int_{G/\gamma} f(g)\, dg,$$

$$\int_{G/\Gamma} \left(\int_{\Gamma/\gamma} f(ga)\, d\bar{a} \right) d\bar{g} = c_2 \int_{G/\gamma} f(g)\, dg,$$

for some positive constants c_1 and c_2.

Next observe that from Lemma 1.4.2,

$$\Gamma/\gamma = \{ a \in \mathbb{Z}^n \mid \gcd(a) = 1 \}. \tag{1.225}$$

So (1.224) says that:

$$c \int_{\mathbb{R}^n} f(x)\, dx = \int_{G/\Gamma} \sum_{\substack{a \in \mathbb{Z}^n \\ \gcd(a)=1}} f(ga)\, d\bar{g}.$$

This implies, by change of variables, that if $t > 0$, we have:

$$ct^{-n} \int_{\mathbb{R}^n} f(x)\, dx = \int_{G/\Gamma} \sum_{\substack{a \in \mathbb{Z}^n \\ \gcd(a)=1}} f(tga)\, d\bar{g}.$$

Now sum over $t = 1, 2, 3, \ldots$ and obtain:

$$c\zeta(n) \int_{\mathbb{R}^n} f(x)\, dx = \int_{G/\Gamma} \sum_{a \in \mathbb{Z}^n - 0} f(ga)\, d\overline{g}. \qquad (1.226)$$

The proof of Siegel's integral formula is completed by showing that $c\zeta(n) = \text{Vol}(G/\Gamma)$. Weil's proof of this fact uses the Poisson summation formula (see Theorem 1.3.2 of Volume I). Let $c^* = c\zeta(n)$ and $V = \text{Vol}(G/\Gamma)$. From (1.226) it follows that:

$$Vf(0) + c^* \int_{\mathbb{R}^n} f(x)\, dx = \int_{G/\Gamma} \sum_{a \in \mathbb{Z}^n} f(ga)\, d\overline{g}. \qquad (1.227)$$

And Poisson tells us that, for $g \in G$, and, for suitable f:

$$\sum_{a \in \mathbb{Z}^n} f(ga) = \sum_{a \in \mathbb{Z}^n} \widehat{f}({}^t g^{-1} a),$$

where $\widehat{f}$ denotes the Fourier transform of f over $\mathbb{R}^n$. Note that

$$\widehat{f}(0) = \int_{\mathbb{R}^n} f(y)\, dy.$$

Therefore

$$Vf(0) + c^* \widehat{f}(0) = \int_{G/\Gamma} \sum_{a \in \mathbb{Z}^n} \widehat{f}({}^t g^{-1} a)\, d\overline{g} = \int_{G/\Gamma} \sum_{a \in \mathbb{Z}^n} \widehat{f}(ga)\, d\overline{g}. \qquad (1.228)$$

Replace f by $\widehat{f}$ in formula (1.227) or formula (1.228) to find that

$$V\widehat{f}(0) + c^* f(0) = Vf(0) + c^* \widehat{f}(0),$$

which says that $(V - c^*)\left(f(0) - \widehat{f}(0)\right) = 0$. Since we can easily find a function $f(x)$ such that $f(0) \neq \widehat{f}(0)$, it follows that $V = c^*$, and we're finished with the proof of Siegel's integral formula.

Corollary 1.4.1. *Suppose that $f : \mathbb{R}^+ \to \mathbb{C}$ is suitably chosen for convergence. Then*

$$\frac{1}{\text{Vol}(\mathcal{SM}_n)} \int_{\mathcal{SM}_n} \sum_{a \in \mathbb{Z}^n - 0} f(W[a])\, dW = \int_{\mathbb{R}^n} f({}^t xx)\, dx.$$

Proof. This corollary follows immediately from (1.221) and Proposition 1.4.2 in the case n is odd, since then $S\mathcal{P}_n/SL(n, \mathbb{Z}) = S\mathcal{P}_n/GL(n, \mathbb{Z}) = S\mathcal{M}_n$ because $GL(n, \mathbb{Z})/SL(n, \mathbb{Z})$ has representatives $I, -I$, both having no effect on $W \in S\mathcal{P}_n$. However, one has to make a more complicated argument when n is even. In that case suppose that I, γ represent $GL(n, \mathbb{Z})/SL(n, \mathbb{Z})$. Then note that

$$S\mathcal{M}_n \cup S\mathcal{M}_n[\gamma]$$

is a fundamental domain for $SL(n, \mathbb{Z})$. Thus, for even n,

$$\text{Vol}\,(S\mathcal{P}_n/SL(n, \mathbb{Z})) = 2\text{Vol}\,(S\mathcal{P}_n/GL(n, \mathbb{Z})).$$

And, setting $\Gamma^0 = SL(n, \mathbb{Z})$ and $\Gamma = GL(n, \mathbb{Z})$, we have

$$\int_{S\mathcal{P}_n/\Gamma^0} \sum_{a \in \mathbb{Z}^n - 0} f(W[a])\, dW = 2 \int_{S\mathcal{P}_n/\Gamma} \sum_{a \in \mathbb{Z}^n - 0} f(W[a])\, dW.$$

The reason for this is the fact that

$$\sum_{a \in \mathbb{Z}^n - 0} f(W[a]) = \sum_{a \in \mathbb{Z}^n - 0} f(W[\gamma a])$$

for any $\gamma \in GL(n, \mathbb{Z})$.

There are many applications of the integral formulas in Proposition 1.4.2 and their generalizations. For example, they give integral tests for the convergence of Eisenstein series. They also imply the existence of quadratic forms with large minima; i.e., the existence of dense lattice packings of spheres in $\mathbb{R}^n$. This result is usually called the Minkowski-Hlawka theorem in the geometry of numbers. But before discussing these applications, let us compute the exact volume of $S\mathcal{M}_n$.

Theorem 1.4.4 (Volume of the Fundamental Domain).

(1)   Using the normalization of measures, given in (1.219), we have

$$\text{Vol}(S\mathcal{M}_n) = \prod_{k=2}^{n} \Lambda(k/2), \quad \Lambda(s) = \pi^{-s}\Gamma(s)\zeta(2s).$$

(2)   Euclidean Volume $\{Y \in \mathcal{M}_n \mid |Y| \le 1\} = 2\,\text{Vol}(S\mathcal{M}_n)/(n + 1)$.

Proof. (2) Note that part (2) is an easy consequence of Lemma 1.4.5.
(1) By Corollary 1.4.1 and the formula for the surface area of the unit sphere in Exercise 1.4.2 of Section 1.4.1, we have the following sequence of equalities for suitable $f : \mathbb{R}^+ \to \mathbb{C}$:

$$\frac{\pi^{n/2}}{\Gamma(n/2)} \int\limits_{r>0} f(r) \, r^{n/2-1} \, dr = \int\limits_{\mathbb{R}^n} f\left({}^t xx\right) dx$$

$$= \frac{\zeta(n)}{\text{Vol}(\mathcal{SM}_n)} \sum_{\substack{a \in \mathbb{Z}^n \\ \gcd(a)=1}} \int\limits_{\mathcal{SM}_n} f(W[a]) \, dW$$

$$= \frac{\zeta(n)}{\text{Vol}(\mathcal{SM}_n)} \sum_{(a\,*)\in\Gamma/\Gamma\cap H} \int\limits_{\mathcal{SP}_n/\Gamma} f(W[a]) \, dW$$

if $\Gamma = GL(n, \mathbb{Z})$, and

$$H = \begin{pmatrix} 1 & * \\ 0 & * \end{pmatrix}.$$

Therefore if $f : \mathbb{R}^+ \to \mathbb{C}$ is suitably chosen for convergence, we have:

$$\frac{\pi^{n/2}}{\Gamma(n/2)} \int\limits_{r>0} f(r) r^{(n/2)-1} \, dr = \frac{\zeta(n)}{\text{Vol}(\mathcal{SM}_n)} \int\limits_{\substack{W=(w_{ij}) \\ W\in\mathcal{SP}_n/\Gamma\cap H}} f(w_{11}) \, dW. \qquad (1.229)$$

It follows from Exercises 1.4.23 and 1.4.24 below and Exercise 1.4.21 of Section 1.1.4 that the integral appearing on the right-hand side of formula (1.229) can be rewritten as:

$$\int\limits_{\mathcal{SP}_n/\Gamma_n\cap H} f(w_{11}) \, dW = \int\limits_{\substack{V \in P_{n-1}/\Gamma_{n-1} \\ t|V| \leq 1, \, h \in [0,1]^{n-1}}} f\left(\left(t\,|V|\right)^{\frac{-1}{n}} t\right)\left(t\,|V|\right)^{\frac{1-n}{2}} t^{n-1} \, dV \, dt \, dh.$$

$$(1.230)$$

Upon setting $U = t^{1/(n-1)}V$, formula (1.230) becomes:

$$\int\limits_{\mathcal{SP}_n/\Gamma_n\cap H} f(w_{11}) \, dW = \int\limits_{\substack{|U|\leq 1 \\ U\in\mathcal{M}_{n-1} \\ 0<t}} f\left(|U|^{-\frac{1}{n}} t\right)|U|^{\frac{1-n}{2}} t^{\frac{n}{2}-1} \, dt \, dU.$$

Therefore if we substitute $x = t|U|^{-1/n}$ and use formulas (1.219) and (1.220), we obtain:

$$\int_{S\mathcal{P}_n/\Gamma_n\cap H} f(w_{11})dW = \int_{\substack{|U|\leq 1 \\ U\in\mathcal{M}_{n-1} \\ 0<t}} f(x)|U|^{1-\frac{n}{2}}x^{\frac{n}{2}-1} \, dt \, dU$$

$$= \mathrm{Vol}(S\mathcal{M}_{n-1})\int_{x>0} f(x)x^{\frac{n}{2}-1} \, dx.$$

Thus we have proved:

$$\int_{S\mathcal{P}_n/\Gamma_n\cap H} f(w_{11}) \, dW = \mathrm{Vol}(S\mathcal{M}_{n-1})\int_{x>0} f(x)x^{\frac{n}{2}-1} \, dt. \tag{1.231}$$

If $f(x)x^{-1+n/2}$ is positive and integrable over $(0, \infty)$, then (1.229) and (1.231) combine to give:

$$\mathrm{Vol}(S\mathcal{M}_n) = \mathrm{Vol}(S\mathcal{M}_{n-1})\pi^{-n/2}\Gamma(n/2)\zeta(n).$$

The theorem follows by induction, using the case $n = 2$ which was obtained in Chapter 3 of Volume I. ∎

Now we need to do the exercises used in the proof.

Exercise 1.4.23. Use formula (1.219) to show that if $f : S\mathcal{P}_n \to \mathbb{C}$ is integrable, then

$$\int_{W\in S\mathcal{P}_n} f(W)dW = \int_{\substack{Y\in\mathcal{P}_n \\ |Y|\leq 1}} f\left(|Y|^{-1/n}Y\right)|Y|^{-(n-1)/2} \, dY,$$

where dY is as in (1.220).

Hint. Let $h(t^{1/n}W) = t\chi_{[0, 1]}(t)f(W)$, for $t > 0$, $W \in S\mathcal{P}_n$ in formula (1.219), where

$$\chi_{[0, 1]}(t) = \begin{cases} 1, & \text{if } t \in [0, 1], \\ 0, & \text{otherwise.} \end{cases}$$

Exercise 1.4.24. Use the partial Iwasawa decomposition

$$Y = \begin{pmatrix} t & 0 \\ 0 & V \end{pmatrix}\begin{bmatrix} 1 & {}^t h \\ 0 & I \end{bmatrix}, \quad t > 0, \quad V \in \mathcal{P}_{n-1}, \quad h \in \mathbb{R}^{n-1},$$

to obtain an explicit fundamental domain for $\mathcal{P}_n/\Gamma \cap H$, with H as it was defined just before formula (1.229).

Three corollaries of Siegel's integral formula (Proposition 1.4.2) can now be derived quite easily.

Corollary 1.4.2 (The Minkowski-Hlawka Theorem I). *There is a matrix $Y \in \mathcal{P}_n$ such that the first minimum m_Y (defined in (1.178) of Section 1.4.1) satisfies:*

$$m_Y > k_n |Y|^{1/n}, \quad \text{with} \quad k_n = \frac{n}{2\pi e},$$

for n sufficiently large.

Proof. Consider a ball $B_n \subset \mathbb{R}^n$ of radius

$$k_n = \left(\frac{n}{2\pi e} \right)^{1/2}.$$

Then

$$\text{Vol}(B_n) = \left(\frac{n}{2\pi e} \right)^{n/2} \frac{\pi^{n/2}}{\Gamma(1 + n/2)}.$$

Stirling's formula implies that $\text{Vol}(B_n)$ tends to zero as n goes to infinity. This is rather surprising, since the radius is blowing up. See the note below for a related paradox.

Now let

$$\chi_{B_n} = \begin{cases} 1, \text{ for } x \in B_n, \\ 0, \text{ otherwise.} \end{cases} \tag{1.232}$$

If this function is plugged into Siegel's integral formula and $V = \text{Vol}(G/\Gamma)$, we obtain:

$$\text{Vol}(B_n) = \frac{1}{V} \int_{G/\Gamma} \sum_{a \in \mathbb{Z}^n - 0} \chi_{B_n}(ga) \, d\bar{g}. \tag{1.233}$$

When n is sufficiently large, formula (1.233) makes it clear that, since $\text{Vol}(B_n) \to 0$, as $n \to \infty$, we must have:

$$\sum_{a \in \mathbb{Z}^n - 0} \chi_{B_n}(ga) < 1$$

for some $g \in G = SL(n, \mathbb{R})$. This means that $({}^t gg)[a] > k_n^2$ for some g and all a in $\mathbb{Z}^n - 0$. Set $Y = {}^t gg$ to finish the proof. ∎

Corollary 1.4.3 (The Minkowski-Hlawka Theorem II). *If d_L from (1.179) is the density of space occupied by spheres of equal radii centered at points of the lattice L, then*

$$d_L \geq \frac{\zeta(n)}{2^{n-1}}.$$

Proof (Siegel [566, pp. 152–154]). Suppose the lattice L is such that the first minimum m_{Y_L} of a quadratic form Y_L corresponding to L is maximal. We know such L exists and $m_{Y_L} = \|a\|^2$ for some $a \in L - 0$. Why?

Let $r_L = \sqrt{m_L}$ and $B_L = \{x \in \mathbb{R}^n \mid \|x\| \leq r_L\}$. By the choice of r_L we know that $B_L \cap L \neq \{0\}$. If χ_{B_L} is as in formula (1.232) and $L = \sum^{\oplus} \mathbb{Z}w_i$ set

$$\xi_L(g) = \sum_{\substack{a = \sum a_i w_i \\ \gcd(a_i) = 1}} \chi_{B_L}(a).$$

Then $\xi_L(g) \geq 2$. Why? Recall that if $a \in B_L$, then so is $\pm pa$, for any p between 0 and 1.

Therefore, if $v_n(1)$ denotes the volume of the unit sphere in $\mathbb{R}^n$, by Siegel's integral formula (Prop. 1.4.2), we have

$$r_L^n v_n(1) = \frac{\zeta(n)}{\mathrm{Vol}\,(G/\Gamma)} \int_{G/\Gamma} \xi_L(g)\, d\bar{g} \geq 2\zeta(n).$$

Using formula (1.180), we see that $d_L = \left(\frac{m_{Y_L}}{4}\right)^{\frac{n}{2}} v_n(1) = \left(\frac{r_L}{2}\right)^n v_n(1)$. The corollary follows.   ∎

Note. There is a mind-boggling paradox associated with balls in high dimensional spaces (cf. Hamming [258, pp. 168–170]). In n-dimensional space, consider a hypercube having a side of length 4 and centered at the origin. Put 2^n unit spheres in each corner of this cube such that each sphere touches all its n neighboring spheres. The distance from the origin to the center of one of these spheres is $\sqrt{n}$. Thus we can put a sphere of radius $\sqrt{n} - 1$ inside all the unit spheres at the corners. See Figure 1.28 for the case $n = 3$. When $n \geq 10$, this inner sphere reaches outside the cube, since $\sqrt{10} - 1 > 2$. Weird.

Our next corollary is a result that we already know, but the proof will be easily generalized to more complicated Eisenstein series.

Corollary 1.4.4 (Integral Test Proof of the Convergence of Epstein's Zeta Function). *The Epstein zeta function*

$$Z(Y, s) = \sum_{a \in \mathbb{Z}^n - 0} Y[a]^{-s}$$

converges absolutely for $\mathrm{Re}\, s > n/2$, $Y \in \mathcal{P}_n$.

Fig. 1.28 The picture in 3d
becomes a paradox in 10d

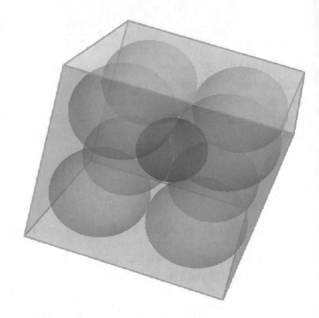

Proof. Siegel tells us that:

$$\frac{1}{\text{Vol}(\mathcal{SM}_n)} \int_{W \in \mathcal{SM}_n} \sum_{\substack{a \in \mathbb{Z}^n - 0 \\ W[a] \geq 1}} W[a]^{-s} \, dW = \int_{\substack{x \in \mathbb{R}^n \\ {}^t xx \geq 1}} (xx)^{-s} \, dx.$$

The integral on the right is easily evaluated as

$$\pi^{n/2} \Gamma(n/2)^{-1} (s - n/2)^{-1},$$

if $\text{Re}\, s > n/2$. Then Fubini's theorem says that the series being integrated converges for almost all W in $\mathcal{SM}_n$. The series differs from Epstein's zeta function $Z(W, s)$ by at most a finite number of terms. Thus $Z(W, s)$ converges for $\text{Re}\, s > n/2$ and almost all W in $\mathcal{SM}_n$. In order to deduce the convergence of $Z(Y, s)$ for all Y in $\mathcal{P}_n$, note that there is a positive constant c depending on Y such that

$$cI[a] \leq Y[a] \leq c^{-1} I[a], \quad \text{for all } a \text{ in } \mathbb{R}^n.$$

Corollary 1.4.5 (The Vanishing of Epstein's Zeta Function in $(0, n/2)$).
   For all s with $0 < \text{Re}\, s < n/2$,

$$\int_{\mathcal{SM}_n} Z(W, s) \, dW = 0.$$

It follows that for any $s \in (0, 1)$ there exist $Y \in \mathcal{P}_n$ such that $Z(Y, s) > 0$. Similarly there are $Y \in \mathcal{P}_n$ such that $Z(Y, s) < 0$, or such that $Z(Y, s) = 0$.

Proof. Use the incomplete gamma expansion of Epstein's zeta function (Theorem 1.4.1 of Volume I):

$$2\pi^{-s}\Gamma(s)Z(Y, s)$$

$$= \frac{|Y|^{-1/2}}{s - \frac{n}{2}} - \frac{1}{s} + \sum_{a \in \mathbb{Z}^n - 0} \left\{ G(s, \pi Y[a]) + |Y|^{-1/2} G\left(\frac{n}{2} - s, \pi Y^{-1}[a]\right) \right\},$$

where the **incomplete gamma function** is:

$$G(s, a) = \int_{t \geq 1} t^{s-1} \exp(-at) \, dt, \quad \text{for} \quad \operatorname{Re} a > 0.$$

Note that we can use this expansion to see that $Z(W, s)$ is integrable over the fundamental domain $\mathcal{SM}_n$ provided that $0 < \operatorname{Re} s < n/2$.

Now apply Siegel's integral formula to see that:

$$\frac{2}{\operatorname{Vol}(\mathcal{SM})} \int_{W \in \mathcal{SM}_n} \pi^{-s}\Gamma(s)Z(W, s) \, dW$$

$$= \frac{1}{s - \frac{n}{2}} - \frac{1}{s} + \int_{\mathbb{R}^n} \left\{ G\left(s, \pi\,{}^t xx\right) + G\left(\frac{n}{2} - s, \pi\,{}^t xx\right) \right\} dx.$$

Use the definition of $G(s, a)$ to write the integral over x in $\mathbb{R}^n$ on the right-hand side of this equality as a double integral over x and t. Then make the change of variables $y = t^{1/2}x$ to see that if $\operatorname{Re} s < n/2$, then we have, for example,

$$\int_{\mathbb{R}^n} G\left(s, \pi\,{}^t xx\right) dx = \frac{1}{\frac{n}{2} - s}.$$

This completes the proof. ∎

When $n = 2$, Corollary 1.4.5 gives the orthogonality of the Eisenstein series and the constants in the spectral decomposition of the Laplacian on $L^2(H/SL(2, \mathbb{Z})$ (see Lemma 3.7.1 in Volume I). When $n = 2$ or 3 there are explicit criteria on Y which tell, for example, whether $Z(Y, (n-1)/2)$ is positive or negative (see Terras [601]). More general results, showing $Z(Y, nu/2) > 0$ if the first minimum $m_Y \leq nu/(2\pi e)$, for $u \in (0, 1)$ and n sufficiently large, can be found in Terras [603]. See Sarnak and Strömbergsson [530] for more information on minima of Epstein zetas and connections with densest lattice packings L of spheres in $\mathbb{R}^n$ and heights of tori $\mathbb{R}^n/L$, where L is a lattice in $\mathbb{R}^n$.

In our studies of Eisenstein series we will need more general integral formulas than the one given in Proposition 1.4.2. The following proposition gives an example which was stated by Siegel [565, Vol. III, p. 46].

Proposition 1.4.3 (Siegel's Integral Formula, Part II, A Generalization). *Let* $G = SL(n, \mathbb{R})$, $\Gamma = SL(n, \mathbb{Z})$ *and if* $1 \leq k < n$, *let* $h : \mathbb{R}^{n \times k} \to \mathbb{C}$ *be integrable. Then*

$$\frac{1}{\text{Vol}(G/\Gamma)} \int_{G/\Gamma} \sum_{\substack{N \in \mathbb{Z}^{n \times k} \\ \text{rk}(N) = k}} h(gN) \, d\bar{g} = \int_{\mathbb{R}^{n \times k}} h(x) \, dx.$$

Proof (of Proposition 1.4.3. A Beginning). We imitate the proof that we gave for Proposition 1.4.2. Let $H = H_k$ be the subgroup of G consisting of matrices of the form:

$$\begin{pmatrix} I_k & * \\ 0 & * \end{pmatrix},$$

where I_k denotes the $k \times k$ identity matrix. Consider an integrable function $f : G/H \to \mathbb{C}$. Then as before, we have the integral formula:

$$\int_{G/H} f(g) \, d\bar{g} = c \int_{G/\Gamma} \int_{a \in \Gamma/\Gamma \cap H} f(ga) \, d\bar{g}.$$

Note that

$$G/H \cong \{x \in \mathbb{R}^{n \times k} \mid \text{rank } x = k\} \doteq S_{n,k}. \tag{1.234}$$

The complement in $\mathbb{R}^{n \times k}$ of the set $S_{n,k}$ on the right in (1.234) has measure 0. Note also that

$$\Gamma/\Gamma \cap H \cong \{A_1 \in \mathbb{Z}^{n \times k} \mid A_1 \text{ can be completed to a matrix in } \Gamma\}.$$

It follows that we have the formula below upon replacing $f(x)$ by $f(xB)$ with B in $\mathbb{Z}^{k \times k}$:

$$\int_{G/\Gamma} \sum_{(A \, *) \in \Gamma} f(gAB) \, d\bar{g} = c|B|^{-n} \int_{\mathbb{R}^{n \times k}} f(x) \, dx. \tag{1.235}$$

Now sum over B in $\mathbb{Z}^{n \times k}$ of rank k modulo $GL(k, \mathbb{Z})$; i.e., B in a complete set of representatives for the equivalence relation:

$$B \sim C \text{ iff } B = CU, \quad \text{for some } U \text{ in } GL(k, \mathbb{Z}).$$

The matrices B are the right greatest common divisors of the matrices AB essentially. Siegel has described the theory of matrix gcd's and other such concepts in his long paper on the theory of quadratic forms in Vol. I of his collected works (see Siegel [565, Vol I, pp. 331–332]). The generalization of concepts from number theory (such as that of class number) to simple algebras (like $\mathbb{Q}^{n \times n}$) is described in Deuring [135] and the end of Weil [661] to some extent.

We need several lemmas to continue the proof. First we want to show that for $\Gamma_k = GL(k, \mathbb{Z})$:

$$\sum_{A \in SL(n,\mathbb{Z})/H_k} \sum_{B \in \mathbb{Z}^{k \times k}/\Gamma_k} f\left(gA\,^tB\right) = \sum_{\substack{N \in \mathbb{Z}^{n \times k} \\ \text{rk} k}} f(gN). \tag{1.236}$$

Here H_k is defined to be the following subgroup of $SL(n, \mathbb{R})$:

$$H_k = \left\{ g \in G = SL(n, \mathbb{R}) \;\middle|\; g = \begin{pmatrix} I_k & * \\ 0 & * \end{pmatrix} \right\}. \tag{1.237}$$

We will finish the proof of Proposition 1.4.3 later in this section—after a couple of lemmas.

Lemma 1.4.6 (A Decomposition for $n \times k$ Integral Rank k Matrices). *If $1 \leq k < n$ and $N \in \mathbb{Z}^{n \times k}$ has rank k, then N has the unique expression:*

$N = A\,^tB$, *with* $(A \;*) \in SL(n, \mathbb{Z})/H_k$, $A \in \mathbb{Z}^{n \times k}$; $B \in \mathbb{Z}^{k \times k}/GL(k, \mathbb{Z})$, *rank* $B = k$.

Here H_k is as defined in (1.237).

Proof.

(a) Existence of the Decomposition.

Since $GL(n, \mathbb{Z})$ consists of matrices generated by those corresponding to elementary row and column operations (as described in the proof of Lemma 1.4.2 in §1.4.2), there is a diagonal integral $k \times k$ matrix D such that

$$N = U \begin{pmatrix} D \\ 0 \end{pmatrix} V, \quad \text{for } U \in GL(n, \mathbb{Z}), \ V \in GL(k, \mathbb{Z}).$$

Set $U = (A \;*)$, with A in $\mathbb{Z}^{n \times k}$. Note that by changing V we can put U in $SL(n, \mathbb{Z})$. Then

$$N = ADV \text{ and we can set } {}^tB = DV.$$

Then reduce B mod $GL(k, \mathbb{Z})$ on the right and modify A to preserve the equality $N = A\,^tB$. This proves the existence of the decomposition.

(b) Uniqueness of the Decomposition.
    Suppose that

$$A\,{}^tB = A'\,{}^tB'.$$

Then let $U = (A \ *)$, $U' = (A' \ *)$ be matrices in $SL(n, \mathbb{Z})$. It follows that

$$U \begin{pmatrix} {}^tB \\ 0 \end{pmatrix} = U' \begin{pmatrix} {}^tB' \\ 0 \end{pmatrix}.$$

Now write

$$\left(U'\right)^{-1} U = \begin{pmatrix} P & Q \\ R & S \end{pmatrix}.$$

It follows that

$$P\,{}^tB = {}^tB' \quad \text{and} \quad R\,{}^tB = 0.$$

Since B is invertible, $R = 0$. So P is in $GL(k, \mathbb{Z})$ and then $B = B'$ so that $P = I$.
But then

$$\left(U'\right)^{-1} U \text{ is in the subgroup } H_k.$$

Then $(A' \ *)$ and $(A \ *)$ are equivalent modulo H_k. This completes the proof.

∎

Next we want to prove a result which was mentioned already in Section 1.4.1,
namely the factorization formula (1.174) for the zeta function of the simple algebra
of $n \times n$ rational matrices.

**Lemma 1.4.7 (A Factorization of the Analogue of the Dedekind Zeta Function
for the Simple Algebra of All $n \times n$ Rational Matrices).** *If* $\mathrm{Re}\,s > k$, *we have
the following factorization of the matrix analogue of Riemann's zeta function into a
product of ordinary Riemann zeta functions:*

$$\zeta_{\mathbb{Q}^{k \times k}}(s) = \sum_{\substack{B \in \mathbb{Z}^{k \times k} \\ \mathrm{rk}\,k / \Gamma}} |B|^{-s} = \prod_{j=0}^{k-1} \zeta(s - j).$$

Here $\Gamma = GL(k, \mathbb{Z})$.

Proof. We need a system of representatives for equivalence classes modulo
$GL(k, \mathbb{Z})$ of $k \times k$ integral matrices B having rank k. That is, we need a complete set
of representatives for the equivalence relation on $B_i \in \mathbb{Z}^{k \times k}$ of rank k defined by:

$$B_1 \sim B_2 \quad \text{iff } B_1 = B_2\gamma \quad \text{for } \gamma \text{ in } GL(k, \mathbb{Z}). \tag{1.238}$$

The complete set of representatives we choose is the following set of upper triangular matrices:

$$\begin{pmatrix} d_{11} & & d_{1k} \\ & \ddots & \\ 0 & & d_{kk} \end{pmatrix}, \tag{1.239}$$

where $d_{ij} \in \mathbb{Z}$, for all i, j, $d_{ii} > 0$, for all i, and $0 \le d_{ij} < d_{ii}$.

To prove this, note that if B in $\mathbb{Z}^{k \times k}$ has rank k, then there exists a matrix U in $GL(k, \mathbb{Z})$ such that

$$BU = \begin{pmatrix} a & c \\ 0 & D \end{pmatrix}, \quad \text{with } a \in \mathbb{Z}^+, \, D \in \mathbb{Z}^{(k-1) \times (k-1)}.$$

To see this, you must solve $k - 1$ homogeneous linear equations in k unknowns. These equations can be solved with relatively prime integers. Those integers can then be made the first column of a matrix in $GL(k, \mathbb{Z})$ by Lemma 1.4.2 from Section 1.4.1. Moreover, we can insure that c is reduced modulo a. Induction finishes the proof that we can choose the representatives for B as given above. And it is not hard to see that, in fact, any pair of distinct upper triangular matrices from (1.239) are inequivalent modulo $GL(n, \mathbb{Z})$ in the sense of (1.238).

But then

$$\zeta_{\mathbb{Q}^{k \times k}}(s) = \sum_{B \in \mathbb{Z}^{k \times k} \text{ rk}k / \Gamma} |B|^{-s} = \sum_{d_{ii} > 0} \prod_{i=1}^{k} d_{ii}^{k-i-s}.$$

Interchange sum and product to see that the term on the right in this last formula is indeed the product of Riemann zeta functions. Thus Lemma 1.4.6 is proved. ∎

Note. Formula (1.239) for a complete set of representatives of the equivalence relation defined on the rank k matrices in $\mathbb{Z}^{k \times k}$ by right multiplication by matrices in $GL(n, \mathbb{Z})$ is the generalization to $k \times k$ matrices of formula (3.117) of Section 3.6.4 of Volume I, a formula which was important in the theory of Hecke operators. We will use formula (1.239) again when we study Hecke operators for $GL(n)$ in the next section.

Proof (of Proposition 1.4.3. The End). First note that Lemma 1.4.6 implies (1.236). Then (1.235) and (1.236) combine to give:

$$\int_{G/\Gamma} \sum_{N \in \mathbb{Z}^{n \times k} \text{ rk } k} f(gN) \, d\overline{g} = c^* \int_{\mathbb{R}^{n \times k}} f(x) \, dx, \tag{1.240}$$

where

$$c^* = \zeta_{\mathbb{Q}^{k \times k}}(n)c, \ V = \mathrm{Vol}(G/\Gamma), \quad G = SL(n, \mathbb{R}), \ \Gamma = SL(n, \mathbb{Z}).$$

Our problem is to determine the constant in Siegel's integral formula by showing that $c^* = V$. It is natural to try to imitate Weil's argument from the proof of Proposition 1.4.2. If we do that, our formula (1.240) plus Poisson summation yields:

$$(c^* - V) \left(\widehat{f}(0) - f(0) \right) = \sum_{j=1}^{k-1} \int_{G/\Gamma} \sum_{N \in \mathbb{Z}^{n \times k} \ \mathrm{rk} \ j} \left(\widehat{f}(gN) - f(gN) \right) \, dg.$$

The claim is that this is zero. But it is not at all obvious from staring at the right-hand side—unless you see something I don't. Perhaps this will be a useful result when we try to analytically continue Eisenstein series using Riemann's method of theta functions and find that terms of lower rank give divergent integrals. Anyway that fact makes one worry slightly about this whole procedure, doesn't it? So, it seems better to try to find a new argument which does not use Poisson summation.

Let us imitate the argument given by Siegel in his original paper for the case $k = 1$. Note that by the definition of the integral, for $g \in G$ and suitable f:

$$\int_{\mathbb{R}^{n \times k}} f(x) \, dx = \lim_{h \to 0+} \left\{ h^{nk} \sum_{N \in \mathbb{Z}^{n \times k} \ \mathrm{rk} \ k} f(hgN) \right\}.$$

To see this, think of what happens if you multiply the points N in the lattice $\mathbb{Z}^{n \times k}$ by a small positive number h. You will get points in a grid with a very small mesh such that each individual hypercube has volume h^{nk}.

Integrate the preceding formula over $\bar{g}$ in G/Γ and obtain:

$$V \int_{\mathbb{R}^{n \times k}} f(x) \, dx = \lim_{h \to 0+} \left\{ h^{nk} \int_{G/\Gamma} \sum_{N \in \mathbb{Z}^{n \times k} \ \mathrm{rk} \ k} f(hgN) \, d\bar{g} \right\}$$

$$= \lim_{h \to 0+} h^{nk} c^* \int_{\mathbb{R}^{n \times k}} f(hx) \, dx, \quad \text{using (1.240)},$$

$$= c^* \int_{\mathbb{R}^{n \times k}} f(x) \, dx, \quad \text{as } |d(hx)/dx| = h^{nk}.$$

This completes the proof that $c^* = V$ and thus finishes (at last) the proof of Siegel's second integral formula. ∎

Next we combine the integral formulas of Siegel (Proposition 1.4.3) and Wishart (which was formula (1.105) in Section 1.2.4).

Corollary 1.4.6 (Siegel Plus Wishart). *Under our usual hypotheses on the function f, we have:*

$$\frac{1}{\text{Vol}(\mathcal{SP}_n/GL(n, \mathbb{Z}))} \int_{\mathcal{SP}_n/GL(n,\mathbb{Z})} \sum_{\substack{N \in \mathbb{Z}^{n \times k} \\ \text{rk}(N)=k}} f(W[N]) \, dW = c_{n,k} \int_{\mathcal{P}_k} f(Y) \, |Y|^{n/2} \, d\mu_k.$$

Here $d\mu_k$ is the invariant volume element on $\mathcal{P}_k$ and $c_{n,k}$ is the constant in Wishart's integral formula; i.e.,

$$c_{n,k} = \prod_{j=n-k+1}^{n} \pi^{j/2} \Gamma(j/2)^{-1}.$$

Proof. Use an argument similar to that given in proof of Corollary 1.4.1. ∎

Exercise 1.4.25 (A Generalization of the Minkowski-Hlawka Theorem I). For $Y \in \mathcal{P}_n$, define the following generalization of the first minimum m_Y in (1.178) if $1 \le k \le n$ by:

$$m_{Y,k} = \min \left\{ |Y[A]| \; \middle| \; A \in \mathbb{Z}^{n \times k}, \quad \text{rank } A = k \right\}.$$

Prove that if k is fixed $1 \le k < n$, n is sufficiently large (depending on k), and $r < (n/(2\pi e))^k$, there exists a matrix $Y \in \mathcal{P}_n$ such that

$$m_{Y,k} > r \, |Y|^{k/n}.$$

Hint. Imitate the proof of Corollary 1.4.2.

Exercise 1.4.26 (Convergence of Koecher's Zeta Function by an Integral Test). Show that Koecher's zeta function $Z_{k,n-k}(Y, s)$ defined by formula (1.173) in Section 1.4.1 will converge absolutely for $\text{Re } s > n/2$. Imitate the proof of Corollary 1.4.4.

Hint. The only new idea that is required is the following. There is a positive constant c such that if $Y[A]$ is in Minkowski's fundamental domain $\mathcal{M}_k$, then $|Y[A]| \le c \, |I[A]|$.

Similarly, there is a positive constant c^* such that if $Y[A] \in \mathcal{M}_k$ then $|Y[A]| \ge c^* \, |I[A]|$. Here c and c^* depend on Y and not on A. You also need to know that the set of matrices $A \in \mathbb{Z}^{n \times k}$ modulo $GL(k, \mathbb{Z})$ such that $|I[A]| \le 1$ is finite. This follows from Exercise 1.4.9(a) in Section 1.4.2.

Koecher's zeta function is another Eisenstein series for $GL(n, \mathbb{Z})$ and thus its analytic continuation is of interest to us. We will consider that problem later and find that our integral formulas are useful in this regard. The analytic continuation of Eisenstein series for $GL(n)$ is much harder than that of Epstein's zeta function.

But we will even want to consider more general Eisenstein series than Koecher's zeta function. Such Eisenstein series are associated with parabolic subgroups of $GL(n)$. Suppose that

$$n = n_1 + \cdots + n_q, \quad \text{with } n_i \in \mathbb{Z}^+,$$

is a **partition of n**. Then the (standard) **parabolic subgroup** $P = P(n_1, \ldots, n_q)$ of $GL(n)$ is defined to be the group of matrices U with block form:

$$\begin{pmatrix} U_1 & * & \cdots & * & * \\ 0 & U_2 & \cdots & * & * \\ \vdots & \vdots & \ddots & \vdots & \vdots \\ 0 & 0 & \cdots & U_{q-1} & * \\ 0 & 0 & \cdots & 0 & U_q \end{pmatrix}, \quad \text{with } U_j \text{ in } GL(n_j). \tag{1.241}$$

Koecher's zeta function is an Eisenstein series associated with a maximal parabolic subgroup (the case $q = 2$).

When $n = 2$ there is only one such standard parabolic subgroup, but for general n there are many such subgroups—as many as there are partitions of n, a number denoted by $p(n)$. The partition function $p(n)$ has been much studied by number theorists. It is a very rapidly increasing function of n. Some examples are:

$$p(10) = 42, \quad p(100) = 190, 569, 292, \quad p(200) = 3, 972, 999, 029, 388.$$

These are asymptotic and exact formulas for $p(n)$ when n is large, thanks to the work of Rademacher, Hardy, Littlewood, and Ramanujan, as well as the fact that $p(n)$ is the nth Fourier coefficient of a modular form of weight $-\frac{1}{2}$, namely $\eta(z)^{-1}$ from formula (3.69) in Section 3.4 of Volume I.

We will develop one version of the general integral formula in some exercises.

Exercise 1.4.27 (Siegel's Integral Formula for a Maximal Parabolic Subgroup $P(k, n - k)$).

(a) Suppose that $P = P(k, n - k)$ is the maximal parabolic subgroup defined in (1.241) and that we have an integrable function $f : \mathcal{P}_k/\Gamma_k \to \mathbb{C}$, with $\Gamma_k = GL(k, \mathbb{Z})$. Show that

$$\int_{\mathcal{SM}_n} \sum_{\substack{(A \ *) \in \Gamma_n/P \\ A \in \mathbb{Z}^{n \times k}}} f(W[A]) \, dW = \text{Vol}(\mathcal{SM}_{n-k}) \int_{\mathcal{M}_k} f(X) \, |X|^{n/2} \, d\mu_k(X).$$

(b) Assuming the necessary integrability conditions on the following functions:

$$f_1 : \mathcal{P}_k/\Gamma_k \to \mathbb{C}, \quad f_2 : \mathbb{R}^+ \to \mathbb{C},$$

show that, defining

$$f(Y) = f_1(Y_1)f_2(|Y|), \quad \text{for } Y = \begin{pmatrix} Y_1 & * \\ * & * \end{pmatrix}, \quad \text{with } Y_1 \in \mathcal{P}_k,$$

we have

$$\int_{\mathcal{M}_n} \sum_{\substack{(A*)\in\Gamma_n/P \\ A\in\mathbb{Z}^{n\times k}}} f(Y[A])\, d\mu_n(Y)$$

$$= \mathrm{Vol}(\mathcal{SM}_{n-k}) \int_{t>0} f_2(t)\, t^{-k/2-1}\, dt \int_{\mathcal{M}_k} f_1(X)|X|^{n/2}\, d\mu_k(X).$$

Hints.

(a) Summing over $(A\ *) \in GL(n, \mathbb{Z})/P(k, n-k)$ is the same as summing over A mod $GL(k, \mathbb{Z})$, where $A \in \mathbb{Z}^{n\times k}$ and A fits into a matrix in $GL(n, \mathbb{Z})$. So we find from similar arguments to those that gave us Corollary 1.4.6 and Lemma 1.4.7 that:

$$\int_{\mathcal{SM}_n} \sum_{\substack{(A*)\in\Gamma_n/P \\ A\in\mathbb{Z}^{n\times k}}} f(W[A])\, dW = c \int_{\mathcal{M}_k} f(X)\, |X|^{n/2}\, d\mu_k(X).$$

Here

$$c = c_{n,k}\mathrm{Vol}(\mathcal{SM}_{n-k}) \prod_{j=0}^{k-1} \zeta(n-j)^{-1},$$

where $c_{n,k}$ is the constant in Wishart's integral formula (see corollary 1.4.6 to Proposition 1.4.3). Now the formula for $\mathrm{Vol}(\mathcal{SM}_n)$ in Theorem 1.4.2 and the formula for $c_{n,k}$ in Corollary 1.4.6 finish this part of the exercise.

(b) Start with part (a) of the exercise and replace $f(X)$ by $f_2(t)f_1(t^{1/n}X)$. Then integrate with respect to $t^{-1}dt$ over $t > 0$ to get the result.

Exercise 1.4.28 (An Integral Formula for an Arbitrary Parabolic Subgroup). Suppose that

$$f(Y) = \prod_{j=1}^{q} f_j(|Y_j|), \quad \text{where } Y_j \in \mathcal{P}_{N_j}, \quad N_j = n_1 + \cdots + n_j,$$

and

$$Y = \begin{pmatrix} Y_j & * \\ * & * \end{pmatrix}.$$

Prove that if $\Gamma_n = GL(n, \mathbb{Z})$ and the parabolic subgroup is $P = P(n_1, \ldots, n_q)$, then

$$\int_{\mathcal{M}_n} \sum_{A \in \Gamma_n/P} f(Y[A]) \, d\mu_n(Y)$$

$$= V_{n_q} \int_{t_q > 0} f_q(t_q) t_q^{(-k/2)-1} \, dt_q \prod_{j=1}^{q-1} V_{n_j} \int_{t_j > 0} f_j(t_j) t_j^{(e_j/2)-1} \, dt_j,$$

where $V_i = \mathrm{Vol}(\mathcal{SM}_{n_i})$, $k = N_{q-1}$, and $e_j = n_j + n_{j+1}$.

Hint. The case $q = 2$ comes from part (b) of Exercise 1.4.27. To prove the general result, use induction on q and write $A \in GL(n, \mathbb{Z})/P$, as

$$A = BC, \quad \text{with} \quad B = (B_1 \; *) \in GL(n, \mathbb{Z})/Q, \; B_1 \in \mathbb{Z}^{n \times k}, \; Q = P(k, n_q), \; k = N_{q-1},$$

$$C = \begin{pmatrix} D & * \\ 0 & * \end{pmatrix} \in Q/P, \quad D \in GL(k, \mathbb{Z})/P^*, \quad P^* = P(n_1, \ldots, n_{q-1}).$$

Note that if $A = (A_1 \; *)$ with $A_1 \in \mathbb{Z}^{n \times k}$, then $A_1 = B_1 D$. Thus

$$\int_{\mathcal{M}_n} \sum_{A \in \Gamma_n/P} f(Y[A]) \, d\mu_n(Y)$$

$$= V_{n_q} \int_{t_q > 0} f_q(t_q) t_q^{(-k/2)-1} \, dt_q \int_{\mathcal{M}_k} f_{q-1}(|Y|) |Y|^{n/2} \sum_{D \in \Gamma_k/P^*} \prod_{j=1}^{q-1} f_j \left(|(Y[D])_j| \right) \, d\mu_k(Y).$$

The proof is completed by induction.

Exercise 1.4.29 (Another Integral Formula for Arbitrary Parabolic Subgroups). Suppose that Y has the partial Iwasawa decomposition:

$$Y = \begin{pmatrix} V_1 & \cdots & 0 \\ \vdots & \ddots & \vdots \\ 0 & \cdots & V_q \end{pmatrix} \begin{bmatrix} \begin{pmatrix} I_{n_1} & \cdots & R_{ij} \\ \vdots & \ddots & \vdots \\ 0 & \cdots & I_{n_q} \end{pmatrix} \end{bmatrix}, \quad \text{with } V_j \in \mathcal{P}_{n_j}, \; R_{ij} \in \mathbb{R}^{n_i \times n_j},$$

and consider a function $g(Y) = h(V_1, \ldots, V_q)$, satisfying suitable integrability conditions. Show that if $P = P(n_1, \ldots, n_q)$ and $\Gamma = GL(n, \mathbb{Z})$, then:

$$\int_{\mathcal{M}_n} \sum_{A \in \Gamma/P} g(Y[A]) \, d\mu_n(Y)$$

$$= \prod_{j=1}^{q} \text{Vol}(\mathcal{S}\mathcal{M}_{n_j}) \int_{V_j \in \mathcal{M}_{n_j}} h(V_1, \ldots, V_q) |V_j|^{f_j} \, d\mu_{n_j}(V_j),$$

where $f_j = (n - N_j - N_{j-1})/2$.

Hint. Actually you should not try to compute the constant in this formula until later. Up to the computation of the constant, the formula follows from:

$$\int_{\mathcal{P}_n/\Gamma} \sum_{\Gamma/P} = c \int_{\mathcal{P}_n/P}$$

and the Jacobian of the partial Iwasawa decomposition (see Maass [426, pp. 149–150] or Varadarajan [623, p. 293]):

$$d\mu_n(Y) = \prod_{j=1}^{q} |V_j|^{f_j} \, d\mu_{n_j}(V_j) \prod_{1 \leq i \leq k \leq q} dR_{ik},$$

where f_j is as given in the problem and dR_{ij} is ordinary Lebesgue measure on $n_i \times n_j$ matrix space.

The preceding exercises are based on what can be viewed as an analogue of the integral formula involved in the Rankin–Selberg method (i.e., formula (3.127) in Section 3.6 of Volume I).

1.5   Maass Forms for $GL(n, \mathbb{Z})$ and Harmonic Analysis on $\mathcal{P}_n/GL(n, \mathbb{Z})$

...and the manuscript was becoming an albatross about my neck. There were two possibilities: to forget about it completely, or to publish it as it stood; and I preferred the second.

From Langlands [392, Preface]

1.5.1   Analytic Continuation of Eisenstein Series by the Method of Inserting Larger Parabolic Subgroups

In order to do harmonic analysis on $\mathcal{P}_n/GL(n, \mathbb{Z})$, as in the case $n = 2$, we need to study Maass forms or automorphic forms for $\Gamma_n = GL(n, \mathbb{Z})$ and, in particular, obtain the analytic continuation of the Eisenstein series which form the continuous spectrum of the $GL(n, \mathbb{R})$-invariant differential operators on the fundamental domain. Selberg [543–545] had already noticed this in the 1950s and the methods we will develop are probably similar to the unpublished methods of Selberg. We will discuss two of Selberg's methods in this and the next section. These methods are also discussed by Maass [426]. There is a third unpublished method of Selberg which makes more use of functional analysis (see Shek-Tung Wong [671]). For other points of view, see the books by Langlands [392] and Osborne and Warner [482] (cf. Langlands [395]) which discuss the analytic continuation of more general Eisenstein series. A more recent reference is Jorgenson and Lang [334].

We define a **Maass form—also known as a modular or automorphic form**—v for $GL(n, \mathbb{Z}) = \Gamma$ to be a function $v : \mathcal{P}_n \to \mathbb{C}$ such that:

$$
\left.
\begin{aligned}
&(1)\ v \text{ is an eigenfunction of all the invariant differential operators } L \text{ in } D(\mathcal{P}_n);\\
&\qquad \text{i.e.,} \quad Lv = \lambda_L v, \text{ for some eigenvalue } \lambda_L;\\
&(2)\ v \text{ is } \Gamma\text{-invariant; i.e.,} \quad v(Y[A]) = v(Y), \text{ for all } Y \in \mathcal{P}_n,\ A \in \Gamma;\\
&(3)\ v \text{ has at most polynomial growth at infinity; i.e.,}\\
&\qquad |v(Y)| \le C|p_s(Y)|, \qquad \text{for some } s \in \mathbb{C}^n \text{ and } C > 0.
\end{aligned}
\right\}
$$

$$(1.242)$$

We shall use the notation $\mathcal{A}(\Gamma, \lambda)$ for the **space of Maass forms or automorphic forms** for a given eigenvalue system λ.

Note that Goldfeld [230] uses slightly different language since he also demands that Maass forms be cuspidal in the sense of the definition in formula (1.243) below for all k. Definition (1.242) is clearly a generalization of the concept of Maass wave form which appeared in Section 3.5 of Volume I. Maass considers these automorphic forms in [426, Section 10] and he calls them "grossencharacters." That name can be explained by the fact that Hecke grossencharacters play the same role in harmonic analysis for $GL(1)$ over a number field that forms in $\mathcal{A}(\Gamma, \lambda)$ play for harmonic analysis on $\mathcal{P}_n/\Gamma$ (see Hecke [268, pp. 215–234, 249–287], Jacquet and Langlands [324], Stark [575], and Weil [660]).

Motivated by the study of representations of semisimple Lie groups, Harish-Chandra has given a much more general definition of automorphic form (see Borel's lecture in Borel and Mostow [68, pp. 199–210]). This definition includes (1.242) above as well as the concept of Siegel modular form for $Sp(n, \mathbb{Z})$ which was introduced by C.L. Siegel in his work on quadratic forms. See Chapter 2 of this volume.

Another motivation for the study of Maass forms for $GL(n)$ is the need to study various kinds of L-functions with many gamma factors in their functional equations.

For L-functions corresponding to Maass forms for $GL(n)$ will indeed have lots of gamma factors and will have an Euler product if they are eigenforms for the algebra of Hecke operators for $GL(n)$. See Sections 1.5.2 and 1.5.4 which follow, also Bump [83] or Goldfeld [230]. An adelic treatment of this subject is part of the Langlands theory (see Gelbart [209], Godement and Jacquet [228], Goldfeld [230], Goldfeld and Hundley [232] or Jacquet et al. [325]). Langlands has conjectured that there is a reciprocity law generalizing the Artin reciprocity law which says that each Artin L-function corresponding to an n-dimensional representation of a Galois group of an extension of number fields is the L-function for some automorphic representation of $GL(n)$. See Langlands [389–395], Arthur [24], Casselman [98], Gelbart [209], and Goldfeld [230, last chapter]. In fact, Artin L-functions do have functional equations involving multiple gamma functions as well as Euler products and this certainly gives good evidence for Langlands' conjecture. A good reference for Artin L-functions is Lang [386].

Booker [60] provides a nice survey on L-functions of number theory, with an emphasis on those corresponding to $GL(3, \mathbb{Z})$. He notes that many number theorists have considered the **Selberg class** of L-functions; i.e., those with Euler product, analytic continuation, and functional equation (and one hopes, someday, to add the Riemann hypothesis on the location of poles). Not every favorite Dirichlet series in this volume is in the Selberg class. For example, the Epstein zeta function does not usually have an Euler product. We will say more about the L-functions for Maass forms f for $GL(3, \mathbb{Z})$ in Sections 1.5.2 and 1.5.4. These are analogues of the L-functions corresponding to Maass wave forms f on the Poincaré upper half plane, assuming f is an eigenform of all the Hecke operators. Such L-functions are in the Selberg class as are Dedekind zeta functions, Dirichlet L-functions, Artin L-functions, the L-functions corresponding to holomorphic modular forms on H which are eigenforms of the Hecke operators, the Hasse–Weil L-functions of elliptic curves, and many more. A survey on L-functions by Iwaniec and Sarnak [321] gives much more information on what is known and what is conjectured.

There are other sorts of L-functions with Euler products and multiple gamma factors in their functional equations—the analogues of the Rankin–Selberg L-functions studied in formula (3.125) of Volume I. These Rankin–Selberg type L-functions have applications to the problem of proving a Ramanujan conjecture for cusp forms for $GL(n, \mathbb{Z})$ (cf. Vol. I, formula (3.87) for the case $n = 2$ and Section 1.5.4 of this volume). Some references are: Bump and Friedberg [86], Elliott et al. [164], Friedberg [190], Goldfeld [230], Jacquet and Shalika [326], Jacquet et al. [325], Moreno and Shahidi [461, 462], Novodvorsky and Piatetski-Shapiro [476], Piatetski-Shapiro [487], and Shahidi [550].

In the preceding section we needed to go backwards in time to commune with Minkowski. The present section unfortunately still demands a time machine that will carry us into the future. Lacking this item, the section will be incomplete.

References for this section include Arthur [24, 25], Ash [28], Baily [32], Bernstein and Gelbart [47], Borel and Casselman [66], Borel and Mostow [68], Bump [83, 84], Casselman [98], Flicker [182], Gelbart [209], Gelbart and Jacquet [210], Gelfand et al. [214], Godement [225, 227], Godement and Jacquet [228], Goldfeld

[230], Goldfeld and Hundley [232], Harish-Chandra [262], Hejhal et al. [272], Jacquet [323], Jacquet et al. [325], Jacquet and Shalika [326, 327], Kazhdan and Patterson [344], Langlands [389–395], Maass [414, 426], Ramanathan [497, 498], Sarnak and Shahidi (Eds.) [529], Selberg [543–546], Tamagawa [587–589], A.B. Venkov [627], and Dorothy Wallace [642–650]. There are websites devoted to Langlands, Sarnak, and Selberg at the Institute for Advanced Study, Princeton. The volume [31] containing proceedings from the conference in honor of Selberg should also be mentioned. This does bring up some unpleasant memories for me. When not invited to this conference, I changed fields.

In earlier sections, we saw how to build up the eigenfunctions of the G-invariant differential operators in $D(\mathcal{P}_n)$ by integrating power functions over orthogonal, abelian, and nilpotent subgroups of $GL(n, \mathbb{R})$, as in formulas (1.60), (1.67), and (1.70) of Section 1.2. The powers s in $p_s(Y)$ provide a way of indexing the eigenvalues of an invariant differential operator $L \in D(\mathcal{P}_n)$ via $Lp_s(Y) = \lambda_L(s)p_s(Y)$. We shall use this sort of indexing when we speak of the dimensionality of the spectrum components. For inversion of the Helgason–Fourier transform on $\mathcal{P}_n$, the spectrum needed was n-dimensional (see Theorem 1.3.1 of Section 1.3.1). The inverse transform required integration over a product of n lines: Re $s_j = -\frac{1}{2}$, $j = 1, 2, \ldots, n-1$, and Re $s_n = (n-1)/4$. We shall see that life is much more complicated in $\mathcal{P}_n/GL(n, \mathbb{Z})$, since there are also discrete and lower dimensional spectra. However, the basic method of constructing $GL(n, \mathbb{Z})$-invariant eigenfunctions in the highest dimensional part of the spectrum is analogous to the construction of spherical and K-Bessel functions. That is, one must sum power functions over $GL(n, \mathbb{Z})$ modulo a parabolic subgroup—perhaps including lower rank cusp forms in the mix. But it is not so simple for general n as it was for $n = 2$ in Section 3.7 of Volume I. In the following discussion, we will sometimes consider only the case of $GL(3, \mathbb{Z})$ in order to simplify the formulas.

Before defining Eisenstein series, we need to consider another sort of Maass form for $GL(n, \mathbb{Z})$—the cusp form. A **Maass cusp form** is a Maass form $f \in \mathcal{A}(\Gamma, \lambda)$, $\Gamma = GL(n, \mathbb{Z})$, with the property that for any k with $1 \leq k \leq n-1$, we have:

$$\int_{X \in (\mathbb{R}/\mathbb{Z})^{k \times (n-k)}} f\left(Y \begin{bmatrix} I & X \\ 0 & I \end{bmatrix}\right) dX = 0, \quad \text{for all } Y \in \mathcal{P}_n. \tag{1.243}$$

This just signifies the vanishing of the **constant terms** in a bunch of Fourier expansions of $f(Y)$ as a periodic function of the X-variable in partial Iwasawa coordinates (see page 268 of Volume I). We shall write $\mathcal{AC}(\Gamma, \lambda)$ for the **space of Maass cusp forms**.

If we knew enough about Fourier expansions for $GL(n)$, we should be able to show that a cusp form is bounded in the fundamental domain. See Goldfeld [230], or the article of Borel and Jacquet in the volume of Borel and Casselman [66, p. 192] or Harish-Chandra [262]. More information on Fourier expansions can be found in

Sections 1.5.3 and 1.5.4, Bump [83], Goldfeld [230], or Goldfeld and Hundley [232, Vol. II, p. 29]. Note that Goldfeld [230] drops the word "cusp" and calls these things "Maass forms" on $GL(n)$.

The cusp forms and constants form the discrete spectrum of the $GL(n, \mathbb{R})$-invariant differential operators on the fundamental domain. It can be proved, using a method of Gelfand and Piatetski-Shapiro, that cusp forms exist (cf. Section 1.5.5 and Theorem 3.7.2 in Volume I or Godement's article in Borel and Mostow [68, pp. 225–234]). We will not be able to give any explicit examples of cusp forms, just as we could not give any examples for the case $n = 2$ in Section 3.5 of Volume I. Recently people have computed computer approximations to cusp forms for $SL(n, \mathbb{Z})$, for $n = 3$ and 4. See Section 1.5.4. There are adelic examples of cusp forms belonging to congruence subgroups of $GL(3, \mathbb{Z})$ corresponding via generalizations of Hecke's correspondence to Hecke L-functions of cubic number fields (see Jacquet et al. [325]). There are also cuspidal examples corresponding via Hecke theory to Rankin–Selberg L-functions for $GL(2, \mathbb{Z})$ (see Gelbart and Jacquet [210] and Moreno and Shahidi [461]). Ash [28] and Ash et al. [29] compute cohomology of $SL(3, \mathbb{Z})$ using Hecke operators and methods of algebraic topology and differential geometry. They show the existence of cusp forms for $SL(3, \mathbb{Z})$ which come from the DeRham cohomology of the fundamental domain and are analogous to holomorphic automorphic forms of weight 2. See also Lee and Schwermer [399], Lee and Szczarba [400], Schwermer [539, 540], and Soulé [571, 572]. Donnelly [138] finds an upper bound for the dimension of the space of cusp forms.

There are also papers proving an $SL(3, \mathbb{Z})$-analogue of the Weyl law for $SL(2, \mathbb{Z})$ to be found in Theorem 3.7.5 of Volume I (see Stephen D. Miller [447] or Eric Stade and Dorothy Wallace [574]). More recently Werner Müller [469] proved Weyl's law for the cusp forms for the congruence subgroup $\Gamma = \Gamma(N)$ of $SL(n, \mathbb{Z})$ by translating the problem to the adeles, then making use of Arthur's trace formula plus the heat kernel. Results of Donnelly [138], Jacquet et al. [325], Luo et al. [413], Colette Mœglin and Jean-Loup Waldspurger [455, 456], Werner Müller and Birgit Speh [470] are used. Lapid and Müller [396] obtain the Weyl law for the cusp forms for the congruence subgroup $\Gamma = \Gamma(N)$ of $SL(n, \mathbb{Z})$ with an error estimate. Müller proves that if d_n is the dimension of the Riemannian manifold $\Gamma \backslash SL(n, \mathbb{R})/SO(n)$ and the eigenvalues of the Laplacian for $L^2(\Gamma \backslash SL(n, \mathbb{R})/SO(n))$ are denoted $\lambda_0 < \lambda_1 \leq \lambda_2 \leq \cdots$, then we have the **Weyl law**:

$$\#\{j \mid \lambda_j \leq x\} \sim \frac{1}{\Gamma\left(1 + \frac{d_n}{2}\right)} \mathrm{Vol}\left(\Gamma \backslash SL(n, \mathbb{R})/SO(n)\right) \left(\frac{x}{4\pi}\right)^{d_n/2}, \quad \text{as } x \to \infty.$$

Lindenstrauss and Venkatesh [405] generalize the result to certain semisimple groups of the sort considered in Chapter 2 of this Volume.

One might expect that the existence of the analogue of odd cusp forms would not be so hard to prove. One would hope to imitate Exercise 3.7.6 from Volume I, replacing $\sin(2\pi nx)$ with a product of $\sin(2\pi nx_i)$ for all the x-variables. Unfortunately it is shown in Goldfeld [230, pp. 162–163] that no odd cusp forms exist for $GL(3, \mathbb{Z})$. Translating into our language, let

$$Y = \begin{pmatrix} y_1 & 0 & 0 \\ 0 & y_2 & 0 \\ 0 & 0 & y_3 \end{pmatrix}, \ X = \begin{pmatrix} 1 & x_1 & x_2 \\ 0 & 1 & x_3 \\ 0 & 0 & 1 \end{pmatrix}, \ \delta = \begin{pmatrix} \delta_1 & 0 & 0 \\ 0 & \delta_2 & 0 \\ 0 & 0 & \delta_3 \end{pmatrix}, \ \delta_i \in \{1, -1\}.$$

Then if φ is a Maass form for $GL(3, \mathbb{Z})$, we have

$$\varphi\left(Y[\delta X \delta]\right) = \varphi\left(Y[X[\delta]]\right), \ \text{and} \ X[\delta] = \begin{pmatrix} 1 & \delta_1 \delta_2 x_1 & \delta_1 \delta_3 x_2 \\ 0 & 1 & \delta_2 \delta_3 x_3 \\ 0 & 0 & 1 \end{pmatrix}.$$

It follows that φ must be an even function of each x_i. Note that $\det \delta = \delta_1 \delta_2 \delta_3$ and thus the matrix δ may not be in $SL(3, \mathbb{Z})$. However you can take any one of the $\delta_i = 1$ and then let the other two δ_j be -1. Thus Goldfeld has proved that any Maass form for $SL(3, \mathbb{Z})$ must be even in each x_i. We will say more about cusp forms in § 1.5.4.

Now we direct our attention to Eisenstein series. We will consider the methods that Selberg and Maass used to continue these series. Before giving our first definition of Eisenstein series, we need to recall some notation concerning the **determinant one surface** $\mathcal{SP}_n$ in $\mathcal{P}_n$. If $Y \in \mathcal{P}_n$, we write

$$Y = t^{1/n} W, \quad t = |Y| > 0, \quad W = Y^0 \in \mathcal{SP}_n. \tag{1.244}$$

Clearly, one can define Maass forms on $\mathcal{SP}_n / \Gamma_n$, $\Gamma_n = GL(n, \mathbb{Z})$, as in (1.242). We will denote **the space of Maass forms on the determinant one surface in** $\mathcal{P}_n$ by $\mathcal{A}^0(\Gamma_n, \lambda)$. See Exercise 1.5.1 below for the relationship between the Laplacian on $\mathcal{P}_n$ and the Laplacian on $\mathcal{SP}_n$.

Now we give our first definition of an Eisenstein series. Suppose that v is a Maass form on a lower rank determinant one surface; i.e., let

$$v \in \mathcal{A}^0(GL(m, \mathbb{Z}), \lambda), \quad 1 \le m < n, \quad \operatorname{Re} s > n/2 \ \text{and} \ Y \in \mathcal{P}_n.$$

Then we define the **Eisenstein series** $E_{m,n-m}(v, s|Y) = E(v, s|Y)$ with lower rank Maass form v by:

$$E(v, s|Y) = \sum_{A=(A_1 \ *) \in \Gamma_n / P(m,n-m)} |Y[A_1]|^{-s} \, v\left(Y[A_1]^0\right). \tag{1.245}$$

Here $A_1 \in \mathbb{Z}^{n \times m}$, $\Gamma_n = GL(n, \mathbb{Z})$, and $P(m, n-m)$ is the parabolic subgroup defined in formula (1.241) of the preceding section.

Note that Exercise 1.4.26 of Section 1.4.4 says that the series in (1.245) converges whenever v is bounded. To see that the series in (1.245) converges when v is an integrable function on the fundamental domain $\mathcal{SM}_m \cong \mathcal{SP}_m / GL(m, \mathbb{Z})$, use the integral formula of Exercise 1.4.27 in Section 1.4.4 and imitate Exercise 1.4.26 of that same section. We can thus compare the series (1.245) with the integral:

$$\int_{\mathcal{S}\mathcal{M}_n} \sum_{\substack{(A\,*)\in\Gamma_n/P \\ |Y[A]|\geq 1}} f(Y[A])\, d\mu_n(Y) = \mathrm{Vol}(\mathcal{S}\mathcal{M}_{n-m}) \int_{\substack{X\in\mathcal{M}_m, \\ |X|\geq 1}} f(X)|X|^{n/2}\, d\mu_m(X),$$

where $f(X) = |X|^{-s}v\left(X^0\right)$. Then formula (1.219) in the preceding section says this last integral is:

$$\mathrm{Vol}(\mathcal{S}\mathcal{M}_{n-m}) \int_{X\in\mathcal{S}\mathcal{M}_m} v(W)\, dW \int_{t\geq 1} t^{n/2-s-1}\, dt,$$

which converges for $\mathrm{Re}\, s > n/2$, if v is integrable on the fundamental domain $\mathcal{S}\mathcal{M}_m$. One can obtain a similar domain of convergence assuming that v is bounded by a power function.

Exercise 1.5.1. Why is the Eisenstein series defined in (1.245) an eigenfunction of all the $GL(n, \mathbb{R})$-invariant differential operators in $D(\mathcal{P}_n)$? Compute the eigenvalue of the Laplacian acting on (1.245) as a function of s and of the eigenvalue of the Laplacian acting on v.

Hint. Maass [426, p. 73] gives formulas for the Laplacian in partial Iwasawa coordinates:

$$Y = \begin{pmatrix} F & 0 \\ 0 & H \end{pmatrix} \begin{bmatrix} I & X \\ 0 & I \end{bmatrix}, \quad F \in \mathcal{P}_m, \quad H \in \mathcal{P}_{n-m}, \quad X \in \mathbb{R}^{m\times(n-m)}.$$

Maass finds that for functions of the form $u(Y) = f(F)h(H)$ the invariant differential operators look like:

$$\mathrm{Tr}\left(\left(Y\frac{\partial}{\partial Y}\right)^k\right) = \mathrm{Tr}\left(\left(F\frac{\partial}{\partial F} + \frac{n-m}{2}I\right)^k\right) + \mathrm{Tr}\left(\left(H\frac{\partial}{\partial H}\right)^k\right)$$

$$-\frac{1}{2}\sum_{j=1}^{k-1}\mathrm{Tr}\left(\left(F\frac{\partial}{\partial F} + \frac{n-m}{2}I\right)^j\right)\mathrm{Tr}\left(H\frac{\partial}{\partial H}\right)^{k-1-j}.$$

We proved the special case $m = 1$ in formula (1.54) of Section 1.2.1. Now in the case under consideration in this exercise our function is

$$u(Y) = |F|^{-s}v\left(F^0\right), \quad \text{where} \quad F^0 = |F|^{-1/m}F \in \mathcal{S}\mathcal{P}_m.$$

So the only term of interest in the formula for the Laplacian in partial Iwasawa coordinates is the first term.

One must also relate the Laplacian on the space $\mathcal{P}_m$ and that on the determinant one surface $\mathcal{SP}_m$. Writing $Y = t^{1/m}W$, for $Y \in \mathcal{P}_m$, $t = |Y| > 0$, $W \in \mathcal{SP}_m$, one can show that $\Delta_Y = m(t\partial/\partial t)^2 + \Delta_W$, where Δ_W is the Laplacian on the determinant one surface induced by the arc length $ds_W^2 = \operatorname{Tr}((W^{-1}dW)^2)$.

When the Maass form v is identically equal to one, the Eisenstein series (1.245) is a quotient of Koecher zeta functions defined in formula (1.173) of the preceding section; i.e.,

$$Z_{m,n-m}(Y,s) = Z_{m,0}(I,s)E(1,s|Y). \tag{1.246}$$

To see this, we need the following lemma.

Lemma 1.5.1 (A Decomposition of the Matrices in the Sums Defining Eisenstein Series). *The quotient $\mathbb{Z}^{n\times m}$ rank m/Γ_m, which means the $n\times m$ rank m integral matrices in a complete set of matrices inequivalent under right multiplication by matrices in $\Gamma_m = GL(m,\mathbb{Z})$, for $1 \le m < n$, can be represented by matrices $A = BC$, where*

$$B \in \mathbb{Z}^{n\times m}, \quad (B *) \in GL(n,\mathbb{Z})/P(m,n-m),$$

$$C \in \mathbb{Z}^{m\times m} \quad \operatorname{rank} m/GL(m,\mathbb{Z}).$$

Proof. First note that by elementary divisor theory, there are matrices $U \in \Gamma_n$ and $V \in \Gamma_m$ such that

$$A = U \begin{pmatrix} D \\ 0 \end{pmatrix} V, \quad \text{with } D \text{ diagonal } m \times m \text{ and nonsingular.}$$

Suppose that $U = (U_1 *)$ with $U_1 \in \mathbb{Z}^{n\times m}$. Then $A = U_1DV$ and U_1 may be taken modulo $GL(m,\mathbb{Z})$, by throwing the difference into DV. The existence of the stated decomposition follows quickly.

To see the uniqueness of the decomposition, suppose

$$B_2C_2 = B_1C_1W, \quad \text{for some } W \in \Gamma_n$$

with

$$V_i = (B_i *) \in \Gamma_n/P(m,n-m) \quad \text{and} \quad C_i \in \mathbb{Z}^{m\times m} \quad \operatorname{rank} m/\Gamma_m, \ i = 1,2.$$

Then

$$V_2 \begin{pmatrix} C_2 \\ 0 \end{pmatrix} = V_1 \begin{pmatrix} C_1W \\ 0 \end{pmatrix} \quad \text{implies} \quad V_1^{-1}V_2 \begin{pmatrix} C_2 \\ 0 \end{pmatrix} = \begin{pmatrix} C_1W \\ 0 \end{pmatrix}.$$

Let

$$Z = V_1^{-1} V_2 \in \Gamma_n \quad \text{and} \quad Z = \begin{pmatrix} X & * \\ Y & * \end{pmatrix},$$

so that $XC_2 = C_1 W$, $YC_2 = 0$, which means that Y must vanish. If $Y = 0$, there are 2 cases. If V_1 is not equivalent to V_2 modulo $P(m, n - m)$, we have a contradiction. If V_1 is equivalent to V_2 modulo $P(m, n - m)$, we may assume $V_1 = V_2$. This gives $C_2 = C_1 W$ and $W = I$, to complete the proof of the lemma.   ∎

Formula (1.246) follows immediately from Lemma 1.5.1. After we have studied Hecke operators for $GL(n, \mathbb{Z})$, we will be able to prove a similar result for a general Eisenstein series of the form (1.245) (see formula (1.278) below). Recall that Lemma 1.4.7 of the previous section expressed Koecher's zeta function $Z_{m,0}(I, s)$ as a product of Riemann zeta functions. Thus formula (1.246) is quite analogous to the formula in part (c) of Exercise 3.5.1 in Section 3.5.1 of Volume I — a result which writes the Eisenstein series for $GL(2, \mathbb{Z})$ as a quotient of Epstein's zeta function divided by Riemann's zeta function.

Exercise 1.5.2. In formula (1.245) for the Eisenstein series $E_{m,n-m}(s, \varphi|Y)$, let $n = 3$, $m = 2$, and $\varphi \in \mathcal{A}^0(GL(2, \mathbb{Z}), \lambda = u(u - 1))$; i.e., φ is a Maass wave form for $GL(2, \mathbb{Z})$ as in Section 3.5 of Volume I. Suppose, in particular, that φ is an Eisenstein series: $\varphi(z) = E_u(z)$, $z \in H$, defined in Equation (3.81) of Section 3.5 of Volume I. Recall the standard identification of the upper half plane H with $\mathcal{SP}_2$ given in Exercise 3.5.1 of Volume I. Then show that

$$E(E_u, s|Y) = \sum_{\substack{(C_i \ *) = C \in \Gamma_3/P_{(3)} \\ C_i \in \mathbb{Z}^{3 \times i}}} Y[C_1]^{-u} \, |Y[C_2]|^{-s + u/2},$$

where $P_{(3)} = P(1, 1, 1)$ is the minimal parabolic subgroup of $\Gamma_3 = GL(3, \mathbb{Z})$ consisting of all upper triangular matrices with ± 1 on the diagonal (see formula (1.241) of the preceding section). Use an integral test argument similar to the ones given after formula (1.245) above and Exercise 1.4.28 of the preceding section to see that the series on the right-hand side of this formula converges for $\mathrm{Re}\, u > 1$ and $\mathrm{Re}(s - u/2) > 1$. Note also that E_u is integrable if $0 < \mathrm{Re}\, u < 1$.

Hint. We can write the sum over $\Gamma_3/P(1, 1, 1)$ in the right-hand side of the formula to be demonstrated as a double sum over $\Gamma_3/P(2, 1)$ and over $P(2, 1)/P(1, 1, 1)$. The latter sum can be identified as a sum over matrices of block form:

$$\begin{pmatrix} \Gamma_2/P(1, 1) & 0 \\ 0 & \pm 1 \end{pmatrix}.$$

This shows that the right-hand side of the formula looks like:

$$\sum_{(A_1 \;*)\in\Gamma_3/P(2,1)} |Y[A_1]|^{-s+u/2} \sum_{(b \;*)\in\Gamma_2/P(1,1)} Y[A_1][b]^{-u},$$

which is $E(E_u, s|Y)$.

Motivated somewhat by the preceding exercise, we define **Selberg's Eisenstein Series for a Parabolic Subgroup** $P = P(n_1, \ldots, n_q)$ to be a function of $Y \in \mathcal{P}_n$ and $s \in \mathbb{C}^q$ given by:

$$E_P(s|Y) = \sum_{\substack{(A_j \;*)=A\in\Gamma_n/P \\ A_j \in \mathbb{Z}^{n \times N_j}}} \prod_{j=1}^{q} |Y[A_j]|^{-s_j} \qquad (1.247)$$

with $N_j = n_1 + \cdots + n_j$. We will also write

$$E_{n_1,\ldots,n_q}(s|Y) = E_P(s|Y).$$

The integral test coming from Exercises 1.4.28 and 1.4.29 of the preceding section generalizes to show that the series above converges absolutely for

$$\operatorname{Re} s_j > (n_j + n_{j+1})/2, \quad j = 1, 2, \ldots, q-1.$$

Exercise 1.5.3. Prove the last statement about the region of convergence of the series (1.247).

Since Selberg proved all the basic properties of the functions (1.247), there is good reason for calling $E_P(s|Y)$ "Selberg's Eisenstein series." Some authors (e.g., Maass [426] and Christian [110]) call (1.247) "Selberg's zeta function." This is confusing since there is another function which has been given that name (namely, the zeta function in formula (3.184) of Volume I).

It is also possible to create Eisenstein series involving Maass forms $v_j \in \mathcal{A}^0(GL(n_j, \mathbb{Z}), \lambda_j)$ by summing a function $f(Y)$ defined by

$$f(Y) = \prod_{j=1}^{q} |Q_j|^{r_j} v(Q_j^0),$$

where

$$Y = \begin{pmatrix} Q_1 & 0 & \cdots & 0 \\ 0 & Q_2 & \cdots & 0 \\ \vdots & \vdots & \ddots & \vdots \\ 0 & \cdots & \cdots & Q_q \end{pmatrix} \begin{bmatrix} I_{n_1} & * & \cdots & * \\ 0 & I_{n_2} & \cdots & * \\ \vdots & \vdots & \ddots & \vdots \\ 0 & \cdots & \cdots & I_{n_q} \end{bmatrix},$$

and the Q_j are positive $n_j \times n_j$ matrices. The **Eisenstein series** is then:

$$E_P(f|Y) = \sum_{\gamma \in \Gamma_n/P} f(Y[\gamma]). \qquad (1.248)$$

Here $\Gamma_n = GL(n, \mathbb{Z})$ and the parabolic subgroup is $P = P(n_1, \ldots, n_q)$. Our integral tests will show that this series converges for Re r_j sufficiently large, assuming that the Maass forms v_j are bounded by power functions or are integrable over the fundamental domain for $GL(n_j, \mathbb{Z})$. In the present subsection we will be studying the special case that all of the v_j are identically 1.

We need to study the analytic continuation, functional equations, residues, Fourier expansions, etc., of these Eisenstein series for $GL(n, \mathbb{Z})$. We will be interested in **two methods for analytic continuation of Eisenstein series.**

The first method is that of inserting larger parabolic subgroups between the modular group $\Gamma_n = GL(n, \mathbb{Z})$ and the parabolic subgroup $P(n_1, \ldots, n_q)$. An example of this method has already appeared in Exercise 1.5.1. The method is also discussed by Maass [426, pp. 275–278]. Maass attributes the idea to Selberg, who announced results of the sort we shall discuss in several places (see Selberg [543–546]). The method was also used by Langlands [392, Appendix I]. Other references are Terras [593, 594].

A **second method for continuing Eisenstein series is the method of theta functions** which goes back to Riemann. Many complications occur, as we shall see in the next section. Again, this method was developed by Selberg, who did not publish his proofs. Selberg did explain this method to various people (including the present author when she was a graduate student in 1969). It is developed in some detail in Maass [426, Section 16] and Terras [593, 594]. The main idea that eliminates the exploding integral is that of making use of differential operators chosen to annihilate the singular terms in the theta series. See also Siegel [565, Vol. III, pp. 328–333].

Other methods of obtaining analytic continuations of Eisenstein series are explained in Harish-Chandra [262], Kubota [377], Langlands [392], Selberg [546], and Wong [671]. These methods are more function-theoretic and apply in a more general context. The method of Wong [671] was outlined by Selberg in a talk with the author and Carlos Moreno in 1984. These methods make use of the Fourier expansion of the Eisenstein series, a topic which we will discuss in a later section. There are also adelic methods (see Jacquet's talk in Borel and Casselman [66, Vol. II, pp. 83–84] and Helen Strassberg [583]). And there is a method which uses Eisenstein series for the Siegel modular group (see Arakawa [18]). See also Christian [110] for the analytic continuation of Eisenstein series for congruence subgroups. And Diehl [136] obtains a relation between Eisenstein series for $GL(n, \mathbb{Z})$ and $Sp(n, \mathbb{Z})$ by methods similar to those of Lemma 1.5.2 and Exercise 1.5.9 below. Other references are Feryâl Alayont [4, 5]. Here it is noted that Selberg's third method of analytic continuation has been further developed by Bernstein in an unpublished manuscript.

More recently a new field has appeared—the field of multiple Dirichlet series. An introduction to the subject is to be found in Bump's paper beginning the volume edited by Bump et al. [88].

There are other sorts of Eisenstein series, created using the heat kernel by Jorgenson and Lang [335]. The introduction contains a brief history of the subject and the authors' attempts to find a treatment of the subject which is accessible to nonexperts.

Many complications arise in higher rank symmetric spaces. For example, there are lots of real variables as well as complex variables in the Eisenstein series, and the trace formula becomes horrendous. This make everyone is a bit grumpy—except perhaps E. Frenkel [188]. Frenkel seems to find joy in his movie involving writing his formulas on a woman's naked body. Well, our $GL(3)$ fundamental domain movie may not provide as much entertainment but it does make me less grumpy about the intricacies of $GL(n)$ Maass forms. Shimura's autobiography may give the reader some more insight into the grumpiness inherent in the field. For example, Shimura has the following comment about one of our heroes C.L. Siegel [557, p. 190]:

> ...we have to know what kind of a man Siegel was. Of course, he established himself as one of the giants in the history of mathematics long ago. He was not known, however, for his good-naturedness.

Now we begin the discussion of the method of inserting larger parabolic subgroups to continue Selberg's Eisenstein series (1.247). We are restricting our attention here to Eisenstein series with the maximal number of complex variables and no lower rank Maass forms. We shall use the notation $P_{(n)}$ for the minimal parabolic subgroup $P(1, \ldots, 1)$ of $\Gamma_n = GL(n, \mathbb{Z})$; i.e., $P_{(n)}$ consists of all upper triangular matrices in Γ_n. And we define, for $Y \in \mathcal{P}_n$, $s \in \mathbb{C}^n$, **Selberg's Eisenstein Series Associated with $P_{(n)}$** by:

$$E_{(n)}(s|Y) = \sum_{\gamma \in \Gamma_n / P_{(n)}} p_{-s}(Y[\gamma]), \quad \text{if} \quad \operatorname{Re} s_j > 1, \quad j = 1, \ldots, n-1. \quad (1.249)$$

First we generalize Exercise 1.5.1.

Note that Goldfeld [230, Ch. 10] names these and the rest of the $GL(n, \mathbb{Z})$-Eisenstein series for Langlands. It seems to me that Selberg was the first to study them and thus I am sticking with my terminology. However, Langlands [392] showed how to do the theory for general reductive groups. Selberg only published summaries of the results and Langlands himself said: "I myself now have difficulty finding my way through it." See [392, p. 284]. This manuscript of Langlands was circulated in partially blue dittoed pages in the early years of my study of this subject. The introduction to the memoir [335] of Jorgenson and Lang details the difficulties of reading expositions of the theory of Eisenstein series. I am following their dictum of trying to write for the nonexpert.

Lemma 1.5.2 (Relations Between $E_{(n)}$ and $E_{(2)}$). *When* Re $s_k > 1$, $k = 1, 2, \ldots, n-1$, *we have the following formula for the Selberg Eisenstein series defined in (1.249), for each $i = 1, 2, \ldots, n-1$:*

$$E_{(n)}(s|Y) = \sum_{\substack{V \in \Gamma_n/P_i^* \\ V = (V_k \, *), V_k \in \mathbb{Z}^{n \times k}}} E_{(2)}(s_i|T) \prod_{\substack{j=1 \\ j \neq 1}}^{n} |Y[V_j]|^{-s_j},$$

if $P_i^* = P(1, \ldots, 1, 2, 1, \ldots, 1)$ with the 2 in the ith position. Here $R \in \mathcal{P}_2$ is defined by the partial Iwasawa decomposition given below (see Exercise 1.1.11 of Section 1.1.3):

$$Y[V_{i+1}] = \begin{pmatrix} Y[V_{i-1}] & 0 \\ 0 & R \end{pmatrix} \begin{bmatrix} I_{i-1} & Q \\ 0 & I_2 \end{bmatrix}, \quad i = 2, \ldots, n-1;$$

and

$$Y[V_2] = R, \quad \text{when } i = 1.$$

Then $T \in \mathcal{P}_2$ is defined by:

$$T = |Y[V_{i-1}]|R, \quad \text{if } i = 2, \ldots, n-1;$$
$$T = R = Y[V_2], \quad \text{if } i = 1.$$

Finally the determinant of T is given by:

$$|T| = \begin{cases} |Y[V_{i-1}]| \, |Y[V_{i+1}]|, & \text{if } i = 2, \ldots, n-2; \\ |Y[V_2]|, & \text{if } i = 1; \\ |Y[V_{n-2}]| \, |Y|, & \text{if } i = n-1. \end{cases}$$

Proof. Observe that $U \in \Gamma_n/P_{(n)}$ can be expressed uniquely as $U = VW$, with $V \in \Gamma_n/P_i^*$ and $W \in P_i^*/P_{(n)}$. Moreover, W can be chosen to have the form:

$$W = \begin{pmatrix} I_{i-1} & 0 & 0 \\ 0 & W^* & 0 \\ 0 & 0 & I_{n-i-1} \end{pmatrix}, \quad W^* \in \Gamma_2/P_{(2)}. \tag{1.250}$$

Here i always denotes our fixed index. If $i = 1$, the top row and first column in (1.250) are not present and if $i = n-1$, the bottom row and last column are absent. Next we write:

$$V = (V_k \, *), \quad V_k \in \mathbb{Z}^{n \times k}, \quad k = 1, 2, \ldots, n; \tag{1.251}$$
$$V_{i+1} = (V_{i-1} \, V_i^*), \quad V_i^* \in \mathbb{Z}^{n \times 2}.$$

Then, according to Exercise 1.5.4 below, we have:

$$|Y[(VW)_j]| = |Y[V_j]|, \quad \text{for } W \text{ as in (1.250), } j \neq i; \tag{1.252}$$

and

$$|Y[(VW)_i]| = R[W_1^*]\, |Y[V_{i-1}]|, \tag{1.253}$$

where R is defined in the lemma being proved and $W_1^* \in \mathbb{Z}^2$ is defined by

$$W^* = (W_1^* \, *), \quad \text{for } W^* \text{ as in (1.250).} \tag{1.254}$$

The lemma follows immediately.   ∎

Exercise 1.5.4. Use the definitions in Lemma 1.5.2.

(a) Show that a complete set of representatives for $W \in P_i^* / P_{(n)}$ can be expressed in the form (1.250).
(b) Prove formula (1.252).
(c) Prove formula (1.253).

Hints. You can do part (a) by multiplying matrices in the appropriate block form. Part (b) follows from the remark that

$$Y\left[(A\,B)\begin{pmatrix} C \\ 0 \end{pmatrix}\right] = Y[AC], \quad \text{for } A \in \mathbb{R}^{n \times k},\ B \in \mathbb{R}^{n \times (n-k)},\ C \in \mathbb{R}^{k \times k},\ 0 \in \mathbb{R}^{(n-k) \times k}$$

where 0 denotes a matrix of zeros. Part (c) is proved using partial Iwasawa decompositions to obtain:

$$
\begin{aligned}
Y[(VW)_i] &= Y\left[V_{i+1}\begin{pmatrix} I_{i-1} & 0 \\ 0 & W_i^* \end{pmatrix}\right] \\
&= \begin{pmatrix} Y[V_{i-1}] & 0 \\ 0 & R \end{pmatrix}\left[\begin{pmatrix} I_{i-1} & Q \\ 0 & I_2 \end{pmatrix}\begin{pmatrix} I_{i-1} & 0 \\ 0 & W_1^* \end{pmatrix}\right] \\
&= \begin{pmatrix} Y[V_{i-1}] & 0 \\ 0 & R[W_1^*] \end{pmatrix}\begin{bmatrix} I_{i-1} & X \\ 0 & 1 \end{bmatrix}.
\end{aligned}
$$

It is now possible to begin the process of analytic continuation of the Selberg Eisenstein series $E_{(n)}$ to a meromorphic function of $s \in \mathbb{C}^n$ by utilizing the analytic continuation of $E_{(2)}$ which was obtained in Section 3.5 of Volume I, using Theorem 1.4.1 in Volume I.

Lemma 1.5.3 (First Step in the Analytic Continuation of Selberg's Eisenstein Series with n Complex Variables). *If $E(n)$ denotes the Selberg Eisenstein series defined in (1.249), set*

$$\Lambda(s) = \pi^{-s}\Gamma(s)\zeta(2s) \quad and \quad \Lambda_i(s|Y) = 2\Lambda(s_i)E_{(n)}(s|Y),$$

for $i = 1, 2, \ldots, n - 1$. Then $\Lambda_i(s|Y)$ can be analytically continued to the region D_i pictured in Figure 1.29. Moreover, $\Lambda_i(s|Y)$ satisfies the functional equation: $\Lambda_i(s|Y) = \Lambda_i(s'|Y)$, where

$$s'_j = \begin{cases} s_j, & j \neq i, i \pm 1, \\ 1 - s_i, & j = i, \\ s_{i\pm 1} + s_i - \frac{1}{2}, & j = i \pm 1. \end{cases}$$

The only poles of Λ_i in the region D_i occur when $s_i = 0$ or 1. Moreover,

$$E_{(n)}(s|Y)\Big|_{s_i=0} = E_{P_i^*}(s_1, \ldots, s_{i-1}, s_{i+1}, \ldots, s_n|Y),$$

with E_p denoting, as usual, the Eisenstein series associated with the parabolic subgroup $P = P_i^*$ of Lemma 1.5.2. If $s_j^* = s_j$ for $j \neq i \pm 1$, $s_{i\pm 1}^* = s_{i\pm 1} + \frac{s_i}{2}$, then

$$\text{Res } E_{(n)}(s|Y)\Big|_{s_i=1} = \frac{E_{P_1^*}(s_1^*, \ldots, s_{i-1}^*, s_{i+1}^*, \ldots, s_n^*|Y)}{2\Lambda_i(1)}.$$

Here $2\Lambda_i(1) = \pi/3$ which is the volume of the fundamental domain for $SL(2, \mathbb{Z})$ acting on the Poincaré upper half plane.

Proof. We can write

$$E_{(2)}(s_i|T) = |T|^{-s_i/2}E_{(2)}(s_i|T^0)$$

where

$$T^0 = |T|^{-1/2}T \in \mathcal{S}\mathcal{P}_2.$$

Substitute the incomplete gamma expansion of $E_{(2)}(s_i|T^0)$ from Theorem 1.4.1 in Volume I into the expression for $E_{(n)}$ given in Lemma 1.5.2. This leads to the formula

$$\Lambda_i(s|Y) = \frac{E_{P_i^*}(s^*|Y)}{s_i - 1} - \frac{E_{P_i^*}(s^*|Y)}{s_i} + \Sigma_3 + \Sigma_4, \tag{1.255}$$

where $E_{P_i^*}(s^*|Y) = E_{P_i^*}(s_1^*, \ldots, s_{i-1}^*, s_{i+1}^*, s_n^*)$, with s^* as defined in the Lemma, and

$$\Sigma_3 = \sum_{V \in \Gamma_n/P_i^*} \prod_{j \neq i} |Y[V_j]|^{-s_j^*} \int_1^{\infty} x^{-s_i} \sum_{a \in \mathbb{Z}^2 - 0} \exp\{-\pi x(T^0)^{-1}[a]\} \, dx,$$

$$\Sigma_4 = \sum_{V \in \Gamma_n / P_i^*} \prod_{j \neq i} |Y[V_j]|^{-s_j^*} \int_1^\infty x^{s_i - 1} \sum_{a \in \mathbb{Z}^2 - 0} \exp\{-\pi x T^0[a]\}\, dx.$$

For any positive ε, we can easily bound the terms Σ_3 and Σ_4 in formula (1.255) by $\Lambda_i(\rho | Y)$, where

$$\rho_j = \begin{cases} \operatorname{Re} s_j, & \text{for } j \neq i \pm 1, \text{ or } i, \\ \operatorname{Re} s_j + \frac{1}{2}\left\{\operatorname{Re} s_j - |\operatorname{Re} s_i| - (1 + \varepsilon)\right\}, & \text{for } j = i \pm 1, \\ |\operatorname{Re} s_j| + 1 + \varepsilon, & \text{for } j = i. \end{cases}$$

Here we are using the fact that

$$(T^0)^{-1} \begin{bmatrix} a_1 \\ a_2 \end{bmatrix} = T^0 \begin{bmatrix} -a_2 \\ a_1 \end{bmatrix},$$

which implies that the sums over $a \in \mathbb{Z}^2 - 0$ in Σ_3 and Σ_4 are the same.

Define the region

$$D_i = \left\{ s \in \mathbb{C}^n \,\middle|\, \operatorname{Re} s_i \geq 0, \operatorname{Re} s_j \geq \tfrac{3}{2}, j \neq i \right\}$$
$$\cup \left\{ s \in \mathbb{C}^n \,\middle|\, \operatorname{Re} s_i \leq 0, \operatorname{Re} s_j \geq \tfrac{3}{2}, j \neq i, i \pm 1, \operatorname{Re}(s_{i\pm 1} + s_i) \geq \tfrac{3}{2} \right\}. \tag{1.256}$$

It follows that $E_{(n)}$ can be analytically continued to the region D_i with the indicated functional equation, poles, residues, and behavior at $s_i = 0$. Note that the transformation $s \to s'$, which appears in the functional equation, maps the region in Figure 1.29 above the lines L_1, L_2, L_3, into itself. In particular, $s \to s'$ takes the line L_1 to L_3 and fixes L_2.   ∎

Fig. 1.29 Real parts of s_i and $s_{i\pm 1}$ for the region D_i of analytic continuation of the Selberg Eisenstein series $E_{(n)}$ defined by formula (1.256). The original region of analyticity for $E_{(n)}$ is that above line L_3. We can enlarge the region D_i to that which lies above the lines L_1, L_2, L_3, but this would lead to more complicated equations. Here the line L_1 has the equation $\operatorname{Re}(s_{i\pm 1} + s_i) = \frac{3}{2}$

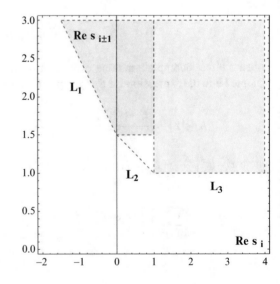

Note that the formulas for values of Eisenstein series at $s_i = 0$ are much simpler than the formulas for the residues at $s_i = 1$. It has often been noticed by number theorists that zeta and L-functions have simpler behavior to the left of the line fixed by the functional equation. For example, the Riemann zeta function has rational values at negative odd integers (cf. Exercise 3.5.7 of Volume I), while the values at positive even integers are more complicated, involving powers of π.

Note also that the zeros of $\Lambda(s_i)$ can produce poles of

$$E_{(n)}(s|Y) = \Lambda_i(s|Y)/2\Lambda(s_i).$$

Thus the zeros of the Riemann zeta function are of interest to us here. Actually they will lie to the left of the line $\mathrm{Re}\, s_i = \frac{1}{2}$, which is the line of interest for our continuation to reach the spectrum of the $GL(n, \mathbb{R})$-invariant differential operators. This happens because $\Lambda(s_i)$ involves $\zeta(2s_i)$ not $\zeta(s_i)$.

Exercise 1.5.5. Prove all the statements made in the proof of Lemma 1.5.3.
Hint. You can use Theorem 2.5.10 in Hörmander [306], which says that every function holomorphic on a connected tube in $\mathbb{C}^n$, $n \geq 2$, can be continued to a function holomorphic on the convex hull of the connected tube. A **tube** $\Omega \subset \mathbb{C}^n$ has the form $\Omega = \{z \in \mathbb{C}^n \mid \mathrm{Re}\, s \in \omega\}$ for some set ω in $\mathbb{R}^n$.

Selberg noticed that the situation in Lemma 1.5.3 is clarified by introducing **new variables** $z \in \mathbb{C}^n$ (cf. Proposition 1.2.1 and formula (1.49) of Section 1.2.1):

$$\left. \begin{array}{l} s_j = z_{j+1} - z_j + \frac{1}{2}, \quad j < n, \quad s_n = -z_n + \frac{1}{2}, \\ z_j = -(s_n + s_{n-1} + \cdots + s_j) + (n - j + 1)/2. \end{array} \right\} \quad (1.257)$$

Note that the z-variables are closely related to the r-variables in part (4) of Theorem 1.2.3.

Exercise 1.5.6. (a) Show that if $z(s)$ is given by formula (1.257) and if $s \to s'$ denotes the transformation appearing in the functional equation of $E_{(n)}$ in Lemma 1.5.3, then $z' = z(s')$ is the transformation σ_i of the z-variables which permutes z_i and z_{i+1}, while leaving the rest of the z_j fixed for $j \neq i$, $i + 1$.
(b) Relate the z-variables of (1.257) with the r-variables in part 4 of Theorem 1.2.3.
Hint. For example,

$$z'_{i-1} = -\left(s_n + s_{n-1} + \cdots + \left(s_{i+1} + s_i - \tfrac{1}{2}\right) + (1 - s_i) + \left(s_{i-1} + s_i - \tfrac{1}{2}\right)\right) + \tfrac{n-i+2}{2}$$
$$= -(s_n + \cdots + s_{i-1}) + \tfrac{n-i+2}{2} = z_{i-1}.$$

The rest of the calculations are easier.

Exercise 1.5.7. Show that in terms of the z-variables the domain D_i in Figure 1.29 contains the following region, after setting $x_j = \mathrm{Re}\, z_j$:

$$R_i = \left\{ z \in \mathbb{C}^n \;\middle|\; x_{j+1} - x_j \geq 1,\ j \neq i,\ x_{i+1} - x_i \geq -\frac{1}{2} \right\}$$

$$\cup \left\{ z \in \mathbb{C}^n \;\middle|\; \begin{array}{l} x_{j+1} - x_j \geq 1,\ j \neq i, i \pm 1,\ x_{i-2} - x_i \geq 1, \\ x_{i+1} - x_{i-1} \geq 1,\ x_i - x_{i+1} \geq -\frac{1}{2} \end{array} \right\}.$$

Note that the second set in the union is the image of the first under the transformation σ_i of Exercise 1.5.6 which permutes z_i and z_{i+1}, leaving the rest of the variables fixed.

Now, as Selberg observed, we have the generators of the group of permutations of the variables $z_1, \ldots, z_n$ (i.e., the Weyl group of $GL(n)$). Thus we should be able to obtain $n!$ functional equations for the Eisenstein series $E_{(n)}$, if we include the identity $E_{(n)} = E_{(n)}$ as a functional equation. Moreover, we will be able to continue $E_{(n)}$ as a meromorphic function in the complex space $\mathbb{C}^n$.

Theorem 1.5.1 (Selberg The Analytic Continuation and Functional Equations of Selberg's Eisenstein Series). *Let $\varphi(s) = \Gamma(s)\zeta(2s)$ and define (using (1.257)):*

$$\Xi(z|Y) = \pi^p E_{(n)}(z(s)|Y) \prod_{1 \leq i < j \leq n} \varphi\left(z_j - z_i + \frac{1}{2}\right),$$

if

$$p = -2 \sum_{j=1}^{n} j z_j.$$

Then

$$E_{(n)}(z(s)|Y) \prod_{1 \leq i < j \leq n} \left(z_j - z_i - \frac{1}{2}\right) \zeta\left(2\left(z_j - z_i + \frac{1}{2}\right)\right)$$

can be continued to a holomorphic function for all $z \in \mathbb{C}^n$. And $\Xi(z|Y)$ satisfies the $n!$ functional equations

$$\Xi(\sigma(z)|Y) = \Xi(z|Y)$$

for every permutation σ of n elements. Here

$$\sigma(z) = (z_{\sigma(1)}, \ldots, z_{\sigma(n)}) \quad \text{if} \quad z = (z_1, \ldots, z_n).$$

Proof. Let us compute $\Xi(\sigma_i(z))$ for $\sigma_i = (i \ \ i+1)$, the transposition of i and $i+1$. The power of π in $\Xi(\sigma_i(z))$ is:

$$p = -2 \sum_{j=1}^{n} j z_j + 2(z_{i+1} - z_i). \tag{1.258}$$

The product of the φ's in $\Xi(\sigma_i(z))$ is:

$$\frac{\varphi\left(z_i - z_{i+1} + \frac{1}{2}\right)}{\varphi\left(z_{i+1} - z_i + \frac{1}{2}\right)} \prod_{1 \le i < j \le n} \varphi\left(z_j - z_i + \frac{1}{2}\right). \tag{1.259}$$

Thus Lemma 1.5.3 completes the proof that $\Xi(\sigma(z) | Y) = \Xi(z|Y)$ for all σ of the form $\sigma = \sigma_i = (i \quad i+1)$. Since these permutations generate the symmetric group of permutations of n elements, it follows that $\Xi(z|Y)$ is invariant under all permutations of the entries of $z \in \mathbb{C}^n$. Thus, in fact, the function $\Xi(z|Y)$ behaves like the spherical function $h_s(Y)$ (see Theorem 1.2.3).

Next we claim that $E_{(n)}$ can be continued as a meromorphic function in the region

$$B^* = \bigcup_{k=1}^{n} B_k, \tag{1.260}$$

where, writing $x_j = \mathrm{Re}\, z_j$,

$$B_k = \left\{ z \in \mathbb{C}^n \mid x_{j+1} - x_j > 2,\ j \ne k, k-1,\ x_{k+1} - x_{k-1} > 2 \right\}. \tag{1.261}$$

Here, we simply drop inequalities that do not make sense; e.g., for $n = 3$, we have

$$B_1 = \left\{ z \in \mathbb{C}^3 \mid x_3 - x_2 > 2 \right\}.$$

To prove that we can continue $E_{(n)}$ to the region (1.260), first continue $E_{(n)}$ to B_1. To do this, note that if $z \in B_1$, there is a permutation σ of n elements such that $\sigma(z) \in R_i$, the region considered in Exercise 1.5.7. For suppose

$$x_j < x_1 \le x_{j+1}, \quad \text{for some} \quad j = 2, \dots, n.$$

Then either $x_{j+1} - x_1 > 1$ or $x_1 - x_j > 1$, otherwise $x_{j+1} - x_j \le 2$, contradicting the definition of B_1.

If $x_{j+1} - x_1 > 1$, then $z \in \sigma^{-1}(R_{j-1})$, where R_{j-1} is defined in Exercise 1.5.7 and

$$\sigma = \begin{pmatrix} 1\ 2 \cdots j-1\ j \\ 2\ 3 \cdots\ \ j\ \ \ 1 \end{pmatrix} = (12)(23) \cdots (j-1\ j).$$

Here we use the standard notation for permutations:

$$\sigma = \begin{pmatrix} 1 & 2 & \cdots & n \\ \sigma(1) & \sigma(2) & \cdots & \sigma(n) \end{pmatrix}.$$

We did not list integers that are fixed and $(i\ j)$ denotes the transposition that interchanges i and j, leaving the other integers fixed. To see that $z \in \sigma^{-1}(R_{j-1})$, note that the following inequalities say that σz with Re $z_j = x_j$ lies in the first subset in the definition of R_{j-1} given in Exercise 1.5.7:

$$x_3 - x_2 \geq 1, \ldots, x_j - x_{j-1} \geq 1, \ x_1 - x_j \geq 0, \ x_{j+1} - x_1 \geq 1, \ldots, x_n - x_{n-1} \geq 1.$$

And, similarly, if $x_1 - x_j > 1$, then $z \in \sigma^{-1}(R_j)$, since the following inequalities say that σz with Re $z_j = x_j$ lies in the first subset of R_j defined in Exercise 1.5.7:

$$x_3 - x_2 \geq 1, \ldots, x_j - x_{j-1} \geq 1, \ x_1 - x_j \geq 1, \ x_{j+1} - x_1 \geq 0, \ldots, x_n - x_{n-1} \geq 1.$$

The functional equations of $\Xi(z|Y)$ allow us to continue it to B_1. And

$$E_{(n)}(z|Y) \prod_{j=2}^{n} \left(z_j - z_1 - \frac{1}{2} \right) \zeta \left(2 \left(z_j - z_i + \frac{1}{2} \right) \right)$$

is holomorphic in B_1. For example, the only poles of

$$\zeta \left(2 \left(z_{j+1} - z_j \right) + 1 \right) \ E_{(n)}(z|Y)$$

in the region R_j of Exercise 1.5.7 occur when $z_{j+1} - z_j = \frac{1}{2}$. This gives $z_{j+1} - z_1 = \frac{1}{2}$ in

$$\sigma^{-1}(R_j) = (j - 1\ \ j) \cdots (23)(12) R_j = \sigma^{-1}(j\ \ j + 1) R_j.$$

Now, in order to continue $E_{(n)}$ to the domain B_k, use the fact that

$$B_k = (k\ \ k - 1) B_{k-1}.$$

So we can use the functional equations of $\Xi(z|Y)$ to continue $E_{(n)}$ to B_k. We find also that

$$E_{(n)}(z|Y) \prod_{1 \leq i < j \leq n} \left(z_j - z_i - \frac{1}{2} \right) \zeta \left(2 \left(z_j - z_i + \frac{1}{2} \right) \right)$$

is holomorphic in the region B^*.

To complete the analytic continuation of $E_{(n)}$, we need to use a theorem from several complex variables mentioned in Exercise 1.5.5. This insures that the function can always be continued to the convex hull of any region in which it is holomorphic (see Theorem 2.5.10 in Hörmander [306]). Thus it suffices to show that B^* is connected with convex hull $\mathbb{C}^n$. This is proved in Exercise 1.5.8. ∎

Exercise 1.5.8. Show that B^* defined in (1.260) is a connected set and then show that B^* has $\mathbb{C}^n$ as its convex hull.
Hint. You can show that n independent lines through the same point lie in the set

$$b^* = \left\{ x \in \mathbb{R}^n \mid x_j = \operatorname{Re} z_j, \ z \in B^* \right\}.$$

You can take, for example, the lines

$$\lambda_i = \{ x \in \mathbb{R}^n \mid x = 3(1, \ 2, \dots, n) + te_i, \ t \in \mathbb{R} \},$$

with $e_i = (0, \dots, 0, \underset{i}{1}, 0, \dots, 0)$, the ith element of the standard basis of $\mathbb{R}^n$. This follows from the fact that the definition of B_i does not restrict the ith coordinate at all.

This completes our discussion of the analytic continuation of $E_{(n)}(s|Y)$ by the method of inserting larger parabolic subgroups. It is possible to generalize our formula relating $E_{(n)}$ and $E_{(2)}$, as the following exercises show.

Exercise 1.5.9 (More General Decompositions Associated with Two Parabolics). Let $P = P(n_1, \ \dots, \ n_q)$ be any parabolic subgroup of $\Gamma_n = GL(n, \mathbb{Z})$. Show that, in the region where the Dirichlet series for $E_{(n)}$ converges absolutely, we have:

$$E_{(n)}(s|Y) = \sum_{V \in \Gamma_n/P} \prod_{i=1}^{q} |Y[V_{N_i}]|^{-s_i} \ E_{(n_i)} \left(s_{N_{i-1}+1}, \dots, s_{N_i-1}, 0 \mid T_i \right),$$

for $N_i = n_1 + \cdots + n_i$, $V = (V_j \ *)$, $V_j \in \mathbb{Z}^{n \times j}$, and $R_i \in \mathcal{P}_{n_i}$, defined by the partial Iwasawa decomposition:

$$Y[V_{N_i}] = \begin{pmatrix} Y[V_{N_{i-1}}] & 0 \\ 0 & R_i \end{pmatrix} \begin{bmatrix} I_{N_{i-1}} & Q \\ 0 & I_{n_i} \end{bmatrix} \quad \text{and} \quad T_i = |Y[V_{N_{i-1}}]| \ R_i.$$

Hints (Terras [593]). Note that we can imitate the proof of Lemma 1.5.2. Write $U \in \Gamma_n/P_{(n)}$ uniquely as $U = VW$ with $V \in \Gamma_n/P$, $W \in P/P_{(n)}$:

$$W = \begin{pmatrix} W_1^* & \cdots & 0 \\ \vdots & \ddots & \vdots \\ 0 & \cdots & W_q^* \end{pmatrix}, \ W_i \in \Gamma_{n_i}/P_{(n_i)}.$$

Then if $N_{i-1} < j < N_i$, we have

$$\left| Y[VW_j] \right| = |Y[V_{N_{i-1}}]| \ \left| R_i[(W_i^*)_{j-N_{i-1}}] \right|.$$

Exercise 1.5.10. Extend Exercise 1.5.9 to relate the Eisenstein series E_P to E_P^* if $P^* \supset P$ are two parabolic subgroups of $GL(n, \mathbb{Z})$. Show that this implies that:

$$E_P \left(s_{N_1}, \ldots, s_{N_q} | Y \right) \Big|_{\substack{s_{N_j}= 0 \\ \text{if } N_j \neq M_i}} = E_{P^*} (s_{M_1}, \ldots, s_{M_{q^*}} | Y),$$

where

$$N_{i_k} = \sum_{\alpha=1}^{i_k} n_\alpha = M_k = \sum_{\beta=1}^{k} m_\beta,$$

if $P = P(n_1, \ldots, n_q)$ and $P^* = P(m_1, \ldots, m_{q^*})$.
Hint (Terras [593]). Use induction on q.

Exercise 1.5.11. Compute

$$\operatorname*{Res}_{\substack{s_j= 1, \\ j = 1, \ldots, n}} E_{(n)}(s|Y).$$

Answer.

$$2^{1-n} \operatorname{Vol}(\mathcal{SP}_n/GL(n, \mathbb{Z}))^{-1}.$$

Exercise 1.5.11 is useful when one seeks to generalize Zagier's argument of Exercise 3.6.17 of Volume I, in order to show that the Hecke points (as seen in Figures 1.18, 1.19, 1.20, 1.21, 1.22, 1.23, 1.24, and 1.25 of Section 1.4.3) actually become dense in the fundamental domain for $SL(3, \mathbb{Z})$ (actually the figures show a union of 2^2 copies of the fundamental domain in which all x-variables run between $-\frac{1}{2}$ and $+\frac{1}{2}$).

The analytic continuation given in Theorem 1.5.1 began with a very concrete formula for $E_{(2)}$—the incomplete gamma expansion found in Section 1.4 of Volume I. But the proof of Theorem 1.5.1 ended in a rather existential way—using the result from several complex variables which extends holomorphic functions of more than one complex variable in a connected tube domain to the convex hull of the domain.

1.5.2   Hecke Operators and Analytic Continuation of L-Functions Associated with Maass Forms by the Method of Theta Functions

Let us now begin the discussion of the second method of analytically continuing Eisenstein series—**the method of theta functions**. In order to relate Eisenstein series and theta functions, we need a generalization of formula (1.246), and thus a generalization of Lemma 1.5.1 in the preceding section. This will require a discussion of Hecke operators for $GL(n, \mathbb{Z})$. We will mostly restrict our attention here to Eisenstein series of the form (1.245), although the method can be generalized much further.

Suppose that $f : \mathcal{SP}_n/GL(n, \mathbb{Z}) \to \mathbb{C}$. Then for any positive integer m, the **mth Hecke operator T_m** is defined by:

$$T_m f(Y) = \sum_{A \in \Xi_m} f(Y[A]^0), \quad \text{for } Y \in \mathcal{P}_n, \ Y^0 = |Y|^{-1/n} Y \in \mathcal{SP}_n. \qquad (1.262)$$

Here Ξ_m denotes any complete system of representatives for $M_m/GL(n, \mathbb{Z})$, where

$$M_m = \{ A \in \mathbb{Z}^{n \times n} | \ |A| = m \} .$$

According to formula (1.239) in the proof of Lemma 1.4.7 in Section 1.4.4, we can take

$$\Xi_m = \left\{ \begin{pmatrix} d_1 & d_{12} & \cdots & d_{1n} \\ 0 & d_2 & \cdots & d_{2n} \\ \vdots & \vdots & \ddots & \vdots \\ 0 & 0 & \cdots & d_n \end{pmatrix} \ \middle| \ \begin{array}{l} d_j > 0 \quad j = 1, \ldots, n, \\ d_{ij} = 0, \quad \text{if } i > j, \\ 0 \leq d_{ij} < d_i, \quad \text{if } i < j, \\ \prod_{i=1}^{m} d_i = m. \end{array} \right\}. \qquad (1.263)$$

Note that the scalar matrices $A = \alpha I$, $\alpha \in \mathbb{Z} - 0$, cancel out when one computes $(Y[A]^0)$ in (1.262). Observe also that we have indeed generalized the Hecke operators given in formula (3.119) of Section 3.6.5, Volume I, although here we use a slightly different normalization in that we do not multiply by $m^{-1/2}$. We have already considered some of the history of Hecke operators in Section 3.6, Vol. I. Maass [414] studied Hecke operators for the Siegel modular group $Sp(n, \mathbb{Z})$. A good reference for the Hecke ring of a general group is Shimura [554, Ch. 3], where there is an exposition of work of Tamagawa [589] connecting Hecke operators with combinatorial results about lattices as well as p-adic convolution operators and a p-adic version of Selberg [543]. Hecke operators for $Sp(n, \mathbb{Z})$ and $SL(n, \mathbb{Z})$ are also considered by Andrianov [9–14] and Freitag [185]. Other references for the Hecke ring of $GL(n)$ over p-adic number fields are Macdonald [428] and Satake [532]. Another general reference is Krieg [371]. See also Goldfeld [230].

The Hecke operator (1.262) appears in many calculations associated with the $GL(n, \mathbb{Z})$. For example, set the function $f(Y) \equiv 1$ identically for all $Y \in \mathcal{SP}_n$. Then, from formulas (1.262) and (1.263), we have:

$$\sum_{m \geq 1} T_m f(Y) m^{-s} = \sum_{\substack{A \in \mathbb{Z}^{n \times n} / \Gamma_n \\ A \neq 0}} |A|^{-s} = Z_{n,0}\left(I, \frac{s}{2}\right) = \prod_{j=0}^{n-1} \zeta(s-j),$$

where $\Gamma_n = GL(n, \mathbb{Z})$ and $Z_{n,0}(Y, s)$ is Koecher's zeta function from formula (1.173) of Section 1.4.1. Solomon [570] considers generalizations of such results and connections with combinatorics. Operators like T_m are also intrinsic to formulas akin to (1.278) below connecting Eisenstein series such as (1.245) defined as sums over $GL(n, \mathbb{Z})$ and zeta functions defined as sums over rank m matrices in $\mathbb{Z}^{n \times m}$. These zeta functions are higher dimensional Mellin transforms of the non-singular terms in a theta function (see formula (1.246)) and we will be able to use a modification of Riemann's method of theta functions to obtain an analytic continuation of these zeta functions for $GL(n, \mathbb{Z})$. Thus it is formula (1.278) and our search for analytic continuations of Eisenstein series that motivate our study of Hecke operators here.

The basic properties of Hecke operators for $GL(n, \mathbb{Z})$ are contained in the following theorem.

Theorem 1.5.2 (Hecke Operators for $GL(n, \mathbb{Z})$).

(1) The Hecke operator T_m maps Maass forms on the determinant one surface, i.e., $f \in \mathcal{A}^0(GL(n, \mathbb{Z}), \lambda)$—as defined in (1.242) and after formula (1.244) of the preceding section—to Maass forms with the same eigenvalue system; i.e., $T_m f \in \mathcal{A}^0(GL(n, \mathbb{Z}), \lambda)$.

(2) The Hecke operator T_m is a Hermitian operator with respect to the inner product:

$$(f, g) = \int_{\mathcal{SP}_n / GL(n, \mathbb{Z})} f(W) \overline{g(W)} \, dW,$$

$dW = $ *the $SL(n, \mathbb{R})$-invariant measure on $\mathcal{SP}_n$, as defined in (1.219).*

(3) The ring of Hecke operators is commutative and thus has a set of simultaneous eigenfunctions which span the space of all Maass forms for $GL(n, \mathbb{Z})$.

(4) If $\gcd(k, m) = 1$, then $T_k T_m = T_{km}$. When the group is $GL(3, \mathbb{Z})$, one has the following formal power series in the indeterminate X for any prime p:

$$\sum_{r \geq 0} T_{p^r} X^r = \left(I - T_p X + \left[(T_p)^2 - T_{p^2}\right] X^2 - p^3 X^3\right)^{-1}.$$

(5) Suppose that $f \in \mathcal{A}^o(GL(n, \mathbb{Z}), \lambda)$ is an eigenfunction for all the Hecke operators; i.e., $T_m f = u_m f$, for some $u_m \in \mathbb{C}$. Form the Dirichlet series:

$$L_f(s) = \sum_{m \geq 1} u_m m^{-s}.$$

This series converges for $\operatorname{Re} s > n/4$ *if f is integrable on the fundamental domain* $\mathcal{S}\mathcal{M}_n$. *Moreover,* L_f *can be analytically continued to a meromorphic function of s with functional equation:*

$$\Lambda(f, s) = \pi^{-ns} L_{f^*}(2s)\Gamma_n(r(f, s)) = \Lambda(f^*, n/2 - s),$$

where $f^*(W) = f(W^{-1})$, $r = r(f, s) \in \mathbb{C}^n$ *is defined via Proposition 1.2.4 of Section 1.2.3 and the formula:*

$$\frac{\Omega\left(\left(|Y|^s f(Y^0)\right)\right)}{|Y|^s f(Y^0)} = \frac{\Omega\left(p_r(Y)\right)}{p_r(Y)}, \text{ for any } \Omega \in D(\mathcal{P}_n),$$

with $D(\mathcal{P}_n)$ *=the algebra of* $GL(n, \mathbb{R})$*-invariant differential operators on* $\mathcal{P}_n$ *and* $p_r(Y)$ *the power function defined in (1.41) of Section 1.2.1. Here* $\Gamma_n(r)$ *is the gamma function defined in (1.44) of Section 1.2.1. For* $GL(3, \mathbb{Z})$ *it follows that* $L_f(s)$ *has the Euler product:*

$$L_f(s) = \prod_{p \text{ prime}} \left(1 - u_p p^{-s} + \left[(u_p)^2 - u_{p^2}\right]p^{-2s} - p^{3-3s}\right)^{-1}.$$

The Euler product in part (5) of Theorem 1.5.2 should be compared with that obtained by Bump [83] using Fourier expansions of Maass forms as sums of Whittaker functions. See also Section 1.5.4 and Goldfeld [230].

Theorem 1.5.2 gives an analogue of much of Theorem 3.6.4 of Volume I. However, the converse result for part (5) is missing. See Goldfeld [230, Chapter 7] for a classical version of the converse theorem for $SL(3, \mathbb{Z})$. Converse theorems have been obtained adelically for $GL(3)$ (see Jacquet et al. [325]). One would expect to need more than one complex variable s in order to be able to invert the Mellin transform over $\mathcal{P}_n/\Gamma$ that leads to $L_f(s)$. Jacquet, Piatetski-Shapiro, and Shalika find that one needs to twist by Maass forms for $GL(n - 2, \mathbb{Z})$ for a $GL(n)$ converse theorem. When $n = 3$, one twists by a **Dirichlet character** $\chi : (\mathbb{Z}/q\mathbb{Z})^* \to \mathbb{T}$, which is a group homomorphism from the multiplicative group of $a(\bmod q)$, with $\gcd(a, q) = 1$ into the multiplicative group of complex numbers of norm 1.

There are many other sorts of Hecke operators and attached Dirichlet series that produce results similar to those which we have stated here. See the references mentioned earlier for some examples.

We will see that the analytic continuation result in (5) is similar to that which we need for the Eisenstein series (1.245).

We break up the proof of Theorem 1.5.2 into the various parts with the proof of Parts (1) and (3) being given as exercises. Mostly we can imitate the arguments given by Maass [414] for the Siegel modular group $Sp(n, \mathbb{Z})$.

Exercise 1.5.12. Prove part (1) of Theorem 1.5.2.
Hint. Imitate the proof of the analogous result for $SL(2, \mathbb{Z})$ to be found in Section 3.6 of Volume I.

Proof of Part (2) of Theorem 1.5.2. One need to only imitate the proof of the analogous result for $SL(2, \mathbb{Z})$ which was part (4) of Theorem 3.6.4 in Volume I. If $\Gamma = GL(n, \mathbb{Z})$, $m \in \mathbb{Z}$, $m \geq 1$, and $A \in M_m$, the set of $n \times n$ integer matrices of determinant m, write $h_A(W) = f\left((W[A])^0\right)$. Note that h_A remains invariant under the **congruence subgroup**:

$$\Gamma(m) = \{B \in \Gamma \mid B \equiv I(\mathrm{mod}\ m)\}. \tag{1.264}$$

For $B \in \Gamma(m)$ and $A \in M_m$ imply that $A^{-1}BA \in \Gamma$. To see this, observe that

$$A^{-1} = (1/m)\ {}^t(\mathrm{adj}\ A) \in (1/m)\mathbb{Z}^{n \times n}.$$

Here adj A denotes the adjoint matrix of A whose ij entry is the determinant of the matrix obtained from A by crossing out the ith row and jth column. Thus $mA^{-1}BA \in \mathbb{Z}^{n \times n}$ and $mA^{-1}BA$ is congruent to $mA^{-1}A = mI$ and thus to 0 modulo m. So $A^{-1}BA$ is an integral matrix of determinant one.

Since the fundamental domain $S\mathcal{P}_n/\Gamma(m)$ consists of $[\Gamma : \Gamma(m)] = \#(\Gamma/\Gamma(m))$ copies of $S\mathcal{P}_n/\Gamma$ (see Exercise 1.5.13 below), one has the following equalities, with Ξ_m as in (1.263):

$$(T_m f, g) = \sum_{A \in \Xi_m} \frac{1}{[\Gamma : \Gamma(m)]} \int_{W \in S\mathcal{P}_n/\Gamma(m)} f((W[A])^0)\ \overline{g(W)}\ dW$$

$$= \sum_{A \in \Xi_m} \frac{1}{[\Gamma : A^{-1}\Gamma(m)A]} \int_{X \in S\mathcal{P}_n/A^{-1}\Gamma(m)A} f(X)\overline{g\left((X[A^{-1}])^0\right)}\ dX$$

$$= (f, T_m g).$$

The second equality is obtained by substituting $X = (W[A])^0$ and noting that $[\Gamma : \Gamma(m)] = [\Gamma : A^{-1}\Gamma(m)A]$. This completes the proof of part (2) of Theorem 1.5.2. ∎

Exercise 1.5.13. Show that the index $[\Gamma : \Gamma(m)] = \#(\Gamma/\Gamma(m))$ is finite, for $\Gamma = GL(n, \mathbb{Z})$ and $\Gamma(m)$ as defined by (1.264).

Exercise 1.5.14. Prove part (3) of Theorem 1.5.2.
Hint. This is proved by Shimura [554, Ch. 3] using the existence of the anti-automorphism $X \mapsto {}^tX$ of $GL(n, \mathbb{Z})$

Proof of Part (4) of Theorem 1.5.2. To see that $T_k T_m = T_{km}$ if $\gcd(k, m) = 1$, one need only multiply the matrices below:

$$
B = \begin{pmatrix} d_1 & d_{12} & \cdots & d_{1n} \\ 0 & d_2 & \cdots & d_{2n} \\ \vdots & \vdots & \ddots & \vdots \\ 0 & 0 & \cdots & d_n \end{pmatrix} \begin{pmatrix} c_1 & c_{12} & \cdots & c_{1n} \\ 0 & c_2 & \cdots & c_{2n} \\ \vdots & \vdots & \ddots & \vdots \\ 0 & 0 & \cdots & c_n \end{pmatrix}
$$

$$
= \begin{pmatrix} d_1 c_1 & d_1 c_{12} + c_2 d_{12} & \cdots & d_1 c_{1n} + d_{12} c_{2n} + \cdots + d_{1n} c_n \\ 0 & d_2 c_2 & \cdots & d_2 c_{2n} + \cdots + d_{2n} c_n \\ \vdots & \vdots & \ddots & \vdots \\ 0 & 0 & \cdots & d_n c_n \end{pmatrix}.
$$

For if d_{ij} runs through a complete set of representatives mod d_i and c_{ij} runs through a complete set of representatives mod c_i, then consider the i, j-entry in the product above for $i < j$:

$$
b_{ij} = d_i c_{ij} + d_{i,i+1} c_{i+1,j} + \cdots + d_{ij} c_j.
$$

Inductively we can assume that the terms $d_{ij'}$ with $j' < j$ and $C_{i'j}$ with $i' > i$ are fixed. Thus what remains is

$$
d_i c_{ij} + d_{ij} c_j + \text{ a fixed number.}
$$

This gives a complete set of representatives modulo $d_j c_j$.

Next we consider the proof of the formula which implies the Euler product for the L-function in part (5) which corresponds to an eigenform of the Hecke operators for $GL(3, \mathbb{Z})$. The proof which follows involves only matrix multiplication but clearly becomes more complicated for $GL(n, \mathbb{Z})$ with $n > 3$. See Exercises 1.5.15 and 1.5.16 for connections with other methods of obtaining such Euler product formulas as well as Shimura [554] and Freitag [185]. See also Goldfeld [230].

Observe that if Ξ_m is as in (1.263),

$$
T_k T_m f(Y) = \sum_{A \in \Xi_m} \sum_{B \in \Xi_k} f\left(Y[BA]^0\right).
$$

It will be helpful to set up the following notation. Suppose that S is a subset of M_m (the set of all $n \times n$ integral matrices of determinant m) and let $T(S)$ denote the operator:

$$
T(S)f(Y) = \sum_{A \in S} f(Y[A]^0). \tag{1.265}
$$

The formal power series identity in part (4) derives from the following two formulas, which are easily checked by multiplying the matrix representatives of the operators involved. The first formula is:

$$T_{p^r} T_p = T_{p^{r+1}} + T(S_1^r) + T(S_2^r), \tag{1.266}$$

where

$$S_1^r = \left\{ \begin{pmatrix} p^e & p(a_1 \bmod p^e) & a_2 \bmod p^e \\ 0 & p^{f+1} & a_3 \bmod p^{f+1} \\ 0 & 0 & p^g \end{pmatrix} \,\middle|\, e \geq 1;\, f, g \geq 0;\ e + f + g = r \right\},$$

$$S_2^r = \left\{ \begin{pmatrix} p^e & a_1 \bmod p^e & p(a_2 \bmod p^e) \\ 0 & p^f & p(a_3 \bmod p^f) \\ 0 & 0 & p^{g+1} \end{pmatrix} \,\middle|\, \begin{array}{l} e \geq 1 \text{ or } f \geq 1 \\ g \geq 0;\ e + f + g = r \end{array} \right\}.$$

The second formula is:

$$T_{p^r}[(T_p)^2 - T_{p^2}] = p^3 T_{p^{r-1}} + T_{p^{r+1}} T_p + T_{p^{r+2}}, \qquad \text{for } r \geq 1. \tag{1.267}$$

To prove formula (1.266), use the following calculations:

$$\begin{pmatrix} p^e & a_1 \bmod p^e & a_2 \bmod p^e \\ 0 & p^f & a_3 \bmod p^f \\ 0 & 0 & p^g \end{pmatrix} \begin{pmatrix} p & b_1 \bmod p & b_2 \bmod p \\ 0 & 1 & 0 \\ 0 & 0 & 1 \end{pmatrix} = \begin{pmatrix} p^{e+1} & c_1 \bmod p^{e+1} & c_2 \bmod p^{e+1} \\ 0 & p^f & c_3 \bmod p^f \\ 0 & 0 & p^g \end{pmatrix},$$

$$\begin{pmatrix} p^e & a_1 \bmod p^e & a_2 \bmod p^e \\ 0 & p^f & a_3 \bmod p^f \\ 0 & 0 & p^g \end{pmatrix} \begin{pmatrix} 1 & 0 & 0 \\ 0 & p & b_3 \bmod p \\ 0 & 0 & 1 \end{pmatrix} = \begin{pmatrix} p^e & p(c_1 \bmod p^e) & c_2 \bmod p^e \\ 0 & p^{f+1} & c_3 \bmod p^{f+1} \\ 0 & 0 & p^g \end{pmatrix},$$

$$\begin{pmatrix} p^e & a_1 \bmod p^e & a_2 \bmod p^e \\ 0 & p^f & a_3 \bmod p^f \\ 0 & 0 & p^g \end{pmatrix} \begin{pmatrix} 1 & 0 & 0 \\ 0 & 1 & 0 \\ 0 & 0 & p \end{pmatrix} = \begin{pmatrix} p^e & c_1 \bmod p^e & p(c_2 \bmod p^e) \\ 0 & p^f & p(c_3 \bmod p^f) \\ 0 & 0 & p^{g+2} \end{pmatrix}.$$

The first set of matrices gives $T_{p^{r+1}}$, except for the $e + 1 = 0$ terms. The second set of matrices gives all of the $e = 0$ terms of $T_{p^{r+1}}$ except the terms with $e = 0$ and $f + 1 = 0$, and it also gives S_1^r. The third set of matrices gives the $e = f = 0$ terms in $T_{p^{r+1}}$ plus S_2^r.

To prove (1.267), use formula (1.266) with $r = 1$ to see that

$$(T_p)^2 - T_{p^2} = T(R_1) + T(R_2) + T(R_3),$$

where

$$R_1 = \left\{ \begin{pmatrix} p & p(b_1 \bmod p) & b_2 \bmod p \\ 0 & p & b_3 \bmod p \\ 0 & 0 & 1 \end{pmatrix} \right\},$$

$$R_2 = \left\{ \begin{pmatrix} p & b_1 \bmod p & p(b_2 \bmod p) \\ 0 & 1 & 0 \\ 0 & 0 & p \end{pmatrix} \right\},$$

$$R_3 = \left\{ \begin{pmatrix} 1 & 0 & 0 \\ 0 & p & p(b_3 \bmod p) \\ 0 & 0 & p \end{pmatrix} \right\}.$$

Then compute the matrix products to find that $T_{p^r} T(R_j) = T(Q_j)$, where

$$Q_1 = \left\{ \begin{pmatrix} p^{e+1} & p(a_1 \bmod p^{e+1}) & a_2 \bmod p^{e+1} \\ 0 & p^{f+1} & a_3 \bmod p^{f+1} \\ 0 & 0 & p^g \end{pmatrix} \right\},$$

$$Q_2 = \left\{ \begin{pmatrix} p^{e+1} & a_1 \bmod p^{e+1} & p(a_2 \bmod p^{e+1}) \\ 0 & p^f & p(a_3 \bmod p^f) \\ 0 & 0 & p^{g+1} \end{pmatrix} \right\},$$

$$Q_3 = \left\{ \begin{pmatrix} p^e & p(a_1 \bmod p^e) & p(a_2 \bmod p^e) \\ 0 & p^{f+1} & p(a_3 \bmod p^{f+1}) \\ 0 & 0 & p^{g+1} \end{pmatrix} \right\}.$$

Now $T(Q_1)$ gives $T(S_1^{r+1})$ from (1.266). And $T(Q_2)$ gives the $e + 1 \neq 0$ part of $T(S_2^{r+1})$ in (1.266). The $e = 0$ part of $T(Q_3)$ gives the remainder of $T(S_2^{r+1})$. The $e \geq 1$ part of $T(Q_3)$ gives $p^3 T_{p^{r-1}}$, since

$$\begin{pmatrix} p^{e-1} & a_1 \bmod p^e & a_2 \bmod p^e \\ 0 & p^f & a_3 \bmod p^{f+1} \\ 0 & 0 & p^g \end{pmatrix}$$

$$= \begin{pmatrix} p^{e-1} & b_1 \bmod p^{e-1} + p^{e-1}(c_1 \bmod p) & b_2 \bmod p^{e-1} + p^{e-1}(c_2 \bmod p) \\ 0 & p^f & b_3 \bmod p^f + p^f(c_3 \bmod p) \\ 0 & 0 & p^g \end{pmatrix}$$

$$= \begin{pmatrix} p^{e-1} & b_1 \bmod p^{e-1} & b_2 \bmod p^{e-1} \\ 0 & p^f & b_3 \bmod p^f \\ 0 & 0 & p^g \end{pmatrix} \begin{pmatrix} 1 & c_1 \bmod p & c_2 \bmod p \\ 0 & 1 & c_3 \bmod p \\ 0 & 0 & 1 \end{pmatrix}.$$

This completes the proof of part (4) of Theorem 1.5.2.   ∎

Exercise 1.5.15. Read Shimura [554, Ch. 3] for another discussion of Hecke operators. If $A \in \mathbb{Z}^{n \times n}$ has rank n, one considers the double coset decomposition:

$$\Gamma A \Gamma = \bigcup_{B \in S_A} B\Gamma \text{ (disjoint).}$$

Here S_A is a set of representatives for $\Gamma A \Gamma / \Gamma$, with $\Gamma = GL(n, \mathbb{Z})$.

Then one can prove a result due to Tamagawa which says that for prime p:

$$\sum_{r \geq 0} T_{p^r} X^r = \left(\sum_{j=0}^{n} (-1)^j p^{j(j-1)/2} T(S_{A_j}) X^j \right)^{-1},$$

where X is an indeterminate and the $n \times n$ diagonal matrix A_j has $n - j$ ones on the diagonal.

$$A_j = \begin{pmatrix} 1 & \cdots & 0 & 0 & \cdots & 0 \\ \vdots & \ddots & \vdots & \vdots & & \vdots \\ 0 & \cdots & 1 & 0 & \cdots & 0 \\ 0 & \cdots & 0 & p & \cdots & 0 \\ \vdots & & \vdots & \vdots & \ddots & \vdots \\ 0 & \cdots & 0 & 0 & \cdots & p \end{pmatrix}.$$

and we use the notation $T(S_A)$ in (1.265). Show that when $\Gamma = GL(3, \mathbb{Z})$, Tamagawa's formal power series above is the same as ours in part (4) of Theorem 1.5.2. In particular, show that

$$(T_p)^2 - T_{p^2} = pT(S_{A_2}).$$

Exercise 1.5.16. Langlands [390] defines a Hecke operator T_A corresponding to a matrix $A \in \mathbb{Q}^{n \times n}$ with $|A| > 0$ by:

$$T_A f(Y) = \sum_{\gamma \in \Gamma / \Gamma \cap (A\Gamma A^{-1})} f\left((Y[\gamma A])^0 \right),$$

for $f : S\mathcal{P}_n / \Gamma \to \mathbb{C}$, $\Gamma = GL(n, \mathbb{Z})$, where $Y^0 = |Y|^{-1/n} Y \in S\mathcal{P}_n$ for $Y \in \mathcal{P}_n$. Show that this definition agrees with that in Exercise 1.5.15.
Hint. Does

$$\sum_{\delta \in \Gamma A \Gamma / \Gamma} f\left((Y[\delta])^0 \right) = \sum_{\gamma \in \Gamma / \Gamma \cap (A\Gamma A^{-1})} f\left(Y[\gamma A]^0 \right)?$$

Bump [83] notes that these operators are no longer self adjoint.

Exercise 1.5.17. Check the Euler product for $GL(3, \mathbb{Z})$ in part (5) of Theorem 1.5.2 by letting f be the the form that is identically one everywhere. Then

$$L_1(s) = \sum_{m \geq 1} T_m(1)m^{-s} = Z_{3,0}\left(I, \frac{s}{2}\right) = \prod_{j=0}^{2} \zeta(s - j),$$

where $Z_{3,0}(I, s/2)$ is Koecher's zeta function from (1.173) of Section 1.4.1 and the factorization into a product of Riemann zeta functions is formula (1.174) of Section 1.4.1. Show that if you substitute the Euler product for Riemann's zeta function (see Exercise 1.4.4 in Volume I) into this formula for $L_1(s)$, you obtain the Euler product given in part (5) of Theorem 1.5.2.

Remarks on Part (5) of Theorem 1.5.2 We will prove part (5) in a slightly more general situation in order to obtain the analytic continuation of the Eisenstein series (1.245) simultaneously. The idea is to imitate Riemann's method of theta functions which gave the analytic continuation of Epstein's zeta function (see the proof of Theorem 1.4.1 of Volume I). However the sailing is not so smooth here because when $k > 1$ there are many singular terms in the sum defining the theta function $\theta(Y, X)$, for $Y \in \mathcal{P}_n$, $X \in \mathcal{P}_k$, $1 \leq k \leq n$, from (1.172) in Section 1.4.1. These terms come from matrices $A \in \mathbb{Z}^{n \times k}$ of rank less than k. When $k = 1$ the only such term comes from the zero vector, but when $k > 1$ there are an infinite number of these terms to deal with.

    This problem has led to gaps in many papers—gaps coming from the subtraction of divergent integrals. We might call this the **curse of the higher rank Eisenstein series**. There was even a gap of this kind in Siegel's first paper on the computation of the volume of the fundamental domain $\mathcal{SM}_n$ (Siegel [565, Vol. I, pp. 459–468]). Such gaps also appear in Koecher's paper [359] on the analytic continuation of his zeta function and other Dirichlet series associated with Siegel modular forms. Käte Hey's thesis [294] on zeta functions of central simple algebras has a similar gap. Siegel makes the following remark at the beginning of his paper that fills the gap (Siegel [565, Vol. III, p. 328]): "Die Korrektur des Fehlschlusses ist dann keineswegs so einfach, wie man zunächst in Gedanken an die Renormalisierung in physikalischen Untersuchungen glauben möchte, und benötigt genauere Abschätzungen unendlicher Reihen."[6] We will not follow Siegel's method here, for it does not seem to give the complete analytic continuation (just the continuation beyond the first pole at $s = n/2$). Siegel's method does, however, allow the computation of the residue at $s = n/2$ of Koecher's zeta function $Z_{m,n-m}(Y, s)$ defined in formula (1.173) of Section 1.4.1. Another reference for Siegel's method is Maass [426, Section 16]. Instead we follow a method due to Selberg which makes use of invariant differential operators to annihilate the lower rank terms in the theta

---

[6]The correction of the wrong deduction is then by no means so simple, as one should like to believe to begin with when thinking of renormalization in the physics literature, and it necessitated more exact estimates of infinite series.

function, generalizing Exercises 1.4.9 and 1.4.10 of Volume I. Other references for Selberg's method are Maass [426, Section 16] and Terras [593, 602, 604]. There are also other ways of accomplishing the analytic continuation, as we mentioned in Section 1.5.1, but we will not discuss them here. In considering Selberg's method we will ignore a perhaps nasty problem: Could the differential operators actually annihilate the automorphic forms?

Proof of Part (5) of Theorem 1.5.2. In order to prove part (5) and complete the proof of Theorem 1.5.2, we need the **theta function** from formula (1.172) of Section 1.4.1, with $1 \leq k \leq n$:

$$\theta(Y, X) = \sum_{A \in \mathbb{Z}^{n \times k}} \exp\{-\pi \operatorname{Tr}(Y[A]X)\}, \qquad \text{for } Y \in \mathcal{P}_n \text{ and } X \in \mathcal{P}_k. \qquad (1.268)$$

For $0 \leq r \leq k$, we set

$$\theta_r(Y, X) = \sum_{\substack{A \in \mathbb{Z}^{n \times k} \\ \operatorname{rank} A = r}} \exp\{-\pi \operatorname{Tr}(Y[A]X)\}. \qquad (1.269)$$

Before proceeding with the proof, we must beg the reader to do the following exercises.

Exercise 1.5.18 (The Transformation Formula of the Theta Function). Show that if the theta function is defined by (1.268), then

$$\theta(Y, X) = |Y|^{-k/2} |X|^{-n/2} \theta(Y^{-1}, X^{-1}).$$

Hint. Imitate the proof of Exercise 1.4.6 in Volume I, using the Poisson summation formula.

Exercise 1.5.19 (A Gamma Integral Associated with a Maass Form for $GL(n)$). Suppose that f is a Maass form on the determinant one surface; i.e., that $f \in \mathcal{A}^0(GL(n, \mathbb{Z}), \lambda)$, where

$$\frac{\Omega(f(Y^0)|Y|^s)}{f(Y^0)|Y|^s} = \frac{\Omega p_r(Y)}{p_r(Y)} = \lambda_\Omega(r) \quad \text{with } r = r(f, s) \in \mathbb{C}^n,$$

for all invariant differential operators $\Omega \in D(\mathcal{P}_n)$. Here $p_r(Y)$ is a power function as in formula (1.41) of Section 1.2.1. Use Proposition 1.2.4 of Section 1.2.3 and the definition of the gamma function Γ_k in formula (1.44) of Section 1.2.1 to prove that:

$$\int_{X \in \mathcal{P}_k} \exp\left\{-\operatorname{Tr}\left(Y^{-1}X\right)\right\} f(X^0) |X|^s \, d\mu_k(X) = \Gamma_k(r(f, s)) f(Y^0) |Y|^s.$$

There is another discussion of this result in Maass [426, Section 7]. Related results were obtained in Exercise 1.2.4 of Section 1.2.1 and in formula (1.108) of Section 1.2.4.

We will write $\Gamma_k = GL(k, \mathbb{Z})$ and hope that the context will make it clear that this is not the gamma function. Defining

$$\Lambda(f, s|Y) = \int_{X \in \mathcal{M}_k} f(X^0) |X|^s \, \theta_k(Y, X) \, d\mu_k(X), \tag{1.270}$$

it follows from Exercise 1.5.19 that

$$\Lambda(f, s|Y) = 2 \int_{\mathcal{P}_k} \sum_{A \in \mathbb{Z}^{n \times k} \text{ rank } k/\Gamma_k} \exp\{-\pi \operatorname{Tr}(Y[A]X)\} f(X^0) |X|^s \, d\mu_k(X)$$

$$= 2\pi^{-ks} \, \Gamma_k(r(f, s)) \sum_{A \in \mathbb{Z}^{n \times k} \text{ rank } k/\Gamma_k} |Y[A]|^{-s} f^*(Y[A]^0) \, .$$

Here $f^*(W) = f(W^{-1})$. The factor 2 comes from the fact that

$$\bigcup_{A \in \Gamma_k} \mathfrak{M}_k[A]$$

represents each point of $\mathcal{P}_n$ exactly twice (see part (3) of Theorem 1.4.1 in Section 1.4.2).

Now suppose that $f(W)$ is an eigenfunction of all the Hecke operators. Then

$$\Lambda(f, s|Y) = \begin{cases} 2\pi^{-ns}\Gamma_n(r(f, s))L_{f^*}(2s)|Y|^{-s}f^*(Y^0), & \text{if } k = n; \\ 2\pi^{-ks}\Gamma_k(r(f, s))L_{f^*}(2s)E_{k,n-k}(f^*, s|Y), & \text{if } 1 \leq k < n. \end{cases} \tag{1.271}$$

The formula for $k = n$ is clear. That for $1 \leq k < n$ will be proved in Corollary 1.5.1. Only the case $k = n$ is needed for Theorem 1.5.2. Here $E_{k,n-k}(f^*, s|Y)$ is the Eisenstein series of formula (1.245).

Exercise 1.5.20. Show that if f is an integrable function on $\mathcal{S}\mathcal{M}_k$ satisfying the rest of the hypotheses of part (5) of Theorem 1.5.2, then the associated L function $L_f(2s)$ converges for Re $s > n/2$.

Hint. Use a similar argument to that which was used to prove the convergence of the Eisenstein series (1.245).

Next we want to consider some differential operators on $\mathcal{P}_k$. As in Section 1.1.5, we write the matrix operator

$$\partial/\partial X = \left(\frac{1}{2}(1 + \delta_{ij})\frac{\partial}{\partial x_{ij}}\right), \qquad \text{if } X = (x_{ij}) \in \mathcal{P}_k. \tag{1.272}$$

See formula (1.29) in Section 1.1.5. Now define the **determinant operator**

$$\partial_X = \det(\partial/\partial X). \tag{1.273}$$

This is a departure from Section 1.1.5, in which we only considered traces of matrix operators. However ∂_X did appear in Section 1.2.5. The property of ∂_X that endears it to us is:

$$\partial_X \exp[\mathrm{Tr}(XY)] = |Y| \exp[\mathrm{Tr}(XY)]. \tag{1.274}$$

This means that the operator ∂_X annihilates the singular terms $\theta_r(Y, X)$ in (1.269) when $r < k$. But ∂_X is not $GL(n, \mathbb{R})$-invariant. To obtain such an operator, consider **Selberg's differential operators**:

$$D_a = |X|^a \partial_X |X|^{1-a}, \quad a \in \mathbb{R}. \tag{1.275}$$

Let $\sigma(X) = X^{-1}$ for $X \in \mathcal{P}_k$ and $D^\sigma f = D(f \circ \sigma) \circ \sigma^{-1}$, for a differential operator D and a function $f : \mathcal{P}_k \to \mathbb{C}$. We know from Theorem 1.2.1 that D^σ is the conjugate adjoint $\overline{D}^*$. This allows you to do part (b) of the following exercise.

Exercise 1.5.21. (a) Show that D_a defined by (1.275) is indeed a $GL(k, \mathbb{R})$-invariant differential operator on $\mathcal{P}_k$; i.e., $D_a \in D(\mathcal{P}_k)$.
(b) Show that if D_a is as in (1.275) and if $\sigma(X) = X^{-1}$, then

$$D_a^\sigma = D_a^* = (-1)^k D_{a^*}, \quad \text{where} \quad a^* = 1 - a + (k+1)/2.$$

Hint. For part (b), use integration by parts to find the adjoint of D_a.

Exercise 1.5.22. (a) Let D_a be as defined in (1.275). Set

$$D = D_a D_1 \text{ for } a = (k - n + 1)/2.$$

Show that

$$(D_1 D_a)^\sigma |X|^{n/2} = |X|^{n/2} D_a D_1.$$

(b) Use part (a) to show that if we write $\theta(Y, X) = \theta^Y(X)$, then

$$D\theta^Y(X) = |Y|^{-k/2} |X|^{-n/2} (D\theta^{Y^{-1}})(X^{-1}).$$

Exercise 1.5.22 shows that the differentiated theta function $D\theta^Y(X)$ satisfies the same transformation formula as $\theta^Y(X)$ itself. By formula (1.274) however, the differentiated theta function is missing all the lower rank terms; i.e.,

$$D\theta^Y(X) = D\theta_k^Y(X).$$

Note that one should really show that $D\theta^Y$ does not vanish. We certainly did this for the special case of Riemann's zeta function in Exercises 1.4.8–1.4.10 of Volume I.

See also Exercise 1.5.24 below for the case of Koecher's zeta function. Are we still under the curse of the higher rank Eisenstein series?

Exercise 1.5.23. Let D be as defined in Exercise 1.5.22 and $\lambda_D^*(r(f, s))$ as defined in Exercise 1.5.19. Show that

$$\lambda_{D^*}(r(f, s)) = \lambda_{D^*}\left(r\left(f^*, \frac{n}{2} - s\right)\right),$$

with $f^*(X) = f(X^{-1})$ and $D^* =$ the adjoint of D.

It is now possible to complete the proof of part (5) of Theorem 1.5.2 by writing:

$$\Lambda(f, s|Y) \, \lambda_{D^*}(r(f, s)) = \int_{\mathcal{M}_k} D\theta^Y(X) \, |X|^s f(X^0) \, d\mu_k(X).$$

Break the fundamental domain into two parts according to whether the determinant is greater than one or not. This gives:

$$\Lambda(f, s|Y) \, \lambda_{D^*}(r(f, s)) = \int_{\substack{X \in \mathcal{M}_k \\ |X| \geq 1}} + \int_{\substack{X \in \mathcal{M}_k \\ |X| \leq 1}}$$

In the second integral replace X by X^{-1} to see that:

$$\Lambda(f, s|Y) \, \lambda_{D^*}(r(f, s))$$
$$= \int_{\substack{X \in \mathcal{M}_k \\ |X| \geq 1}} \left(|X|^s f(X^0) D\theta^Y(X) + |Y|^{-k/2} |X|^{n/2-s} f^*(X^0) D\theta^{Y^{-1}}(X) \right) d\mu_k(X).$$

It is clear from this formula that we have the analytic continuation of $\Lambda(f, s|Y)$ to all values of s along with the functional equation:

$$\Lambda(f, s \mid Y) = |Y|^{-k/2} \Lambda\left(f^*, \frac{n}{2} - s \mid Y^{-1}\right), \qquad (1.276)$$

using Exercises 1.5.22 and 1.5.23. Set $n = k$ and $Y = I$ to obtain the functional equation in part (5) of Theorem 1.5.2, thus completing the proof of that theorem. ∎

Exercise 1.5.24 (The Eigenvalue for Koecher's Zeta Function).

(a) Let the differential operator D_1 be defined by (1.275) and the polynomial $h(s)$ be defined by

$$D_1^* |X|^s = (-1)^k h(s) |X|^s, \quad \text{for } X \in \mathcal{P}_k, \ s \in \mathbb{C}.$$

Show that

$$h(s) = \prod_{j=0}^{k-1} \left(s - \frac{j}{2} \right).$$

(b) Use part (a) to show that if ∂_X is defined by (1.273) then

$$\partial_X |X|^s = h(s + (k-1)/2)|X|^{s-1}.$$

(c) Let D be the operator defined in Exercise 1.5.22 and let $\lambda_D^*(r(f, s))$ be the eigenvalue defined in Exercise 1.5.19. Suppose that f is the Maass form that is identically one. Show that

$$\lambda_D^*(r(1, s)) = h(s)h(n/2 - s).$$

Hint. Compare with Exercise 1.2.31 of Section 1.2.5. We find that:

$$(-1)^k \Gamma_k(0, \ldots, 0, s)h(s) = \int_{\mathcal{P}_k} \exp\{-\mathrm{Tr}(X)\}\, D_1^* |X|^s \, d\mu_k(X)$$

$$= \int_{\mathcal{P}_k} (D_1 \exp\{-\mathrm{Tr}(X)\})\, |X|^s \, d\mu_k(X)$$

$$= (-1)^k \Gamma_k(0, \ldots, 0, s+1).$$

Now we can consider the analytic continuation of the Eisenstein series defined in formula (1.245).

Corollary 1.5.1 (Analytic Continuation and Functional Equation of the Eisenstein Series $E_{k,n-k}(f, s|Y)$).
Suppose that $f \in \mathcal{A}^0(GL(k, \mathbb{Z}), \lambda)$ is a Maass form on the determinant one surface such that f is integrable over the fundamental domain $\mathcal{SM}_k$ and let f be an eigenfunction of the Hecke operators T_m for all positive integers m. Suppose also that $L_f(s)$ is the L-function associated with f in part (5) of Theorem 1.5.2. Then the Eisenstein series $E_{k,n-k}(f, s|Y)$, $1 \leq k < n$, defined in formula (1.245) of Section 1.5.1 has analytic continuation to all $s \in \mathbb{C}$ as a meromorphic function. Moreover it satisfies the functional equation below, using the notation of part (5) of Theorem 1.5.2:

$$\Lambda(f, s|Y) = 2\pi^{-ks} \Gamma_k(r(f, s)) L_f^*(2s) E_{k,n-k}\left(f^*, s| Y\right)$$

$$= |Y|^{-k/2} \Lambda\left(f^*, \frac{n}{2} - s \middle| Y^{-1}\right),$$

where $f^*(W) = f(W^{-1}).$

Proof. The only chore that remains is the proof of (1.271) when $1 \leq k < n$. To do this, we need to show that the **zeta function** defined by:

$$Z(f, s|Y) = \sum_{A \in \mathbb{Z}^{n \times k} \text{ rank } k/GL(k,\mathbb{Z})} |Y[A]|^{-s} f(Y[A]^0), \quad \text{Re}s > n/2, \quad (1.277)$$

has the factorization

$$Z(f, s|Y) = L_f(2s) \, E(f, s|Y). \quad (1.278)$$

Use Lemma 1.5.1 in Section 1.5.1 to see that summing over rank k matrices $A \in \mathbb{Z}^{n \times k}$ modulo $GL(k, \mathbb{Z})$ is equivalent to summing $A = BC$ over

$$B \in \mathbb{Z}^{n \times k}, \quad (B \, *) \in GL(k, \mathbb{Z})/P(k, n - k),$$
$$C \in \mathbb{Z}^{k \times k}, \quad \text{rank } k/GL(k, \mathbb{Z}).$$

Here $P(k, n - k)$ is the parabolic subgroup defined in formula (1.241) of Section 1.4.4. It follows that

$$Z(f, s|Y) = \sum_{B,C} |Y[BC]|^{-s} f(Y[BC]^0)$$
$$= \sum_{B} |Y[B]|^{-s} \sum_{r \geq 1} r^{-2s} \sum_{|C|=r} f((Y[B]^0)[C]^0),$$

which completes the proof of (1.278). The corollary follows from the proof of part (5) of Theorem 1.5.2. In particular, the functional equation of the Eisenstein series is formula (1.276).

Remarks. (1) The method of analytic continuation using differentiated theta functions is magical (and perhaps problematic in general), and it does not appear to allow one to find the residues of the zeta function (1.277) at $s = n/2$, for example. For this, one must investigate the divergent integrals, as Siegel did in [565, Vol. III, pp. 328–333]. There is also an approach using Fourier expansions of Maass forms (see Terras [601]) and we will consider this method in the next section (see (1.308)–(1.310) and Exercise 1.5.35 in Section 1.5.3).
(2) The gamma factors $\Gamma_k(r(f, s))$ appearing in the functional equations of the Dirichlet series $L_f(2s)$ resemble those in the functional equations of Hecke L-functions with grossencharacter (see Hecke [268, pp. 215–234, 249–287]). Of course, for proper congruence subgroups of $GL(2)$, this fact allowed Maass [417] to prove the existence of nonholomorphic cusp forms. For $GL(n, \mathbb{Z})$, one must generalize Theorem 1.5.5 to congruence subgroups and prove some kind of a converse theorem in order to obtain results similar to those of Maass. Jacquet et al. [325] manage this in the language of representations of the

adelic version of $GL(3)$ and thus prove the existence of an adelic automorphic representation of $GL(3)$ corresponding to L-functions for cubic number fields.

(3) There is a problem with the method of analytic continuation which was presented in this section since we don't know where $L_f(2s)$ vanishes. In the case of Epstein's zeta function, the L-function was Riemann's zeta function. For a general Maass form f for $GL(n, \mathbb{Z})$ there is much less known about L_f than there is about Riemann's zeta function. The main reference producing a nonvanishing theorem for the adelic version of L-functions associated with Maass forms for $GL(n)$ is the paper of Jacquet and Shalika [327]. See also Goldfeld [230].

(4) We should perhaps say something about the more general Eisenstein series defined by (1.248). Let us consider the case when $q = 2$ and the parabolic subgroup is maximal. When $n = n_1 + n_2$ write:

$$Y = a[u] \quad \text{with } a = \begin{pmatrix} a_1(Y) & 0 \\ 0 & a_2(Y) \end{pmatrix}, \quad a_i(Y) \in \mathcal{P}_{n_i}, \ i = 1, 2,$$

$$u = \begin{pmatrix} I_{n_1} & X \\ 0 & I_{n_2} \end{pmatrix}, \qquad\qquad X \in \mathbb{R}^{n_1 \times n_2}. \tag{1.279}$$

Let $f_i \in \mathcal{A}(GL(n_i, \mathbb{Z}), \lambda_i)$, $i = 1, 2$, $Y \in \mathcal{P}_n$, $n = n_1 + n_2$, and define the **Eisenstein series**:

$$E_{n_1,n_2}(f_1, f_2 | Y) = \sum_{A \in GL(n,\mathbb{Z})/P(n_1,n_2)} f_1(a_1(Y[A]))f_2(a_2(Y[A])). \tag{1.280}$$

If $f_i(Y) = |Y|^{r_i} \varphi_i(Y^0)$, with $\varphi_i \in \mathcal{A}^0(GL(n_i, \mathbb{Z}), \eta_i)$, the series (1.280) will converge if φ_i is integrable on $\mathcal{SM}_{n_i}$ and $\mathrm{Re}\, r_i$ is sufficiently large. For one can use an integral test based on the integral formula in Exercise 1.4.27 of Section 1.4.4. In order to obtain an analytic continuation similar to that given in Corollary 1.5.1, one must relate the Eisenstein series (1.280) to a zeta function:

$$Z_{n_1,n_2}(f_1, f_2 | Y) = \sum_{A \in \mathbb{Z}^{n \times n} \text{ rank } n/P(n_1,n_2)} f_1(a_1(Y[A]))f_2(a_2(Y[A])). \tag{1.281}$$

Exercise 1.5.25. Suppose that the Maass forms f_i in (1.280) can be written:

$$f_i(W) = |W|^{r_i} \varphi_i(W^0), \quad \text{with } \varphi_i \in \mathcal{A}^0(GL(n_i, \mathbb{Z}), \eta_i), \ r_i \in \mathbb{C},$$

with $\mathrm{Re}\, r_i$ sufficiently large for convergence of the Eisenstein series. Assume that the Maass forms φ_i are eigenfunctions of all the Hecke operators for $GL(n_i, \mathbb{Z})$. Let $L_{\varphi_i}(s)$ be the L function associated with such a Maass form as in part (5) of Theorem 1.5.2. Show that

$$Z(f_1, f_2 | Y) = E(f_1, f_2 | Y) \, L_{\varphi_1}(2r_1) \, L_{\varphi_2}(2r_2).$$

Hint. You need an analogue of Lemma 1.5.1 in Section 1.5.1 to be able to write a rank n matrix $A \in \mathbb{Z}^{n \times n}$ modulo $P(n_1, n_2)$ in the form $A = BC$ with

$$B \in GL(n, \mathbb{Z})/P(n_1, n_2),$$

$$C = \begin{pmatrix} C_1 & D \\ 0 & C_2 \end{pmatrix}, \quad C_i \in \mathbb{Z}^{n_i \times n_i} \text{ rank } n_i/GL(n_i, \mathbb{Z}).$$

(5) There are many generalizations of the zeta functions and Eisenstein series considered here. For example, Maass [420] deals with functions generalizing $Z(f, s|Y)$ by adding a spherical function $u : \mathbb{R}^{n \times k} \to \mathbb{C}$. See also Maass [421, 422] and Christian [111]. In introducing these Maass zeta functions, Maass was motivated by the problem of studying the number of representations of a positive matrix $T \in \mathcal{P}_k$ in the form $T = S[G]$ for $G \in \mathbb{Z}^{n \times k}$, with $n > k$ and $S \in \mathcal{P}_n$ fixed. Maass wanted to study the zeta functions for $GL(n)$ with the additional variable coming from spherical functions for $\mathbb{R}^{n \times k}$ in order to employ a method analogous to that used by Hecke for similar problems in algebraic number fields. The last theorem in the paper of Maass [420] gives the analytic continuation and functional equations of these zeta functions. Jorgenson and Lang [335] consider heat kernel Eisenstein series.

The Eisenstein series (1.245) are eigenfunctions of the Hecke operators (1.262). The following proposition gives an explicit expression for the eigenvalue.

Proposition 1.5.1 (Eigenvalues for the Action of Hecke Operators on Eisenstein Series). *Using the definitions* (1.245) *of the Eisenstein series $E(\varphi, s|Y)$ and* (1.262) *for the Hecke operator $T_k^{(n)} = T_k$, we have:*

$$T_k^{(n)} E(\varphi, s|Y) = u_k^{(n)} E(\varphi, s|Y),$$

where

$$u_k^{(n)} = k^{2ms/n} \sum_{t|k} d_{n-m}\left(\frac{k}{t}\right) t^{n-m-2s} u_t^{(m)}.$$

Here $T_t^{(m)} \varphi = u_t^{(m)} \varphi$ and

$$d_r(v) = \sum_{v = d_1 \cdots d_r} d_1^{r-1} d_2^{r-2} \cdots d_{r-1}$$

where the d_j are all nonnegative integers.

Proof. Let us change our notation slightly from (1.262) and let $M_k^{(n)}$ denote the set of all $n \times n$ matrices of determinant k. Clearly we need representatives of $M_k^{(n)}$ modulo $P(m, n - m)$. One can write $A \in M_k^{(n)}$ as $A = BC$, with

$$B = (B_1 \ *) \in \Gamma_n / P(m, n - m),$$

$$C = \begin{pmatrix} F & H \\ 0 & G \end{pmatrix} \in M_k^{(n)}, \quad F \in \mathbb{Z}^{m \times m}, \quad G \in \mathbb{Z}^{(n-m) \times (n-m)}.$$

Here $P(m, n - m)$ is the maximal parabolic subgroup defined in formula (1.241) of Section 1.4.4.

It follows that the sum over $A \in M_k^{(n)} / P(m, n - m)$ is the same as the double sum over

$$(B_1 \ *) \in \Gamma_n / P(m, n - m) \quad \text{and} \quad F \in M_t^{(m)} / \Gamma_m,$$

$$G \in M_{k/t}^{(n-m)} / \Gamma_{n-m}, \quad H \bmod F,$$

for all divisors t of k. The notation "$H \bmod F$" denotes a complete set of representatives for the equivalence relation:

$$H \sim H' \Leftrightarrow H' = FU + H, \quad \text{for some } U \in \mathbb{Z}^{m \times (n-m)}.$$

The number of $H \bmod F$ is easily seen to be $|F|^{n-m} = t^{n-m}$. The number of $G \in M_{k/t}^{(n-m)} / \Gamma_{n-m}$ is $d_{n-m}(k/t)$, using the definition of $d_r(\nu)$ to be found in the statement of the proposition.

Putting all this together and setting $P = P(m, n - m)$, we see that

$$T_k^{(n)} E_{m,n-m}(\varphi, s | Y) = \sum_{\substack{B \in M_k^{(n)} / \Gamma_n}} \sum_{\substack{A = (A_1 \ *) \in \Gamma_n / P \\ A_1 \in \mathbb{Z}^{n \times m}}} \varphi\left(Y [BA_1]^0\right) \left|Y\left[k^{-1/n} BA_1\right]\right|^{-s}$$

$$= k^{2ms/n} \sum_{\substack{(A_1 \ *) \in M_k^{(n)} / P \\ A_1 \in \mathbb{Z}^{n \times m}}} \varphi\left(Y [A_1]^0\right) |Y [A_1]|^{-s}$$

$$= k^{2ms/n} \sum_{\substack{(B_1 \ *) \in \Gamma_n / P \\ B_1 \in \mathbb{Z}^{n \times m}}} \sum_{t/k} d_{n-m}\left(\tfrac{k}{t}\right) t^{n-m-2s} u_t \, \varphi\left(Y [B_1]^0\right) |Y [B_1]|^{-s}.$$

This is easily seen to be equal to $u_k^{(n)} E_{m,n-m}(\varphi, s | Y)$. ∎

The next exercise shows that the Eisenstein series (1.280) always satisfy a trivial functional equation.

Exercise 1.5.26 (Trivial Functional Equation of the Eisenstein Series). Consider the Eisenstein series (1.280). Define $f^*(Y) = f(Y^{-1})$. Prove that

$$E_{k,n-k}(f_1, f_2 | Y) = E_{n-k,k}\left(f_2^*, f_1^* | Y^{-1}\right).$$

There is a similar argument relating Eisenstein series for the parabolic subgroup $P(n_1, \ldots, n_q)$ to those for the **associated parabolic subgroup** $P(n_{\sigma(1)}, \ldots, n_{\sigma(q)})$, for any permutation σ of q elements.

Hint. If Y is expressed in the form

$$Y = \begin{pmatrix} a_1 & 0 \\ 0 & a_2 \end{pmatrix} \begin{bmatrix} I_{n_1} & X \\ 0 & I_{n_2} \end{bmatrix},$$

then we have

$$Y^{-1} \begin{pmatrix} 0 & I \\ I & 0 \end{pmatrix} = \begin{pmatrix} a_2^{-1} & 0 \\ 0 & a_1^{-1} \end{pmatrix} \begin{bmatrix} I_{n_2} & {}^t X \\ 0 & I_{n_1} \end{bmatrix}.$$

Next note that if

$$\omega = \begin{pmatrix} 0 & I \\ I & 0 \end{pmatrix},$$

and we set $\gamma = {}^t\omega \, {}^t\tau^{-1}\omega$, then γ runs through $\Gamma_n/P(k, n-k)$ as fast as τ runs through $\Gamma_n/P(n-k, k)$.

The functional equation in Exercise 1.5.26 does not extend the domain of convergence. See Exercise 1.5.39 below for a similar result.

1.5.3 Fourier Expansions of Eisenstein Series

Generalities on Fourier Expansions of Eisenstein Series

Next we plan to study Fourier expansions of Maass forms for $GL(n, \mathbb{Z})$ using methods modeled on those of Siegel [565, Vol. II, pp. 97–137]. Before proceeding further, the reader should review the Fourier expansions of Maass wave forms for $SL(2, \mathbb{Z})$ in Exercise 3.5.3 of Volume I. The main results in the present section are Fourier expansions of Eisenstein series for $GL(n)$ given in Theorems 1.5.3–1.5.5 from the papers Terras [606] and Kaori Imai (Ota) and Terras [318]. Similar results are obtained in Terras [596], Takhtadzhyan and Vinogradov [586], and Proskurin [494]. These results should also be compared with those of Bump [83] and Goldfeld [230] and we will do so in Section 1.5.4.

Why look at Fourier expansions of Maass forms for $GL(n)$? There are many reasons beyond simple curiosity. We will see in Section 1.5.5 that harmonic analysis on the fundamental domain $\mathcal{SP}_n/GL(n, \mathbb{Z})$ requires knowledge of the "constant term" in the Fourier expansion of Eisenstein series just as it did for $SL(2, \mathbb{Z})$ in

Lemma 3.7.1 of Volume I. We will discuss the constant term in a rather simple-minded way. A more elegant theory of the constant term has been developed (see Arthur's talk in Borel and Casselman [66, Vol. I, pp. 253–274], Harish-Chandra [262], and Langlands [392]) as well as Section 1.5.4. Of course the constant term will not be a constant in our case. It was not even a constant in the case of Maass wave forms for $SL(2, \mathbb{Z})$.

There are also reasons for considering the "nonconstant terms" in these Fourier expansions. For example, we saw in Section 3.6 of Volume I that, in the case of $SL(2, \mathbb{Z})$, a knowledge of the exact form of the nonconstant terms was useful in our quest to understand Maass's extension of Hecke's correspondence between modular forms and Dirichlet series (Theorem 3.6.2, Vol. I). In particular, the Fourier expansion of the Eisenstein series for $SL(2, \mathbb{Z})$, given in Exercise 3.5.4, Vol. I, has had many applications in number theory. Many of these applications stem from the Kronecker limit formula (to be found in Exercise 3.5.6, Vol. I). Hecke [268, pp. 198–207] noticed that the Fourier expansion of Epstein's zeta function (see Proposition 1.5.2 below) gives a Kronecker limit formula for zeta functions of number fields. See Theorem 1.4.2 of Vol. I and Bump and Goldfeld [89]. However the analogue of the Dedekind eta function $\eta(z)$ has yet to be completely understood for $GL(n)$, $n > 2$. Efrat [153] obtains an analogue of $|\eta(z)|$ which is a harmonic Maass form for $SL(3, \mathbb{Z})$ of weight $\frac{1}{2}$ and considers the consequences for cubic number fields. Bill Duke and Özlem İmamoḡlu [148] consider analogues of $|\eta(z)|$ for $GL(n)$. Siegel [563] found that Hecke's result could be generalized to Hecke L-functions with grossencharacter. Takhtadzhyan and Vinogradov [586] have also announced applications of Fourier expansions of Eisenstein series for $GL(3)$ to the theory of divisor functions.

There has, in fact, been much work on Fourier expansions of Maass forms for general discrete groups Γ acting on symmetric spaces $X = K\backslash G$ such that the fundamental domain X/Γ has "cusps." For example, Siegel considered $X = SU(n)\backslash Sp(n, \mathbb{R})$ and $\Gamma = Sp(n, \mathbb{Z})$, the **Siegel modular group**. In this case, X can be identified with the Siegel upper half space $\mathcal{H}_n$, consisting of $X + iY$, with X and Y real $n \times n$ symmetric matrices and positive Y, to be considered in Chapter 2. Siegel obtained the Fourier expansions of holomorphic Eisenstein series (see Siegel [565, Vol. II, pp. 97–137], Baily [32, pp. 228–240], and Chapter 2). Baily [32, p. 238] uses the rationality and bounded denominators of the Fourier coefficients in these expansions to show that the Satake compactification of X/Γ is defined over the field of rational numbers as an algebraic variety. Fourier expansions of non-holomorphic Eisenstein series for $Sp(n, \mathbb{Z})$ have been obtained by Maass [426, Section 18]. The arithmetic parts of both the holomorphic and non-holomorphic Eisenstein series are "singular series" or divisor-like functions (see the discussion in Exercise 3.5.4, Vol. I). The non-arithmetic or analytic part in the holomorphic case is $\exp\{-\text{Tr}(NY)\}$, for $Y \in \mathcal{P}_n$, N a nonnegative symmetric half-integral $n \times n$ matrix. Here "half-integral" means that $2n_{ij} \in \mathbb{Z}$, when $i \neq j$ and $n_{ii} \in \mathbb{Z}$. And N "nonnegative" means that $N[x] \geq 0$ for all $x \in \mathbb{R}^n$. In the non-holomorphic case, the non-arithmetic part is a matrix analogue of a confluent hypergeometric function of the sort which was studied in Section 1.2.2. We will obtain similar results for $GL(n)$.

It is also possible to obtain Fourier expansions of Eisenstein series for congruence subgroups of $SL(2, \mathfrak{O}_K)$, where $\mathfrak{O}_K$ is the ring of integers of a number field. This will be discussed in the next chapter. Such Fourier expansions have been used by number theorists to study Gauss sums and elliptic curves for example (see Kubota [376], Heath-Brown and Patterson [267], and Goldfeld et al. [231]).

Many authors take an adelic representation-theoretic approach to the subject of Fourier expansions for $GL(n)$. See, for example, Jacquet et al. [325]—a paper which makes use of Whittaker models for representations. Stark [578] indicates a way to bridge the gap between the classical and adelic points of view. See also Rhodes [510].

From our earlier study of the case of $SL(2)$, in view of work of Harish-Chandra and Langlands, we should be willing to believe that the spectral measure in the spectral resolution of the Laplacian on $L^2(S\mathcal{P}_n/GL(n, \mathbb{Z}))$ comes from the asymptotics and functional equations of the Eisenstein series. Knowing the asymptotic behavior of the Eisenstein series as the argument $Y \epsilon \mathcal{P}_n$ approaches the boundary of the fundamental domain is the same as knowing the "constant term" in the Fourier expansion. Let us attempt to find a simple-minded way of obtaining this constant term in the region where the Dirichlet series defining $E_P(\varphi, s|Y)$ converges. That is we want to find a generalization of the method we had in mind for part (b) of Exercise 3.5.2 of Volume I.

As our first example, consider the Eisenstein series $E_{m,n-m}(\varphi, s|Y)$ defined in formula (1.245) of Section 1.5.1. Let $Y \in \mathcal{P}_n$ have partial Iwasawa decomposition:

$$Y = \begin{pmatrix} V & 0 \\ 0 & W \end{pmatrix} \begin{bmatrix} I & X \\ 0 & I \end{bmatrix}, \quad V \in \mathcal{P}_m, \ W \in \mathcal{P}_{n-m}, \ X \in \mathbb{R}^{m \times (n-m)}.$$

Recall that the Eisenstein series $E_{m,n-m}(\varphi, s|Y)$ was defined in formula (1.245) of Section 1.5.1 as:

$$E_{m,n-m}(\varphi, s|Y) = \sum_{A = (A_1 *) \epsilon \Gamma_n/P} \varphi\left(Y[A_1]^0\right) \ |Y[A_1]|^{-s}, \quad \text{for } \operatorname{Re} s > n/2,$$

where $\Gamma_n = GL(n, \mathbb{Z})$ and $P = P(m, n - m)$. Here we assume that the Maass form $\varphi \in \mathcal{A}^0(GL(m, \mathbb{Z}), \lambda)$ is bounded on the determinant one surface of the fundamental domain. Write

$$A_1 = \begin{pmatrix} B \\ C \end{pmatrix}, \quad B \in \mathbb{Z}^{m \times m}, \quad \text{so that} \quad Y[A] = V[B + XC] + W[C].$$

If W approaches infinity in the sense that the diagonal entries in its Iwasawa decomposition all approach infinity, then it is not too hard to see that when $\operatorname{Re} s > n/2$, the term $|Y[A]|^{-s}$ must approach zero unless $C = 0$. It follows that $A \in P$ and so $A = I$. Thus we find that for fixed s with $\operatorname{Re} s > n/2$

$$E_{m,n-m}(\varphi, s|Y) \sim \varphi\left(V^0\right) \ |V|^{-s},$$

as W goes to infinity in the sense described above. When s is not in the region of convergence of the Dirichlet series, the functional equation can add in other terms to this asymptotic formula—a phenomenon that we saw already in the case of $SL(2, \mathbb{Z})$ in Section 3.5 of Volume I.

The preceding example is a little too simple-minded perhaps. So let us try to be a little more explicit about the approach to the boundary of the fundamental domain while attempting to consider the more general Eisenstein series defined by formula (1.280) in Section 1.5.2. We also want to relate all this to the theory of roots, parabolic subgroups, and Bruhat decompositions.

These considerations lead us to examine the possible ways of approaching the boundary of our fundamental domain. For concreteness, let us consider only the case of $GL(4)$. In this case we can write

$$Y = A[n], \quad \text{where } A = \begin{pmatrix} a_1 & 0 & 0 & 0 \\ 0 & a_2 & 0 & 0 \\ 0 & 0 & a_3 & 0 \\ 0 & 0 & 0 & a_4 \end{pmatrix} \quad \text{and } n = \begin{pmatrix} 1 & * & * & * \\ 0 & 1 & * & * \\ 0 & 0 & 1 & * \\ 0 & 0 & 0 & 1 \end{pmatrix}.$$

The ways of approaching the boundary of $\mathcal{SP}_4/GL(4, \mathbb{Z})$ can be described by subsets of the set of quotients a_i/a_j, $i < j$, which are allowed to approach zero. These quotients are the multiplicative version of "roots" to be discussed in Chapter 2. See Figure 1.30 for a diagram of the various sets of roots giving ways of approaching the boundary, as well as the corresponding parabolic subgroups.

The general theory of parabolic subgroups says that they correspond to sets J of simple roots $\alpha_i(a) = a_i/a_{i+1}$ (see Borel's article in Borel and Mostow [68, pp. 1–19]) via:

$$\bigcap_{\alpha \in J} \ker \alpha = S_J, \quad P_J = Z(S_J)N, \tag{1.282}$$

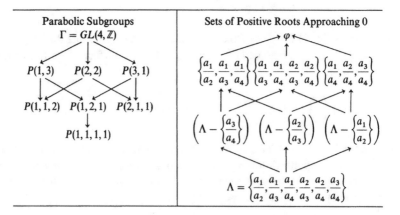

Fig. 1.30 Subgroups of $GL(4)$ and ways to approach the boundary of $\mathcal{P}_4/GL(4, \mathbb{Z})$. Arrows indicate containment

where N is the nilpotent or unipotent subgroup of upper triangular matrices with ones on the diagonal and $Z(S_J)$ denotes the centralizer of S_J. Let W be the Weyl group of all permutation matrices in K (cf. Exercise 1.5.30 below). Denote by w_i the Weyl group element (which is almost the $n \times n$ identity matrix except for a 2×2 matrix on the diagonal at row i):

$$
w_i = \begin{pmatrix}
1 & \cdots & 0 & 0 & 0 & 0 & \cdots & 0 \\
\vdots & \ddots & \vdots & \vdots & \vdots & \vdots & \cdots & \vdots \\
0 & \cdots & 1 & 0 & 0 & 0 & \cdots & 0 \\
0 & \cdots & 0 & 0 & 1 & 0 & \cdots & 0 \\
0 & \cdots & 0 & -1 & 0 & 0 & \cdots & 0 \\
0 & \cdots & 0 & 0 & 0 & 1 & \cdots & 0 \\
\vdots & \cdots & \vdots & \vdots & \vdots & \vdots & \ddots & \vdots \\
0 & \cdots & 0 & 0 & 0 & 0 & \cdots & 1
\end{pmatrix},
$$

where there are $i-1$ ones on the diagonal, then a 2×2 matrix, $\begin{pmatrix} 0 & 1 \\ -1 & 0 \end{pmatrix}$ then $(n-i-1)$ ones on the diagonal, with the other entries being 0.

$$(1.283)$$

This matrix represents the permutation that interchanges i and $i+1$, leaving all else fixed. Then set

$$W_J = \quad \text{the group generated by the } w_i \text{ for } \alpha_i \in J.$$

One can show that $P_J = BW_JB$, where B is the **Borel or minimal parabolic subgroup** $B = P(1, 1, \ldots, 1)$, using the notation (1.241) from Section 1.4.4. The **Bruhat decomposition** of $GL(n, \mathbb{R})$ corresponding to parabolic subgroups P_J and $P_{J'}$, is:

$$GL(n, \mathbb{R}) = \bigcup_{w \in W_J \backslash W / W_{J'}} P_J w P_{J'} \quad \text{(disjoint)}. \tag{1.284}$$

This result was discussed by Curtis [127]. We will consider only some special cases in what follows (see Exercises 1.5.29 and 1.5.36 as well as Lemma 1.5.5 below).

Now let's look at some simple examples of (1.282).

Example 1.5.1. $P(3, 1)$ corresponds to the group generated by w_1, w_2 and the set

$$J = \{a_1/a_2, a_2/a_3\}$$

of simple roots.

The roots in Figure 1.30 which correspond to the parabolic subgroup $P(3, 1)$ are $\{a_1/a_4, a_2/a_4, a_3/a_4\}$. The complement of this set is the set formed from roots in J and products of roots in J.

Example 1.5.2. $P(2,2)$ corresponds to the group generated by w_1 and w_3 and the set

$$J = \{a_1/a_2, a_3/a_4\}$$

of simple roots.

Now the roots in Figure 1.30 which correspond to $P(2,2)$ form the set $\{a_1/a_3, a_1/a_4, a_2/a_3, a_2/a_4\}$ which is the complement of J.

Now we want to know how the Eisenstein series $E_{2,2}$ behaves as Y approaches the boundary in the direction corresponding to the parabolic subgroup $P(2,2)$. We use the notation:

$$A_i = \begin{pmatrix} a_i & 0 \\ 0 & a_{i+1} \end{pmatrix} \begin{pmatrix} 1 & x_i \\ 0 & 1 \end{pmatrix}, \quad i = 1, 3;$$

$$Y = I \left[\begin{pmatrix} A_1 & 0 \\ 0 & A_3 \end{pmatrix} \begin{pmatrix} I_2 & X_3 \\ 0 & I_2 \end{pmatrix} \right].$$

Regard again the Eisenstein series $E_{2,2}(f_1, f_2 | Y)$ in (1.280) with $f_i(W) = |W|^{r_i} v_i(W^0)$, $v_i \in \mathcal{A}^0 (GL(n_i, \mathbb{Z}), \eta_i)$. It will be shown that for $\mathrm{Re}\,(r_1 - r_2) > 2$, there is the following asymptotic formula:

$$E_{2,2}(f_1, f_2 | Y) \sim f_1(A_1) f_2(A_3), \quad \text{as } A_1 A_3^{-1} \to 0. \tag{1.285}$$

Here "$A_1 A_3^{-1} \to 0$" means that a_1/a_3, a_1/a_4, a_2/a_3, a_2/a_4 all approach zero.

To prove (1.285), note that upon setting

$$a = \begin{pmatrix} A_1 & 0 \\ 0 & A_3 \end{pmatrix}, \quad n' = \begin{pmatrix} I_2 & X_3 \\ 0 & I_2 \end{pmatrix},$$

$$\gamma = \begin{pmatrix} \gamma_{11} & \gamma_{12} \\ \gamma_{21} & \gamma_{22} \end{pmatrix}, \quad \gamma_{ij} \in \mathbb{Z}^{2 \times 2},$$

then $Y = I[an]$ and

$$Y[\gamma] = I[an\gamma] = I[ana^{-1} a \gamma a^{-1} a]$$

$$= I \left[\begin{pmatrix} I_2 & A_1 X_3 A_3^{-1} \\ 0 & I_2 \end{pmatrix} \begin{pmatrix} A_1 \gamma_{11} A_1^{-1} & A_1 \gamma_{12} A_3^{-1} \\ A_3 \gamma_{21} A_1^{-1} & A_3 \gamma_{22} A_3^{-1} \end{pmatrix} a \right].$$

The first matrix in the last bracket approaches the identity as $A_1 A_3^{-1}$ approaches zero (see the discussion following for more details). In order for the 2nd matrix in brackets to remain finite (as well as the corresponding term in the Eisenstein series), the block entry γ_{21} must vanish. Thus γ must be in $P(2,2)$. So we must be looking at a term of the Eisenstein series which corresponds to the coset of the identity matrix. To see this in more detail, note that

$$A_1 X_3 A_3^{-1} = \begin{pmatrix} 1 & 0 \\ x_1 & 1 \end{pmatrix} \begin{pmatrix} a_1 & 0 \\ 0 & a_2 \end{pmatrix} X_4 \begin{pmatrix} a_3^{-1} & 0 \\ 0 & a_4^{-1} \end{pmatrix} \begin{pmatrix} 1 & 0 \\ -x_3 & 1 \end{pmatrix},$$

where

$$X_4 = \begin{pmatrix} u & v \\ y & z \end{pmatrix} = \begin{pmatrix} 1 & x_1 \\ 0 & 1 \end{pmatrix} X_3 \begin{pmatrix} 1 & -x_3 \\ 0 & 1 \end{pmatrix}.$$

It follows that

$$A_1 X_3 A_3^{-1} = \begin{pmatrix} 1 & 0 \\ x_1 & 1 \end{pmatrix} \begin{pmatrix} a_1 u a_3^{-1} & a_1 v a_4^{-1} \\ a_2 y a_3^{-1} & a_2 z a_4^{-1} \end{pmatrix} \begin{pmatrix} 1 & 0 \\ -x_3 & 1 \end{pmatrix}.$$

This certainly approaches zero as a_1/a_3, a_1/a_4, a_2/a_3, a_2/a_4 all approach zero. Similarly, each entry in $A_3 \gamma_{21} A_1^{-1}$ must approach infinity unless γ_{21} vanishes. Note that the corresponding term of the Eisenstein series involves

$$\left| I \begin{bmatrix} A_1(\gamma_{11} + X_3 \gamma_{21}) A_1^{-1} \\ A_3 \gamma_{21} A_1^{-1} \end{bmatrix} \right|^{r_2 - r_1}.$$

So we are looking at a term $|I[B] + I[C]|^s$, where $\mathrm{Re}\, s < 0$, and the eigenvalues of B remain bounded while those of C become infinite unless γ_{21} vanishes. Therefore this term goes to zero unless γ_{21} is zero and then $\gamma = I$. So we obtain (1.285).

To complete the discussion of $E_{2,2}$, one should note that if we let Y approach the boundary in a direction dictated by some other parabolic subgroup such as $P(3, 1)$, then we will see it approach zero, under the hypothesis that v_2 is a cusp form. For example, consider the behavior of $E_{2,2}$ as Y approaches the boundary in the direction corresponding to $P(3, 1)$; i.e., as $a_i/a_4 \to 0$, for $i = 1, 2, 3$. Note that we can let a_i/a_4 approach zero so that the quotients:

$$\frac{a_1}{a_3} = \frac{a_1/a_4}{a_3/a_4} \quad \text{and} \quad \frac{a_2}{a_3} = \frac{a_2/a_4}{a_3/a_4}$$

both approach zero. This puts us in the situation that we just encountered, except that, in addition, we are letting a_3/a_4 approach zero. Now if v_2 is a cusp form

$$v_2(A_3^0) \to 0 \quad \text{as } a_3/a_4 \to 0,$$

because

$$A_3^0 = \begin{pmatrix} \sqrt{a_3/a_4} & 0 \\ 0 & \sqrt{a_4/a_3} \end{pmatrix} \begin{pmatrix} 1 & x_2 \\ 0 & 1 \end{pmatrix}.$$

The usual discussions of the asymptotics of Eisenstein series involve the computation of integrals like the following integral which represents the zeroth Fourier coefficient of $E_{m,n-m}$ with respect to $P(m, n-m)$:

$$\int\limits_{X \in (\mathbb{R}/\mathbb{Z})^{m \times (n-m)}} E_{m,n-m} \left(f \, \middle| \, Y \begin{bmatrix} I & X \\ 0 & I \end{bmatrix} \right) dX.$$

See, for example, Langlands [392] and Goldfeld [230], plus Section 1.5.4. The general result obtained by Langlands says that the constant term for E_P involves a sum over the Weyl group W_J associated with $P = P_J$ by (1.282). The terms in this sum are of the sort obtained in (1.285) multiplied by factors appearing in the functional equations of the Eisenstein series. In order to connect the two methods, one must understand the asymptotic behavior of the terms in the Fourier expansions of the Eisenstein series.

It would also be interesting to clarify the connection between the various notions of approach to the boundary of the fundamental domain and the compactifications of the fundamental domain that have been considered by authors such as Satake [532], Baily and Borel [33], Borel and Serre [69], and Ash et al. [30]. See also Freitag [185] and Goresky [236]. To create a compactification, Satake adjoins to a fundamental domain $\mathcal{F}_n$ for $GL(n, \mathbb{Z})$ a union of lower rank fundamental domains $\mathcal{F}_m$ and defines approach of $Y \in \mathcal{F}_n$ to an element V^* of the lower rank fundamental domain $\mathcal{F}_m$ if

$$Y = \begin{pmatrix} V & 0 \\ 0 & W \end{pmatrix} \begin{bmatrix} I & X \\ 0 & I \end{bmatrix}, \quad V \in \mathcal{P}_m, \ W \in \mathcal{P}_{n-m}, \ X \in \mathbb{R}^{m \times (n-m)}, \qquad (1.286)$$

and V approaches V^* while W approaches infinity in the sense that the diagonal part in the Iwasawa decomposition of W goes to infinity.

It would also be useful to clarify the connections with the truncations used by Langlands and Arthur [25] in the continuous spectrum integrals of Eisenstein series appearing in the trace formula. See Section 1.5.4.

We will discuss several methods of obtaining Fourier expansions. The first method comes from Terras [595, 606] and goes back to the paper of Chowla and Selberg [107] which makes use of theta functions. The other methods are analogous to that used by Siegel [565, Vol. II, pp. 97–137] for $Sp(n, \mathbb{Z})$ and come from Kaori Imai (Ota) and Terras [318].

First note that, for any Maass form $f \in \mathcal{A}(GL(n, \mathbb{Z}), \lambda)$, the function $f(Y)$ must be a periodic function of each entry of the matrix X when Y has the partial Iwasawa decomposition (1.286) above. Therefore $f(Y)$ must have a **Fourier expansion**:

$$f(Y) = \sum_{N \in \mathbb{Z}^{n \times (n-m)}} c_N(V, W) \, \exp\{2\pi i \, \mathrm{Tr}(^t NX)\}. \tag{1.287}$$

Many authors have obtained such expansions. The case that $|Y| = 1$, $n = 2$, is considered in Exercise 3.5.3, Volume I. Here we are mainly motivated by the work of Siegel [565, Vol. II, pp. 97–137] and Maass [426] for $Sp(n, \mathbb{Z})$. See also Freitag [185].

The terms $c_N(V, W)$ in the Fourier expansion (1.287) always have two parts. For Eisenstein series, one part is **arithmetic**—either a singular series (for an Eisenstein series defined as a sum over Γ_n/P) or a divisor-type function (for an Eisenstein series defined as a sum over rank m matrices in $\mathbb{Z}^{n \times m}/\Gamma_m$). When $n = 2$, the arithmetic part of the kth Fourier coefficient of the Eisenstein series is:

$$k^s \sum_{0 < t | k} t^{1-2s} = k^s \sigma_{1-2s}(k), \tag{1.288}$$

where σ_s denotes the divisor function. The **singular series** version of this is:

$$\sum_{\substack{c > 0, d \bmod c \\ \gcd(d,c)=1}} c^{-2s} \exp(2\pi i k d/c) = \sigma_{1-2s}(k)/\zeta(2s). \tag{1.289}$$

We found in Terras [596] (see Exercise 1.5.36) that the arithmetic part of the term corresponding to $N \in \mathbb{Z}^{m \times (n-m)}$ of rank m in $E_{m,n-m}(1, s|Y)$ is essentially the **singular series**:

$$\sum_{R \in (\mathbb{Q}/\mathbb{Z})^{m \times m}} \nu(R)^{-2s} \exp\left\{2\pi i \, \mathrm{Tr}\left(^t R \, N\right)\right\}. \tag{1.290}$$

Here $\nu(R)$ is the product of the reduced denominators of the elementary divisors of R. Siegel [565, Vol. II, pp. 97–137] obtains an analogous result for holomorphic Eisenstein series for $Sp(n, \mathbb{Z})$. Maass [426, Section 18] does the non-holomorphic case for $Sp(n, \mathbb{Z})$. Lower rank terms are more complicated to describe. For example, the "most singular" term listed in Maass [426, p. 307] required quite a long computation. We will find here that the arithmetic part of Fourier expansions of $E(\varphi, s|Y)$ cannot be separated out so easily for general $\varphi \in \mathcal{A}^0(\Gamma_m, \eta)$.

The terms in the Fourier expansion (1.287) of a Maass form $f \in \mathcal{A}^0(\Gamma_n, \lambda)$ will have a second part which is **analytic**—a matrix argument confluent hypergeometric function. For $GL(n, \mathbb{Z})$ one obtains analogues either of K-Bessel or Whittaker functions. We work mostly with K-Bessel functions because we are attempting to stay close to the Siegel-type Fourier expansions. We should caution the reader that most researchers use a slightly different formulation (see Jacquet [323], Bump [83], Goldfeld [230], Proskurin [494]). We will discuss the connections between the K-Bessel function expansions and the Whittaker function expansions in Section 1.5.4. See also Grenier [243].

As we mentioned, the Fourier expansion (1.287) is analogous to that for Siegel modular forms. Koecher [359] proved that in the definition of holomorphic Siegel modular forms for $Sp(n, \mathbb{Z})$, one needs no hypothesis on the behavior of the form at infinity when $n \geq 2$ to rule out terms $\exp\{-\text{Tr}(NY)\}$, where N is not nonnegative. See also Maass [426, pp. 185–187] and Chapter 2. It would be interesting to know whether a similar phenomenon occurs for $GL(n, \mathbb{Z})$—ruling out the analogues of I-Bessel functions in Fourier expansions (1.287) without explicitly assuming the third hypothesis in the definition (1.242) of Maass form for $GL(n, \mathbb{Z})$.

I. Method of Chowla and Selberg

This method is a generalization of that of Chowla and Selberg [107]. See Volume I, p. 263. In Proposition 1.5.2 we consider the method for the special case of Epstein's zeta function. Then in Theorem 1.5.3 we consider only the terms of maximal rank in Fourier expansions of Eisenstein series $E_{m,n-m}$ with respect to the parabolic subgroup $P(m, n - m)$.

Proposition 1.5.2 (Fourier Expansions of Epstein's Zeta Function). *Let* $Z_{1,n-1}(1, s|Y)$ *be Epstein's zeta function, defined for* $\text{Re } s > n/2$ *and* $Y \in \mathcal{P}_n$ *by:*

$$Z_{1,n-1}(1, s|Y) = \frac{1}{2} \sum_{a \in \mathbb{Z}^n - 0} Y[a]^{-s}.$$

Here we use the notation (1.277) in Section 1.5.2 rather than (1.173) of Section 1.4.1. If Y has the partial Iwasawa decomposition (1.286), then we have the following Fourier expansion of the normalized Eisenstein series:

$$\begin{aligned}
\Lambda_{1,n-1}(1, s|Y) &\doteq 2\pi^{-s}\Gamma(s)Z_{1,n-1}(1, s|Y) \\
&= \Lambda_{1,m-1}(1, s|V) + \frac{1}{\sqrt{|V|}}\Lambda_{1,n-m-1}\left(1, s - \frac{m}{2} \mid W\right) \\
&+ \frac{2}{\sqrt{|V|}} \sum_{\substack{0 \neq b \in \mathbb{Z}^m \\ 0 \neq c \in \mathbb{Z}^{n-m}}} e^{2\pi i \, {}^t bXc} \left(\frac{V^{-1}[b]}{W[c]}\right)^{\frac{2s-m}{4}} K_{s-\frac{m}{2}}\left(2\pi \sqrt{V^{-1}[b]W[c]}\right).
\end{aligned}$$

Here $K_s(y)$ is the usual K-Bessel function defined by the formula in Exercise 3.2.1 on p. 166 of Volume I.

Proof (Compare Epstein [169], Terras [595]). We know that (as in part (a) of Exercise 1.4.7 on p. 65 of Volume I):

$$\Lambda_{1,n-1}(1, s|Y) = \int_{t>0} t^{s-1}\theta_1(Y, t) \, dt,$$

where

$$\theta_1(Y, t) = \sum_{0 \neq a \in \mathbb{Z}^n} \exp\{-\pi Y[a]t\}.$$

Write

$$a = \begin{pmatrix} b \\ c \end{pmatrix}, \quad \text{with } b \in \mathbb{Z}^m.$$

Then $Y[a] = V[b + Xc] + W[c]$. We can split the sum defining θ_1 into two parts—that consisting of terms with $c = 0$ and the rest. The part of the sum with $c = 0$ gives $\Lambda_{1,m-1}(1, s|V)$. The rest of the terms have $c \neq 0$ and thus b is summed over all of $\mathbb{Z}^m$ and we can apply Poisson summation to obtain:

$$\sum_{b \in \mathbb{Z}^m} \exp\{-\pi V[b + Xc]t\} = |V|^{-1/2} t^{-m/2} \sum_{b \in \mathbb{Z}^m} \exp\left\{2\pi i\,{}^t bXc - \pi V^{-1}[b]t^{-1}\right\}.$$

Substitute this into our integral and obtain the result, since the $b = 0$ term gives

$$|V|^{-1/2} \Lambda_{1,n-m-1}(1, s - m/2|W)$$

and the rest of the terms come from summing over nonzero b and c:

$$|V|^{-1/2} \exp\left\{2\pi i\,{}^t bXc\right\} \int_{t>0} t^{s-1-m/2} \exp\left\{-\pi(V^{-1}[b]t^{-1} + W[c]t)\right\}\, dt.$$

The integral is easily evaluated in terms of the ordinary K-Bessel function to complete the proof. ∎

Exercise 1.5.27. (a) Prove that if Y has the partial Iwasawa decomposition (1.286) with $v > 0$, $W \in \mathcal{P}_{n-1}$, $X \in \mathbb{R}^{1 \times (n-1)}$ in the case $m = 1$, and if $\operatorname{Re} s > n/2$, then

$$E_{1,n-1}(1, s|Y) \sim v^{-s},$$

as W approaches infinity in the sense that the diagonal elements in the Iwasawa decomposition of W all approach infinity.

(b) Use Proposition 1.5.2 to deduce the functional equation of Epstein's zeta function.

(c) Use Proposition 1.5.2 to find an analogue of the Kronecker limit formula from Exercise 3.5.6 on p. 264 of Volume I.

Hint. See Terras [595] and Efrat [153].

Theorem 1.5.3 (Highest Rank Terms in the Fourier Expansion of an Eisenstein Series). *Consider the Eisenstein series $E_{m,n-m}(\varphi, s|Y)$ defined in (1.245)* *of* Section 1.5.1 *in the special case that φ is Selberg's Eisenstein series; i.e.,* $\varphi(W) = E_{(m)}(r|W)$, $r \in \mathbb{C}^{n-1}$, *as defined in (1.249) of Section 1.5.1. Suppose that $m < n/2$. Then when $N \in \mathbb{Z}^{m \times m}$ is non-singular, the Nth Fourier coefficient in the expansion (1.287) of the normalized Eisenstein series*

$$\Lambda(\varphi, s|Y) = 2\pi^{-ms}\Gamma_m(r(\varphi, s))L_{\varphi^*}(2s)E_{m,n-m}(\varphi^*, s|Y)$$

as in (1.271) and Exercise 1.5.19 of Section 1.5.2 is:

$$2|V|^{-m/2} \sum_{\substack{D|Z \\ D \in \mathbb{Z}^{m \times m}/\Gamma_m}} p_{-r}(I[D]) \sum_{a \in P_{(m)} \backslash \Gamma_m} K_m\left(-r, s - m/2 \left|\pi W\left[\,{}^t N\,{}^t a\right], \pi V^{-1}\left[a^{-1}\right]\right.\right),$$

where "$D|N$" means that there is a matrix C in $\mathbb{Z}^{(n-m)\times m}$ such that $N = D\,{}^t C$. This says that each elementary divisor in D divides the corresponding elementary divisor in N.

Proof (Terras [606]). The proof will be similar to that of Proposition 1.5.2. Begin with formulas (1.270) and (1.271) for the normalized Eisenstein series:

$$\Lambda(f, s|Y) = \int_{H \in \mathcal{M}_m} \varphi(H^0)|H|^s \theta_m(Y, H)\, d\mu_m(H). \qquad (1.291)$$

Since we have the partial Iwasawa decomposition (1.286), we find that the partial theta function can be written in the following way for $B \in \mathbb{Z}^{m \times m}$:

$$\theta_m(Y, H) = \sum_{\left(\begin{smallmatrix} B \\ C \end{smallmatrix}\right) \in \mathbb{Z}^{n \times m},\ \text{rank } m} \exp\left\{-\pi \operatorname{Tr}\left(Y\begin{bmatrix} B \\ C \end{bmatrix}H\right)\right\}$$

$$= \sum_{\left(\begin{smallmatrix} B \\ C \end{smallmatrix}\right) \in \mathbb{Z}^{n \times m},\ \text{rank } m} \exp\{-\pi \operatorname{Tr}(V[B + XC]H + W[C]H)\}.$$

The terms of maximal rank in the Fourier expansion of $E_{m,n-m}$ correspond to the matrices C of rank m in $\mathbb{Z}^{(n-m)\times m}$. We shall not consider the other terms here. For such terms B is summed over the full lattice $\mathbb{Z}^{m \times m}$ and we can use Poisson summation on B. This leads to the following formula for the terms with C of rank m in $\theta_m(Y, H)$:

$$|H|^{-m/2}|V|^{-m/2} \sum_{B \in \mathbb{Z}^{m \times m}} \exp\left\{2\pi i \operatorname{Tr}\left({}^t BXC\right) - \pi \operatorname{Tr}\left(V^{-1}[B]H^{-1}\right)\right\}.$$

Substitute this into (1.291) to find that the terms with C of rank m are:

$$|V|^{-m/2} \sum_{\substack{B \in \mathbb{Z}^{m \times m} \text{ rank } m/\Gamma_m \\ C \in \mathbb{Z}^{(n-m) \times m} \text{ rank } m/\Gamma_m}} \exp\left\{2\pi i \operatorname{Tr}\left(C\,{}^t BX\right)\right\} I\left(\varphi, s - \frac{m}{2} \,\middle|\, \pi W[C],\ \pi V^{-1}[B]\right),$$

(1.292)

where for $F, G \in \mathcal{P}_m$, $s \in \mathbb{C}$:

$$I(\varphi, s | F, G) = \int_{H \in \mathcal{P}_m} |H|^s\, \varphi\left(H^0\right) \exp\left\{-\operatorname{Tr}(FH + GH^{-1})\right\}\, d\mu_m(H).$$

(1.293)

From Lemma 1.5.4 below, we see that (1.292) is equal to

$$2|V|^{-m/2} \sum_{\substack{B \in \mathbb{Z}^{m \times m} \text{ rank } m/\Gamma_m \\ C \in \mathbb{Z}^{(n-m) \times m} \text{ rank } m}} \exp\left\{2\pi i \operatorname{Tr}\left(C\,{}^t BX\right)\right\}$$

$$\times \sum_{u \in P_{(m)} \backslash \Gamma_m} K_m\left(-r, s - \frac{m}{2} \,\middle|\, \pi W\left[C\,{}^t u\right], \pi V^{-1}[Bu^{-1}]\right).$$

To finish this proof, we need to make use of another kind of Hecke operator to move the B around to be next to C in

$$I\left(\varphi, s\ \middle|\ \pi W[C], \pi V^{-1}[B]\right).$$

These Hecke operators are associated with a matrix $N \in \mathbb{Z}^{m \times (n-m)}$ and defined for $f : \mathcal{P}_m \to \mathbb{C}$ by;

$$T_N f(Y) = \sum_{\substack{B \in \mathbb{Z}^{m \times m}/\Gamma_m \\ N = B\,{}^t C, \text{ for some} \\ C \in \mathbb{Z}^{(n-m) \times m}}} f(Y[B]).$$

(1.294)

Then we have exercise 1.5.28 below to finish the proof of Theorem 1.5.3.   ∎

Lemma 1.5.4. *Suppose $A, B \in \mathcal{P}_m$, $r \in \mathbb{C}^m$, $\operatorname{Re} r_i > 1$. If $E_{(m)}(r|Y)$ is the Eisenstein series (1.247), in Section 1.5.1, we have the following expression for the integral from formula (1.293) above involving K-Bessel functions from formula (1.61) of Section 1.2.2:*

$$I\left(E_{(m)}(r|*), s\,\middle|\,F, G\right) = \int_{H \in \mathcal{P}_m} |H|^s\, E_{(m)}(r|H)\, \exp\left\{-\mathrm{Tr}\left(FH + GH^{-1}\right)\right\}\, d\mu_m(H)$$

$$= 2 \sum_{u \in P_{(m)} \backslash \Gamma_m} K_m\left(-r + (0, s)\,\middle|\,F\left[{}^t u\right], G\left[u^{-1}\right]\right).$$

Proof. Let N be the nilpotent subgroup of $GL(n, \mathbb{R})$ consisting of all upper triangular matrices with ones on the diagonal. We find that

$$2 \int_{H \in \mathcal{P}_m} E_{(m)}(r|H)\, \exp\left\{-\mathrm{Tr}(FH + GH^{-1})\right\}\, d\mu_m(H)$$

$$= \int_{Y \in \mathcal{P}_m / \Gamma_m} \sum_{\tau \in \Gamma_m / \Gamma_m \cap N} p_{-r}(Y[\tau]) \sum_{\gamma \in \Gamma_m} \exp\left\{-\mathrm{Tr}\left(FY[\gamma] + G(Y[\gamma])^{-1}\right)\right\}\, d\mu_m(Y)$$

$$= \int_{Y \in \mathcal{P}_m / N} p_{-r}(Y) \int_{n \in N / N \cap \Gamma_m} \sum_{\gamma \in \Gamma_m} \exp\left\{-\mathrm{Tr}\left(FY[n\gamma] + G(Y[n\gamma])^{-1}\right)\right\}\, dn\, d\mu_m(Y).$$

Here the factor of "2" comes from the order of the center of Γ_m. On the other hand,

$$K_m(-r|A, B) = \int_{Y \in \mathcal{P}_m / N} p_{-r}(Y) \int_{n \in N} \exp\left\{-\mathrm{Tr}\,(AY[n] + B(Y[n])^{-1}\right\}\, dn\, d\mu_m(Y).$$

Thus it suffices to show the easily verified identity:

$$\int_{n \in N / N \cap \Gamma_m} \sum_{\gamma \in \Gamma_m} \exp\left\{-\mathrm{Tr}\,(FY[n\gamma] + G(Y[n\gamma])^{-1})\right\}\, dn$$

$$= \sum_{u \in \Gamma_m \cap N \backslash \Gamma_m} \int_{n \in N} \exp\left\{-\mathrm{Tr}\left(F\left[{}^t u\right] Y[n] + G\left[u^{-1}\right] (Y[n])^{-1}\right)\right\}\, dn.$$

This completes the proof of the lemma.   ∎

Exercise 1.5.28. (a) Show that if T_N is as in formula (1.294), for $N \in \mathbb{Z}^{m \times (n-m)}$ of rank m (assuming that $m > n - m$),

$$T_N E_{(m)}(r|Y) = a_N(r) E_{(m)}(r|Y),$$

where

$$a_N(r) = \sum_{\substack{D|N \\ D \in \mathbb{Z}^{m \times m} / \Gamma_m}} p_{-r}(I[D]).$$

Here $D|N$ means that there is a matrix $C \in \mathbb{Z}^{(n-m) \times m}$ such that $N = D^{\,t}C$.
(b) Use part (a) to finish the proof of Theorem 1.5.3.

It would be interesting to use similar methods to find the lower rank Fourier coefficients and to deal with the case that φ is a cusp form rather than an Eisenstein series, but we shall not do that here.

Now let us consider other methods for finding Fourier expansions of Eisenstein series.

II. Methods of Imai and Terras [318] and Terras [596]

The prerequisite for Fourier expansions of Eisenstein series using the method of Siegel (see Baily [32, pp. 228–240], Maass [426, pp. 300–308], Siegel [565, Vol. II, pp. 97–137], and Terras [596]) is the Bruhat decomposition or some related matrix decomposition.

Exercise 1.5.29 (The Bruhat Decomposition of $SL(n, \mathbb{Q})$ with Respect to $P(n-1, 1)$).
  Let $P(n - 1, 1) = P \subset G = SL(n, \mathbb{Q})$ be defined as in (1.241) of Section 1.4.4 by:

$$P = \left\{ \begin{pmatrix} A & b \\ 0 & c \end{pmatrix} \in G \,\middle|\, A \in GL(n-1, \mathbb{Q}),\ b \in \mathbb{Q}^{n-1},\ c \in \mathbb{Q} - 0 \right\}.$$

Show that we have the following disjoint union:

$$SL(n, \mathbb{Q}) = P \cup (P \sigma P),$$

where

$$\sigma = \begin{pmatrix} 0 & 0 & 1 \\ 0 & I_{n-2} & 0 \\ -1 & 0 & 0 \end{pmatrix}.$$

Hint. Note that a matrix

$$\begin{pmatrix} E & F \\ g & h \end{pmatrix} \in SL(n, \mathbb{Q}) \quad \text{with} \quad E \in \mathbb{Q}^{(n-1) \times (n-1)},\ h \in \mathbb{Q},$$

lies in $P\sigma P$ if and only if the rank of g is one.

The result of Exercise 1.5.29 should be compared with the general Bruhat decomposition (1.284). Next let us consider another view of the Weyl group W of all permutation matrices in K. Define the following subgroups of $K = O(n)$:

$$M = \text{the centralizer of } A \text{ in } K = \left\{ k \in K \mid kak^{-1} = a, \quad \text{for all } a \in A \right\},$$
$$M' = \text{the normalizer of } A \text{ in } K = \left\{ k \in K \mid kak^{-1} \in A, \text{ for all } a \in A \right\}. \tag{1.295}$$

Here A is the group of all positive diagonal $n \times n$ matrices, as usual. The **Weyl group** W of $GL(n)$ can then be defined to be

$$W = M'/M. \tag{1.296}$$

We can identify W with the group of permutations of n elements, as the following exercise shows.

Exercise 1.5.30. (a) Show that M defined by (1.295) consists of all diagonal matrices with entries ± 1 (see the definition of the boundary of G/K in (1.19) of Section 1.1.4).
(b) Show that M' in (1.295) consists of all matrices such that each row or column has exactly one nonzero entry equal to ± 1.
(c) Show that the Weyl group of $GL(n, \mathbb{R})$ can be identified as the group of permutations of n elements.

Exercise 1.5.31. Show that formula (1.284) agrees with the decomposition in Exercise 1.5.29 when $n = 3$.
Hint. When $P = P(2,1)$, what is J? You need to look at

$$\begin{pmatrix} b_1 & b_2 & b_3 \\ 0 & b_4 & b_5 \\ 0 & 0 & b_6 \end{pmatrix} \begin{pmatrix} g_1 & g_2 & g_3 \\ h_1 & h_2 & h_3 \\ 0 & 0 & j_3 \end{pmatrix}.$$

To put this matrix into $B = P(1,1,1)$, the Borel or minimal parabolic subgroup, requires the ability to interchange the rows of g's with the row's of h's. So J corresponds to the permutation matrix w_1 in the notation (1.283).

What is $W_J \backslash W / W_J$? Clearly representatives are the permutation matrices corresponding to the identity and the transposition (13).

Lemma 1.5.5 (The Bruhat Decomposition for the Minimal Parabolic or Borel Subgroup $B = P(1, \ldots, 1)$).
Let k denote any field. Then we have the disjoint union:

$$GL(n, k) = \bigcup_{w \in W} BwB,$$

where W is the Weyl group of all permutation matrices.

Proof. Note that given $g \in GL(n, k)$, there is an element $b \in B$ such that the first nonzero entry in each row of the matrix bg must occur in a different position for each row. Thus we can find an element $w \in W$ to put the rows of wbg in the correct order to form an element of B. ∎

Exercise 1.5.32.   Use Lemma 1.5.5 to obtain part of Lemma 1.3.2 of Section 1.3.3 which says that $G = SL(n, \mathbb{R})$ has the form:

$$G = (\overline{N}B) \cup (\text{something lower dimensional}),$$

where $\overline{N}$ denotes the subgroup of all lower triangular matrices with ones on the diagonal and B is the Borel or minimal parabolic subgroup of all upper triangular matrices in G.

Hint. Note that

$$\begin{pmatrix} 0 & \cdots & 1 \\ & & \cdot \\ \vdots & \cdot & \vdots \\ & \cdot & \\ 1 & \cdots & 0 \end{pmatrix} N \begin{pmatrix} 0 & \cdots & 1 \\ & & \cdot \\ \vdots & \cdot & \vdots \\ & \cdot & \\ 1 & \cdots & 0 \end{pmatrix} = \overline{N}.$$

Lemma 1.5.6 (Coset Representatives à la Bruhat). *Suppose that $P = P(n-1, 1)$. The cosets in $SL(n, \mathbb{Z})/P$ can be represented by*

$$S_1^* \cup S_2^*,$$

where $S_1^* = \{I\}$ and

$$S_2^* = \left\{ \begin{pmatrix} {}^t A^{-1} & 0 \\ 0 & 1 \end{pmatrix} n_q \sigma p_q \,\middle|\, \begin{array}{l} A \in SL(n-1, \mathbb{Z})/P(1, n-2) \\ q = e/f,\ f \geq 1,\ \gcd(e,f) = 1 \\ e, f \in \mathbb{Z} \end{array} \right\},$$

with

$$n_q = \begin{pmatrix} 1 & 0 & q \\ 0 & I_{n-2} & 0 \\ 0 & 0 & 1 \end{pmatrix}, \quad \sigma = \begin{pmatrix} 0 & 0 & 1 \\ 0 & I_{n-2} & 0 \\ -1 & 0 & 0 \end{pmatrix}, \quad p_q = \begin{pmatrix} f & 0 & g \\ 0 & I_{n-2} & 0 \\ 0 & 0 & 1/f \end{pmatrix},$$

if $eg \equiv 1 \pmod{f}$, $0 \leq g < f$.

Proof (Kaori Imai (Ota)). The general idea is to use the method of Baily [32] and Terras [596]. This requires Exercises 1.5.29 and 1.5.33. We shall write $P_{\mathbb{Q}}$ when we wish to consider the parabolic subgroup in $SL(n, \mathbb{Q})$ and $P_{\mathbb{Z}}$ when we want the parabolic subgroup of $SL(n, \mathbb{Z})$. Define

$$T : P_{\mathbb{Q}} \rightarrow (SL(n, \mathbb{Z}) \cap P_{\mathbb{Q}}\sigma P_{\mathbb{Q}})/P_{\mathbb{Z}},$$

by $T(p) = p\sigma p' (\mathrm{mod}\ P_{\mathbb{Z}})$, for $p \in P_{\mathbb{Q}}$. Here p' is chosen in $P_{\mathbb{Q}}$ to put $p\sigma p'$ in $SL(n, \mathbb{Z})$. This is possible by Exercise 1.5.33.

Then matrix multiplication shows $\sigma p = p'\sigma$ is equivalent to:

$$p = \begin{pmatrix} a & 0 & 0 \\ c & D & e \\ 0 & 0 & g \end{pmatrix}, \quad \text{with } D \in \mathbb{Q}^{(n-2) \times (n-2)}.$$

So define

$$P_{\mathbb{Q}}^* = \left\{ p = \begin{pmatrix} a & 0 & 0 \\ c & D & 0 \\ 0 & 0 & g \end{pmatrix} \Bigg| \ p \in SL(n, \mathbb{Q}),\ D \in \mathbb{Q}^{(n-2) \times (n-2)} \right\}.$$

Finding the coset representatives for $SL(n, \mathbb{Z})/P_{\mathbb{Z}}$ is the same as reducing $p \in P_{\mathbb{Q}}$ modulo $P_{\mathbb{Q}}^*$. Representatives for $P_{\mathbb{Q}}/P_{\mathbb{Q}}^*$ are:

$$\begin{pmatrix} {}^t A^{-1} & {}^t A^{-1} c \\ 0 & 1 \end{pmatrix}, \quad A \in SL(n-1, \mathbb{Z})/P(1, n-1)_{\mathbb{Z}}, \quad c = \begin{pmatrix} q \\ 0 \end{pmatrix}, \ q \in \mathbb{Q}.$$

The equality $p\sigma p' = p_1 \sigma p_1'$ with $p, p_1, p', p_1' \in P_{\mathbb{Q}}$ implies $p_1^{-1}p \in P_{\mathbb{Q}}^*$. Thus if

$$p = \begin{pmatrix} {}^t A^{-1} & 0 \\ 0 & 1 \end{pmatrix} n_q,$$

then $T(p) = p\sigma p'$ gives a complete set of representatives for

$$SL(n, \mathbb{Z}) \cap (P_{\mathbb{Q}}\sigma P_{\mathbb{Q}})/P_{\mathbb{Z}}.$$

Finally it must be proved that if $q = e/f$, $f > 1$, $\gcd(e,f) = 1$, then

$$p' = \begin{pmatrix} f & 0 & g \\ 0 & I_{n-2} & 0 \\ 0 & 0 & 1/f \end{pmatrix}, \quad \text{with}\quad eg \equiv 1 (\mathrm{mod}\ f).$$

To see this, write

$${}^t A^{-1} = \begin{pmatrix} a & b \\ c & D \end{pmatrix}, \quad D \in \mathbb{Z}^{(n-2) \times (n-2)}.$$

Then

$$
\begin{pmatrix} {}^t A^{-1} & 0 \\ 0 & 1 \end{pmatrix}
\begin{pmatrix} 1 & 0 & q \\ 0 & I_{n-2} & 0 \\ 0 & 0 & 1 \end{pmatrix}
\begin{pmatrix} 0 & 0 & 1 \\ 0 & I_{n-2} & 0 \\ -1 & 0 & 0 \end{pmatrix} p'
$$

$$
= \begin{pmatrix} -ae & b & a(-qg + 1/f) \\ -ce & d & c(-qg + 1/f) \\ -f & 0 & -g \end{pmatrix}.
$$

Clearly the matrix on the right lies in $SL(n, \mathbb{Z})$, assuming that $q = e/f$ is as stated in the lemma. According to Exercise 1.5.29, the proof of Lemma 1.5.6 is now complete.   ∎

Exercise 1.5.33. Show that $SL(n, \mathbb{Q}) = SL(n, \mathbb{Z}) B_{\mathbb{Q}}$, where $B_{\mathbb{Q}} = P(1, 1, \ldots, 1)_{\mathbb{Q}}$ is the minimal parabolic or Borel subgroup of upper triangular matrices in $SL(n, \mathbb{Q})$.
Hint. Use induction. First observe that for $A \in SL(n, \mathbb{Q})$ there is a matrix $U \in SL(n, \mathbb{Z})$ such that the first $n - 1$ elements in the last row of UA vanish. For $n - 1$ homogeneous linear equations in n unknowns with rational coefficients have relatively prime integral solutions. Recall Lemma 1.4.2 of Section 1.4.2.

Kaori Imai (Ota) used Lemma 1.5.6 to give the Fourier expansion of $E_{2,1}(f, s|Y)$ by a method like the third method of Exercise 3.5.4, Vol. I for the Eisenstein series in the case of $GL(2, \mathbb{Z})$. Now we want to consider a decomposition that leads to an analogue of the second method in that same exercise from Volume I.

Exercise 1.5.34 (Coset Representatives sans[7] Bruhat). Show that the cosets of $\mathbb{Z}^{n \times (n-1)}$ rank $(n-1)/GL(n-1, \mathbb{Z})$ can be represented by $S_1 \cup S_2$, where

$$
S_1 = \left\{ \begin{pmatrix} B \\ 0 \end{pmatrix} \,\middle|\, B \in \mathbb{Z}^{(n-1)\times(n-1)} \text{ rank } (n-1)/GL(n-1, \mathbb{Z}) \right\},
$$

$$
S_2 = \left\{ \begin{pmatrix} B \\ {}^t c \end{pmatrix} \,\middle|\, \begin{array}{l} B = HD, \ H \in GL(n-1, \mathbb{Z}) / {}^t P, \\ d_j > 0, \ j = 2, \ldots, n \\ d_{ij} \bmod d_j, \ j = 2, \ldots, n, \\ {}^t c = (c_1 0 \cdots 0), \ c_1 > 0 \end{array} \quad D = \begin{pmatrix} d_1 & \cdots & 0 \\ \vdots & \ddots & \vdots \\ d_{ij} & \cdots & d_{n-1} \end{pmatrix} \right\}.
$$

Here ${}^t P$ denotes the lower triangular subgroup of $GL(n-1, \mathbb{Z})$. The point of the inequalities on the lower triangular matrix D in S_2 is that d_1 is an arbitrary integer while $d_2, \ldots, d_n$ are all positive integers and the d_{ij} can be taken to lie between 0 and $d_j - 1$.

---

[7] Without.

Hint. Write $A \in \mathbb{Z}^{n \times (n-1)}$ of rank $(n-1)$ as

$$A = \begin{pmatrix} B \\ {}^t c \end{pmatrix} \quad \text{with } B \text{ in } \mathbb{Z}^{(n-1) \times (n-1)} \text{ and } c \in \mathbb{Z}^{n-1}.$$

If $c = 0$, then A lies in the set of representatives S_1. Otherwise there is a matrix W in $GL(n-1, \mathbb{Z})$ such that

$$AW = \begin{pmatrix} b_1 & B_2 \\ c_1 & 0 \end{pmatrix} \quad \text{with } c_1 > 0, \ B_2 \in \mathbb{Z}^{(n-1) \times (n-2)}, \ b_1 \in \mathbb{Z}^{n-1}.$$

Moreover we can write:

$$\begin{pmatrix} b_1 & B_2 \\ c_1 & 0 \end{pmatrix} = \begin{pmatrix} b_1' & B_2' \\ c_1' & 0 \end{pmatrix} \begin{pmatrix} x & y \\ V & W \end{pmatrix} \quad \text{if and only if } x = 1 \text{ and } y = 0 \in \mathbb{Z}^{n-1}.$$

So we need to take $A = \begin{pmatrix} B \\ {}^t c \end{pmatrix}$ modulo the subgroup of $GL(n-1, \mathbb{Z})$ of matrices of the form

$$\begin{pmatrix} 1 & 0 \\ v & W \end{pmatrix}.$$

Thus W must be in $GL(n-2, \mathbb{Z})$. It must be shown that this puts A in S_2. Elementary divisor theory writes $B \in \mathbb{Z}^{(n-1) \times (n-1)}$ in the form $B = HD$ with H in $GL(n-1, \mathbb{Z})$ and

$$D = \begin{pmatrix} d_1 & 0 \\ d_{12} & D_2 \end{pmatrix},$$

with d_1 in $\mathbb{Z}$ and with a lower triangular, non-singular D_2 in $\mathbb{Z}^{(n-2) \times (n-2)}$. And we can reduce H modulo the lower triangular group ${}^t P$.

Next we use the preceding matrix decompositions to obtain some explicit Fourier expansions for Eisenstein series $E_{2,1}$ from Kaori Imai (Ota) and Terras [318]. Let $\varphi \in \mathcal{A}^0(SL(2, \mathbb{Z}), \chi)$ have the Fourier expansion:

$$\varphi(U) = \alpha_0 \, k_{1,1}(1 - r|U, 0) + \alpha_0' \, k_{1,1}(r|U, 0) + \sum_{n \neq 0} \alpha_n k_{1,1}(r|U, n), \qquad (1.297)$$

if

$$U = \begin{pmatrix} y & 0 \\ 0 & 1/y \end{pmatrix} \begin{bmatrix} 1 & 0 \\ x & 1 \end{bmatrix}, \quad y > 0, \quad x \in \mathbb{R},$$

and $k_{1,1}(r|U, a)$ denotes the k-Bessel function defined in formula (1.60) of Section 1.2.2. Formula (1.297) is just a restatement of Exercise 3.5.3 of Volume I, using the notation of Section 1.2.2.

Theorem 1.5.4 (Kaori Imai (Ota)). *Let $\varphi_r \in \mathcal{A}^0(SL(2, \mathbb{Z}), r(r-1))$ be a cusp form having Fourier expansion (1.297) with Fourier coefficients α_k, $k \neq 0$, and $\alpha_0 = \alpha'_0 = 0$. Then when*

$$Y = \begin{pmatrix} U & 0 \\ 0 & w \end{pmatrix} \begin{bmatrix} I_2 & X \\ 0 & 1 \end{bmatrix}, \quad U \in \mathcal{P}_2, \quad w = |U|^{-1}, \quad x \in \mathbb{R}^2,$$

we have the following Fourier expansion of the Eisenstein series $E_{2,1}(\varphi_r, s|Y)$ defined in (1.245)

$$E_{2,1}(\varphi_r, s|Y) = |U|^{-s} \varphi_r(U^0)$$
$$+ \sum \exp\{2\pi i \,^t xAm\} \, c_f(n) \, \alpha_k \, f^{1-2s-r} \, k_{2,1}\left(s - \frac{r}{2}, r \middle| U(A, w), m\right),$$

where the sum runs over $k \in \mathbb{Z} - 0$, $n \in \mathbb{Z}$, $f \geq 1$, $A \in SL(2, \mathbb{Z})/P(1, 1)$,

$$m = \begin{pmatrix} n \\ -kf \end{pmatrix}, \quad U(A, w) = \begin{pmatrix} U[\,^t A^{-1}] & 0 \\ 0 & w \end{pmatrix},$$

*and $P(1, 1)$ is the minimal parabolic subgroup of $SL(2, \mathbb{Z})$. Here $c_f(n)$ is **Ramanujan's sum**:*

$$c_f(n) = \sum_{\substack{0 < e < f \\ \gcd(e, f) = 1}} \exp(2\pi i n e/f),$$

and $k_{2,1}$ is the Bessel function from (1.60) in Section 1.2.2.

Proof. Set

$$\widetilde{Y} = Y \begin{bmatrix} \,^t A^{-1} & 0 \\ 0 & 1 \end{bmatrix} = \begin{pmatrix} \widetilde{U} & 0 \\ 0 & \widetilde{w} \end{pmatrix} \begin{bmatrix} I_2 & \widetilde{x} \\ 0 & 1 \end{bmatrix}. \tag{1.298}$$

It is easily seen that

$$\widetilde{U} = U[\,^t A^{-1}], \quad \widetilde{U}^{-1} = U^{-1}[A], \quad \widetilde{w} = w, \tag{1.299}$$

$$\tilde{x} = \begin{pmatrix} \widetilde{x}_1 \\ \widetilde{x}_2 \end{pmatrix} = \,^t Ax = \begin{pmatrix} \,^t a_1 x \\ \,^t a_2 x \end{pmatrix}, \quad \text{if } A = (a_1 a_2).$$

Using Lemma 1.5.6 with $n = 3$, we are led to compute $Y[(n_q \sigma p_q)_1]$, where the subscript "1" means that we must take the first two columns of the 3×3 matrix $n_q \sigma p_q$. Recall that $q = e/f$ and

$$
(n_q \sigma p_q)_1 = \begin{pmatrix} -qf & 0 \\ 0 & 1 \\ -f & 0 \end{pmatrix}.
$$

So we set

$$
Y^\# = \widetilde{Y}[(n_q \sigma p_q)_1] = \widetilde{U} \begin{bmatrix} -f(q + \tilde{x}_1) & 0 \\ -f\,\tilde{x}_2 & 1 \end{bmatrix} + \begin{pmatrix} wf^2 & 0 \\ 0 & 0 \end{pmatrix}.
$$

In order to use the Fourier expansion (1.297), we must set

$$
Y^\# = |Y^\#|^{1/2} \begin{pmatrix} y & 0 \\ 0 & y^{1/2} \end{pmatrix} \begin{bmatrix} 1 & 0 \\ x & 1 \end{bmatrix} = \begin{pmatrix} * & * \\ x\,|Y^\#|^{1/2}/y & |Y^\#|^{1/2}/y \end{pmatrix} \tag{1.300}
$$

and

$$
\widetilde{U} = \begin{pmatrix} t & 0 \\ 0 & v \end{pmatrix} \begin{bmatrix} 1 & 0 \\ p & 1 \end{bmatrix} = \begin{pmatrix} * & * \\ vp & v \end{pmatrix}. \tag{1.301}
$$

It follows that

$$
Y^\# = \begin{pmatrix} t & 0 \\ 0 & v \end{pmatrix} \begin{bmatrix} -f(q + \tilde{x}_1) & 0 \\ -f\{p(q + \tilde{x}_1) + \tilde{x}_2\} & 1 \end{bmatrix} + \begin{pmatrix} f^2 w & 0 \\ 0 & 0 \end{pmatrix}.
$$

Putting all this together, we find that

$$
|Y^\#| = vf^2 \{t(q + \tilde{x}_1)^2 + w\}, \quad y = \sqrt{\tfrac{f^2}{v}} \sqrt{t(q + \tilde{x}_1)^2 + w},
$$
$$
x = -f \{p(q + \tilde{x}_1) + \tilde{x}_2\}.
$$

By Lemma 1.5.6, q runs over all of the field of rational numbers. So we break this sum up into a sum over $q \in \mathbb{Q}/\mathbb{Z}$ and a sum over $n \in \mathbb{Z}$. Then use Poisson summation on the variable n to see that

$$
E_{2,1}(\varphi_r, s | Y) = |U|^{-s} \varphi_r(U^0) + \sum_{A,k,q,n} \alpha_k T(s, r | A, k, q, n),
$$

where the sum is over $A \in SL(2, \mathbb{Z})/P(1, 1)$, $k \neq 0$, $q \in \mathbb{Q}/\mathbb{Z}$, $q = e/f$, $f \geq 1$, $\gcd(e, f) = 1$, $n \in \mathbb{Z}$, and

$$T = T(s, r|A, k, q, n)$$

$$= \int\limits_{z \in \mathbb{R}} \left(vf^2 \{ t(z + q + \tilde{x}_1)^2 + w \} \right)^{-s} \left(\frac{f^2}{v} \{ t(z + q + \tilde{x}_1)^2 + w \} \right)^{(1-r)/2}$$

$$\times k_{1,1} \left(r \,\Big|\, I_2, \frac{kf}{\sqrt{v}} \sqrt{t(z + q + \tilde{x}_1)^2 + w} \right)$$

$$\times \exp\left(-2\pi i k f \{ p(z + q + \tilde{x}_1) + \tilde{x}_2 \} - 2\pi i n z \right) \, dz.$$

Next let $u = (t/w)^{1/2}(z + q + \tilde{x}_1)$ and use part (5) of Theorem 1.2.2 of Section 1.2.2 to obtain:

$$T = \exp\{2\pi i (nq + n\tilde{x}_1 - kf\,\tilde{x}_2)\} \, f^{-2s+1-r} v^{-s-(1-r)/2} t^{-1/2} w^{-s+(2-r)/2}$$
$$\times k_{2,1} \left(s - \tfrac{r}{2}, r \,\Big|\, I_3, \left((kpf + n)\sqrt{\tfrac{w}{t}}, -kf\sqrt{\tfrac{w}{v}} \right) \right).$$

Now the last argument of $k_{2,1}$ is the vector:

$$\sqrt{w} \begin{pmatrix} t^{-1/2} & 0 \\ 0 & v^{-1/2} \end{pmatrix} \begin{pmatrix} 1 & -p \\ 0 & 1 \end{pmatrix} \begin{pmatrix} n \\ -kf \end{pmatrix} = \sqrt{w} M \begin{pmatrix} n \\ -kf \end{pmatrix}, \quad \text{with}$$

$$M = \begin{pmatrix} t^{-1/2} & 0 \\ 0 & v^{-1/2} \end{pmatrix} \begin{pmatrix} 1 & -p \\ 0 & 1 \end{pmatrix}.$$

And ${}^t M M = \tilde{U}^{-1} = U^{-1}[A]$. Part (4) of Theorem 1.2.2 in Section 1.2.2 says that if ${}^t m = (n, -kf)$:

$$k_{2,1} \left(s - \frac{r}{2}, r \,\Big|\, I_3, \sqrt{w} M m \right) = p_{s-r/2, r-3/2}(U^{-1}[A]) \, k_{2,1} \left(s \,\Big|\, \begin{pmatrix} U\,[^tA^{-1}] & 0 \\ 0 & w \end{pmatrix}, m \right).$$

Next note that

$$v^{-s-(1-r)/2} w^{-s+(2-r)/2} t^{-1/2} = p_{s-r/2, r-3/2}(U^{-1}[A])^{-1}.$$

Thus the power functions cancel and we find that

$$T = \exp\{2\pi i (nq + n\tilde{x}_1 - kf\,\tilde{x}_2)\} \, f^{-2s+1-r} \, k_{2,1} \left(s - \frac{r}{2}, r \,\Big|\, \begin{pmatrix} U\,[^tA^{-1}] & 0 \\ 0 & w \end{pmatrix}, m \right).$$

This completes the proof of Theorem 1.5.4. ∎

Next we want to use Exercise 1.5.34 to obtain an alternate Fourier expansion.

Theorem 1.5.5. *Let $\varphi_r \in \mathcal{A}^0(SL(2, \mathbb{Z}), r(r - 1))$ be a cusp form having Fourier expansion (1.297) with Fourier coefficients α_k, $k \neq 0$, and $\alpha_0 = \alpha'_0 = 0$. Suppose that*

$$Y = \begin{pmatrix} U & 0 \\ 0 & w \end{pmatrix} \begin{bmatrix} I_2 & x \\ 0 & 1 \end{bmatrix}, \quad U \in \mathcal{P}_2, \ w = |U|^{-1}, \ x \in \mathbb{R}^2.$$

Then the Eisenstein series $E_{2,1}(\varphi_r, s|Y)$ defined by (1.245) has the following Fourier expansion with respect to the parabolic subgroup $P(2,1) \subset GL(3, \mathbb{Z})$:

$$L_{\varphi_r}(2s)E_{2,1}(\varphi_r, s|Y) = L_{\varphi_r}(2s)\varphi_r(U^0)|U|^{-s}$$
$$+ \sum \alpha_k \, c^{2-2s-r} d_2^{r-2s} \exp\left(2\pi i \, {}^t xAm\right) k_{2,1}\left(s - \tfrac{r}{2}, r \, \middle| \begin{pmatrix} U\begin{bmatrix} {}^t A^{-1} \end{bmatrix} & 0 \\ 0 & w \end{pmatrix}, m \right)$$

where ${}^t m = c(d_1, k/d_2) \in \mathbb{Z}^2$ and the sum is over $A \in SL(2, \mathbb{Z})/P(1,1)$, $c > 0$, $d_1 \in \mathbb{Z}$, $0 < d_2 | k$, $k \neq 0$. The parabolic subgroup $P(1,1)$ of $SL(2, \mathbb{Z})$ consists of the upper triangular matrices of determinant one. The L function $L_{\varphi_r}(2s)$ is the one that is associated with φ_r by part (5) of Theorem 1.5.2 in Section 1.5.2 (see also (1.278) in Section 1.5.2). Here $k_{2,1}$ is the Bessel function from (1.60) in Section 1.2.2.

Proof. Everything goes as it did in Theorem 1.5.4, except that we use Exercise 1.5.34 rather than Lemma 1.5.6. Define $\widetilde{Y}$ as in (1.298) and (1.299). Using Exercise 1.5.34 we must set

$$Y^\# = \widetilde{Y}\begin{bmatrix} D \\ {}^t g \end{bmatrix}, \quad \text{where} \quad \begin{cases} {}^t g = (c\ 0), \ c > 0, \ D = \begin{pmatrix} d_1 & 0 \\ d_{12} & d_2 \end{pmatrix}, \\ d_1 \in \mathbb{Z}, \ d_2 > 0, \ d_{12} \bmod d_2. \end{cases} \qquad (1.302)$$

Suppose that $\widetilde{U}$ is again given by (1.301). Then

$$Y^\# = \begin{pmatrix} t & 0 \\ 0 & v \end{pmatrix}\begin{bmatrix} d_1 + \tilde{x}_1 c & 0 \\ p(d_1 + \tilde{x}_1 c) + d_{12} + \tilde{x}_2 c & d_2 \end{bmatrix} + \begin{pmatrix} wc^2 & 0 \\ 0 & 0 \end{pmatrix}.$$

We compute $|Y^\#|$, x, y in (1.300) to be:

$$\begin{rcases} |Y^\#| = (vd_2^2)\left\{t(d_1 + \tilde{x}_1 c)^2 + wc^2\right\}, \\ y = \frac{\sqrt{t(d_1 + \tilde{x}_1 c)^2 + wc^2}}{\sqrt{v}d_2}, \quad x = \frac{p(d_1 + \tilde{x}_1 c) + d_{12} + \tilde{x}_2 c}{d_2}. \end{rcases} \qquad (1.303)$$

Since Exercise 1.5.34 says that the sum defining

$$Z(\varphi_r, s|Y) = L_{\varphi_r}(2s)E(\varphi_r, s|Y)$$

in (1.277) and (1.278) runs over all $d_1 \in \mathbb{Z}$, we can use Poisson summation to find that:

$$Z(\varphi_r, s|Y) = L_{\varphi_r}(2s)\varphi_r(U^0)|U|^{-s} + \sum \alpha_k T(s, r|A, c, D, k),$$

where the sum is over:

$$D = \begin{pmatrix} d_1 & 0 \\ d_{12} & d_2 \end{pmatrix}, \quad d_1 \in \mathbb{Z}, \ d_2 > 0, \ d_{12} \bmod d_2,$$

$A \in SL(2, \mathbb{Z})/P(1,1)$, $c > 0$, and $k \neq 0$. We define

$$T = T(s, r|A, c, D, k) =$$
$$\int_{z \in \mathbb{R}} \exp \left\{ 2\pi i \left(\tfrac{k}{d_2} (p(z + \tilde{x}_1 c) + d_{12} + \tilde{x}_2 c) - zd_1 \right) \right\} \left(v d_2^2 \{ t(z + \tilde{x}_1 c)^2 + wc^2 \} \right)^{-s}$$
$$\times \left(\frac{t(z+x_1 c)^2 + wc^2}{v d_2^2} \right)^{\frac{1-r}{2}} k_{1,1} \left(r \ \bigg| \ I_2, k \sqrt{\frac{t(z + \tilde{x}_1 c)^2 + wc^2}{v d_2^2}} \right) dz.$$

$$(1.304)$$

Now use the fact that

$$\sum_{0 \leq d_{12} < d_2} \exp\{2\pi i k d_{12}/d_2\} = \begin{cases} 0 & \text{if } d_2 \nmid k \\ d_2 & \text{if } d_2 | k \end{cases} \div \chi(d_2, k). \tag{1.305}$$

Therefore

$$T = \chi(d_2, k)(v d_2^2)^{-s-(1-r)/2} \exp\{2\pi i (p\tilde{x}_1 + \tilde{x}_2)kc/d_2\} \int_{z \in \mathbb{R}} \exp\{-2\pi i z (d_1 - kp/d_2)\}$$
$$\times (t(z + \tilde{x}_1 c)^2 + wc^2)^{-s+(1-r)/2} k_{1,1} \left(r \ \bigg| \ I_2, k \sqrt{\frac{t(z + \tilde{x}_1 c)^2 + wc^2}{v d_2^2}} \right) dz.$$

As in the proof of Theorem 1.5.4, set $u = (wc^2/t)^{-1/2}(z + \tilde{x}_1 c)$ and use part (5) of Theorem 1.2.2 of Section 1.2.2 to obtain the following formula for T defined by (1.304):

$$T = \chi(d_2, k) d_2^{-2s+r-1} c^{-2s-r+3/2} \exp\{2\pi i c(d_1 \tilde{x}_1 + \tilde{x}_2 k/d_2)\}$$
$$\times v^{-s-(1-r)/2} w^{1-s-r/2} t^{-1/2} k_{2,1} \left(s - \tfrac{r}{2}, r \ \bigg| \ I_3, \sqrt{w} \left(\tfrac{c}{\sqrt{t}} \left(\tfrac{kp}{d_2} - d_1 \right), \tfrac{kc}{\sqrt{v} d_2} \right) \right).$$

Set $M = \begin{pmatrix} t^{-1/2} & 0 \\ 0 & v^{-1/2} \end{pmatrix} \begin{pmatrix} 1 & -p \\ 0 & 1 \end{pmatrix}$. Then ${}^t MM = \tilde{U}^{-1}$ and if we set ${}^t m = c(d_1, k/d_2)$, part (4) of Theorem 1.2.2 of Section 1.2.2 says that:

$$T = \chi(d_2, k) d_2^{r-2s-1} c^{2-2s-r} \exp\{2\pi i \, {}^t x A m\} k_{2,1} \left(s - \tfrac{r}{2}, r \ \bigg| \ \begin{pmatrix} U^{-1} [\,{}^t A^{-1}] & 0 \\ 0 & w \end{pmatrix}, m \right).$$

For, again the power functions of $\tilde{U}^{-1}$ cancel. This completes the proof of Theorem 1.5.5.

Next we consider the case that φ_r in $E_{2,1}(\varphi_r, s|Y)$ is itself an Eisenstein series. By Exercise 1.5.2 in Section 1.5.1, using the notation of (1.245) and (1.249) in that section, we know that:

$$E_{2,1}(E_r, s|Y) = E_{(3)}(r, s - r/2, 0|Y). \tag{1.306}$$

Instead of considering the Eisenstein series itself, we obtain the Fourier expansion of the zeta function $Z_{2,1}(\varphi_r, s|Y)$ of formula (1.277) in Section 1.5.2, where $\varphi_r = \pi^{-r}\Gamma(r)Z_{1,1}(1, r|Y)$ and $Z_{1,1}(1, r|Y)$ is Epstein's zeta function from Proposition 1.5.2. Of course, the Eisenstein series $E_{2,1}$ is related to the zeta function $Z_{2,1}$ by formula (1.278) in Section 1.5.2.

Theorem 1.5.6 (Fourier Expansion of Selberg's Eisenstein Series for $GL(3, \mathbb{Z})$).
Suppose that $\varphi_r(W) = \pi^{-r}\Gamma(r)Z_{1,1}(1, r|W)$ where $Z_{1,1}$ denotes Epstein's zeta function for $GL(2, \mathbb{Z})$ from Proposition 1.5.2. If

$$Y = \begin{pmatrix} U & 0 \\ 0 & w \end{pmatrix}\begin{bmatrix} I_2 & x \\ 0 & 1 \end{bmatrix}, \quad U \in \mathcal{P}_2, \ w > 0, \ x \in \mathbb{R}^2,$$

the Fourier expansion of the normalized Eisenstein series $E_{2,1}(\varphi_r, s|Y)$ defined by (1.245) in Section 1.5.2 as a periodic function of $x \in \mathbb{R}^2$ is:

$$\pi^{-(s-r/2)}\Gamma(s - r/2)\pi^{-(s-(1-r)/2)}\Gamma(s - (1-r)/2)Z_{2,1}(\varphi_r, s|Y)$$

$$= c(s, r) + c((6 - 2s - 3r)/4, s - r/2) + c((3 + 3r - 2s)/4, \ s - (1 - r)/2)$$

$$+ \sum_{\substack{k=0 \\ A,c,d_1 \neq 0, d_2}} \alpha_0' c^{2-2s-r} d_2^{r-2s} \exp\left\{2\pi i\, {}^t x A m\right\}$$

$$\times k_{2,1}\left(s - \frac{r}{2}, r \ \middle| \ \begin{pmatrix} U[{}^t A^{-1}] & 0 \\ 0 & w \end{pmatrix}, \pi m\right)$$

$$+ \sum_{\substack{k=0 \\ A,c,d_1 \neq 0, d_2}} \alpha_0 c^{1-2s+r} d_2^{1-r-2s} \exp\left\{2\pi i\, {}^t x A m\right\}$$

$$\times k_{2,1}\left(s - \frac{1-r}{2}, 1 - r \ \middle| \ \begin{pmatrix} U[{}^t A^{-1}] & 0 \\ 0 & w \end{pmatrix}, \pi m\right)$$

$$+ \sum_{\substack{k \neq 0 \\ A,c,d_1,d_2}} \alpha_k c^{2-2s-r} d_2^{r-2s} \exp\left\{2\pi i\, {}^t x A m\right\}$$

$$\times k_{2,1}\left(s - \frac{r}{2}, r \ \middle| \ \begin{pmatrix} U[{}^t A^{-1}] & 0 \\ 0 & w \end{pmatrix}, \pi m\right).$$

Here the zeta function $Z_{2,1}$ is defined by (1.277) in Section 1.5.2 and relates to the Eisenstein series $E_{2,1}$ via (1.278) in that section. And we define the following quantities:

$$\alpha_0 = \Lambda(s,r) \Big/ B\left(\frac{1}{2}, \frac{1}{2} - r\right), \quad \alpha_0' = \Lambda(s,r) \Big/ B\left(\frac{1}{2}, r - \frac{1}{2}\right),$$

$$B(x,y) = \Gamma(x)\Gamma(y)/\Gamma(x+y), \quad \text{the beta function,}$$

$$\alpha_k = \Lambda(s,r)\sigma_{1-2r}(k)/\zeta(2r),$$

$$\Lambda(s,r) = \pi^{-(s-r/2)}\Gamma(s-r/2)\pi^{-(s-(1-r)/2)}\Gamma(s-(1-r)/2),$$

$$c(s,r) = \Lambda(r)\Lambda(s-r/2)\Lambda(s-(1-r)/2)E_r(U^0)|U|^{-s},$$

$$\Lambda(r) = \pi^{-r}\Gamma(r)\zeta(2r),$$

$$E_r(U^0) = \text{the Eisenstein series for } GL(2,\mathbb{Z}) \text{ defined in formula (3.81)}$$

$$\text{in Section 3.5 of Volume I.}$$

The three sums in the formula above are over $A \in SL(2,\mathbb{Z})/P(1,1)$, where $P(1,1)$ is the subgroup of upper triangular matrices of determinant one, $c > 0$, $d_1 \in \mathbb{Z}$ (with $d_1 \neq 0$ in the first two sums), $d_2 > 0$, $d_2|k$, $k \in \mathbb{Z}$ ($k \neq 0$ in the third sum). And the vector $m \in \mathbb{Z}^2$ is defined by ${}^t m = c(d_1, k/d_2)$. Here $k_{2,1}$ denotes the Bessel function from formula (1.60) in Section 1.2.2.

Proof. The proof is the same as that of Theorem 1.5.5 except that α_0 and α_0' are not zero. We need to use formula (1.297) with the Fourier coefficients given by Exercise 3.5.4 of Volume I. The constant term in the Fourier expansion of:

$$\pi^{-(s-r/2)}\Gamma(s-r/2)\pi^{-(s-(1-r)/2)}\Gamma(s-(1-r)/2)Z_{2,1}(\varphi_r, s|Y)$$

is:

$$\Lambda(s,r)|U|^{-s}E(r|U^0)L_{\varphi_r}(2s) + \alpha_0 k_{2,1}(s-(1-r)/2)|I_3, 0)$$
$$\times \sum_{A,c,d_2} d_2^{-2s+1-r} c^{-2s+1+r} |U|^{3(1-r)/4+s/2-3/2} U^{-1}[a_1]^{-s+(1-r)/2}$$
$$+ \alpha_0' k_{2,1}(s-r/2, r|I_3, 0) \sum_{A,c,d_2>0} d_2^{r-2s} c^{2-2s-r} |U|^{3r/4+s/2-3/2} U^{-1}[a_1]^{-s+r/2}.$$

The computation of Harish-Chandra's c-function (i.e., the calculation after formula (1.155) in Section 1.3.3) shows that

$$k_{2,1}(s-r/2, r|I_3, 0) = B\left(\frac{1}{2}, \frac{r}{2} + s - 1\right) B\left(\frac{1}{2}, r - \frac{1}{2}\right). \tag{1.307}$$

In order to compute the L-function corresponding to the form φ_r, recall that Exercise 3.5.4, Vol. I, showed that the nth Fourier coefficient of φ_r is:

$$c_n = 4\pi^{-r}\Gamma(r)\zeta(2r)|n|^{r-1/2}\sigma_{1-2r}(n).$$

The theory of Hecke operators for $SL(2,\mathbb{Z})$, to be found in part (5) of Theorem 3.6.4 of Volume I, shows that if T_k denotes the kth Hecke operator for $SL(2,\mathbb{Z})$ and

$$T_k\varphi_r = u_k\varphi_r, \ k \geq 1,$$

then (since our Hecke operators T_k are $k^{1/2}$ times those of Section 3.6, Vol. I):

$$u_n = \frac{c_n}{c_1}n^{1/2} = n^r\sigma_{1-2r}(n).$$

Thus

$$L_{\varphi_r}(2s) = \sum_{n\geq 1}n^{-2s}u_n = \sum_{n\geq 1}\sum_{0<d|n}n^{r-2s}d^{1-2r}$$

$$= \sum_{m\geq 1}\sum_{d\geq 1}(md)^{r-2s}d^{1-2r} = \zeta(2s-r)\zeta(2s+r-1).$$

This shows that the first part of the constant term is indeed $c(s,r)$. The third part of the constant term is:

$$\Lambda(s,1-r)\, B\left(\tfrac{1}{2},\tfrac{r}{2}+s-1\right)\, B\left(\tfrac{1}{2},r-\tfrac{1}{2}\right)\, \zeta(2s-r)\, \zeta\left(2s+\tfrac{3}{2}-r\right)$$
$$\times|U|^{-(3/2-s/2-3r/4)}E(s-r/2|U^0)\,/B\left(\tfrac{1}{2},r-\tfrac{1}{2}\right)$$
$$= c((6-2s-3r)/4, s-r/2).$$

So the second part of the constant term must be:

$$c((3-2s+3r)/4, s-(1-r)/2).$$

The rest of the proof of Theorem 1.5.6 proceeds as in Theorem 1.5.5. ∎

Exercise 1.5.35 (Remarks on the Constant Term). Let $E_r(U)$ be the Eisenstein series for $GL(2,\mathbb{Z})$ defined in formula (3.81) of Section 3.5, Volume I. Recall that by Exercise 1.5.1,

$$E_{2,1}(E_r, s|Y) = E_{(3)}(r, s-r/2, 0|Y),$$

using definitions (1.245) and (1.249) in Section 1.5.1. Consider Selberg's change of variables (1.257) in Section 1.5.1:

$$r = z_2 - z_1 + \frac{1}{2}, \ s - r/2 = z_3 - z_2 + \frac{1}{2}.$$

Show that the three parts of the constant term in the Fourier expansion of $E_{2,1}(E_r, s|Y)$ in Theorem 1.5.6 correspond to the permutations (1), (23), and (13) of the z-variables. This is a special case of a very general phenomenon described by Langlands [392]. See also Section 1.5.4.

Exercise 1.5.36 (Another Bruhat Decomposition and Its Consequences).

(a) Suppose that $n \geq 2m$ and $P = P(m, n-m)$. Show that we have the following disjoint union (Bruhat decomposition):

$$SL(n,\mathbb{Q}) = \bigcup_{r=0}^{m} P_{\mathbb{Q}}\sigma_r P_{\mathbb{Q}}, \quad \text{with } \sigma_r = \begin{pmatrix} 0 & 0 & I_r & 0 \\ 0 & I_{m-r} & 0 & 0 \\ -I_r & 0 & 0 & 0 \\ 0 & 0 & 0 & I_{n-m-r} \end{pmatrix}.$$

(b) Obtain a complete set of representatives for $(SL(n,\mathbb{Z}) \cap (P_{\mathbb{Q}}\sigma_r P_{\mathbb{Q}}))/P_{\mathbb{Z}}$ of the form $p\sigma_r p'$ where σ_r is as in part (a),

$$p = \begin{pmatrix} {}^tA^{-1} & 0 \\ 0 & B \end{pmatrix} \begin{pmatrix} I_r & 0 & U & 0 \\ 0 & I_{m-r} & 0 & 0 \\ 0 & 0 & I_r & 0 \\ 0 & 0 & 0 & I_{n-m-r} \end{pmatrix},$$

$U \in \mathbb{Q}^{r \times r}$, $A \in SL(m,\mathbb{Z})/P(r, m-r)$, $B \in SL(n-m,\mathbb{Z})/P(r, n-m-r)$. The element $p' \in P_{\mathbb{Q}}$ is fixed, once p is.

(c) Use part (b) to obtain the Fourier expansion of the Eisenstein series $E_{m,n-m}(1, s|Y)$ with respect to $P(m, n-m)$. Here the Maass form f in $\mathcal{A}^0(GL(m,\mathbb{Z}), \lambda)$ is chosen to be identically one in formula (1.245) of Section 1.5.1.

Hint (Terras [596]). If

$$Y = \begin{pmatrix} V & 0 \\ 0 & W \end{pmatrix} \begin{bmatrix} I & Q \\ 0 & I \end{bmatrix}, \quad V \in \mathcal{P}_m, \ W \in \mathcal{P}_{n-m}, \ Q \in \mathbb{R}^{m \times (n-m)},$$

the Fourier coefficient corresponding to $N \in \mathbb{Z}^{m \times (n-m)}$ of rank r, $0 < r \leq m$, involves the Bessel function K_r or $k_{r,r}$ from §1.2.2, plus the *singular series*

$$\sigma(C, s) = \sum_{R \in \mathbb{Q}^{r \times r}/\mathbb{Z}^{r \times r}} \nu(R)^{-2s} \exp(2\pi i \operatorname{Tr}({}^tRC)),$$

where $\nu(R)$ denotes the product of the reduced denominators of the elementary divisors of R. This is an analogue of the singular series appearing in the Fourier coefficients of Eisenstein series for the Siegel modular group (see Siegel [565, Vol. II, pp. 97–137]).

Next let us consider another application of these Fourier expansions—an application to the analytic continuation of Eisenstein series $E_{m,n-m}(f, s|Y)$ for Maass forms $f \in \mathcal{A}^0(GL(m, \mathbb{Z}), \lambda)$. See Jacquet and Shalika [326] for a similar adelic argument. According to formulas (1.270) and (1.271) in Section 1.5.2, we have:

$$\Lambda_{m,n-m}(f, s|Y) = \Lambda(f, s|Y) = 2\pi^{-ms}\Gamma_m(r(f, s))L_f^*(2s)E_{m,n-m}(f^*, s|Y)$$

$$= \int_{\substack{X \in \mathcal{M}_m, \\ |X| \geq 1}} \theta_m(Y, X) f(X^0)|X|^s \, d\mu_m + \int_{\substack{X \in \mathcal{M}_m, \\ |X| \geq 1}} \theta_m(Y, X^{-1}) f^*(X^0)|X|^{-s} \, d\mu_m.$$

Therefore, by the transformation formula of theta (from Exercise 1.5.18 in Section 1.5.2), we see that

$$\left.\begin{aligned}
\Lambda_{m,n-m}(f, s|Y) = \int_{X \in \mathcal{M}_m, |X| \geq 1} \theta_m(Y, X) f(X^0)|X|^s \, d\mu_m \\
+ |Y|^{-m/2} \int_{\substack{X \in \mathcal{M}_m, \\ |X| \geq 1}} \theta_m(Y^{-1}, X) f^*(X^0)|X|^{n/2-s} \, d\mu_m + \sum_{k=0}^{m-1} I_k(f, s|Y),
\end{aligned}\right\}$$

$$(1.308)$$

where

$$I_k(f, s|Y) = \int_{X \in \mathcal{M}_m, |X| \geq 1} \left(|Y|^{-\frac{m}{2}} |X|^{\frac{n}{2}-s} f^*(X^0) \, \theta_k(Y^{-1}, X) - |X|^s f(X^0) \, \theta_k(Y, X)\right) d\mu_m.$$

$$(1.309)$$

The term $I_0(f, s|Y)$ is no problem:

$$I_0(f, s|Y) = \begin{cases} 0, & \text{if } f \text{ is orthogonal to the constants} \\ |Y|^{-m/2} \left(\frac{n}{2} - s\right)^{-1} - s^{-1}, & \text{if } f \text{ is identically 1.} \end{cases}$$

$$(1.310)$$

To study I_k, for $0 < k < m$, we need the following exercise.

Exercise 1.5.37 (Study of the Integrals Occurring in the Analytic Continuation of Eisenstein Series by the Method of Theta Functions Terras [604]).

(a) Show that we can express every rank k matrix $A \in \mathbb{Z}^{n \times m}$, for $1 \leq k \leq m - 1$, uniquely in the form:

$$A = B\,{}^tC, \quad B \in \mathbb{Z}^{n \times k}, \quad \text{rank } k,$$

$$(C *) \in GL(m, \mathbb{Z})/P(k, m - k), \quad C \in \mathbb{Z}^{m \times k}.$$

(b) Obtain the Jacobian of the following change of variables for $Y \in \mathcal{P}_n$:

$$t^{-1/n}Y = \begin{pmatrix} u^{-1}T & 0 \\ 0 & uV \end{pmatrix}\begin{bmatrix} I & Q \\ 0 & I \end{bmatrix}, \quad u > 0, \quad T \in \mathcal{SP}_p,$$

$$V \in \mathcal{SP}_q, \ Q \in \mathbb{R}^{p \times q}, \ n = p + q.$$

Answer. $d\mu_n(Y) = (2pq/n)u^{-pq-1}t^{-1}dt \, du \, dT \, dV$. Compare Exercise 1.4.21 of Section 1.4.3.

(c) Rewrite the integral $I_k(f, s|Y)$ in (1.309) using parts (a) and (b).

(d) Now suppose that $f \in \mathcal{A}^0(GL(m, \mathbb{Z}), \lambda)$ has the Fourier expansion:

$$f(W) = \sum_{N \in \mathbb{Z}^{k \times (m-k)}} A_{N,f}\left(u^{-1}T, uV\right) \exp\left\{2\pi i \operatorname{Tr}\left({}^t NQ\right)\right\},$$

if W is expressed as in part (b). Show that

$$\frac{m}{2k(m-k)}I_k(f, s|Y)$$
$$= \int\limits_{T,V,t,u} \left(A_{0,f^*}\left(u^{-1}T, uV\right) t^{n/2-s} |Y|^{-m/2}\right.$$
$$\times \sum_{B \in \mathbb{Z}^{n \times k} \text{ rank } k} \exp\left\{-\pi \operatorname{Tr}\left(Y^{-1}[B]t^{1/m}u^{-1}T\right)\right\}$$
$$\left. - A_{0,f}(u^{-1}T, uV)\, t^{-s} \sum_{B \in \mathbb{Z}^{n \times k} \text{ rank } k} \exp\left\{-\pi \operatorname{Tr}(Y[B]t^{-1/m}u^{-1}T\right\}\right)$$
$$\times u^{-\tau}t^{-1} \, du \, dt \, dT \, dV,$$

where $\tau = k(m - k) + 1$ and the integral is over $T \in \mathcal{SM}_k$, $V \in \mathcal{SM}_{m-k}$, $t \geq 1$, $u \geq 0$.

(e) What hypotheses on f are necessary to justify the preceding arguments? Be careful. It was just this sort of divergent integral problem that led to gaps in many papers, as we mentioned in the remarks on part (5) of Theorem 1.5.2 in Section 1.5.2.

(f) In the special case $n = 3$, $m = 2$, $f(X) = E_r(X)$, $X \in \mathcal{SP}_2$, the Eisenstein series from formula (3.81) of Section 3.5, Volume I and Exercise 1.5.2 of Section 1.5.1, let

$$E_{2,1}(E_r, s|Y) = E_{(3)}(r, s - r/2, 0|Y)$$

be continued as above and show that then the integral I_1 is:

$$I_1(E_r, s|Y) = |Y|^{-1}\frac{\Lambda_3(Y^{-1}, 1-r)}{s-1-r/2} - \frac{\Lambda_3(Y, 1-r)}{s+(r-1)/2} + \frac{c(r)|Y|^{-1}\Lambda_3(Y^{-1}, r)}{s+(r-3)/2}$$
$$- \frac{c(r)\Lambda_3(Y, r)}{s-r/2},$$

where $\Lambda_3(Y, r) = \Lambda(r)E_r(Y)$, $c(r) = \Lambda(1-r)/\Lambda(r)$, $\Lambda(r) = \pi^{-r}\Gamma(r)\zeta(2r)$.
(g) What happens when $f(X) \in \mathcal{A}^0(GL(m, \mathbb{Z}), \lambda)$ is a cusp form?

This ends our discussion of Fourier expansions of Maass forms for $GL(n, \mathbb{Z})$. Many questions remain. In particular, we have certainly not obtained the most general sorts of Fourier expansions. One would also like to build up a Hecke correspondence by making use of the Fourier expansions. In this regard, note that we already have the Mellin transform of the K-Bessel functions from Exercise 1.2.18 in Section 1.2.2. However it is not useful to attempt to do the Mellin transforms of the k-Bessel functions in the following exercise.

Exercise 1.5.38 (Mellin Transforms That Diverge).

(a) Show that the following integral diverges in general

$$\int\limits_{t>0} t^r k_{1,1}\left(s \left| \begin{pmatrix} t & 0 \\ 0 & 1 \end{pmatrix}, n \right.\right) dt.$$

(b) Do the same for

$$\int\limits_{U \in \mathcal{M}_2} |U|^{-r} f(U^0) k_{2,1}\left(s \left| \begin{pmatrix} U & 0 \\ 0 & w \end{pmatrix}, n \right.\right) \exp(-\mathrm{Tr}(U)) \, dU.$$

when $r \in \mathbb{C}$, $f \in \mathcal{A}^0(GL(2, \mathbb{Z}), \lambda)$.

The next exercise should be compared with Exercise 1.5.26 in Section 1.5.2.

Exercise 1.5.39 (Another Functional Equation for Eisenstein Series).

(a) Use part (4) of Proposition 1.2.1 of Section 1.2.1 to show that Selberg's Eisenstein series $E_{(n)}(s|Y)$ defined by (1.249) of Section 1.5.1 satisfies the functional equation:

$$E_{(n)}(s|Y^{-1}) = E_{(n)}(s^*|Y), \quad s^* = \left(s_{n-1}, \ldots, s_1, -\sum_{j=1}^n s_j\right), \text{ for } s \in \mathbb{C}^n.$$

(b) Apply Exercise 1.5.10 of Section 1.5.1 to the Eisenstein series $E_{k,n-k}(1, s_k|Y)$, in the notation (1.245), to show that this Eisenstein series is essentially a specialization of $E_{(n)}(s|Y)$ arrived at by setting all but one variable equal to zero; more specifically,

$$E_{k,n-k}(1, s|Y) = E_{(n)}(s|Y)\Big|_{\substack{s_j=0 \\ \forall\, j \neq k}}.$$

(c) Show that the Eisenstein series $E_{k,n-k}(1, s|Y)$ satisfies the functional equation:

$$E_{k,n-k}(1, s|Y^{-1}) = |Y|^s E_{n-k,k}(1, s|Y).$$

In particular, this means that if $Z_{k,n-k}(1, s|Y)$ denotes Koecher's zeta function, using the notation (1.277) of Section 1.5.2 rather than (1.173) of Section 1.4.1, then the following equality holds:

$$\begin{aligned}
Z_{2,1}(1, s|Y) &= \zeta(2s - 1)\zeta(2s)|Y|^{-s}E_{1,2}(1, s|Y^{-1}) \\
&= \zeta(2s - 1)|Y|^{-s}Z_{1,2}(1, s|Y^{-1}).
\end{aligned}$$

Thus, in this special case, Koecher's zeta function is just a product of Riemann and Epstein zeta functions. Use this result to check the formula obtained in part (f) of Exercise 1.5.37.

Hint. (c) You will need to take the limit of the quantity

$$r(r - 1)\Lambda_{2,1}(E_r, s|Y)$$

in (1.308) and Exercise 1.5.37 as r approaches zero. In particular, this leads to the formula:

$$\lim_{r \to 0} r(r - 1)\Lambda(r)I_1(E_r, s|Y)$$
$$= \frac{|Y|^{-1}\Lambda_{1,2}(1,1|Y)}{s-1} - \frac{\Lambda_{1,2}(1,1|Y)}{s-1/2} + \frac{\Lambda(1)|Y|^{-1}}{s-3/2} - \frac{\Lambda(1)}{s}.$$

Thus Exercise 1.5.37 gives the analytic continuation of Koecher's zeta function and agrees with formula (3.16) in Koecher [359].

The last exercise in this section shows that one must be careful in obtaining the analytic continuation of Koecher's zeta function.

Exercise 1.5.40 (Double Poles). Consider the special case of Koecher's zeta function $Z_{m,0}(1, s|Y)$, in the notation of (1.277) in Section 1.5.2—a case in which the zeta function factors into a product of Riemann zeta functions as in (1.174) of Section 1.4.1. Then form the normalized function from (1.271) of Section 1.5.2:

$$\Lambda_{m,0}(1, s|Y) = 2\pi^{-ms}\Gamma_m(s)Z_{m,0}(1, s|Y) = 2|Y|^{-s}\prod_{j=0}^{m-1}\Lambda(s - j/2),$$

where $\Lambda(s) = \pi^{-s}\Gamma(s)\zeta(2s)$. Show that $\Lambda_{m,0}(1, s|Y)$ has simple poles at $s = 0, m/2$ and has double poles at $s = \frac{1}{2}, 1, \frac{3}{2}, \ldots, \frac{m-2}{2}, \frac{m-1}{2}$.

This exercise demonstrates that Koecher [359] formula (3.16) is incorrect, in general, as Koecher's formula says that only simple poles occur. Another reference on the location of poles of Eisenstein series is Feit [175]. Other references on L-functions and Eisenstein series are: Böcherer [54], Duke [144], Garrett [203], and Goldfeld [230].

Remarks on Maass Cusp Forms

Recall that in Section 1.5.1 we defined a Maass form $f \in \mathcal{A}^0(GL(n, \mathbb{Z}), \lambda)$ (where $\mathcal{A}^0$ means f lives on $\mathcal{SP}_n$) to be a **cusp form** if for every $k = 1, \ldots, n-1$:

$$a_0^k(Y) = \int\limits_{X \in (\mathbb{R}/\mathbb{Z})^{k \times (n-k)}} f\left(Y\begin{bmatrix} I & X \\ 0 & I \end{bmatrix}\right) dX = 0, \qquad \text{for all } Y \in \mathcal{P}_n. \tag{1.311}$$

Thus a cusp form has a zero constant term for each one of the Fourier expansions (1.287) with respect to maximal parabolic subgroups $P(k, n-k)$, $1 \le k \le n-1$.

A reader familiar with the definition of cusp form for the Siegel modular group might ask whether our definition implies that all of the Fourier coefficients $a_N(U, V)$ vanish for $N \in \mathbb{Z}^{k \times (n-k)}$ not of maximal rank in (1.287).

Another question raised by Siegel's approach to the definition of cusp form is that of defining cusp forms for $GL(n, \mathbb{Z})$ to be those that are in the kernel of some analogue of the Siegel Φ-operator which we will consider in Section 2.2 below. In [243], Grenier defines a ϕ-operator. Assume that $f \in \mathcal{A}^0(\Gamma_n, \lambda)$, where $\Gamma_n = GL(n, \mathbb{Z})/\{\pm I\}$ and the eigenvalues of the invariant differential operators on f agree with those of the power function p_{-s}. If this is so, we write $\lambda = \lambda(s)$. Note that we can write the power function for $W \in \mathcal{SP}_{n-1}$, $v > 0$, $x \in \mathbb{R}^{n-1}$ as:

$$p_{-s}\begin{pmatrix} v^{-1} & 0 \\ 0 & v^{1/(n-1)}W \end{pmatrix} = v^{s_1 + \xi_1} p_{-s'}(W), \quad \text{where}$$

$$\text{if } {}^t s = (s_1, \ldots, s_{n-1}), \ {}^t s' = (s_2, \ldots, s_{n-1}) \ \text{ then } \xi_1 = \frac{1}{n-1}\sum_{k=2}^{n-1} (n-k) s_k.$$

The **Grenier ϕ-operator** is defined by:

$$(f|\phi)(W) = \lim_{v \to \infty} v^{-s_1 - \xi_1} f\left(\begin{pmatrix} v^{-1} & 0 \\ 0 & v^{1/(n-1)}W \end{pmatrix}\begin{bmatrix} 1 & {}^t x \\ 0 & I_{n-1} \end{bmatrix}\right),$$

$$\text{for } W \in \mathcal{SP}_{n-1}, \ v > 0, x \in \mathbb{R}^{n-1} \text{ as.} \tag{1.312}$$

Grenier's ϕ-operator is an analogue of Siegel's Φ-operator defined in formula (2.57) of Section 2.2. Grenier proves that it maps $f \in \mathcal{A}^0(\Gamma_n, \lambda(s))$ to $f|\phi \in \mathcal{A}^0(\Gamma_{n-1}, \lambda(s'))$. To do this, he obtains the Fourier expansion of $f \in \mathcal{A}^0(\Gamma_n, \lambda)$ with respect to the parabolic subgroup $P(1, n-1)$ from the Whittaker–Fourier expansion of Bump to be discussed in Section 1.5.4 and then uses properties of K-Bessel functions. Grenier also begins a discussion of Maass–Selberg relations for Maass forms $f, g \in \mathcal{A}^0(\Gamma_n, \lambda)$. Sadly, these investigations were not continued, as far as I know.

Siegel would define f to be cuspidal if $f|\phi = 0$ (see Chapter 2, Maass [426, pp. 187–198] or Freitag [185]). Does the analogous statement hold for the Grenier ϕ-operator and Maass cusp forms?

Sadly we leave these questions open but we do say a bit more about cusp forms and their Fourier and Fourier–Whittaker expansions in the next section.

1.5.4   Update on Maass Cusp Forms for $SL(3, \mathbb{Z})$ and L-Functions Plus Truncating Eisenstein Series

Maass Cusp Forms for $SL(3, \mathbb{Z})$ and L-Functions

First we want to sketch some of the work on cusp forms for $SL(3, \mathbb{Z})$ that has occurred since I was writing the old edition. Indeed some of the things mentioned here were done earlier but in adelic language, which I never managed to translate. Most of these things here could be done for $SL(n, \mathbb{Z})$ but we stick to $n = 3$ for simplicity. We mostly follow Dorian Goldfeld [230] in this section since he provides a translation from adelic to classical language as well as proofs of many of the earlier results of authors such as Jacquet, Piatetski-Shapiro, and Shalika.

Adelic computations involve infinite vectors with real, complex, and p-adic entries for each prime p. You will not find Mathematica programs to do this. Goldfeld's book [230] includes Mathematica programs to do many things, such as computing Whittaker functions. Moreover, Goldfeld's has summed up our view in the following quotation.

> In line with the philosophy of understanding by simple example, we have avoided the use of adeles, and as much as possible the theory of representations of Lie groups. This very explicit language appears particularly useful for analytic number theory where precise growth estimates of L-functions and Maass forms play a major role. (from Goldfeld [230, pp. xi–xii].)

It is a bit difficult to sync with Goldfeld's notation as he replaces $Y \in \mathcal{SP}_3$ with $z \in T_3$ the group of upper triangular matrices with positive diagonal entries:

$$z(x, y) = \begin{pmatrix} 1 & x_2 & x_3 \\ 0 & 1 & x_1 \\ 0 & 0 & 1 \end{pmatrix} \begin{pmatrix} y_1 y_2 & 0 & 0 \\ 0 & y_1 & 0 \\ 0 & 0 & 1 \end{pmatrix}. \tag{1.313}$$

Since the determinant of g is not 1, the map $T_3 \to \mathcal{SP}_3$ is

$$z = z(x,y) \longmapsto \left(z\,^t z\right)^0. \tag{1.314}$$

Here we use our notation $Y^0 = |Y|^{-1/3}\,Y$, for $Y \in \mathcal{P}_3$. Note that this map takes the *left* $GL(n,\mathbb{R})$ action on $z \in T_3$ to the *right* action of $GL(n,\mathbb{R})$ on $\mathcal{SP}_3$.

However there is still a problem with formula (1.314), if we want to use our usual Iwasawa decomposition in the definition of power functions. Thus we make use of a Weyl group element which will change the x-variables from upper to lower triangular, and vice versa. The Weyl group element is

$$\omega = \begin{pmatrix} 0 & 0 & 1 \\ 0 & -1 & 0 \\ 1 & 0 & 0 \end{pmatrix}. \tag{1.315}$$

This matrix $\left(z\,^t z\right)^0 [\omega]$ has our usual Iwasawa decomposition with diagonal part equal to

$$\left(y_1^{-4/3} y_2^{-2/3},\, y_1^{2/3} y_2^{-2/3},\, y_1^{2/3} y_2^{4/3}\right).$$

Our power function is

$$p_s\left(\left(z\,^t z\right)^0 [\omega]\right) = \left(y_1^{-2/3} y_2^{-4/3}\right)^{s_2} \left(y_1^{-4/3} y_2^{-2/3}\right)^{s_1} = y_1^{-(2s_2+4s_1)/3} y_2^{-(4s_2+2s_1)/3} = y_1^a y_2^b,$$

where $a = \frac{-2}{3}(s_2 + 2s_1)$ and $b = \frac{-2}{3}(2s_2 + s_1)$. Goldfeld replaces the power function $p_s(Y)$ with his function

$$I_\nu(z) = \prod_{1 \le i,j \le 2} y_i^{b_{ij}\nu_j}, \quad \text{where} \quad b = \begin{pmatrix} 1 & 2 \\ 2 & 1 \end{pmatrix}. \tag{1.316}$$

So $I_\nu(z) = y_1^{\nu_1+2\nu_2} y_2^{2\nu_1+\nu_2}$. It follows that $\frac{-2}{3}s_2 = \nu_1$ and $\frac{-2}{3}s_1 = \nu_2$, if we make the identification (1.314) and throw in the Weyl group element to identify Goldfeld's $z \in T_3$ with our $\left(z\,^t z\right)^0 [\omega] \in \mathcal{SP}_3$.

Goldfeld defines f to be a **Maass form for $SL(3,\mathbb{Z})$ of type ν** if f is $SL(3,\mathbb{Z})$-invariant and the differential operators in $\mathcal{D}(\mathcal{SP}_3)$ share the same eigenvalues with f as the power function I_ν in formula (1.316). Goldfeld also assumes that his Maass form f is cuspidal. Here we do not always make that assumption.

In the case of a Maass cusp form for $SL(3,\mathbb{Z})$ of type ν, Bump (see Goldfeld [230, p. 160]) obtains the **Whittaker–Fourier** expansion:

$$f(z) = \sum_{\gamma \in P(1,1) \backslash SL(2,\mathbb{Z})} \sum_{m_1 \geq 1} \sum_{m_2 \neq 0} \frac{A(m_1, m_2)}{|m_1 m_2|}$$

$$\times W \left(\left(\begin{pmatrix} |m_1 m_2| & 0 & 0 \\ 0 & m_1 & 0 \\ 0 & 0 & 0 \end{pmatrix} \begin{pmatrix} \gamma & 0 \\ 0 & 1 \end{pmatrix} z; v, \psi_{\left(1, \frac{m_2}{|m_2|}\right)} \right) \right), z \in T_3. \qquad (1.317)$$

Here $A(m_1, m_2) \in \mathbb{C}$ and $W(s|Y, r)$ denotes the **Jacquet–Whittaker function** defined for the character $\psi_{(a,b)}$ of N (the group of upper triangular real 3×3 matrices with 1's on the diagonal)

$$\psi_{(a,b)} \left(\begin{pmatrix} 1 & u_2 & u_3 \\ 0 & 1 & u_1 \\ 0 & 0 & 1 \end{pmatrix} \right) = \exp\left(2\pi i \left(au_1 + bu_2\right)\right),$$

by:

$$W\left(z; v, \psi_{(a,b)}\right) = \int_{u \in N} I_v(\omega u z)\overline{\psi_{(a,b)}(u)}du, \quad \text{with} \quad du = du_1 du_2 du_3, \qquad (1.318)$$

where the Weyl group element ω is as defined in formula (1.315). Compare our definition of Whittaker functions at the end of Section 1.2.2.

The expansion (1.317) should be compared with the Fourier expansions (1.287). The Whittaker–Fourier expansion is obtained by starting with our usual sort of expansion (1.287) for the parabolic subgroup $P(2, 1)$:

$$f(Y) = \sum_{r \in \mathbb{Z}^2} f_r(Y), \quad \text{where } f_r(Y) = \int_{x \in \mathbb{R}^2} f\left(Y \begin{bmatrix} I_2 & X \\ 0 & 1 \end{bmatrix}\right) \exp\left(-2\pi i \, {}^t rx\right) dx.$$

This means that

$$f_r \left(Y \begin{bmatrix} I_2 & X \\ 0 & 1 \end{bmatrix}\right) = \exp\left(2\pi i \, {}^t rx\right) f_r(Y), \quad \text{for all } x \in \mathbb{R}^2.$$

Then note that for $A \in SL(2, \mathbb{Z})$ and $r \in \mathbb{R}^2$, we have:

$$f_{{}^t Ar} \left(Y \begin{bmatrix} A & 0 \\ 0 & 1 \end{bmatrix}\right) = f_r(Y). \qquad (1.319)$$

If $r \in \mathbb{Z}^2$, then there is a matrix $A \in SL(2, \mathbb{Z})$ such that ${}^t Ar = {}^t(0, m)$ for some positive integer m. Furthermore, if $a \in \mathbb{Z}$, one has

$$f_r \left(Y \begin{bmatrix} 1 & a & 0 \\ 0 & 1 & 0 \\ 0 & 0 & 1 \end{bmatrix}\right) = f_{{}^t(r_1, r_2 - r_1 a)}(Y). \qquad (1.320)$$

This shows that when $r_1 = 0$, the coefficient f_r must be invariant under

$$Y \mapsto Y \begin{bmatrix} 1 & a & 0 \\ 0 & 1 & 0 \\ 0 & 0 & 1 \end{bmatrix}, \quad \text{for } a \in \mathbb{Z}.$$

Thus one can find the Fourier expansion with respect to the x_{12} variable in N, the group of upper triangular matrices with ones on the diagonal. That takes us from Bessel to Whittaker functions. It is the multiplicity one theorem of Shalika [552] which says that the resulting functions must be multiples of Whittaker functions. See Goldfeld [230, p. 155]. A summary of work on Maass forms for $GL(3)$ in the language of Bump [83] can also be found in Friedberg [189].

Exercise 1.5.41.   Give the relation between the Jacquet–Whittaker function (1.318) and that of the $k_{2,1}$-Bessel function from formula (1.60) of Section 1.2.2. Deduce the K-Bessel expansion of $f \in \mathcal{A}^0 (SL(3, \mathbb{Z}), \lambda)$ from the Fourier–Whittaker expansion of f. See Grenier [243].

Once you have the Whittaker–Fourier expansion (1.317) of a Maass cusp form for $SL(3, \mathbb{Z})$, then you can define the **Godement–Jacquet L-function** for $\mathrm{Re}\, s > 2$ by:

$$L_f(s) = \sum_{n \geq 1} \frac{A(1, n)}{n^s}, \tag{1.321}$$

as in Goldfeld [230, p. 174]. It can be proved that if f is a **Maass cusp form for $GL(3, \mathbb{Z})$ of type v**, then the Godement–Jacquet L-function can be extended to a holomorphic function of all $s \in \mathbb{C}$ and satisfies a functional equation

$$G(v, s)L_f(s) = G(\widetilde{v}, 1 - s)L_{\widetilde{f}}(1 - s), \tag{1.322}$$

where

$$\left. \begin{aligned} G(v, s) &= \pi^{-3s/2} \Gamma\left(\tfrac{s+1-2v_1-v_2}{2}\right) \Gamma\left(\tfrac{s+v_1-v_2}{2}\right) \Gamma\left(\tfrac{s-1+v_1+2v_2}{2}\right), \\ G(\widetilde{v}, s) &= \pi^{-3s/2} \Gamma\left(\tfrac{s+1-v_1-2v_2}{2}\right) \Gamma\left(\tfrac{s-v_1+v_2}{2}\right) \Gamma\left(\tfrac{s-1+2v_1+v_2}{2}\right). \end{aligned} \right\} \tag{1.323}$$

Here the **dual Maass form** $\widetilde{f}$ is defined by $\widetilde{f}(z) = f\left(\omega \,{}^t z^{-1} \omega\right)$ with ω defined by (1.315). Goldfeld [230, p. 161] shows that $\widetilde{f}$ is a Maass form of type $\widetilde{v} = (v_2, v_1)$. It turns out that if $A(m_1, m_2)$ are the Whittaker–Fourier coefficients of f, then $A(m_2, m_1)$ are the Whittaker–Fourier coefficients of $\widetilde{f}$. Bump (see also Goldfeld) has a method of obtaining the functional equations (1.322) by using those for the Eisenstein series with the same v-type.

Exercise 1.5.42. Suppose that ω is defined by (1.315) and $Y_z = (I \, [^t z])^0$, for $z = z(x, y)$ from (1.313). Suppose $\widetilde{z} = \omega \, {}^t z^{-1} \omega$. Show that $Y_{\widetilde{z}} = (Y_z)^{-1} [\omega]$.

Mellin transforms of Whittaker functions are used by Goldfeld to prove the multiplicity one theorem (see [230, p. 155]). I would expect that this idea would work for K-Bessel multiplicity one as well. The use of Mellin transforms of Maass forms to obtain functional equations for the L-functions attached to Maass forms as presented in Goldfeld's book seems a bit more complicated than that which we have presented. See Goldfeld [230, Ch. 6]. Bump [83] and Bump and Friedberg [85] give more information on Mellin transforms of Whittaker functions.

Grenier [243] notices that one can look at the Whittaker–Fourier expansion in a different way since the Whittaker function is the Fourier transform of the $k_{2,1}$-Bessel function, this means that one can start with the Whittaker–Fourier expansion of a Maass form for $GL(3, \mathbb{Z})$ and proceed to derive an ordinary Fourier expansion of Maass forms for $\mathcal{SP}_3$ in K_2-Bessel functions. This would give a simpler Mellin transform approach to the functional equations of L-functions.

The Bump–Goldfeld version of Hecke operators is essentially the same as ours. So we won't bother to summarize. It's the Fourier coefficients that differ.

It in addition to being a Maass cusp form for $SL(3, \mathbb{Z})$, f is an eigenform of all the Hecke operators, then, assuming f is not the 0-form, by multiplying f by a constant, we may assume that the Whittaker–Fourier coefficient $A(1, 1) = 1$. It follows that $T_n f = A(n, 1)f$ for all $n = 1, 2, \ldots$. From this and the multiplicative properties of the Hecke operators one obtains an Euler product for the Godement–Jacquet L-function which is similar to our earlier **Euler product** Theorem 1.5.2:

$$L_f(s) = \prod_{p \text{ prime}} \left(1 - A(1, p)p^{-s} + A(p, 1)p^{-2s} - p^{-3s}\right)^{-1}. \tag{1.324}$$

See Goldfeld [230, p. 174].

The converse theorem in Hecke theory was proved by Jacquet, Piatetski-Shapiro and Shalika [325] in 1979. One must twist by primitive Dirichlet characters mod q for all positive integers q. See Goldfeld [230, p. 195 ff] for a proof of Goldfeld and Meera Thillainatesan.

It is possible to use the converse theorem and Rankin–Selberg L-functions to obtain the Gelbart–Jacquet lift of a cuspidal Maass wave form for $SL(2, \mathbb{Z})$ (which is a Hecke eigenform) to a Maass form for $SL(3, \mathbb{Z})$ which is self-dual of type $(2v_f/3, 2v_f/3)$. See Goldfeld [230, Ch 7]. Here $v = v_f$ means that $\Delta f = (v(v - 1))f$. This does not give an explicit $SL(3, \mathbb{Z})$-example until one has an explicit example for a Maass form for $SL(2, \mathbb{Z})$.

The **Ramanujan conjecture** for a Maass cusp form f for $SL(3, \mathbb{Z})$ which is eigen for all the Hecke operators says that if we factor the polynomial in the Euler product (1.324),

$$\left(1 - A(1,p)x + A(p,1)x^2 - x^3\right) = \prod_{i=1}^{3} \left(1 - \alpha_{p,i}x\right), \quad \text{then all } \left|\alpha_{p,i}\right| = 1.$$

$$(1.325)$$

Equivalently this gives a bound on the Whittaker–Fourier coefficients $|A(p,1)| \le 3$. A related conjecture is the **Selberg eigenvalue conjecture** which says that if f is a Maass cusp form of type ν for $SL(3,\mathbb{Z})$ then $\text{Re}\,\nu_i = \frac{1}{3}$, for $i = 1,2$. Both conjectures are still open for the case we are discussing. See [230, p. 384] for a discussion of the Luo et al. [412, 413] result from the late 1990s which gives a bound on the $\alpha_{p,i}$ in formula (1.325) of the form $\left|\alpha_{p,i}\right| \le p^e$, where $e = \frac{1}{2} - \frac{1}{10}$. A similar result is obtained for the ν_i. Sarnak discusses progress on the Ramanujan conjecture for general groups in [528].

Why is one interested in all this? Much of the interest comes from the **Langlands conjectures** which basically say that all the number theorists' L-functions in the Selberg class come from $GL(n)$. This would hopefully end up proving things like the Artin conjecture on the holomorphicity of nontrivial Artin L-functions. See [230, last chapter].

Now we want to consider the computational results for Maass cusp forms for $SL(3,\mathbb{Z})$. As with $SL(2,\mathbb{Z})$ Maass forms, there are only computer approximations— no explicit examples. In his thesis, Stephen D. Miller [446] found a region with no Maass cusp forms f of type ν. He writes the power function, for $t \in T_3$, as $\varphi_\mu(I[t]) = t_1^{2\mu_1-1} t_2^{2\mu_2} t_3^{2\mu_3+1}$ with the eigenvalue λ of Δ such that $|\lambda| = \left(\mu_1^2 + \mu_2^2 + \mu_3^2\right) - 1/2$. He finds that $|\lambda| > 80$. Using some representation theory about the irreducible subspaces of $L^2(\Gamma \backslash G)$ coming from cusp forms and a description of the unitary dual from Birgit Speh [573], he gets the fact that $\mu_1 + \mu_2 + \mu_3 = 0$, also $\{\mu_i\} = \{-\overline{\mu_i}\}$. There are 2 possible cases—either the μ_j are all imaginary and add to 0 or 2 of the μ_j are not imaginary and the 3rd μ_j is imaginary. The first case is called the **tempered** case and the 2nd the **non-tempered**. Selberg's eigenvalue conjecture says that cusp forms are tempered. In the tempered case then $\mu_1 = iy_1$, $\mu_2 = iy_2$ and $\mu_3 = -i(y_1 + y_2)$. By the explicit formula for the Rankin–Selberg L-function of f with a carefully chosen set of kernel functions, Miller obtains a figure for the excluded region for the tempered unitary (y_1, y_2) pairs.

More recently David Farmer, Sally Koutsoliotas & Stefan Lemurell [172] have done numerical computations of L-functions of Maass cusp forms which are eigenforms of the Hecke operations for $SL(3,\mathbb{Z})$. They find a slightly larger excluded region using the method of approximate functional equations for the L-functions. See Figure 1.31 in which the black region is that of Stephen D. Miller's thesis. They found more than 2000 spectral parameters (y_1, y_2) and associated Dirichlet coefficients. These are on the web at

http://www.LMFDB.org/L/degree3.

Other references for such computations are Ce Bian [49] and Borislav Mezhericher [445]. In his thesis, Bian looked at the L-functions twisted by a Dirichlet character and confirmed the Ramanujan conjecture for the computed

Fig. 1.31 From Farmer et al. [172], this figure shows the parameters in the functional equation of the L-functions for cuspidal Maass eigenforms for $SL(3, \mathbb{Z})$ as *purple dots*. The *black region* is Miller's region containing no such parameters [446]. The *purple region* also contains no parameters, assuming the Ramanujan conjecture

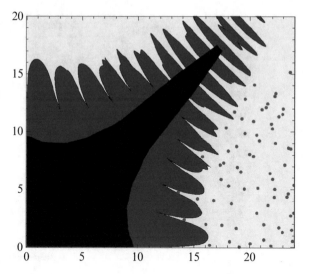

Maass cusp forms.   Bian also checked whether Whittaker–Fourier coefficients satisfy the Sato-Tate or Wigner semi-circle distribution for the first Gelbart–Jacquet lift of a Maass cusp form for $SL(2, \mathbb{Z})$. He looked as well at the distribution of the Fourier coefficients of a non-Gelbart–Jacquet lift and compared it with the distribution of the absolute trace on $SU(3)$ with respect to Haar measure. See Figure 3 of Bian's paper which is an earlier version of Figure 1.32 below. In both cases, the agreement between the conjectural and computed histograms is good.

Andrew R. Booker, who was Bian's thesis advisor, notes in [60] that Bian's data passes three tests for Maass cusp forms on $GL(3, \mathbb{Z})$—at least to computer precision. Test 1 checks whether the Whittaker–Fourier coefficients are multiplicative. Test 2 asks for the check provided by Figure 1.32, which was just emailed to me by Booker in August, 2015. The non-lift is the same as that appearing in Figure 3 of [49] and Figure 2 of [60]. The lift comes from the first Maass cusp form for $SL(2, \mathbb{Z})$. Figure 1.32 shows that the data from a non-lift Maass cusp form matches Langlands' conjecture on the distribution of the Fourier coefficients $A(p, 1)$ for prime p. This figure simultaneously shows that the data from lifts have distributions ruled by the underlying $GL(2, \mathbb{Z})$ Maass form. Test 3 asks that the L-function obey the Riemann hypothesis. Booker notes that Michael Rubinstein numerically verified the Riemann hypothesis for the first few zeros of the L-functions corresponding to Bian's data while at an American Institute of Mathematics meeting in 2008.

Mezhericher [445] uses the Fourier expansions of Maass forms. For this, he develops his own algorithms to compute the Whittaker functions.

Kowalski and Ricotta [369] obtain a central limit theorem for the Whittaker–Fourier coefficients of Maass forms using Voronoi summation formulas. Krötz and Opdam [372] use knowledge of the holomorphic extension of eigenfunctions of the invariant differential operators to the crown domain of the symmetric space to show that Maass cusp forms have exponential decay.

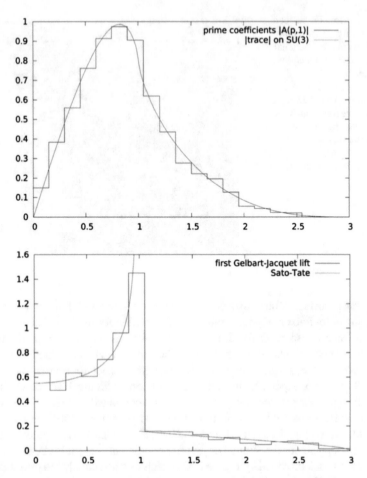

Fig. 1.32 The *top figure* is a histogram for the first 2312 values $|A(p, 1)|$ for Bian's first Maass cusp form for $SL(3, \mathbb{Z})$ compared with the conjectural distribution. The *lower figure* is a similar comparison for the first Gelbart–Jacquet lift. The figures were based on Bian's thesis and were provided by Andrew Booker

Langlands' Inner Product Formulas for Truncated Eisenstein Series

Now we wish to address the Langlands formulas for inner products of truncated Eisenstein series for $SL(3, \mathbb{Z})$. Finding a non-adelic discussion of these results which is not so general and obscured with notation from representation theory is difficult. We have been looking at Paul Garrett's articles on his website [204], Feryâl Alayont [4, 5], and Stephen D. Miller [446, 447]. Other references are Robert P. Langlands [392], James Arthur [21, 22], and Colette Mœglin and Jean-Loup Waldspurger [455, 456].

For $\Gamma = GL(3, \mathbb{Z})$ there are three parabolics: $P_0 = P(1,1,1)$, $P_1 = P(2,1)$, and $P_2 = P(1,2)$. Then we say that P_1 and P_2 are **associated** (writing $P_1 \sim P_2$) and P_0 is **self-associate**. If $P = P(n_1, n_2)$ is a maximal parabolic subgroup of $G = GL(3, \mathbb{R})$, M^P the Levi subgroup of matrices of the form $\begin{pmatrix} m_1 & 0 \\ 0 & m_2 \end{pmatrix}$, with $m_i \in GL(n_i, \mathbb{R})$, and N^P the nilpotent or unipotent subgroup of matrices of the form $\begin{pmatrix} I_{n_1} & * \\ 0 & I_{n_2} \end{pmatrix}$, then we have the **Langlands decomposition** of the parabolic subgroup along with a decomposition of G itself:

$$P = N^P M^P \quad \text{and} \quad G = PK, \quad \text{for } K = O(3). \tag{1.326}$$

Given a function f on G which is left $P \cap \Gamma$-invariant, we can form the **incomplete theta series** (alias a Poincaré or Eisenstein series) as in Volume I, page 321:

$$T^P f(g) = \sum_{\gamma \in \Gamma \cap P \backslash \Gamma} f(\gamma g).$$

As we saw on pages 322–323 of Volume I, these incomplete theta series have some important relations with the constant coefficient operator and the Eisenstein series. Here we only sketch a few similar results. It would be useful to do more. We leave that to the poor abused reader.

Again given a function f on G which is left $N^P \cap \Gamma$-invariant, we can form the **constant term**

$$c_P f(g) = \int_{N^P \cap \Gamma \backslash N^P} f(ng)\,dn. \tag{1.327}$$

Suppose that f_i are Maass cusp forms for $GL(n_i, \mathbb{Z})$, meaning that when $n_i = 1$, then f_i is identically 1. The **Eisenstein series** are special incomplete theta series of the form

$$T^P \varphi = E_\varphi^P(g) = \sum_{\gamma \in P \backslash \Gamma} \varphi(\gamma g), \quad \text{where } \varphi(nmk) = \|m_1\|^{n_2 s} \|m_2\|^{-n_1 s} f_1(m_1) f_2(m_2).$$

One finds that if say $P = P_1 = P(2,1)$ and $Q = P_2 = P(1,2)$ (or vice versa), then the constant terms for P and Q are:

$$c_P T^P \varphi = \varphi \quad \text{and} \quad c_Q T^P \varphi = \xi_s \varphi. \tag{1.328}$$

Here ξ_s^φ denotes the quotient of L-functions and gamma functions appearing in the **functional equation** of the Eisenstein series for the maximal parabolic subgroup

$$E_\varphi^P(g) = \xi_s^\varphi E_{\varphi^w}^Q, \tag{1.329}$$

where

$$\varphi^w(nmk) = \|m_1\|^{n_2(1-s)} \|m_2\|^{-n_1(1-s)} f_1(m_1) f_2(m_2) = \varphi\left(wmw^{-1}\right) \frac{\|m_2\|^{n_1}}{\|m_1\|^{n_2}}.$$
(1.330)

Here w is a Weyl group element which exchanges M^{P_1} and M^{P_2}. We can take $w = \omega$ from formula (1.315).

Now we can define the truncation operators Λ^A, for A large and positive, and again with respect to the maximal parabolic subgroups. First define a **height function** and corresponding **truncation set**, using the decomposition (1.326):

$$h_P(g) = \frac{\|m_1\|^{n_2}}{\|m_2\|^{n_1}}, \text{ if } g = nmk \text{ and } S_P^A = \left\{g \in G \mid h^P(g) \geq A\right\}.$$

Define for a left $\left(N^P \cap \Gamma\right)$-invariant function F

$$c_P^A F = (c_P F) \cdot 1_{S_P^A},$$

where 1_S denotes the indicator function of S. Then if $P = P_1 = P(2,1)$ and $Q = P_2 = P(1,2)$, the **truncation operator** is

$$\Lambda^A E_\varphi^P = E_\varphi^P - T^P\left(c_P^A E_\varphi^P\right) - T^Q\left(c_Q^A E_\varphi^P\right).$$
(1.331)

Langland's inner product formula involves a second truncated Eisenstein series E_ψ^P with ψ defined as follows using the decomposition (1.326), assuming the h_i are Maass cusp forms for $GL(n_i, \mathbb{Z})$,

$$\psi(nmk) = \|m_1\|^{n_2 r} \|m_2\|^{-n_1 r} h_1(m_1) h_2(m_2).$$

The **Langlands inner product formula** is then

$$\left\langle \Lambda^A E_\varphi^P, \Lambda^A E_\psi^P \right\rangle = \langle f, h \rangle \frac{A^{s+\bar{r}-1}}{s + \bar{r} - 1} + \langle f^w, h^w \rangle \xi_s^\varphi \overline{\xi_r^\psi} \frac{A^{(1-s)+(1-\bar{r})-1}}{(1-s)+(1-\bar{r})-1}.$$
(1.332)

Here $\langle f, h \rangle$ is the inner product on $GL(n_1) \times GL(n_2)$ mod the center for $f(m_1, m_2) = f_1(m_1) f_2(m_2)$ and h defined similarly. See Garrett [204] for a proof of (1.332). In the case of a self-associate parabolic subgroup P the number of terms in the Langland's inner product formula is the square of the order of the Weyl group $W(P, P)$, which is 36 for $P(1, 1, 1)$.

This formula is the analogue of the Maass–Selberg relations for $SL(2, \mathbb{Z})$ and has been used by Stephen D. Miller [447] in his pre-trace formula proof of the Weyl law for cusp forms for $SL(3, \mathbb{Z})$. He also had to use the analogous formula for the minimal parabolic $P_0 = P(1, 1, 1)$ with its 36 terms. Other uses are in the proof of the analytic continuation of Eisenstein series. See Feryâl Alayont [4]. There is reputed to be a manuscript of Bernstein on this subject. Hopefully one could use the result along with the analogue for P_0 to prove a trace formula for $SL(3, \mathbb{Z})$. But the number of Langlands inner product terms for $SL(3, \mathbb{Z})$, plus the number of orbital integrals associated with the conjugacy classes in $SL(3, \mathbb{Z})$ have kept us from working on this in our old age. So in the next section we still do not tell you what the trace formula for $SL(3, \mathbb{Z})$ is. Sorry. You can find adelic versions of the trace formula in Arthur's papers [21–23, 25].

Deitmar and Pavey [132] obtain a prime geodesic theorem for rank 1 closed geodesics in the fundamental domain for a discrete cocompact subgroup Γ of $SL(4, \mathbb{R})$ with the goal of applying the result to the asymptotics of class numbers of quartic number fields.

1.5.5   Remarks on Harmonic Analysis on the Fundamental Domain

Recall that the Euclidean Poisson summation formula for a Schwartz function f : $\mathbb{R}^n/\mathbb{Z}^n \to \mathbb{C}$ says:

$$f(x + a) = \sum_{a \in \mathbb{Z}^n} \widehat{f}(a) \exp\left(2\pi i \, {}^t ax\right),$$

where $\widehat{f}$ denotes the Fourier transform on $\mathbb{R}^n$ defined by

$$\widehat{f}(a) = \int_{y \in \mathbb{R}^n} f(y) \exp\left(-2\pi i \, {}^t ay\right) \, dy.$$

We have seen in Section 1.4 of Volume I that there are many applications. For example, setting $f(x) = \exp(-Y[x])$ for $Y \in \mathcal{P}_n$, we obtain the transformation formula for the theta function. This transformation formula allows one to prove the analytic continuation and functional equation of Epstein's zeta function. It also gives information about the asymptotics of the fundamental solution of the heat equation on $\mathbb{R}^n/\mathbb{Z}^n$.

In Theorem 3.7.3 of Volume I, we found an analogue of Poisson summation for functions on the fundamental domain for $SL(2, \mathbb{Z})$ in the Poincaré upper half plane. Now we want to examine a generalization of this result to $GL(n, \mathbb{Z})$. The result we seek to prove follows easily from the generalization of the Roelcke–Selberg spectral decomposition of the Laplacian to $GL(n, \mathbb{Z})$, a result stated by Arthur [25]. For a proof of this highly nontrivial theorem, one needs the discussion of Langlands [392] or Harish-Chandra [262], or Osborne and Warner [482]. Some of the basics are to be found in the preceding subsection. Other references are Goldfeld [230] as well as Jorgenson and Lang [334]. Here we suppose for simplicity that f lies in $C_c^\infty(\mathcal{SP}_n)$ and that f is $O(n)$-invariant. This implies that the Helgason–Fourier transform $\widehat{f}$

defined in (1.121) of Section 1.3.1 is independent of the rotation variable $k \in O(n)$. Thus we can write $\widehat{f}$ as a function of $s \in \mathbb{C}^n$ alone:

$$\widehat{f}(s) = \int_{Y \in \mathcal{P}_n} f(Y) \overline{p_s(Y)} \, d\mu_n(Y).$$ (1.333)

Poisson summation for $\Gamma_n = GL(n, \mathbb{Z})$ says:

$$\left. \begin{array}{l} \displaystyle\sum_{\gamma \in \Gamma_n/\{\pm I\}} f(I[a\gamma b^{-1}]) = \sum_{m \geq 0} \widehat{f}(s_m) w_m(I[a]) \overline{w_m(I[b])} \\[2mm] \displaystyle \qquad + \sum_P \sum_{v \in A(P)} \kappa_P \int_{\substack{r \in \mathbb{C}^{|P|-1} \\ \operatorname{Re} r = \text{constant}}} \widehat{f}(s_P(v, r)) E_P(v, r | I[a]) \overline{E_P(v, r | I[b])} \, dr. \end{array} \right\}$$

(1.334)

Here the sum over P runs over all nonassociated parabolic subgroups $P = P(n_1, \ldots, n_q)$ as in (1.241) and we write $|P| = q$. We say that two parabolic subgroups $P(n_1, \ldots, n_q)$ and $P(m_1, \ldots, m_q)$ are **associated** if the m_j are a permutation of the n_i. The sum over $m \geq 0$ corresponds to a sum over the discrete spectrum of the G-invariant differential operators in $D(\mathcal{SP}_n)$ on $L^2(\mathcal{SP}_n/\Gamma_n)$, $\Gamma_n = GL(n, \mathbb{Z})$. We will use the notation:

$$A(GL(n)) = \{w_m | m \geq 0\}$$ (1.335)

to represent an orthonormal basis of $L^2(\mathcal{SP}_n/\Gamma_n)$ consisting of eigenfunctions of $D(\mathcal{SP}_n)$. We will let w_0 be the constant function. This may be viewed as a residue of an Eisenstein series. There will also be cusp forms as defined in (1.311) of Section 1.5.3. Thus $A(GL(n))$ has both "cuspidal" and "residual" parts. It is somewhat confusing to refer to the corresponding parts of the spectrum as the "cuspidal spectrum" and the "residual spectrum" since the term "residual spectrum" has a different meaning in spectral theory. See, for example Reed and Simon [501, pp. 188, 194] where it is proved that self-adjoint operators have *no* residual spectrum in the sense of functional analysis. So let us just say that the spectrum has "cuspidal" and "non-cuspidal" components. The "non-cuspidal" part has proved to be problematical for higher rank versions of the trace formula, as we noted at the end of the last subsection.

To continue with the definitions of the terms in (1.334), now suppose that we are given the parabolic subgroup $P = P(n_1, \ldots, n_q)$ and write the corresponding partial Iwasawa decomposition of $Y \in \mathcal{SP}_n$ as:

$$Y = \begin{pmatrix} a_1 & \cdots & 0 \\ \vdots & \ddots & \vdots \\ 0 & \cdots & a_q \end{pmatrix} \begin{bmatrix} I_{n_1} & \cdots & * \\ \vdots & \ddots & \vdots \\ 0 & \cdots & I_{n_q} \end{bmatrix}, \quad a_j \in \mathcal{P}_{n_j}.$$ (1.336)

Here

$$\prod_{j=1}^{q} |a_j| = 1, \quad \text{assuming } |Y| = 1.$$

If $n_j > 1$, we have a nonempty discrete spectrum $A(GL(n_j))$ as in (1.335). We define

$$A\left(P(n_1, \ldots, n_q)\right) = \left\{v = (v_1, \ldots, v_q) \mid v_j \in A\left(GL\left(n_j\right)\right)\right\}, \qquad (1.337)$$

where if $n_j = 1$, we write $A(GL(1)) = \{1\}$.

As in (1.248) of Section 1.5.1, the **Eisenstein series** in the continuous spectrum are given by:

$$E_P(v, r|Y) = \sum_{\gamma \in \Gamma_n/P} \varphi(Y[\gamma]), \qquad (1.338)$$

for

$$\varphi(Y) = \prod_{j=1}^{q} v_j\left(a_j^0\right) |a_j|^{-r_j}, \quad r_j \in \mathbb{C},$$

with $v = (v_1, \ldots, v_q) \in A(P)$, Re r_j suitably restricted for convergence of (1.338). As usual, we are using the notation:

$$a_j^0 = |a_j|^{-1/n_j} a_j \in \mathcal{SP}_{n_j}.$$

Actually, since the determinant of Y is 1, we have only $q - 1$ independent r-variables. So we just integrate over the first $q - 1$ of them. Also, we must continue the Eisenstein series outside of the region of convergence of (1.338) to reach the spectrum of the invariant differential operators. Since, $E_P(v, r|Y)$ is an eigenfunction of all the G-invariant differential operators on $\mathcal{SP}_n$, it determines, as in Proposition 1.2.4 of Section 1.2.3, a vector of powers $s_P(v, r) = s \in \mathbb{C}^n$ to put in the Helgason–Fourier transform $\widehat{f}(s(r))$ in (1.333). That is, E_P and p_{-s} have the same eigenvalues for the G-invariant differential operators L:

$$LE_P(v, r|Y) = \lambda_L E_P(v, r|Y) \quad \text{and} \quad Lp_{=s} = \lambda_L p_{-s} \text{ for all } L \in D(\mathcal{SP}_n),$$
$$(1.339)$$

where $p_s(Y)$ denotes the power function defined in (1.41) of Section 1.2.1. Finally, the κ_P in formula (1.334) denotes a positive constant.

The proof of Poisson's summation formula comes from the spectral decomposition of the G-invariant differential operators on $L^2(\mathcal{SP}_n/GL(n, \mathbb{Z}))$ plus Selberg's basic lemma that eigenfunctions of G-invariant differential operators are eigenfunctions of G-invariant integral operators (Proposition 1.2.4 of Section 1.2.3). We spell out the details later in the section.

One of the first applications of the Poisson summation formula is to study an analogue of the circle problem on the asymptotics of the number of lattice points in a circle in $\mathbb{R}^2$ of radius x as x approaches infinity. In our case the statistical function is

$$\#\left\{\gamma \in GL(n,\mathbb{Z}) \;\middle|\; \mathrm{Tr}({}^t\gamma\gamma) \le x\right\}, \qquad (1.340)$$

and one should obtain the asymptotic character of this function as x approaches infinity. In order to imitate the discussion for $n = 2$ that appears in pages 334–337 of Volume I, it is necessary to study the Helgason–Fourier transform of the function $\exp(-p\,\mathrm{Tr}(Y))$ for $p > 0$ and Y in $\mathcal{SP}_n$. When $n = 2$, this is a K-Bessel function (though it looks like a gamma function). We have not carried out the details for $GL(n)$. Results have been obtained by many authors, going back to Delsarte in the 1940's for cocompact subgroups. Results for general compact $K\backslash G/\Gamma$ are obtained by Bartels [38]. The case of $G = SL(2,\mathbb{C})$, coming from imaginary quadratic fields is considered by Elstrodt et al. [168]. We will say more about this in Chapter 2.

The Poisson summation formula can also be used to study the cusp forms themselves. Stephen D. Miller [446] uses this to obtain a pre-trace formula for cocompact Γ as well as $SL(3,\mathbb{Z})$ and then deduce a Weyl law as we noted earlier.

We should also mention that the spectral decomposition of the $GL(n,\mathbb{R})$-invariant differential operators on $\mathcal{SP}_n/GL(n,\mathbb{Z})$ can be used to obtain a converse theorem for an analogue of Hecke's correspondence that takes place between Siegel modular forms and Dirichlet series in several variables. This is worked out for $Sp(2,\mathbb{Z})$ by Kaori Imai (Ota) [317], extending work of Koecher [359] and Maass [426]. We will say a little more about the subject of Hecke's correspondence for Siegel modular forms in Chapter 2. Weissauer [663] obtains a very general converse theorem for congruence subgroups of $Sp(n,\mathbb{Z})$. Bill Duke and Özlem Imamoḡlu [146] obtain a lifting of certain holomorphic cusp forms on the upper half plane to Siegel cusp forms using Kaori Imai's converse theorem.

After this brief introduction, let us attempt to understand the discrete spectrum better. To do this we need to develop an analogue of Theorem 3.7.2 on p. 331 of Volume I which gave the existence of a complete orthonormal set of cusp forms for $SL(2,\mathbb{Z})$. So we define:

$$L_0^2\left(\mathcal{SP}_n/GL(n,\mathbb{Z})\right) = \left\{h \in L^2(\mathcal{SP}_n/GL(n,\mathbb{Z})) \;\middle|\; \begin{array}{l} a_0^k(Y) = 0, \text{ for almost all} \\ Y \in \mathcal{SP}_n, \; 1 \le k \le n-1 \end{array}\right\},$$
$$(1.341)$$

where $a_0^k(v)$ is defined by (1.311) in § 1.5.3 as the 0th Fourier coefficient of h with respect to $P(k, n - k)$. Gelfand and Piatetski-Shapiro have proved the $GL(n)$-analogue of Theorem 3.7.2, Volume I. Let us give a sketch of Godement's proof of this result (see Borel and Mostow [68, pp. 225–234]). The argument is very close to the one we gave in Volume I, but it is complicated by the non-abelian nature of the nilpotent subgroup $N \subset G = SL(n,\mathbb{R})$, where N consists of upper triangular matrices with ones on the diagonal.

We want to prove the compactness of the convolution operators C_h, defined in (1.342) below, acting on $L^2_0(\mathcal{SP}_n/SL(n, \mathbb{Z}))$, for $h : \mathcal{SP}_n \to \mathbb{C}$ infinitely differentiable with compact support. We identify functions on $\mathcal{SP}_n$ with functions on $G = SL(n, \mathbb{R})$ via $f\left((I[g])^0\right) = f(g)$ for all $g \in G$ and write the convolution operator as in (1.24) of Section 1.1.4:

$$C_h f(a) = \int_{b \in G = SL(n,\mathbb{R})} h(ab^{-1}) f(b) \, db. \tag{1.342}$$

Theorem 1.5.7 (Gelfand and Piatetski-Shapiro). *The convolution operator C_h defined in (1.342) gives a compact operator on the space of square integrable functions on the fundamental domain with vanishing constant terms; i.e., on $L^2_0(\mathcal{SP}_n/GL(n, \mathbb{Z}))$ defined in (1.341).*

Proof. This discussion follows a similar path to that laid out in the proof of Theorem 3.7.2 in Volume I. Here we will consider only the case that $n = 3$.

Just as for $n = 2$, it suffices, by the Theorem of Arzelà and Ascoli (see Kolmogorov and Fomin [362]) to show that if $f \in L^2_0(\mathcal{SP}_n/SL(n, \mathbb{Z}))$,

$$|C_h f(a)| \leq k \|f\|_2, \quad \text{for some positive constant } k,$$

where

$$\|f\|^2_2 = \int_{\mathcal{SP}_3/\Gamma_3} |f(W)|^2 dW, \quad \Gamma_3 = GL(3, \mathbb{Z}).$$

We assume that $h(b) = h(b^{-1})$ so that:

$$C_h f(x) = \int_{\mathcal{SP}_3} f(W) h\left(W\left[x^{-1}\right]\right) dW$$

$$= \int_{\mathcal{SP}_3/N_{\mathbb{Z}}} f(W) \sum_{n \in N_{\mathbb{Z}}} h\left(W\left[nx^{-1}\right]\right) dW,$$

where $N_{\mathbb{Z}}$ denotes the integral upper triangular 3×3 matrices with ones on the diagonal.

Now apply ordinary Poisson summation as stated at the beginning of this section to rewrite the sum over $N_{\mathbb{Z}}$. You find that this sum is equal to:

$$\sum_{R \in N_{\mathbb{Z}}} \widehat{h}(R), \quad \text{where } \widehat{h}(R) = \int_{n \in N_{\mathbb{R}}} h\left(W\left[nx^{-1}\right]\right) \exp\left(2\pi i \sum_{1 \leq j < k \leq 3} n_{jk} r_{jk}\right) dn.$$

Here $N_{\mathbb{R}}$ denotes the real 3×3 upper triangular matrices with ones on the diagonal.

We may assume that h lies in a Dirac delta sequence at the identity, in particular, that h is $K = O(3)$ bi-invariant. Write the Iwasawa decomposition of $x \in G$ as $x = kav$, where $k \in K = O(3)$, $v \in N_\mathbb{R}$, and a is positive diagonal with jth diagonal entry a_j. We want to bound $C_h f(x)$ as a_i/a_{i+1} approaches zero for $i = 1, 2$. Let

$$W = b[t] = \begin{pmatrix} b_1 & 0 & 0 \\ 0 & b_2 & 0 \\ 0 & 0 & b_3 \end{pmatrix} \begin{bmatrix} 1 & t_{12} & t_{13} \\ 0 & 1 & t_{23} \\ 0 & 0 & 1 \end{bmatrix}, \quad \text{for } b_j > 0.$$

Then

$$\widehat{h}(R) = \int_{n \in N_\mathbb{R}} h\left(b\left[tnv^{-1}a^{-1}\right]\right) \exp\left(2\pi i \sum_{j<k} n_{jk} r_{jk}\right) dn.$$

Next change variables via $m = tnv^{-1}$, $n = t^{-1}mv$. Recall that

$$t = \begin{pmatrix} 1 & t_{12} & t_{13} \\ 0 & 1 & t_{23} \\ 0 & 0 & 1 \end{pmatrix} \text{ implies } t^{-1} = \begin{pmatrix} 1 & -t_{12} & t_{12}t_{23} - t_{13} \\ 0 & 1 & -t_{23} \\ 0 & 0 & 1 \end{pmatrix}$$

and

$$\begin{pmatrix} 1 & m_{12} & m_{13} \\ 0 & 1 & m_{23} \\ 0 & 0 & 1 \end{pmatrix} \begin{pmatrix} 1 & v_{12} & v_{13} \\ 0 & 1 & v_{23} \\ 0 & 0 & 1 \end{pmatrix} = \begin{pmatrix} 1 & m_{12} + v_{12} & m_{13} + v_{13} + m_{12}v_{23} \\ 0 & 1 & m_{23} + v_{23} \\ 0 & 0 & 1 \end{pmatrix}.$$

From this, we see that, if $n = t^{-1}mv$ and we use the same notation for entries of n as for entries of t, then we obtain:

$$n_{12} = -t_{12} + m_{12} + v_{12},$$

$$n_{23} = -t_{23} + m_{23} + v_{23},$$

$$n_{13} = -t_{13} + m_{13} + v_{13} + m_{12}v_{23} + t_{12}t_{23} - t_{12}(m_{23} + v_{23}).$$

It follows that, with the n_{ij} as above:

$$\widehat{h}(R) = \int_{m \in N_\mathbb{R}} h\left(b\left[ma^{-1}\right]\right) \exp\{2\pi i \left(r_{12}n_{12} + r_{13}n_{13} + r_{23}n_{23}\right)\} \, dm.$$

Then we must change variables via $n = ama^{-1}$, with Jacobian:

$$\alpha(a) = \prod_{i<j} a_j/a_i.$$

This gives:

$$\widehat{h}(R) = \alpha(a) \int_{n \in N_{\mathbb{R}}} h\left(b\left[a^{-1}n\right]\right) \exp\{2\pi i T(n)\} \, dn,$$

where

$$T(n) = r_{12}\left(\frac{a_2}{a_1}n_{12} - t_{12} + v_{12}\right) + r_{23}\left(\frac{a_3}{a_2}n_{23} - t_{23} + v_{23}\right)$$

$$+r_{13}\left\{v_{13} + \frac{a_2}{a_1}n_{12}v_{23} + \frac{a_3}{a_1}n_{13} - t_{12}\left(\frac{a_3}{a_2}n_{23} + v_{23}\right) + t_{12}t_{23} - t_{13}\right\}.$$

It will be easier if we approximate h by functions such that:

$$h\left(b[t]\right) = h_0(b)h_{12}(t_{12})h_{13}(t_{13})h_{23}(t_{23}).$$

Then we have:

$$\widehat{h}(R) = \alpha(a)h_0\left(b\left[a^{-1}\right]\right) \exp\{2\pi i\, L(r, v, t)\}\, \widehat{h}_{12}\left(\frac{a_2}{a_1}(r_{12} + r_{13}v_{23})\right)$$

$$\times \widehat{h}_{13}\left(\frac{a_3}{a_1}r_{13}\right)\widehat{h}_{23}\left(\frac{a_3}{a_2}(r_{23} - r_{13}t_{12})\right),$$

where

$$L(r, v, t) = r_{12}\left(v_{12} - t_{12}\right) + r_{13}\left(v_{13} - t_{12}v_{23} + t_{12}t_{23} - t_{13}\right) + r_{23}\left(v_{23} - t_{23}\right).$$

If $R = (r_{12}, r_{13}, r_{23}) = 0$, then the fact that $f \in L_0^2(\mathcal{SP}_n/SL(n, \mathbb{Z}))$ says that the integral over $t_{ij} \in [0, 1]$ is zero. Thus we need to only consider $R \neq 0$.

Since h is infinitely differentiable with compact support (i.e., $h \in C_c^\infty$), we can apply differential operators to h and stay in C_c^∞. But then we can obtain the following bound:

$$\sum_{R \neq 0} \widehat{h}(R) \leq \alpha(a)h_0(b[a^{-1}])Z(W(a, v, t), s),$$

where $Z(W, s)$ is Epstein's zeta function from Section 1.4 of Volume I and its first argument is the positive matrix:

$$W(a, v, t) = I \begin{bmatrix} a_2/a_1 & v_{23}a_2/a_1 & 0 \\ 0 & a_3/a_1 & 0 \\ 0 & -t_{12}a_3/a_2 & a_3/a_2 \end{bmatrix}.$$

Here s is any sufficiently large integer. Epstein's zeta function can in turn be bounded by

$$\max(\delta_1^{-s}, \delta_2^{-s}, \delta_3^{-s}) \sum_{R \in \mathbb{Z}^3 - 0} Y(v, t)[R]^{-s},$$

where $\delta_1 = a_2/a_1$, $\delta_2 = a_3/a_1$, $\delta_3 = a_3/a_2$, and

$$Y(v, t) = I \begin{bmatrix} 1 & v_{23} & 0 \\ 0 & 1 & 0 \\ 0 & -t_{12} & 1 \end{bmatrix}.$$

Since v and t can be assumed to be bounded, our only problem is to deal with $h_0\left(b\left[a^{-1}\right]\right)$ for a_i/a_{i+1} near zero. But, because h_0 has compact support, $h_0\left(b\left[a^{-1}\right]\right) \neq 0$ implies

$$c_1 < \frac{b_i}{b_{i+1}} \frac{a_{i+1}}{a_i} < c_2.$$

If we assume that the quotient a_i/a_{i+1} is sufficiently near zero, then

$$b_i/b_{i+1} < c,$$

for as small a constant c as we like.

Thus we have bounded $|C_h f(x)|$ by a constant multiplied by the product of some powers of a_i/a_{i+1} and the following integral, if $x = kav$, as above:

$$\int\limits_{|t_{ij}| \leq 1/2} \int\limits_{b_i \leq cb_{i+1}} f\left(b[t]\right) \, db_1 \, db_1 \, db_2 \, dt_{12} \, dt_{13} \, dt_{23}.$$

To complete the proof of the bound on $C_h f(x)$, one must note that when $c = 1$, the domain of integration in the integral above must lie in the fundamental domain $\mathcal{SF}_3$ of Section 1.4.3. For this, see Exercise 1.4.19 of Section 1.4.3. Therefore the desired inequality follows from the Cauchy–Schwartz inequality.   ∎

There is another question one might raise at this point, as Kaori Imai (Ota) pointed out to me. Our discussion of the discreteness of the spectrum of the cusp forms for $SL(3, \mathbb{Z})$ used the vanishing of the integral defined by:

$$\iiint\limits_{t_{ij}\in[0,1]} f(Y[t]) \, dt_{12} \, dt_{13} \, dt_{23}, \quad t = \begin{pmatrix} 1 & t_{12} & t_{13} \\ 0 & 1 & t_{23} \\ 0 & 0 & 1 \end{pmatrix}. \qquad (1.343)$$

The definition of cusp form given in (1.311) requires the vanishing of both of the integrals below:

$$\left. \begin{aligned} \iint\limits_{t_{ij}\in[0,1]} f(Y[t]) \, dt_{12} \, dt_{13}, t = \begin{pmatrix} 1 & t_{12} & t_{13} \\ 0 & 1 & 0 \\ 0 & 0 & 1 \end{pmatrix}, \\ \iint\limits_{t_{ij}\in[0,1]} f(Y[t]) \, dt_{13} \, dt_{23}, \quad t = \begin{pmatrix} 1 & 0 & t_{13} \\ 0 & 1 & t_{23} \\ 0 & 0 & 1 \end{pmatrix}. \end{aligned} \right\} \qquad (1.344)$$

Note that (1.344) implies (1.343) but the converse is not clear.

Next let us say a little more about the proof of (1.334). Note that we saw in Section 1.5.3 for a special case that if P and Q are nonassociated parabolic subgroups of $GL(n)$, then:

$$E_P(v, r|Y) \sim \prod_{j=1}^{q} v_j\left(a_j^0\right) \left|a_j\right|^{-r_j}, \quad \text{as } Y \to \partial \mathcal{SF}_n,$$

in the direction of P. Here $\partial \mathcal{SF}_n$ denotes the boundary of the fundamental domain $\mathcal{SF}_n$ for $\mathcal{SP}_n/GL(n, \mathbb{Z})$. However, because P and Q are nonassociated parabolic subgroups and if v is a cusp form:

$$E_Q(v, r|Y) \sim 0,$$

as Y approaches the boundary of the fundamental domain in the direction corresponding to Q. One expects that this implies that E_P and E_Q are orthogonal provided that P and Q are nonassociated and at least one of the Eisenstein series has a cusp form floating around in it. An argument similar to that of Exercise 1.5.37, Section 1.5.3 should work here provided that one has proved the vanishing of the 0th Fourier coefficient of E_Q with respect to the parabolic subgroup P.

Of course, if P and Q are associated parabolic subgroups, then one can see that E_P and E_Q are really the same function after a trivial change of variables. See (1.329) in the preceding section or Exercise 1.5.26 in Section 1.5.2.

Thus it appears likely that a generalization of the principle of asymptotics and functional equations from Section 1.3 should lead to the spectral resolution of the G-invariant differential operators on $L^2(\mathcal{SP}_n/GL(n, \mathbb{Z}))$. One could attempt to imitate the approach given for Theorem 3.7.1 in Volume I for $SL(2, \mathbb{Z})$, using incomplete theta series attached to the nonassociated parabolic subgroups. We will not go into this here beyond what was said in the preceding subsection. We would, for example, need to see that someone orthogonal to all the Eisenstein series E_P for all P must be in L_0^2 and thus in the span of the discrete spectrum $A(GL(n))$. The ultimate result would be that any $g \in L^2(\mathcal{SP}_n/GL(n, \mathbb{Z}))$ has the **spectral decomposition**:

$$\left. \begin{aligned} g(Y) = &\sum_{w \in A(GL(n))} (g, w) w(Y) \\ &+ \sum_{P} \sum_{v \in A(P)} \kappa_P \int_{\substack{r \in \mathbb{C}^{q-1} \\ \text{Re } r_j = \text{constant}}} (g, E_P) \, E_P(v, r | Y) \, dr, \end{aligned} \right\} \qquad (1.345)$$

where (f, g) is the inner product on $L^2(\mathcal{SP}_n/GL(n, \mathbb{Z}))$ with respect to the G-invariant measure. We use the notation of (1.334)–(1.339) above. See Langlands [392], Harish-Chandra [262], Osborne and Warner [482], and Arthur's lecture in Borel and Casselman [66, Vol. I, pp. 253–274]. The proof of formula (1.345) by Meera Thillainatesan is in Goldfeld [230, p. 324 ff].

In order to derive the Poisson summation formula (1.334) from (1.345), one must use Selberg's basic lemma which says that eigenfunctions of G-invariant differential operators are eigenfunctions of G-invariant integral operators (see Proposition 1.2.4 in Section 1.2.3). This shows that if we write $f(I[a]) = f(a)$ for all $a \in G$ and

$$g(a) = \sum_{\gamma \in \Gamma_n/\pm I} f(a\gamma b^{-1}), \quad \text{with } f \text{ as in (1.334),}$$

then by Proposition 1.2.4 of Section 1.2.1, since we are assuming that $f(a) = f(a^{-1})$, we have:

$$\begin{aligned} (g, E_P) &= \int_{a \in G/\Gamma_n} \sum_{\gamma \in \Gamma_n/\pm I} f\left(a\gamma b^{-1}\right) \overline{E_P(I[a])} \, da \\ &= \int_G f\left(ab^{-1}\right) \overline{E_P(I[a])} \, da \\ &= \int_G f\left(ba^{-1}\right) \overline{E_P(I[a])} \, da = C_f(\overline{E}_P)(b) = \widehat{f}(s_P) \overline{E_P(I[b])}. \end{aligned}$$

Here $s_P = s_P(v, r)$ is chosen as in (1.339). A similar argument applies to the discrete spectrum terms in (1.334). This completes our very sketchy discussion of Poisson summation for $GL(n, \mathbb{Z})$.

The main application of the Poisson summation formula (1.334) is to the trace formula. If we are looking at a discrete subgroup Γ of $GL(n, \mathbb{R})$ acting on $\mathcal{SP}_n$ without fixed points and so as to have a **compact** fundamental domain, then the trace formula is easily obtained, since (1.334) then involves no continuous or residual spectrum. See Mostow [466] for an example of such a group Γ. In what follows we give a sketchy discussion of the Selberg trace formula, imitating what we said in the last part of Section 3.7 in Volume I. Another reference for Selberg's trace formula is Hejhal [269].

Stage 1.

$$\int_{K\backslash G/\Gamma} \sum_{\gamma \in \Gamma/\{\pm I\}} f\left(b\gamma b^{-1}\right) dg = \sum_{m\geq 0} \widehat{f}\left(s_m\right). \tag{1.346}$$

Formula (1.346) is obtained by integrating (1.334) and using the orthonormality of the various elements of the spectrum of the G-invariant differential operators on $\mathcal{SP}_n/GL(n, \mathbb{Z})$.

If the fundamental domain for Γ is not compact, divergent integrals arise on the left in (1.346) and must be cancelled against integrals from components of the continuous spectrum, as in the case of $SL(2, \mathbb{Z})$ in Section 3.7 of Volume I. This is reminiscent of renormalization methods in quantum field theory. There are even problems showing that one can take the trace. Some references are Arthur [21–25], A.B. Venkov [627], Dorothy Wallace [642–650], and Warner's paper in the volume of Hejhal et al. [272, pp. 529–534].

Stage 2. Let $\{\gamma\} = \{a\gamma a^{-1} \mid a \in \Gamma/\pm I\} = $ the **conjugacy class** of γ in $\Gamma/\pm I$. And let Γ_γ be the **centralizer** of γ in Γ. Then it is easily shown, as in (3.162) of Section 3.7, Volume I, that

$$\sum_{m\geq 0} f(s_m) = \sum_{\{\gamma\}} c_f(\gamma), \tag{1.347}$$

where the sum on the right is over all distinct conjugacy classes $\{\gamma\}$ in $\Gamma/\pm I$, and the **orbital integral** is:

$$c_f(\gamma) = \int_{K\backslash G/\Gamma_\gamma} f\left(b\gamma b^{-1}\right) db. \tag{1.348}$$

Stage 3. Now one needs to evaluate orbital integrals. If γ is in the **center** of G, this is easy, since it is clear that:

$$c_f(\gamma) = f(\gamma) \operatorname{Vol}(K\backslash G/\Gamma), \quad \text{for } \gamma \text{ in the center of } \Gamma.$$

This can be evaluated in terms of $\widehat{f}(s)$ using the inversion formula from Theorem 1.3.1 in Section 1.3.1.

Next suppose that γ is **hyperbolic**; i.e., γ is conjugate in $G = SL(n, \mathbb{R})$ to a matrix with n distinct real eigenvalues d_i none of which equal one. Let us assume for simplicity that $n = 3$. In this case, one finds that the orbital integral for γ can be replaced by one in which the argument of f has the following form:

$$
\begin{pmatrix} y_1 & 0 & 0 \\ 0 & y_2 & 0 \\ 0 & 0 & y_3 \end{pmatrix}
\begin{pmatrix} 1 & x_1 & x_3 \\ 0 & 1 & x_2 \\ 0 & 0 & 1 \end{pmatrix}
\begin{pmatrix} d_1 & 0 & 0 \\ 0 & d_2 & 0 \\ 0 & 0 & d_3 \end{pmatrix}
\begin{pmatrix} 1 & x_1 & x_3 \\ 0 & 1 & x_2 \\ 0 & 0 & 1 \end{pmatrix}^{-1}
\begin{pmatrix} 1/y_1 & 0 & 0 \\ 0 & 1/y_2 & 0 \\ 0 & 0 & 1/y_3 \end{pmatrix}
$$

$$
= \begin{pmatrix} d_1 & x_1 y_1 (d_2 - d_1)/y_2 & x_3 y_1 (d_3 - d_1)/y_3 + x_1 x_2 y_1 (d_1 - d_2)/y_3 \\ 0 & d_2 & x_2 y_2 (d_3 - d_2)/y_3 \\ 0 & 0 & d_3 \end{pmatrix}.
$$

Now look at the fundamental domain $K \backslash G / \Gamma_\gamma$ for the centralizer of γ. We can assume that we have conjugated everything so that Γ_γ consists of diagonal matrices with diagonal entries c_j, $j = 1, 2, 3$. The c_j must be units in a totally real cubic number field. The elements of $K \backslash G$ are represented by matrices of the form:

$$
\begin{pmatrix} y_1 & 0 & 0 \\ 0 & y_2 & 0 \\ 0 & 0 & y_3 \end{pmatrix}
\begin{pmatrix} 1 & x_1 & x_3 \\ 0 & 1 & x_2 \\ 0 & 0 & 1 \end{pmatrix}, \quad y_3 = \frac{1}{y_1 y_2}.
$$

Then γ acts on g by multiplying the y_j by c_j. So we can take the y_j in a fundamental domain for this action. From number theory (Dirichlet's theorem on units in algebraic number fields which was mentioned in Section 1.4 of Volume I), one knows that there is a compact fundamental domain for the action of the c's on the y's. The volume of this compact fundamental domain is measured by what number theorists call a regulator. See of Section 1.4 of Volume I for the definition and properties of the regulator.

Now make a change of variables in the orbital integral via:

$$
u_1 = x_1 y_1 (d_2 - d_1)/y_2, \quad u_2 = x_2 y_2 (d_3 - d_2)/y_3,
$$
$$
u_3 = x_3 y_1 (d_3 - d_1)/y_3 + x_1 x_2 y_1 (d_1 - d_2)/y_3.
$$

You obtain

$$
H(d) \int_{\substack{y \bmod c \\ u \in \mathbb{R}^3}} f \begin{pmatrix} d_1 & u_1 & u_3 \\ 0 & d_2 & u_2 \\ 0 & 0 & d_3 \end{pmatrix} y_1^{-3} y_2^{-2} \, dy \, du,
$$

where

$$
H(d) = \prod_{i<j} |d_i - d_j|^{-1}.
$$

Thus we end up with the Harish transform of f over N as in the hint to Exercise 1.3.3 in Section 1.3.1. The Helgason–Fourier transform (1.333) is a composition of Mellin transforms over the y-variables and the Harish-transform over the x-variables. See Dorothy Wallace [644] for more details on this calculation.

Since we have seen that units in cubic number fields appear in these formulas, one expects to generalize Sarnak's result (Theorem 3.7.6 in Volume I) to cubic and higher degree totally real number fields once the full trace formula for $SL(n, \mathbb{Z})$ is worked out in excruciating detail.

We should also note that there will be other sorts of orbital integrals, for example, from loxodromic elements of $SL(n, \mathbb{Z})$ with mixed eigenvalues (not all real), none of which are equal to each other or to ± 1. Wallace computes the contribution to the trace formula for $SL(3, \mathbb{Z})$ from loxodromic conjugacy classes in [649]. She notes that: "For $SL(4, \mathbb{Z})$ and higher dimensional groups, ... there are many more kinds of terms that, for the purposes of computing anything, must be distinguished from each other."

Stage 4. In this stage one must deal with the problems which arise when the fundamental domain for Γ is not compact. The orbital integrals for parabolic $\gamma \in \Gamma$ (i.e., γ some of whose eigenvalues are 1) require truncation, as do the inner products of continuous spectrum terms. We will not say more about this, beyond what was said in the preceding subsection. One needs to truncate and use the Langlands inner product formula. Dorothy Wallace [650] has a version of the result. There may be some terms missing. This is a constant problem with higher rank Selberg trace formulas. There are just too many terms. For an example, one has only to read Howard Resnikoff's review in *Math. Reviews, Vol. 53* #2841, of a paper computing dimensions of spaces of Siegel modular forms.

As is the case in Dorothy Wallace's calculations in [650], it is likely that one must use the trace formula for $SL(n - 1, \mathbb{Z})$ in order to prove that the cancellation occurs and we get a trace formula for $SL(n, \mathbb{Z})$. Even when $n = 3$, complicated calculations are needed, to obtain an explicit useful result. Many very large sheets of paper are required.

Of course, we should mention that various cases of the trace formula have already been worked out; for example in the case of discrete or discontinuous group actions on the quaternionic upper half space $SL(2, \mathbb{C})/SU(2)$, or actions on products of upper half planes, or actions on Siegel's upper half space which is the quotient $Sp(n, \mathbb{R})/SU(n)$. See Chapter 2 for more information on such examples. References for this and related topics are: Arakawa [18], Arthur [21–25], Christian [109], Efrat [152], Eie [160], Elstrodt et al. [165–168], Gangolli and Warner [199, 200], Godement [226], Hashimoto [264], Hejhal et al. [272, pp. 253–276], Langlands [389–395], Mennicke [442, 443], Morita [463], Müller [469], Petra Ploch [489], Sarnak [526], Selberg [543, 545, 546] Shimizu [553], Tanigawa [591], A.B. Venkov [627], Marie-France Vignéras [630–632], Dorothy Wallace [642–650], Warner [656], and Zograf [677].

Langlands [389] has obtained information on dimensions of spaces of automorphic forms by noting that since it is not too hard to compute some of the terms in the trace formula, e.g., the identity and hyperbolic terms, then one can make use of a good vanishing result for the remaining terms or a way of showing them asymptotically negligible. See Langlands' talk in Borel and Mostow [68, pp. 251–257] and Warner's discussion of the Selberg principle [655, Vol. II, p. 370] as well as Section 2.2.3.

Adelic versions of the trace formula have also been obtained. See Arthur [21–25], and Flicker [182]. There is also a trace formula for $N_\mathbb{R}/N_\mathbb{Z}$, where N is the nilpotent group of 3×3 upper triangular matrices with ones on the diagonal (see Osborne's second paper in Hejhal et al. [272, pp. 375–385]). It would be natural to use this result in our derivation of the trace formula for $SL(3, \mathbb{Z})$, but, so far as I know, no one has done this. Finally there are twisted trace formulas, which are needed for base change and the work of Langlands [394] on Artin's conjecture.

Let us just close this section by saying that if you want more exercises, you could try to translate anything in Section 3.7 in Volume I over to $GL(n, \mathbb{Z})$. Many have slaved on this project. But there is still much to do.

1.5.6 Finite and Other Analogues

Finite analogues of the results in this chapter have been studied but not as exhaustively as the finite upper half planes in §3.3.8 of Vol. I. For example, one might hope to write down the trace formula for $GL(n, \mathbb{F}_q)$. In this setting, the trace formula would require one to understand the character table of $GL(n, \mathbb{F}_q)$. I could only find the character tables for $n = 3$ and 4 in print, when writing [609]. Of course you can use the computer software GAP to give you these character tables for explicit finite fields. In a less ambitious vein, one can simply look for analogues of finite upper half planes, finite upper half plane graphs, and finite Eisenstein series considered in Volume I. Let us sketch what we know about this work.

In attempting to find analogues of finite upper half plane graphs, Nancy Tufts Allen [6] considered Cayley graphs attached to the group G of matrices of the form

$$\begin{pmatrix} y & x & z \\ 0 & 1 & 0 \\ 0 & 0 & 1 \end{pmatrix} = (y, x, z), \text{ for } x, y, z \text{ in the finite field } \mathbb{F}_p, \ y \neq 0,$$

where p is a prime greater than 3. The vertices of the graph are the elements of G. The edges in the graph are between an element g in G and an element gs, for s in the edge set S. Here, for fixed δ, a, c which are nonzero elements of the finite field $\mathbb{F}_p$, the edge set $S = S_p(\delta, a, c)$ consists of elements $(y, x, z) \in G$ such that $x^2 + cz^2 = ay + \delta(y-1)^2$. Call this graph $X_p(\delta, a, c)$. Allen asks whether these graphs are Ramanujan as defined in Vol. I, p. 93. She finds that there is much evidence for her conjecture that the graphs $X_p(\delta, a, c)$ are Ramanujan if $\left(\frac{-c}{p}\right) = \left(\frac{a(a-4\delta)}{p}\right) = 1$, where $\left(\frac{a}{p}\right)$ denotes the Legendre symbol. Allen also generalizes these graphs to analogous subgroups of $GL(n, \mathbb{F}_p)$ and shows that for $n > 3$, the analogous graphs are not Ramanujan.

A **hypergraph** $X = X(V, E)$ consists of a vertex set $V = V(X)$ and an edge set $E = E(X)$ such that each element of $E(X)$ is a nonempty subset of $V(X)$. The hypergraphs considered here are (d, r)-regular and finite. This means that any given edge contains r elements of $V(X)$ and each vertex of X is contained in d edges. For a Cayley graph $r = 2$. The **adjacency matrix** of a hypergraph X is a $|V| \times |V|$ matrix with diagonal entries $A_{x,x} = 0$, for $x \in V$, and off-diagonal entries

$$A_{x,y} = \#\{e \in E \mid x, y \in E\}, \quad \text{for } x, y \in V \text{ with } x \neq y.$$

Winnie Li and Patrick Solé [404] have defined a finite connected (d, r)-regular hypergraph X to be a **Ramanujan hypergraph** if every eigenvalue λ of A, $\lambda \neq d(r-1)$, satisfies

$$|\lambda - (r-2)| \leq 2\sqrt{(d-1)(r-1)}.$$

María Martínez [434] has found hypergraphs attached to a finite upper half-space for $GL(n, \mathbb{F}_q)$. Martínez proceeds as follows. She replaces the finite upper half plane with the finite upper half space $\mathcal{H}_q^3$ defined using a root θ of an irreducible cubic polynomial $\theta^3 + a\theta^2 + b\theta + c = 0$, where $a, b, c \in \mathbb{F}_q$. Then $\mathcal{H}_q^3$ consists of vectors

$$Z = \begin{pmatrix} W_1 \\ W_2 \\ 1 \end{pmatrix}, \quad \begin{array}{l} \text{where } W_1, W_2 \in \mathbb{F}_q(\theta), \\ W_1, W_2, 1 \text{ linearly independent over } \mathbb{F}_q \end{array}$$

The action of $GL(n, \mathbb{F}_q)$ on points in $\mathcal{H}_q^3$ is that from viewing these points as elements of projective 2-space. That is $gZ = gZ\left(\frac{1}{(gZ)_3}\right)$, where $(gZ)_3$ denotes the third component of the vector gZ. For $Z \in \mathcal{H}_q^3$, define $\det Z = \det \alpha_Z$, where

$$\alpha_Z = \begin{pmatrix} u_{11} & u_{12} & u_{13} \\ u_{21} & u_{22} & u_{23} \\ 0 & 0 & 1 \end{pmatrix} \quad \text{if } Z = \begin{pmatrix} u_{11}\theta^2 + u_{12}\theta + u_{13} \\ u_{21}\theta^2 + u_{22}\theta + u_{23} \\ 1 \end{pmatrix}.$$

Let $Nx = N_{\mathbb{F}_q(\theta)/\mathbb{F}_q}$ denote the **norm** in the finite field extension $\mathbb{F}_q(\theta)/\mathbb{F}_q$. Martínez replaces the point-pair invariant for points in the finite upper half plane with a **point-triple invariant**

$$k(Z_1, Z_2, Z_3) = \left(\frac{N \left(\det (Z_1 \, Z_2 \, Z_3) \right)}{\det(Z_1) \det(Z_2) \det(Z_3)} \right)^2 \quad \text{for} \quad Z_1, Z_2, Z_3 \in \mathcal{H}_q^3.$$

Martínez defines the **hypergraph** $\mathcal{H}_q^3(a)$ to have vertex set $\mathcal{H}_q^3$ and a hyperedge between $Z_1, Z_2, Z_3 \in \mathcal{H}_q^3$ if $k(Z_1, Z_2, Z_3) = a^2$. Then she shows that for small p these hypergraphs are indeed Ramanujan. Martínez also considers the analogues for $GL(n, \mathbb{F}_q)$.

María Martínez, Harold Stark, and I found other Ramanujan hypergraphs associated with $GL(n, \mathbb{F}_q)$ (see [435]). For example we considered the subgroup $A(1, n-1)$ of $GL(n, \mathbb{F}_q)$ defined by

$$A(1, n-1) = \left\{ \begin{pmatrix} 1 & {}^t b \\ 0 & c \end{pmatrix} \middle| c \in GL(n-1, \mathbb{F}_q), b \in \mathbb{F}_q^{n-1} \right\}.$$

Then let

$$\mathcal{A}_q^n = GL(n, \mathbb{F}_q)/A(1, n-1) \cong \mathbb{F}_q^n - \{0\}.$$

The identification here sends a coset $gA(1, n-1)$ to the first column of g. The $\mathcal{M}_q^n$ **space** is defined for $n \geq 3$ as

$$\mathcal{M}_q^n = \left\{ \begin{pmatrix} 1 \\ x \end{pmatrix} \middle| x \in \mathbb{F}_q^{n-1} \right\}.$$

An element $g \in A(1, n-1)$ acts on a column vector $v \in \mathcal{M}_q^n$ by $v \mapsto {}^t g v$, where ${}^t g$ denotes the transpose of g. Define the **n-point invariant function**

$$D(x_1, \ldots, x_n) = \det (x_1 \ \ldots \ x_n)^2 .$$

It is easily seen that $D(g x_1, \ldots, g x_n) = D(x_1, \ldots, x_n)$ if $\det g = \pm 1$.

The **α-hypergraph** $\alpha_q^n(a)$ has as its vertex set $\mathcal{A}_q^n$ and an edge between $x_1, \ldots, x_n \in \mathcal{A}_q^n$ (assumed to be pairwise distinct) if $D(x_1, \ldots, x_n) = a^2$. We proved that, for $n \geq 3$ and $a \neq 0$, the hypergraph $\alpha_q^n(a)$ is Ramanujan only when $n = 3$ and $q = 2, 3, 4$ or when $n = 4$ and $q = 2$.

The **μ-hypergraph** $\mu_q^n(a)$ has vertex set $\mathcal{M}_q^n$ and an edge between the n distinct elements $x_1, \ldots, x_n \in \mathcal{M}_q^n$ if the n-point invariant $D(x_1, \ldots, x_n) = a^2$. We found that, for $a \neq 0$, these hypergraphs are Ramanujan only if $n = 3$, for all q, or if $n = 4$, but then only when $q = 2$.

Winnie Li [402] has found many examples of finite hypergraphs associated with finite quotients of the Bruhat–Tits building of $GL(n, F)$ over a local field F. Examples of local fields F include the field of p-adic numbers, $\mathbb{Q}_p$, which is the completion of the rationals at the prime p. Another example is the completion of $\mathbb{F}_p(x)$ at some prime (i.e., irreducible) polynomial in $\mathbb{F}_p[x]$. Much more is known for the 2nd example and it is this second type of field that Winnie Li considers.

We briefly sketch some of her results. Let $X = PGL(n, F)/PGL(n, \mathcal{O}_F)$ be the Bruhat–Tits building associated with $PGL(n, F) = GL(n, F)/Z$, where Z is the center of $GL(n, F)$. It is assumed that F is a non-Archimedean local field with q elements in its residue field. One can view points $x \in X$ as equivalence classes of rank n lattices L_x over $\mathcal{O}_F$. Suppose that π is the uniformizer of F. To obtain a hypergraph, one says that the points $x_1, \dots, x_n \in X$ are in a **hyperedge** if the points x_i can be represented by lattices L_{x_i} such that

$$L_{x_1} \supset L_{x_2} \supset \cdots L_{x_n} \supset \pi L_{x_1}.$$

One can also view X as a finite simplicial complex. Hecke operators A_i, $i = 1, \dots, n-1$, act on functions on X by convolution with the indicator function of the $PGL(n, \mathcal{O}_F)$ double coset of the diagonal matrix with diagonal $(\pi, \dots, \pi, 1, \dots, 1)$, where π occurs i times. The operator spectrum of A_i acting on $L^2(X)$ is known. See Macdonald [428]. Suppose that Γ is a discrete subgroup of $PGL(n, F)$ such that $\Gamma \backslash X$ is finite. Then we can view $\Gamma \backslash X$ as a hypergraph with operators A_i giving analogues of the Laplacian or Hecke operators. The general definition of a **Ramanujan hypergraph** (or complex) covered by X asks that the nontrivial eigenvalues of the operators A_i, $i = 1, \dots, d-1$, lie in the spectrum of A_i acting on $L^2(X)$. Since for $n > 2$, the A_i's are normal but not necessarily self-adjoint, it follows that these spectra are subsets of $\mathbb{C}$ not $\mathbb{R}$. Li [403] uses the multiplicative group D of a division algebra of dimension n^2 over a function field to construct for each prime power q and positive integer $n \geq 3$, an infinite family of Ramanujan $(q + 1)$-regular n-hypergraphs. She proves her result using results of Laumon, Rapoport, and Stuhler on the Ramanujan conjecture for adelic automorphic representations of D, by suitably choosing two division algebras and comparing their trace formulae. Alex Lubotzky, Beth Samuels & Uzi Vishne [411] prove similar results to those of Winnie Li differently, using work of Lafforgue, who generalized the Ramanujan conjecture from $GL(2)$ to $GL(n)$ over fields of positive characteristic. At the end of their paper very explicit examples are given. Note also that, in order to prove the Ramanujanicity of their examples, they need the global Jacquet–Langlands correspondence from adelic division algebras to adelic $GL(n)$ for function fields such as $\mathbb{F}_p(x)$. In an earlier related paper, Cristina Ballantine [35] constructed finite 3-hypergraphs as quotients of the building attached to PGL_3 over the archimedian field of p-adic numbers, $\mathbb{Q}_p$, and proved that they are Ramanujan using representation theory.

Chris Storm [580] has generalized the Ihara zeta function of a graph to hypergraphs. Recall that the Ihara zeta is a graph theory analogue of Selberg's zeta. See my book [611] for more information on Ihara zetas. Storm proves that for (d, r)-regular hypergraphs, a modified Riemann hypothesis is true if and only if the hypergraph is Ramanujan in the sense of Winnie Li and Patrick Solé. He notes the connection between a hypergraph $X = X(V, E)$ and its associated bipartite graph B. The bipartite graph has vertices indexed by V and E. The vertices $v \in V$ and $e \in E$ are adjacent if $v \in e$. Storm shows that the zeta function of the hypergraphcan be represented in terms of the Ihara zeta function of the bipartite graph. But he also shows that there are infinitely many hypergraphs whose generalized zeta function is never the Ihara zeta function of a graph. Ren et al. [504] have applied zeta functions of hypergraphs to the processing of visual information extracted from images. The main reason for using such methods according to them "is that hypergraph representations allow nodes to be multiply connected by edges, and can hence capture multiple relationships between features." They look at some data that appears to show that the zeta function method is superior to the method of computing Laplacian spectra. Deitmar and Hoffman [131] study an analogue of the Ihara zeta function of $PGL(3)$ and find a determinant formula for it.

The study of automorphic forms/representations over global function fields such as $\mathbb{F}_p(x)$ seems to be much easier than that over global number fields such as $\mathbb{Q}$. For example, consider the $\mathbb{F}_p(x)$-analogue of Riemann's zeta function, for whom the Riemann hypothesis is proved. See Rosen [515] for more information on function fields and their zeta functions. The function field analogue of the Langlands correspondence was proved for $GL(2)$ by Drinfeld (see [139–142]) and for $GL(n)$ by Lafforgue [381]. Frenkel [187] discusses the Langlands program for both number fields and function fields. Such results have led to something called "the geometric Langlands program" in which $\mathbb{F}_p(x)$ is replaced by $\mathbb{C}(x)$. This is of interest to physicists such as Witten working on quantum field theory. See Witten's interview [480]. This connects with work on string theory, supermanifolds and supertrace formulas. See Alice Rogers [513] and Christian Grosche [245].

Chapter 2
The General Noncompact Symmetric Space

"These things will become clear to you," said the old man gently, "at least," he added with a slight doubt in his voice, "clearer than they are at the moment."

From *The Hitchhiker's Guide to the Galaxy*, by Douglas Adams, Pocket Books, NY, 1981. Reprinted by permission of The Crown Publishing Group.

2.1 Geometry and Analysis on G/K

2.1.1 Symmetric Spaces, Lie Groups, and Lie Algebras

Volume I [612] and the first chapter of this tome considered various examples and applications of symmetric spaces X, along with harmonic analysis on X and X/Γ for discrete groups Γ of isometries of X. Here we consider some aspects of analysis on a general noncompact symmetric space $X = G/K$. Our discussion will be very sketchy. The main goal is to lay the groundwork for extension of the results of the preceding chapters to other symmetric spaces which are of interest for applications; in particular, the Siegel upper half space $\mathcal{H}_n$ [which can be identified with $Sp(n, \mathbb{R})/U(n)$] and hyperbolic three space $\mathcal{H}^c$ [which can be viewed as $SL(2, \mathbb{C})/SU(2)$]. We will also be interested in the fundamental domains $\mathcal{H}_n/Sp(n, \mathbb{Z})$ for the Siegel modular group as well as the fundamental domain $\mathcal{H}^c/SL(2, \mathbb{Z}[i])$ for the Picard modular group. It is possible to generalize just about everything we did in the earlier chapters for such examples; e.g., the Selberg trace formula. And our main motivations for doing so come from number theory. Because it is time consuming and sometimes not so enlightening to do each of these examples separately, we have decided to present some results on the general symmetric space. Those interested in number theoretic applications may find this equally tedious and attempt to jump to the next section. But I think it is useful to know what a general

© Springer Science+Business Media New York 2016
A. Terras, *Harmonic Analysis on Symmetric Spaces—Higher Rank Spaces, Positive Definite Matrix Space and Generalizations*, DOI 10.1007/978-1-4939-3408-9_2

Iwasawa decomposition is, for example, in order to find the right coordinates to use in solving a given problem on the symmetric space. Of course, others will say that the discussion which follows is neither sufficiently general, detailed, nor rigorous. We refer those characters to the texts of other authors which are listed below.

Some topics in physics that lead one to study these other symmetric spaces are: quantum statistical mechanics and quantum field theory (see Hurt [312], Frenkel [187], and Ooguri's interview of Witten [480]), particle physics (see Wybourne [672, Ch. 21]), coherent states (see Hurt [312], Monastyrsky and Perelomov [457], and Perelomov [484]), boson fields (see Shale [551] and Cartier's article in Borel and Mostow [68, pp. 361–386]), solitons (see Dubrovin et al. [143], McKean and Trubowitz [440], Lonngren and Scott [407], and Novikov [474]), rotating tops (see Sofya Kovalevskaya [aka Sonya Kovalevsky] [368], Linda Keen [345], Pelageya Kochina [358], and Cooke [124]), and string theory (see Polyakov [491]).

Many branches of number theory steer one into these realms; e.g., the theory of quadratic forms (see Siegel [561–565]) and algebraic number theory (see Hecke [268] and Siegel [563]). The study of the ring $\mathbb{Z}[i]$ of Gaussian integers and similar rings for various algebraic number fields leads one to think that anything one can do for $\mathbb{Z}$ should be generalizable to $\mathbb{Z}[i]$. In particular, we will see that the theory of Maass wave forms for $SL(2, \mathbb{Z})$ has an analogue for $SL(2, \mathbb{Z}[i])$. This leads to some interesting formulas for the Dedekind zeta function of $\mathbb{Q}(i)$, among other things. See the Corollaries to Theorem 2.2.1 in Section 2.2 which follows. There is also an analogue of Selberg's trace formula (see the last section in this volume or Elstrodt et al. [168]).

Finally electrical engineering has many applications of these symmetric spaces as well (see Blankenship [51] and Helton [284–286]). We saw in Section 3.1 of Volume I that 2-port microwave circuits lead to quantities in $SL(2, \mathbb{R})$. Similarly, more complicated circuits lead to higher rank Lie groups.

References for this section include: Baily [32], Barut and Rączka [39], Broecker and tom Dieck [81], Chevalley [104], Yvonne Choquet-Bruhat, Cécile DeWitt-Morette, & Margaret Bleick [106], Dieudonné [137], Gangolli [195–198], Harish-Chandra [262, 263], Helgason [273–282], Hermann [289, 290], Hua [308], Loos [408], Maass [426], Piatetski-Shapiro [485], Sagle and Walde [524], Séminaire Cartan [547], Siegel [561–565], Varadarajan [623–625], Wallach [651–653], Warner [655], and Wybourne [672].

We will assume that the reader has had a decent course in multivariable calculus. Our favorite books for this are Lang [385, 388]. The notions of differential, tangent space, matrix exponential, Taylor's formula are all covered there. You may also need to refer to a book like that of Sagle and Walde [524] for more details on various arguments. The true story of everything is found in Helgason's big green books. Varadarajan [623] is also useful, for example, as a source for all the details of the root space calculations.

Élie Cartan obtained the basic theory of symmetric spaces between 1914 and 1927. Then, beginning in the 1950s, Harish-Chandra, Helgason, and others developed harmonic analysis and representation theory on these spaces and their Lie groups of isometries.

A **symmetric space** M is a connected Riemannian manifold (as in the discussion at the beginning of Chapter 2, Volume I) such that at each point $P \in M$ there is a geodesic-reversing isometry

$$s_P : M \to M;$$

i.e., s_P preserves the Riemannian metric and flips geodesics about the fixed point P.

Our first goal is to produce a multitude of examples of symmetric spaces. We always start with a **Lie group** G; i.e., a real analytic manifold which is also a group such that the mapping

$$
\begin{aligned}
G \times G &\to G \\
(g, h) &\mapsto gh^{-1}
\end{aligned}
\tag{2.1}
$$

is analytic. We will only consider Lie groups of real or complex matrices here. As we have said earlier, it is often useful to replace $\mathbb{R}$ or $\mathbb{C}$ with a finite field, or a local field such as $\mathbb{Q}_p$, the field of p-adic numbers. Mostly, we will avoid doing this.

The **Lie algebra** $\mathfrak{g}$ of a Lie group G is the tangent space to G at the identity, once it has been provided with an additional operation called "the Lie bracket." It is traditional that the Lie algebra is written as the lowercase German letter (fraktur) corresponding to the uppercase Latin letter which is the group. The fraktur letters used here should be recognizable—except $\mathfrak{k}$, which is k.

The Lie bracket operation is defined by identifying the Lie algebra

$$\mathfrak{g} = T_e(G) = \text{ the tangent space to } G \text{ at the identity } e \in G,$$

with the space of left-invariant vector fields on G. These vector fields are first order differential operators on G (with real analytic coefficients) which commute with left translation. This identification is achieved by making use of the left translation $L_g(x) = gx$, for $x, g \in G$. If $\tilde{X}$ is a left-invariant vector field, then

$$\tilde{X}_g = dL_g(X), \qquad \text{for } g \in G, \ X \in \mathfrak{g}.$$

Here dL_g denotes the differential of left multiplication on G. Now we define the **Lie bracket** of two left invariant vector fields $\tilde{X}, \tilde{Y}$ by;

$$[\tilde{X}, \tilde{Y}] = \tilde{X}\tilde{Y} - \tilde{Y}\tilde{X},$$

which is also a left invariant vector field; i.e., the bracket of two first order differential operators is actually a first order and not a second order differential operator. Write $[X, Y]$ for the corresponding bracket of elements X, Y in the Lie algebra $\mathfrak{g}$.

What makes $\mathfrak{g}$ a **Lie algebra**? The answer is that the bracket can be shown to have the following defining properties of such an algebra:

(1) $[X, Y]$ is a bilinear map of $\mathfrak{g} \times \mathfrak{g}$ into $\mathfrak{g}$;
(2) $[X, Y] = -[Y, X]$;
(3) $[X, [Y, Z]] + [Y, [Z, X]] + [Z, [X, Y]] = 0$ (**Jacobi's identity**).

Then one defines subalgebra, ideal, homomorphism, isomorphism, etc., for Lie algebras in the usual way (see the references). For example, an ideal $\mathfrak{a} \subset \mathfrak{g}$ is a vector subspace $\mathfrak{a}$ of $\mathfrak{g}$ such that $[\mathfrak{a}, \mathfrak{g}] \subset \mathfrak{a}$.

Exercise 2.1.1. Prove that if $G = GL(n, \mathbb{R})$, then the corresponding Lie algebra can be identified with the vector space $\mathbb{R}^{n \times n}$ of all $n \times n$ real matrices with bracket defined by

$$[A, B] = AB - BA, \qquad \text{for } A, B \in \mathbb{R}^{n \times n}.$$

Here AB denotes the usual matrix product. Thus $\mathfrak{gl}(n, \mathbb{R})$ is identified with $\mathbb{R}^{n \times n}$.
Hint. See Dieudonné [137, Vol. VI, pp. 145–146] or Sagle and Walde [524, pp. 117–118]. The vector space $\mathbb{R}^{n \times n}$ can certainly be identified with the tangent space to $GL(n, \mathbb{R})$, since $GL(n, \mathbb{R})$ is an open subset of $\mathbb{R}^{n \times n}$. In fact, using the matrix exponential, we can make the identification as follows. Suppose $A \in \mathbb{R}^{n \times n}, g \in GL(n, \mathbb{R})$, and $f : GL(n, \mathbb{R}) \to \mathbb{C}$. Then

$$(\tilde{A}f)(g) = \left.\frac{d}{dt}f(g \exp tA)\right|_{t=0}.$$

One has for $A, B \in \mathbb{R}^{n \times n}$ and $g \in GL(n, \mathbb{R})$:

$$(\tilde{A}\tilde{B}f)(g) = \left.\frac{\partial^2}{\partial t\,\partial s}f(g \exp tA \exp sB)\right|_{t=s=0}.$$

Use the chain rule to see that at $g = e$ this is $f_e''(A, B) + f_e'(AB)$. If you interchange A and B and then subtract, the second order terms cancel and you get $f_e'(AB - BA)$. This shows that the identification of $\mathfrak{g}$ with $\mathbb{R}^{n \times n}$ does preserve brackets.

There is a representation of any (real) Lie algebra $\mathfrak{g}$ in $\mathfrak{gl}(n, \mathbb{R})$, where $n = \dim_{\mathbb{R}} \mathfrak{g}$. This representation is called the **adjoint representation** defined as follows, thinking of $\mathfrak{gl}(n, \mathbb{R})$ as the space of linear transformations of $\mathfrak{g}$ into itself:

$$\begin{aligned} \text{ad} : \mathfrak{g} \to \mathfrak{gl}(n, \mathbb{R}), \quad n = \dim_{\mathbb{R}} \mathfrak{g}, \\ (\text{ad } X)Y = [X, Y], \quad \text{for } X, Y \in \mathfrak{g}. \end{aligned} \tag{2.2}$$

Exercise 2.1.2. Show that the adjoint representation defined by (1.1) above does indeed preserve brackets; i.e., $[ad\, X, ad\, Y] = ad[X, Y]$.

The **Killing form** of a Lie algebra $\mathfrak{g}$ is defined to be the bilinear form:

$$B_{\mathfrak{g}} = B(X, Y) = \text{Tr}(\text{ad } X \text{ ad } Y), \quad \text{for } X, Y \in \mathfrak{g}. \tag{2.3}$$

A Lie algebra is called **semisimple** if the Killing form B is nondegenerate; i.e., $B(X, Y) = 0$ for all $Y \in \mathfrak{g}$ implies $X = 0$. A Lie algebra $\mathfrak{g}$ is **simple** if it is semisimple, and if, in addition, it has no ideals but $\{0\}$ and itself.

Example 2.1.1 $(GL(n, \mathbb{R}))$.
   Since matrices of the form aI, $a \in \mathbb{R}$, commute with $n \times n$ matrices, it is clear that $\text{ad}(aI) = 0$ and thus that $B(aI, Y) = 0$ for all $Y \in \mathbb{R}^{n \times n}$. Thus $\mathfrak{gl}(n, \mathbb{R}) \cong \mathbb{R}^{n \times n}$ is not semisimple.

   It will be useful to compute the Killing form for $\mathfrak{gl}(n, \mathbb{R})$. One can do this as follows. Let E_{ij} be the $n \times n$ matrix with i, j entry equal to one and the rest zero. Let H be the diagonal matrix with ith diagonal entry h_i. Then $\text{ad}(H)E_{ij} = (h_i - h_j)E_{ij}$. Therefore

$$B(H, H) = \text{Tr}(\text{ad} H \text{ ad } H) = \sum_{i,j=1}^{n} (h_i - h_j)^2 = 2n \, \text{Tr}(H^2) - 2(\text{Tr } H)^2.$$

Note that it suffices to compute the Killing form on diagonal matrices. For the map $X \mapsto gXg^{-1}$, with $g \in GL(n, \mathbb{R})$ and $X \in \mathfrak{gl}(n, \mathbb{R})$, leaves the Killing form invariant. Moreover matrices conjugate to a diagonal matrix are dense in $\mathfrak{gl}(n, \mathbb{R})$.

Exercise 2.1.3. (a) Show that if σ is a Lie algebra automorphism of $\mathfrak{g}$, then

$$B(X, Y) = B(\sigma X, \sigma Y), \quad \text{for all } X, Y \in \mathfrak{g}.$$

(b) Show that

$$B(X, [Y, Z]) = B(Y, [Z, X]) = B(Z, [X, Y]), \quad \text{for all } X, Y, Z \in \mathfrak{g}.$$

There is an analogue of the matrix exponential for any Lie group G. It is, appropriately enough, called the **exponential map** and it maps the Lie algebra into the Lie group such that if $X \in \mathfrak{g}$, $g \in G$, and $f : G \to \mathbb{C}$ is infinitely differentiable, then

$$\tilde{X}_g f = \frac{d}{dt} f(g \exp tX)|_{t=0}. \tag{2.4}$$

For matrix groups the matrix exponential is the Lie group exponential map. For general Lie groups, the existence of $\exp : \mathfrak{g} \to G$ comes from standard results in ordinary differential equations.

Let us list a few properties of exp. The curve

$$\mathbb{R} \to G$$
$$t \mapsto \exp(tX), \text{ for } X \in \mathfrak{g},$$

is a **one-parameter subgroup** of G; i.e., $\exp(0) = e$, the identity in G, and

$$\exp(tX)\exp(sX) = \exp(t + s)X, \quad \text{for all real numbers } s, t. \tag{2.5}$$

Taylor's formula for G says that:

$$f(g \exp X) = \sum_{n \geq 0} \frac{1}{n!}(\tilde{X}^n f)(g), \quad \text{for } g \in G, \ X \in \mathfrak{g}, \tag{2.6}$$

where f is a real analytic function on G. **The exponential map allows one to relate multiplication on the Lie group with bracket on the Lie algebra via**:

$$\exp tX \exp tY = \exp\left\{ t(X + Y) + \frac{1}{2}t^2[X, Y] + O(t^3) \right\}, \tag{2.7}$$

for $X, Y \in \mathfrak{g}$, and $t \in \mathbb{R}$. It is possible to continue the expansion inside the braces in (2.7) and the result is called the Campbell–Hausdorff formula.

Exercise 2.1.4. Prove formula (2.7).
Hint. First consider the case of $GL(n, \mathbb{R})$. The same sort of proof works in general using Taylor's formula (2.6).

It is possible to compute the **differential of exp** and obtain:

$$(d\exp)_X Y = (dL_{\exp X})_e \left(\frac{1 - e^{-\operatorname{ad} X}}{\operatorname{ad} X} \right)(Y), \quad \text{for} \ X, Y \in \mathfrak{g}. \tag{2.8}$$

Formula (2.8) implies in particular that the mapping from X to $\exp X$ is a diffeomorphism from an open neighborhood of 0 in $\mathfrak{g}$ onto an open neighborhood of the identity e in G.

Let us prove (2.8) in the case of matrix exp. First note that

$$\lim_{t \to 0}(e^{X+tY} - e^X)/t = \lim_{t \to 0}\sum_{n \geq 0}\frac{1}{n!t}\{(X + tY)^n - X^n\}$$

$$= \sum_{n \geq 0}\frac{1}{(n + 1)!}\{X^n Y + X^{n-1}YX + \cdots + YX^n\}.$$

Beware that $XY \neq YX$, in general, so that you cannot blindly use the binomial theorem. However, it is possible to be clever (although that is unworthy of a Vulcan), since right and left multiplication by X do commute as operators. Define $R_X Y = YX$ and $L_X Y = XY$. Observe that $\text{ad} X = L_X - R_X$. The three operators R_X, L_X, and ad X will commute. Thus we can apply the binomial theorem to obtain:

$$R_X^m = (L_X - \text{ad} X)^m = \sum_{k=0}^{m} \binom{m}{k} L_X^{m-k}(-\text{ad} X)^k.$$

This allows us to write:

$$X^n Y + X^{n-1} YX + \cdots + YX^n = \sum_{i=0}^{n} X^i \sum_{k=0}^{n-i} \binom{n-i}{k} X^{n-i-k}(-\text{ad} X)^k Y$$

$$= \sum_{k=0}^{n} \sum_{i=0}^{n-k} \binom{n-i}{k} X^{n-k}(-\text{ad} X)^k Y,$$

upon reversing sums. It is an **exercise** in the properties of binomial coefficients to show that

$$\sum_{i=0}^{n-k} \binom{n-i}{k} = \binom{n+1}{k+1}. \tag{2.9}$$

Therefore

$$(d\exp)_X Y = \sum_{n \geq 0} \frac{1}{(n+1)!} \sum_{k=0}^{n} \binom{n+1}{k+1} X^{n-k}(-\text{ad} X)^k Y$$

$$= \sum_{k \geq 0} \sum_{n \geq k} \frac{1}{(k+1)!(n-k)!} X^{n-k}(-\text{ad} X)^k Y$$

$$= \sum_{k \geq 0} \sum_{r \geq 0} \frac{1}{(k+1)!r!} X^r(-\text{ad} X)^k Y$$

$$= e^X \sum_{k \geq 0} \frac{1}{(k+1)!}(-\text{ad} X)^k Y.$$

This completes the proof of (2.8) in the case of the matrix exponential. The general result is proved in a similar way (see Helgason [275]).

Exercise 2.1.5. Prove formula (2.8) for a general Lie group.

One of the most important tools in Lie group theory is the **dictionary** that allows one to translate between Lie groups and Lie algebras. We list a few results from the dictionary. For the proofs, see references such as Helgason's big green books, Sagle and Walde [524], or Varadarajan [623].

For each Lie group G with Lie algebra $\mathfrak{g}$ and for each Lie subalgebra $\mathfrak{h}$ of $\mathfrak{g}$, there is a unique connected Lie subgroup H of G with Lie algebra $\mathfrak{h}$. However, H may not have the induced topology; e.g., consider the densely wound line in the torus:

$$\mathbb{R} \to (\mathbb{R}/\mathbb{Z})^2 = \mathbb{T}^2$$

$$t \mapsto (e^{it}, e^{iat}), \quad \text{when} \quad a \in \mathbb{R} \text{ is irrational.}$$

If $f : G_1 \to G_2$ is a Lie group homomorphism of connected Lie groups, then the differential $(df)_e : \mathfrak{g}_1 \to \mathfrak{g}_2$ is a Lie algebra homomorphism. Moreover, we have the following relations between images and kernels:

$$\text{Lie Algebra } (f(G_1)) = (df)_e \, \mathfrak{g}_1,$$
$$\text{Lie Algebra } (\ker f) = \ker(df)_e.$$

If $\lambda : \mathfrak{g}_1 \to \mathfrak{g}_2$ is a Lie algebra homomorphism and G_1, G_2 are connected Lie groups with Lie algebras $\mathfrak{g}_1, \mathfrak{g}_2$, respectively, and if, in addition, G_1 is simply connected, then there exists a unique Lie group homomorphism $f : G_1 \to G_2$ such that $(df)_e = \lambda$.

The hypothesis that G_1 be simply connected cannot be removed in the preceding result. For example, $\mathbb{R}/\mathbb{Z}$ and $\mathbb{R}$ have the same Lie algebra. But the identity mapping of $\mathbb{R}$ onto itself cannot be the differential of a Lie group homomorphism from $\mathbb{R}/\mathbb{Z}$ to $\mathbb{R}$.

Exercise 2.1.6. (a) Show that the exponential map $\mathfrak{g} \to G$ need not be onto.
(b) Show that $\exp : \mathfrak{gl}(n, \mathbb{C}) \to GL(n, \mathbb{C})$ is onto.

Hints.

(a) Take $G = SL(2, \mathbb{R})$ and consider

$$A = \begin{pmatrix} r & 0 \\ 0 & 1/r \end{pmatrix} \quad \text{for} \quad r < -1.$$

If $A = \exp(X)$, consider the eigenvalues of X.
(b) Use the Jordan canonical form.

The final dictionary entry that we list here concerns a closed subgroup H of a Lie group G. Then H must have the induced topology and

$$\text{Lie Algebra } (H) = \mathfrak{h} = \{X \in \mathfrak{g} \mid \exp(tX) \in H, \text{ for all } t \in \mathbb{R}\}. \tag{2.10}$$

Formula (2.10) provides a quick way to compute Lie algebras. For example, since $\det(e^X) = e^{\mathrm{Tr}X}$, for matrices X, it follows that the Lie algebra $\mathfrak{sl}(n, \mathbb{R})$ consists of all $n \times n$ real matrices of trace zero. As we said, we use the notation that the Lie algebra of a group G is in lowercase German Fraktur letters so that $\mathfrak{sl}(n, \mathbb{R})$ is the Lie algebra of $SL(n, \mathbb{R})$. One can show that the Killing form of $\mathfrak{sl}(n, \mathbb{R})$ is:

$$B_{\mathfrak{sl}(n,\mathbb{R})}(X, Y) = 2n\mathrm{Tr}(XY), \quad \text{for } X, Y \in \mathfrak{sl}(n, \mathbb{R}).$$

Therefore $\mathfrak{sl}(n, \mathbb{R})$ is a semisimple Lie algebra. In fact, it is actually a simple Lie algebra.

Exercise 2.1.7.   (a)   Verify the comments made in the last paragraph.
(b)   Find the Lie algebra of the **symplectic group**[1]:

$$Sp(n, \mathbb{R}) = \left\{ g \in \mathbb{R}^{2n\times 2n} \mid {}^t g J_n g = J_n \right\}, \quad \text{for } J_n = \begin{pmatrix} 0 & I_n \\ -I_n & 0 \end{pmatrix}.$$

(c)   Find the Lie algebra of the Lorentz-type group

$$O(p, q) = \left\{ g \in \mathbb{R}^{n\times n} \mid {}^t g I_{p,q} g = I_{p,q} \right\},$$

where $n = p + q$, and

$$I_{p,q} = \begin{pmatrix} I_p & 0 \\ 0 & -I_q \end{pmatrix}.$$

Hint. The answer to part (b) is given in formula (2.12).

There is an analogue of the adjoint representation on the group level, denoted Ad. To obtain it, proceed as follows. If $g \in G$, define

$$\mathrm{Int}(g)x = gxg^{-1}, \quad \text{for all } x \in G \text{ and } \mathrm{Ad}(g) = (d\,\mathrm{Int}(g))_e, \text{ where } e \text{ is the identity of } G.$$
$$(2.11)$$

Then we have a commutative diagram:

$$\begin{array}{ccc}
 & \mathrm{Ad}(g) & \\
\mathfrak{g} & \to & \mathfrak{g} \\
\exp \downarrow & & \downarrow \exp \\
G & \stackrel{\mathrm{Int}(g)}{\to} & G.
\end{array}$$

It can be proved that $(d\,\mathrm{Ad})_e X = \mathrm{ad}X$, for all $X \in \mathfrak{g}$. Thus we have another commutative diagram:

$$\begin{array}{ccc}
 & \mathrm{ad} & \\
\mathfrak{g} & \to & \mathfrak{gl}(\mathfrak{g}) \\
\exp \downarrow & & \downarrow \text{ matrix exp} \\
G & \stackrel{\mathrm{Ad}}{\to} & G.
\end{array}$$

---

[1] Beware! Some authors write $Sp(2n, F)$ instead of $Sp(n, F)$.

If G is a matrix group already, then the matrix $\mathrm{Ad}(g)$ is the matrix of $\mathrm{Int}(g)$, since $\mathrm{Int}(g)x$ is a linear function of x. If $\mathfrak{g}$ is semisimple, then the kernel of ad is $\{0\}$ and the kernel of Ad is the center of G, which must then be discrete.

Exercise 2.1.8. Prove that $(d\,\mathrm{Ad})_e X = \mathrm{ad}X$, for all $X \in \mathfrak{g}$.

It is possible to classify all the simple Lie algebras over the complex numbers. Except for a finite number of exceptional Lie algebras, the **simple Lie algebras over** $\mathbb{C}$ are in the following list (with J_n as in Exercise 2.1.7 above):

$$
\left.
\begin{aligned}
\mathfrak{a}_n &= \mathfrak{sl}(n+1, \mathbb{C}), \quad n \geq 1; \\
\mathfrak{b}_n &= \mathfrak{so}(2n+1, \mathbb{C}) = \left\{ X \in \mathbb{C}^{(2n+1)\times(2n+1)} \mid {}^tX = -X \right\}, \quad n \geq 2; \\
\mathfrak{c}_n &= \mathfrak{sp}(n, \mathbb{C}) = \left\{ X \in \mathbb{C}^{(2n)\times(2n)} \mid {}^tXJ_n + J_nX = 0 \right\} \\
&= \left\{ \begin{pmatrix} A & B \\ C & -{}^tA \end{pmatrix} \,\middle|\, A, B, C \in \mathbb{R}^{n\times n}, B, C \text{ symmetric} \right\}, \quad n \geq 3; \\
\mathfrak{d}_n &= \mathfrak{so}(2n, \mathbb{C}) = \left\{ X \in \mathbb{C}^{(2n)\times(2n)} \mid {}^tX = -X \right\}, \quad n \geq 4.
\end{aligned}
\right\}
\tag{2.12}
$$

The indices n are restricted because in low dimensions some strange things happen; e.g., $\mathfrak{d}_1$ and $\mathfrak{d}_2$ are not simple, since $\mathfrak{d}_1$ is abelian and $\mathfrak{d}_2 \cong \mathfrak{a}_1 \oplus \mathfrak{a}_1$. Also $\mathfrak{a}_1 \cong \mathfrak{b}_1 \cong \mathfrak{c}_1$, $\mathfrak{b}_2 \cong \mathfrak{c}_2$, $\mathfrak{d}_3 \cong \mathfrak{a}_3$. These things can be proved using Dynkin diagrams. You can find the details in Varadarajan [623]. The Lie groups corresponding to the Lie algebras in this list are $SL(n, \mathbb{C})$, the special linear group of $n \times n$ complex matrices of determinant one, $SO(n, \mathbb{C})$, the special orthogonal group of $n \times n$ complex matrices g of determinant one such that ${}^tgg = I$, and $Sp(n, \mathbb{C})$, the complex symplectic group of $(2n) \times (2n)$ matrices g with the property that ${}^tgJ_ng = J_n$, for J_n as in Exercise 2.1.7 above. We should perhaps note again that some authors write $Sp(2n, \mathbb{C})$ instead of $Sp(n, \mathbb{C})$. This is rather confusing.

Cartan's classification of symmetric spaces makes use of the preceding classification of complex simple Lie algebras. It also uses the surprising, but simple, observation that **the group** $I(M)$ **of isometries of a symmetric space** M **acts transitively on** M. To see this fact, it helps to recall the **Hopf–Rinow theorem** in differential geometry (see Helgason [273, p. 56]) which says that if M is a Riemannian manifold, then the following are equivalent:

(a) M is a complete metric space;
(b) each maximal geodesic $\gamma(t)$ in M can be extended to all $t \in \mathbb{R}$;
(c) each bounded closed subset of M is compact.

If M is a complete Riemannian manifold, then any two points P, Q in M can be joined by a geodesic whose length is the metric space distance between P and Q. To see that a symmetric space M must be complete, note that if a point P lies on the geodesic γ of M and s_P denotes the geodesic-reversing isometry at P, then $s_P\gamma$ is an extension of γ. Thus each maximal geodesic of a symmetric space must have domain the set of all real numbers. Then to see that the group $I(M)$ of isometries of M acts transitively on M, note that if P and Q are in M, then the geodesic-reversing isometry at the midpoint of the geodesic connecting P to Q will exchange P and Q.

It is possible to prove that $I(M)$ is a Lie group such that the connected component of the identity in $I(M)$ still acts transitively on M (see Helgason [273, Ch. 4]). **Let G be the connected component of the identity in $I(M)$.** Now fix a point o to be called the **origin** of the symmetric space M. And let K denote the subgroup of G consisting of elements which fix o. Then K is compact and we can identify M with G/K. Suppose next that s_o denotes the **geodesic-reversing isometry at the origin.** Now consider the map:

$$\sigma : G \to G$$

$$g \mapsto s_o g s_o.$$

Note that σ is an involutive automorphism of G (i.e., σ is an automorphism in the sense of Lie groups such that σ^2 is the identity). Moreover, setting

$$K_\sigma = \{g \in G \mid \sigma g = g\}$$

and

$$(K_\sigma)_o = \text{the connected component of the identity in } K_\sigma,$$

we have

$$(K_\sigma)_o \subset K \subset K_\sigma.$$

This means that K and K_σ have the same Lie algebra.

Now consider the consequences of the preceding remarks about symmetric spaces and Lie groups of isometries on the Lie algebras of these groups, using the dictionary relating Lie groups and Lie algebras. One sees that:

$$(d\sigma)_e : \mathfrak{g} \to \mathfrak{g}$$

is an involutive Lie algebra automorphism which fixes $\mathfrak{k}$, the Lie algebra of K. Moreover, the eigenspace decomposition of $(d\sigma)_e$ on $\mathfrak{g}$ is:

$$\mathfrak{g} = \mathfrak{k} \oplus \mathfrak{p},$$

where $\mathfrak{k} = \{X \in \mathfrak{g} \mid (d\sigma)_e X = X\}$ and $\mathfrak{p} = \{X \in \mathfrak{g} \mid (d\sigma)_e X = -X\}$; that is, $\mathfrak{k}$ is the space of eigenvectors corresponding to the eigenvalue $+1$ while $\mathfrak{p}$ consists of eigenvectors corresponding to the eigenvalue -1.

If $\pi : G/K \to M$ is the natural identification, then $(d\pi)_o$ maps $\mathfrak{k}$ to $\{0\}$ and identifies $\mathfrak{p}$ with the tangent space $T_o(M)$.

To proceed further with the classification of symmetric spaces, one must reduce to semisimple Lie algebras, using the following result of E. Cartan (see Helgason [273, Ch. 5]).

Symmetric Space Decomposition Suppose that M is a simply connected symmetric space. Then M is a product:

$$M = M_e \times M_c \times M_n,$$

where M_e is of **Euclidean** type, M_c is **compact with semisimple Lie group** of isometries, and M_n is **noncompact with semisimple Lie group** of isometries having a Lie algebra with a Cartan decomposition described below.

We say that a semisimple Lie algebra is of **compact type** if its Killing form is negative definite (see Helgason [273, p. 122]).

A **Cartan decomposition** of a noncompact semisimple Lie algebra $\mathfrak{g}$ is a vector space direct sum decomposition $\mathfrak{g} = \mathfrak{k} \oplus \mathfrak{p}$ such that the Killing form of $\mathfrak{g}$ is negative definite on $\mathfrak{k}$ and positive definite on $\mathfrak{p}$. Also the mapping $\theta : \mathfrak{g} \to \mathfrak{g}$ with $\theta(X+Y) = X - Y$, for $X \in \mathfrak{k}$ and $Y \in \mathfrak{p}$, must be an automorphism of $\mathfrak{g}$. We call θ the **Cartan involution**.

Example 2.1.2. Consider the simple Lie algebra $\mathfrak{sl}(n, \mathbb{R})$ of all trace zero $n \times n$ real matrices. Set

$$\mathfrak{so}(n) = \{\text{skew-symmetric } n \times n \text{ real matrices}\}$$

and

$$\mathfrak{p}_n = \{\text{symmetric } n \times n \text{ real matrices of trace } 0\}.$$

Clearly we have the direct sum decomposition:

$$\mathfrak{sl}(n, \mathbb{R}) = \mathfrak{so}(n) \oplus \mathfrak{p}_n,$$

with Cartan involution $\theta(X) = -{}^t X$. The Killing form on $\mathfrak{sl}(n, \mathbb{R})$ is $B(X, Y) = 2n\mathrm{Tr}(XY)$, and it is easy to see that this is negative definite on $\mathfrak{so}(n)$ and positive definite on $\mathfrak{p}_n$. It is also easy to check that the Cartan involution preserves the Lie bracket in $\mathfrak{sl}(n, \mathbb{R})$, which is $[X, Y] = XY - YX$.

Exercise 2.1.9. Prove all the claims made in the preceding example.

There is a mirror image of the **Cartan decomposition on the Lie group level**:

$$G = KP,$$

where $P = \exp \mathfrak{p}$. For the example above, we have

$$SL(n, \mathbb{R}) = SO(n)\,\mathcal{SP}_n, \tag{2.13}$$

where, as usual, $\mathcal{SP}_n$ denotes the positive $n \times n$ real matrices of determinant one. The proof of (2.13) is easy (see Exercise 1.1.5 of Section 1.1.2).

We have not given more than a rough sketch of the preceding arguments on classification of symmetric spaces because we are more interested in studying the examples. Thus we will give more attention to the question: **How does one obtain symmetric spaces out of Cartan decompositions of semisimple Lie algebras?** Suppose that $\mathfrak{g} = \mathfrak{k} \oplus \mathfrak{p}$ is a Cartan decomposition of the semisimple Lie algebra $\mathfrak{g}$ with Cartan involution θ. Let G be a connected real semisimple Lie group with Lie algebra $\mathfrak{g}$, and let K be a Lie subgroup of G having Lie algebra $\mathfrak{k}$. Then G/K has a unique analytic manifold structure such that the mapping of $\mathfrak{p}$ into G/K defined by sending X to $(\exp X)K$ is a diffeomorphism. If $\mathfrak{g}$ is of noncompact type, it can be proved that K is closed, connected, and equal to the fixed point set of an involutive automorphism $t : G \to G$ such that $(dt)_e$ is the Cartan involution θ. Such a map t clearly exists if G is simply connected (making use of the dictionary between group and algebra). But one does not really have to assume that G is simply connected in the noncompact case. Moreover K is compact if and only if the center of G is finite and then K is a **maximal compact subgroup** of G. For proofs of these results, see Helgason [273, Ch. 6].

To make G/K a symmetric space, we use the Killing form B of $\mathfrak{g}$. Let $\pi : G \to G/K$ be defined by $\pi(g) = gK$. Define the Riemannian metric Q on G/K by translating the Killing form on the space $\mathfrak{p}$:

$$Q_{gK}((d\pi)_g \tilde{X}_g, (d\pi)_g \tilde{Y}_g) = B(X, Y), \quad \text{for all } X, Y \in \mathfrak{p}. \tag{2.14}$$

Here $\tilde{X}$ denotes the left invariant vector field corresponding to $X \in \mathfrak{p}$. The metric Q is well defined because the Killing form is invariant under $\mathrm{Ad}(k)$, for $k \in K$. It is clear that the metric is positive from the definition of the Cartan decomposition. And it is easily seen that the metric is G-invariant.

The **geodesic-reversing isometry** s_o at the origin o, which is the coset K in G/K, is obtained from the involutive automorphism $t : G \to G$ as follows:

$$s_o : G/K \to G/K$$

$$gK \mapsto t(g)K.$$

Translate by elements of G to obtain the geodesic-reversing isometries at other points of G/K.

Example 2.1.3. The Riemannian structure on $SL(n, \mathbb{R})/SO(n)$ obtained from (2.14) above is just the same as that defined in Chapter 1 of this volume. To see this, first note that one has an identification:

$$SL(n, \mathbb{R})/SO(n) \to \mathcal{SP}_n$$

$$gSO(n) \mapsto g\,{}^t g.$$

The action of $g \in SL(n, \mathbb{R})$ on $Y \in \mathcal{SP}_n$ is given by $a_g(Y) = Y[{}^t g]$. The differential is $(da_g)_I = a_g$ since $a_g(Y)$ is a linear function of Y. So we find that if $Y = g\,{}^t g$, for

$g \in SL(n, \mathbb{R})$, and if u, v are in $T_Y(\mathcal{SP}_n)$, the tangent space to $\mathcal{SP}_n$ at the point Y, then:

$$Q_Y(u, v) = 2n \operatorname{Tr}\left((da_g)_I^{-1} u \cdot (da_g)_I^{-1} v\right) = 2n \operatorname{Tr}\left(g^{-1} u \, {}^t g^{-1} \cdot g^{-1} v \, {}^t g^{-1}\right)$$
$$= 2n \operatorname{Tr}(Y^{-1} u Y^{-1} v).$$

This is exactly the Riemannian structure of Chapter 1 of this volume.

Before considering more examples, let us record a few more general facts. Suppose again that we have a Cartan decomposition $\mathfrak{g} = \mathfrak{k} \oplus \mathfrak{p}$ of a semisimple noncompact Lie algebra over the real numbers. Assume that the Lie group $\tilde{G}$ is the universal covering group of G. Then there is a unique involutive automorphism $\tilde{\iota} : \tilde{G} \to \tilde{G}$ such that the differential $(d\tilde{\iota})_e$ is θ, the Cartan involution. It can be proved that the center $\tilde{Z}$ of $\tilde{G}$ is contained in $\tilde{K}$, where $\tilde{K}$ is the analytic subgroup of $\tilde{G}$ with Lie algebra $\mathfrak{k}$ (see Helgason [273, p. 216]). Now G is a quotient $\tilde{G}/N$ for some $N \subset \tilde{Z}$. Thus $\tilde{\iota}$ induces an involution automorphism of G. Setting $K = \tilde{K}/N$, we have:

$$G/K \cong (\tilde{G}/N)/(\tilde{K}/N) \cong \tilde{G}/\tilde{K}.$$

So the symmetric space G/K is independent of the choice of Lie group G with Lie algebra $\mathfrak{g}$. So we may assume that G is simply connected whenever we need this. Furthermore, it can be proved that all the K's are conjugate (see Helgason [273, p. 256]). Note, however, that the K's need not be semisimple.

The preceding arguments fail for symmetric spaces of compact type. For example, the center of G need not lie in K; e.g., consider $G = SU(n)$, $K = SO(n)$. Also K need not be connected; e.g., $SO(3)/K = \mathbb{P}^2$, the real projective plane, with K the subgroup of $SO(3)$ leaving a line through the origin invariant. Finally, the Cartan involution need not correspond to an automorphism of G in the compact case.

Another difference between compact and noncompact symmetric spaces is that the noncompact ones are topologically (though not geometrically) identifiable with the Euclidean space $\mathfrak{p}$. However, the compact symmetric spaces are not topologically trivial (see Greub et al. [244]). This fact makes the compact and noncompact symmetric spaces very different. However, there is a duality between the two types, as we shall see.

2.1.2   Examples of Symmetric Spaces

Now we intend to manufacture many examples of symmetric spaces by exploring the connection between real forms of complex simple Lie algebras and Cartan decompositions of real Lie algebras.

A **real form** $\mathfrak{g}$ of a complex simple Lie algebra $\mathfrak{g}^c$ is defined by the equality of the complexification of $\mathfrak{g}$ with $\mathfrak{g}^c$; i.e.,

$$\mathfrak{g}^c = \mathfrak{g} \otimes_{\mathbb{R}} \mathbb{C} = \mathfrak{g} \oplus i\mathfrak{g}.$$

It is possible to list the real forms $\mathfrak{g}$ of $\mathfrak{g}^c$ by listing the conjugations of $\mathfrak{g}^c$. By a **conjugation** of $\mathfrak{g}^c$, we mean a mapping $C : \mathfrak{g}^c \to \mathfrak{g}^c$ which is conjugate linear, bracket preserving, and such that C^2 is the identity. See Helgason [273, Chs. 3, 10] or Loos [408, Vol. II, Ch. VII].

It can also be shown that any complex semisimple Lie algebra $\mathfrak{g}^c$ has a compact real form $\mathfrak{u}$; i.e., the Killing form of $\mathfrak{u}$ is negative definite. Then to make a *list of symmetric spaces of noncompact type coming from complex simple Lie algebras $\mathfrak{g}^c$, one must follow through the following plan of action.*

Plan for Construction of Noncompact Symmetric Spaces of Type III

I. List the conjugations ; i.e., involutive automorphisms of $\mathfrak{g}^c$. The fixed points will be real forms of $\mathfrak{g}^c$. One of these real forms $\mathfrak{u}$ will be compact.

II. For the noncompact real forms $\mathfrak{g}$ of $\mathfrak{g}^c$, the Cartan decomposition is:

$$\mathfrak{g} = (\mathfrak{g} \cap \mathfrak{u}) \oplus (\mathfrak{g} \cap i\mathfrak{u}).$$

Note that the Killing form of $\mathfrak{g}$ has the correct behavior on the decomposition since $\mathfrak{u}$ is compact; i.e., the Killing form is negative definite on $\mathfrak{u}$. Furthermore, if τ is the conjugation of $\mathfrak{g}^c$ corresponding to the compact real form $\mathfrak{u}$, then the restriction of τ to $\mathfrak{g}$ is θ, the Cartan involution corresponding to this Cartan decomposition.

III. Form the symmetric space G/K by taking Lie groups $G \supset K$ with Lie algebras $\mathfrak{g}, \mathfrak{k}$, respectively. Here $\mathfrak{k} = \mathfrak{g} \cap \mathfrak{u}$.

Type $\mathfrak{a}$ Examples

I. **Real Forms of $\mathfrak{sl}(n,\mathbb{C})$.**

1. $\mathfrak{sl}(n, \mathbb{R})$=normal real form = fixed points of the conjugation $\tau(X) = \overline{X}$.
2. $\mathfrak{su}(n, \mathbb{R})$=compact real form = fixed points of the conjugation $\tau(X) = -\,{}^t\overline{X}$.
3. $\mathfrak{su}(p, q)$=fixed points of the conjugation $\tau(X) = -I_{p,q}\,{}^t\overline{X}\,I_{p,q}$, where

$$I_{p,q} = \begin{pmatrix} I_p & 0 \\ 0 & -I_q \end{pmatrix}, \quad n = p + q.$$

4. $\mathfrak{su}^*(2m)$=fixed points of the conjugation $\tau(X) = J_m \overline{X} J_m^{-1}$, where

$$J_m = \begin{pmatrix} 0 & I_m \\ -I_m & 0 \end{pmatrix}, \quad n = 2m \quad \text{(for even } n\text{)}.$$

II. Cartan Decompositions of Noncompact Real Forms of $\mathfrak{sl}(n,\mathbb{C})$.

1. $\mathfrak{sl}(n, \mathbb{R}) = \mathfrak{so}(n) \oplus \mathfrak{p}_n$, where

$$\mathfrak{so}(n) = \{X \in \mathfrak{sl}(n, \mathbb{R}) \mid {}^tX = -X\},$$
$$\mathfrak{p}_n = \{X \in \mathfrak{sl}(n, \mathbb{R}) \mid {}^tX = X\}.$$

The Cartan involution is $\theta(X) = -\,{}^tX, \; X \in \mathfrak{sl}(n, \mathbb{R})$.

2. $\mathfrak{su}(p, q) = \mathfrak{k} \oplus \mathfrak{p}$, where

$$\mathfrak{k} = \left\{ \begin{pmatrix} A & 0 \\ 0 & B \end{pmatrix} \;\middle|\; A \in \mathfrak{u}(p), \; B \in \mathfrak{u}(q), \; \mathrm{Tr}(A + B) = 0 \right\},$$

$$\mathfrak{p} = \left\{ \begin{pmatrix} 0 & Z \\ {}^t\overline{Z} & 0 \end{pmatrix} \;\middle|\; Z \in \mathbb{C}^{p \times q} \right\}.$$

The Cartan involution is $\theta(X) = I_{p,q} X I_{p,q}, \; X \in \mathfrak{su}(p, q)$.

3. $\mathfrak{su}^*(2m) = \mathfrak{k} \oplus \mathfrak{p}$, where

$$\mathfrak{k} = \mathfrak{sp}(m, \mathbb{C}) \cap \mathfrak{u}(2m) \doteq \mathfrak{sp}(m) \quad \text{(by definition)},$$
$$\mathfrak{p} = \mathfrak{su}^*(2m) \cap (i\, \mathfrak{u}(2m)).$$

The Cartan involution is $\theta(X) = -J_m\, {}^tX J_m^{-1}$.

III. The Noncompact Symmetric Spaces Corresponding to the Noncompact Real Forms.

1. $SL(n, \mathbb{R})/SO(n)$.

2. $SU(p, q)/S(U_p \times U_q)$, where $n = p + q$ and

$$SU(p, q) = \left\{ g \in SL(n, \mathbb{C}) \mid {}^t\overline{g} I_{p,q} g = I_{p,q} \right\},$$
$$U(p) = \left\{ g \in \mathbb{C}^{p \times p} \mid {}^t\overline{g} g = I_p \right\} = \text{ the unitary group},$$
$$S(U_p \times U_q) = \left\{ g \in SL(n, \mathbb{C}) \;\middle|\; g = \begin{pmatrix} A & 0 \\ 0 & B \end{pmatrix}, \; A \in U(p), B \in U(q) \right\}.$$

3. $SU^*(2n)/Sp(n)$, where

$$SU^*(2n) = \left\{ g \in SL(2n, \mathbb{C}) \mid g = J_n \overline{g} J_n^{-1} \right\},$$
$$Sp(n) = Sp(n, \mathbb{C}) \cap U(2n),$$
$$Sp(n, \mathbb{C}) = \left\{ g \in \mathbb{C}^{2n \times 2n} \mid {}^t g J_n g = J_n \right\} = \text{the complex symplectic group}.$$

IV. The Corresponding Compact Symmetric Spaces.

4. $SU(n)/SO(n)$.
5. $SU(p+q)/S(U_p \times U_q)$.
6. $SU(2n)/Sp(n)$.

Type c Examples

I. Real Forms of $sp(n,\mathbb{C})$.

1. $\mathfrak{sp}(n, \mathbb{R})$ =normal real form=fixed points of the conjugation $\tau(X) = \overline{X}$.
2. $\mathfrak{sp}(n)$ =compact real form=fixed points of the conjugation $\tau(X) = -\,{}^t\overline{X}$. Note that $\mathfrak{sp}(n) = \mathfrak{sp}(n, \mathbb{C}) \cap \mathfrak{u}(2n)$.
3. $\mathfrak{sp}(p, q)$ = fixed points of the conjugation $\tau(X) = -K_{p,q}\,{}^t\overline{X}K_{p,q}$, where

$$
K_{p,q} = \begin{pmatrix} I_p & 0 & 0 & 0 \\ 0 & -I_q & 0 & 0 \\ 0 & 0 & I_p & 0 \\ 0 & 0 & 0 & -I_q \end{pmatrix}, \quad p+q = n.
$$

II. Cartan Decompositions of Noncompact Real Forms of $sp(n,\mathbb{C})$.

1. $\mathfrak{sp}(n, \mathbb{R}) = \mathfrak{k} \oplus \mathfrak{p}$, where

$$
\mathfrak{k} = \left\{ \begin{pmatrix} A & B \\ -B & A \end{pmatrix} \,\middle|\, A, B \in \mathbb{R}^{n \times n},\ B = {}^tB,\ A = -\,{}^tA \right\} \cong \mathfrak{u}(n),
$$

$$
\mathfrak{p} = \left\{ X \in \mathfrak{sp}(n, \mathbb{R}) \,\middle|\, X = {}^tX \right\}.
$$

To see that $\mathfrak{k} \cong \mathfrak{u}(n)$, map

$$
\begin{pmatrix} A & B \\ -B & A \end{pmatrix} \in \mathfrak{k} \quad \text{to} \quad A + iB \in \mathfrak{u}(n).
$$

The Cartan involution is $\theta(X) = -\,{}^tX$.

2. $\mathfrak{sp}(p, q) = \mathfrak{k} \oplus \mathfrak{p}$, where

$$
\mathfrak{k} = \left\{ \begin{pmatrix} X_{11} & 0 & X_{13} & 0 \\ 0 & X_{22} & 0 & X_{24} \\ -X_{13} & 0 & X_{11} & 0 \\ 0 & -X_{24} & 0 & X_{22} \end{pmatrix} \,\middle|\, \begin{array}{l} X_{11} \in \mathfrak{u}(p),\ X_{22} \in \mathfrak{u}(q), \\ X_{13} \in \mathbb{C}^{p \times p},\ {}^tX_{13} = X_{13} \\ X_{24} \in \mathbb{C}^{q \times q},\ {}^tX_{24} = X_{24} \end{array} \right\}
$$

$\cong \mathfrak{sp}(p) \times \mathfrak{sp}(q)$.

The Cartan involution is $\theta(X) = K_{p,q}XK_{p,q}$.

III.   The Corresponding Noncompact Symmetric Spaces.

1. $Sp(n, \mathbb{R})/U(n)$.

Here $G = Sp(n, \mathbb{R})$ is the symplectic group defined in Exercise 2.1.7 while $U(n)$ is really the subgroup $K = G \cap O(2n)$ which is isomorphic to the unitary group

$$U(n) = \{g \in \mathbb{C}^{n \times n} \mid {}^t\overline{g}g = I\},$$

by part (b) of Lemma 2.1.1 below.

There are two equivalent but rather different ways to view this symmetric space. The first is as the space $\mathcal{P}_n^*$ of **positive symplectic $2n \times 2n$ real matrices**.

The second version of $Sp(n, \mathbb{R})/U(n)$ is the **Siegel upper half space** $\mathcal{H}_n$ defined by:

$$\mathcal{H}_n = \{Z \in \mathbb{C}^{n \times n} \mid {}^tZ = Z, \ \mathrm{Im}\, Z \in \mathcal{P}_n\}.$$

This example is the most important one for the rest of this book. We will discuss the various identifications of $Sp(n, \mathbb{R})/U(n)$ below. Sometimes the space $\mathcal{H}_n$ is called the "Siegel upper half plane," despite the fact that it is definitely not two-dimensional for $n > 1$. We must also apologize for our abusive use of the letter H. In Volume I, H was the Poincaré upper half plane. Now it should be $\mathcal{H}_1$. Then there is the Helgason–Fourier transform. Help! I need more alphabets!

2. $Sp(p, q)/Sp(p) \times Sp(q)$.

Here $Sp(p, q) = \{g \in SL(p + q, \mathbb{C}) \mid {}^tgK_{p,q}\overline{g} = K_{p,q}\}$, where, as before

$$K_{p,q} = \begin{pmatrix} -I_p & & & 0 \\ & I_q & & \\ & & -I_p & \\ 0 & & & I_q \end{pmatrix}.$$

Exercise 2.1.10.   Check the computations for the type A and C noncompact symmetric space examples above.

This is just about all the examples of symmetric spaces that we shall discuss. In particular, we are avoiding the exceptional Lie groups and their symmetric spaces. Table 2.1 lists some other examples of symmetric spaces. We will also be interested in the symmetric space $SL(2, \mathbb{C})/SU(2)$, which can be identified with the quaternionic upper half space or hyperbolic 3-space. It is considered at the end of this section. It is the symmetric space of a complex Lie group considered as a real group. We do not discuss compact symmetric spaces here, except to note that there is a duality between symmetric spaces U/K' of compact type and symmetric spaces G/K of noncompact type. This duality is obtained on the Lie algebra level by writing

$$\mathfrak{g} = \mathfrak{k} \oplus \mathfrak{p}$$

and

$$\mathfrak{u} = \mathfrak{k} \oplus i\mathfrak{p},$$

where $\mathfrak{u}$ is a compact real form of the complexification of $\mathfrak{g}$. See Helgason [273, Ch. 5] for more details. Helgason [273, p. 321] gives a global duality result for bounded symmetric domains (which will be defined below) allowing them to be viewed as open submanifolds of a compact Hermitian space. This is the Borel embedding theorem (see Borel [61, 62]). Such results can be applied to compute dimensions of spaces of automorphic forms via the Hirzebruch–Riemann–Roch theorem (see Hirzebruch [297, pp. 162–165] and Section 2.2.3). Healy [266] provides an example of the implications of this duality for harmonic analysis on $SU(2)$ and hyperbolic 3-space.

Example 2.1.4 (The Duality Between Compact and Noncompact Symmetric Spaces).

This example shows that hyperbolic geometry is dual to spherical geometry. We begin with the two Cartan decompositions:

$$\mathfrak{sl}(2, \mathbb{R}) = \mathfrak{so}(2) \oplus \mathfrak{p}_2, \quad \mathfrak{su}(2) = \mathfrak{so}(2) \oplus i\mathfrak{p}_2,$$

where

$$\mathfrak{p}_2 = \left\{ X \in \mathbb{R}^{2\times 2} \mid {}^t X = X, \ \mathrm{Tr}\, X = 0 \right\}.$$

The symmetric space $SL(2, \mathbb{R})/SO(2)$ can be viewed as the hyperbolic upper half plane of Chapter 3, Vol. I, while $SU(2)/SO(2)$ can be viewed as the sphere S^2 in $\mathbb{R}^3$, which is the symmetric space considered in Chapter 2, Vol. I.

The Siegel upper half space $\mathcal{H}_n$ is a **Hermitian symmetric space**; i.e., a symmetric space with a complex structure, invariant under each geodesic-reversing symmetry. It turns out that the Hermitian symmetric spaces of the compact or noncompact type have non-semisimple maximal compact subgroups K (see Helgason [273, p. 281] or Loos [408, Vol. II, p. 161]). Such is indeed the case for $G = Sp(n, \mathbb{R})$, $K = U(n)$. It also turns out that the Hermitian symmetric spaces of noncompact type are the **bounded symmetric domains** D in complex n-space. Here **symmetric** means that for every $z \in D$ there is a biholomorphic involutive map on D having z as an isolated fixed point (see Helgason [273, pp. 311–322] or Loos [408, Vol. II, p. 164]). Koecher [361] found a way of constructing all the Hermitian symmetric spaces from Jordan algebras. We shall see in Exercise 2.1.11 that the

Siegel upper half space is identifiable with a bounded symmetric domain, namely the **generalized unit disc**:

$$\mathcal{D}_n = \left\{ W \in \mathbb{C}^{n \times n} \mid {}^t W = W,\ I - \overline{W}W \in \mathcal{P}_n \right\}. \qquad (2.15)$$

The identification map is the **generalized Cayley transform**:

$$\begin{aligned} \alpha : \mathcal{H}_n &\to \mathcal{D}_n \\ Z &\mapsto (Z - iI)(Z + iI)^{-1}. \end{aligned} \qquad (2.16)$$

Exercise 2.1.11 (The Cayley Transform). Show that $W = (Z - iI)(Z + iI)^{-1}$ maps $Z \in \mathcal{H}_n$ into W in the generalized unit disc defined by

$$\mathcal{D}_n = \left\{ W \in \mathbb{C}^{n \times n} \mid {}^t W = W,\ I - W\overline{W} \in \mathcal{P}_n \right\}.$$

This mapping allows us to view the symmetric space of the symplectic group as a bounded symmetric domain. What is the image of $\mathcal{P}_n$, viewing $Y \in \mathcal{P}_n$ as the element $iY \in \mathcal{H}_n$?

Cartan proved in 1935 that there are only six types of irreducible homogeneous bounded symmetric domains (see Helgason [278, p. 518]). Here **irreducible** means that the corresponding Lie group is simple. It is possible to generalize many results from analysis and number theory to these classical domains (see Hua [308], Piatetski-Shapiro [485], and Siegel [564], [565, Vol. II, pp. 274–369]).

We could also have differentiated between the three types of symmetric spaces M according to their **sectional curvature**. The sectional curvature is defined as $-g(R(u, v)u, v)$, where g is the Riemannian metric for M, R is the curvature tensor, and u, v are orthonormal tangent vectors in $T_P(M)$, the tangent space to M at a point P. For a symmetric space, the curvature tensor at the origin is:

$$R_o(X, Y)Z = -[[X, Y], Z], \quad \text{for} \quad X, Y, Z \in \mathfrak{p}$$

(see Helgason [273, p. 180]).

Then one has the **classification of types of symmetric spaces M by sectional curvature** (see Helgason [273, p. 205]):

> M is of noncompact type $\Leftrightarrow$ the sectional curvature of M is ≤ 0;
> M is of compact type $\Leftrightarrow$ the sectional curvature of M is ≥ 0;
> M is of Euclidean type $\Leftrightarrow$ the sectional curvature of M is $= 0$.

It is possible to prove the **conjugacy of all maximal compact subgroups of noncompact semisimple real Lie groups** G using Cartan's fixed point theorem, which says that if a compact group K_1 acts on a simply connected Riemannian manifold of negative curvature such as G/K, there must be a fixed point. And $x^{-1}K_1 x \subset K$ means xK is fixed. See Helgason [273, p. 75] for more details on Cartan's theorem.

Table 2.1 Irreducible Riemannian symmetric spaces of types I and III for the non-exceptional groups

	Noncompact	Compact
AI	$SL(n, \mathbb{R})/SO(n)$	$SU(n)/SO(n)$
AII	$SU^*(2n)/Sp(n)$	$SU(2n)/Sp(n)$
AIII	$SU(p, q)/S\left(U_p \times U_q\right)$	$SU(p + q)/S\left(U_p \times U_q\right)$
BDI	$SO_o(p, q)/SO(p) \times SO(q)$	$SO(p + q)/SO(p) \times SO(q)$
DIII	$SO^*(2n)/U(n)$	$SO(2n)/U(n)$
CI	$Sp(n, \mathbb{R})/U(n)$	$Sp(n)/U(n)$
CII	$Sp(p, q)/Sp(p) \times Sp(q)$	$Sp(p + q)/Sp(p) \times Sp(q)$

The grand finale of the classification theory is the listing of the four types of irreducible symmetric spaces given in Helgason [273, Ch. 9] and [278, pp. 515–518]. Once more, irreducible means that the corresponding Lie group is simple.
The Four Types of Irreducible Symmetric Spaces are:

I. G/K, where G is a compact connected simple real Lie group and K is the subgroup of points fixed by an involutive automorphism of G.

II. G is a compact, connected simple Lie group provided with a left and right invariant Riemannian structure unique up to constant factor.

III. G/K where G is a connected noncompact simple real Lie group and K is the subgroup of points fixed by an involutive automorphism of G (a maximal compact subgroup).

IV. G/U, where G is a connected Lie group whose Lie algebra is a simple Lie algebra over $\mathbb{C}$ viewed as a real Lie algebra, and U is a maximal compact subgroup of G.

The irreducible symmetric spaces of types I and III which come from non-exceptional Lie groups are in Table 2.1. In this table $SO^*(2n) = \{g \in SO(2n, \mathbb{C}) \mid {}^t\overline{g}J_ng = J_n\}$, where J_n is defined in part (4) of the list of real forms of $\mathfrak{sl}(n, \mathbb{C})$ and $SO_o(p, q)$ is the identity component of $SO(p, q) = \{g \in SL(n, \mathbb{R}) \mid {}^tgI_{p,q}g = I_{p,q}\}$, where $n = p + q$ and $I_{p,q}$ is defined in part (3) of the list of real forms of $\mathfrak{sl}(n, \mathbb{C})$.

2.1.3 Cartan, Iwasawa, and Polar Decompositions, Roots

From now on, our emphasis will be upon the symmetric space $Sp(n, \mathbb{R})/U(n)$. Our first task is to study the various realizations of this space. We begin with the realization as the space of **positive symplectic matrices**:

$$Sp(n, \mathbb{R})/U(n) \cong \mathcal{P}_n^* = \{Y \in Sp(n, \mathbb{R}) \mid Y \in \mathcal{P}_{2n}\}. \tag{2.17}$$

The proof of (2.17) involves the global or group level Cartan decomposition. Let $\mathfrak{g}$ be a noncompact semisimple (real) Lie algebra with Cartan decomposition $\mathfrak{g} = \mathfrak{k} \oplus \mathfrak{p}$. Suppose that K and G are the corresponding connected Lie groups with Lie algebras $\mathfrak{k}$ and $\mathfrak{g}$, respectively. Then we have the **global Cartan decomposition**:

$$G = KP, \qquad P = \exp \mathfrak{p},$$

and G is diffeomorphic to $K \times \mathfrak{p}$. Note that G and K are Lie groups but P is not. The main idea of the proof of (2.17) is to use the Adjoint representation of G to deduce the Cartan decomposition of G from that for $GL(n, \mathbb{R})$, which is:

$$GL(n, \mathbb{R}) = O(n) \cdot \mathcal{P}_n \tag{2.18}$$

(see Exercise 1.1.5 of Section 1.1.2). Proofs of the general Cartan decomposition can be found in Helgason [273, p. 215] or Loos [408, Vol. I, p. 156]. We shall only consider the special case of interest.

Lemma 2.1.1 (The Cartan Decomposition for the Symplectic Group).

(a) *The Cartan decomposition for $G = Sp(n, \mathbb{R})$ comes from the Cartan decomposition (2.18) for $GL(2n, \mathbb{R})$ by taking intersections; i.e.,*

$$G = Sp(n, \mathbb{R}) = K \cdot \mathcal{P}_n^*, \quad \text{with } \mathcal{P}_n^* = \mathcal{P}_{2n} \cap G \text{ and } K = O(2n) \cap G.$$

(b) *The maximal compact subgroup K of G given in part (a) can be identified with the unitary group $U(n) = \{g \in \mathbb{C}^{n \times n} \mid {}^t\overline{g}g = I\}$. It follows also that the symmetric space $G/K = Sp(n, \mathbb{R})/U(n)$ can be identified with $\mathcal{P}_n^*$.*

Proof. (a) See Helgason [273, p. 345] and [278, p. 450].

Observe that (2.18) says that $g \in G$ can be written as $g = up$ with $u \in O(2n)$ and $p \in \mathcal{P}_{2n}$. We need to show that both u and p lie in $Sp(n, \mathbb{R})$. To see this, note that $p^2 = {}^tgg$. Moreover, $g \in G$ implies that ${}^tg^{-1}$ and ${}^tg$ both also lie in G (a situation really brought about by the existence of an involution of G with differential the Cartan involution of $\mathfrak{g}$). Thus $p^2 \in G$.

Now we must show that $p^2 \in G$ implies that p lies in G. To do this, note that G is a pseudoalgebraic group, meaning that there is a finite set of polynomials

$$f_j \in \mathbb{C}[X_1, \ldots, X_{4n^2}]$$

such that a matrix g lies in G if and only if g is a root of all the f_j. Now there is a rotation matrix $k \in O(2n)$ such that

$$k^{-1}p^2k = \begin{pmatrix} e^{h_1} & & 0 \\ & \ddots & \\ 0 & & e^{h_{2n}} \end{pmatrix}.$$

And $k^{-1}Gk$ is also a pseudoalgebraic group. Thus the diagonal matrices

$$k^{-1}p^{2r}k = \begin{pmatrix} e^{rh_1} & & 0 \\ & \ddots & \\ 0 & & e^{rh_{2n}} \end{pmatrix}$$

satisfy a certain set of polynomial equations for any integer r. But if an exponential polynomial

$$F(t) = \sum_{j=1}^{B} c_j \exp(b_j t)$$

vanishes for all integers t, then it must vanish for all real numbers t as well. Thus, in particular, p must lie in the group G, as will all elements

$$p_t = \exp(tX), \quad \text{for } t \in \mathbb{R}, \quad \text{if } p^2 = \exp(2X).$$

But then $p \in G$ implies that $u = gp^{-1} \in G$. This completes the proof of part (a)—except to show the uniqueness of the expression $g = up$ and the fact that G is diffeomorphic to $K \times \mathfrak{p}$. We leave these proofs as an **exercise**.

(b) To see that K is isomorphic to $U(n)$, proceed as follows. First recall that we have $K = G \cap O(2n)$. Thus if $J = J_n$ is as defined in Exercise 2.1.7, then

$$M = \begin{pmatrix} A & B \\ C & D \end{pmatrix} \in K \Leftrightarrow JM = MJ \text{ and } {}^tMJM = J$$

$$\Leftrightarrow C = -B, \ D = A, \ {}^tAB = {}^tBA \ \text{ and } \ {}^tAA + {}^tBB = I.$$

The last statement is equivalent to saying that $A + iB \in U(n)$. Thus the identification of K and $U(n)$ on the group level is the same as that on the Lie algebra level which was discussed when we listed the Cartan decompositions corresponding to noncompact real forms of $\mathfrak{sp}(n, \mathbb{C})$. In fact,

$$\sigma \begin{pmatrix} A & B \\ -B & A \end{pmatrix} = A + iB$$

defines a mapping which preserves matrix multiplication as well as addition. The map σ identifies K with $U(n)$. A good reference for these things is Séminaire Cartan [547, Exp 3]. The proof of Lemma 2.1.1 is now complete.   ∎

Exercise 2.1.12. Fill in all the details in the proof of Lemma 2.1.1.

Note that most calculations are far easier on the Lie algebra level than on the group level. For an example of the difference between the algebra and the group, note that it is clear that $\mathfrak{sp}(n, \mathbb{R})$ is contained in $\mathfrak{sl}(2n, \mathbb{R})$, but it is not obvious that $Sp(n, \mathbb{R})$ is contained in $SL(2n, \mathbb{R})$, though it is true.

Exercise 2.1.13. Prove the last statement.
Hint. Show that $Sp(n, \mathbb{R})$ is connected. See Chevalley [104, p. 36] for the useful result which says that H and G/H connected implies G connected, where H is a closed subgroup of the topological group G.

Next we seek to generalize the **Iwasawa decomposition** from Exercise 1.2.12 in Section 1.1.3:

$$G = GL(n, \mathbb{R}) = KAN, \tag{2.19}$$

where K is the compact group $O(n)$, A is the abelian group of positive diagonal matrices in G, and N is the nilpotent group of upper triangular matrices in G with ones on the diagonal. In order to obtain such an Iwasawa decomposition for any noncompact semisimple (real) Lie group G, one must discuss the **root space decomposition** of the Lie algebra of G. We do not give a detailed discussion of root spaces, except for several examples. The details for the general case can be found in Helgason [273, 278] or Loos [408].

Some definitions are needed to discuss the root space decomposition of the Lie algebra $\mathfrak{g}$ of G. Define $\mathfrak{a}$ to be a **maximal abelian subspace of** $\mathfrak{p}$. Here $\mathfrak{p}$ comes from the Cartan decomposition of $\mathfrak{g}$. Then for any **real linear functional (root)** $\alpha : \mathfrak{a} \to \mathbb{R}$, define the **root space**:

$$\mathfrak{g}_\alpha = \{X \in \mathfrak{g} \mid (\operatorname{ad} H)X = \alpha(H)X, \ \text{for all} \ H \in \mathfrak{a}\}.$$

If $\mathfrak{g}_\alpha \neq \{0\}$, then we say that the linear functional α is a **restricted root**. Let Λ denote the set of all **nonzero restricted roots**. When we are considering a normal real form such as $\mathfrak{sl}(n, \mathbb{R})$, the **restricted roots** are restrictions of roots of the complexification of $\mathfrak{g}$.

Next set $\mathfrak{m}$ equal to the **centralizer** of $\mathfrak{a}$ in $\mathfrak{k}$, where $\mathfrak{k}$ comes from the Cartan decomposition of $\mathfrak{g}$; i.e.,

$$\mathfrak{m} = \{X \in \mathfrak{k} \mid [X, \mathfrak{a}] = 0\}.$$

In fact, $\mathfrak{m}$ will always be zero for normal or split real forms such as $\mathfrak{sp}(n, \mathbb{R})$.

Finally, **the root space decomposition of the real noncompact semisimple Lie algebra $\mathfrak{g}$ is**:

$$\mathfrak{g} = \mathfrak{a} \oplus \mathfrak{m} \oplus \sum_{\alpha \in \Lambda}^{\oplus} \mathfrak{g}_\alpha.$$

To prove the validity of this decomposition, consider the positive definite bilinear form F on $\mathfrak{g}$ defined as follows, using the Killing form B and the Cartan involution θ of $\mathfrak{g}$:

$$F(X, Y) = -B(X, \theta Y), \quad \text{for} \quad X, Y \in \mathfrak{g}. \tag{2.20}$$

If $X \in \mathfrak{p}$, then $\mathrm{ad}X$ is symmetric with respect to F and thus is a diagonalizable linear transformation of $\mathfrak{g}$. Therefore the commuting family of all the $\mathrm{ad}X$ for $X \in \mathfrak{a}$ is simultaneously diagonalizable with real eigenvalues. It remains to show that the eigenspace corresponding to the zero functional is:

$$\mathfrak{g}_0 = (\mathfrak{g}_0 \cap \mathfrak{k}) \oplus (\mathfrak{g}_0 \cap \mathfrak{p}) = \mathfrak{m} \oplus \mathfrak{a}.$$

This comes from the definitions.

Note that if $\mathfrak{g}$ is $\mathfrak{sl}(n, \mathbb{R})$ or $\mathfrak{sp}(n, \mathbb{R})$, then $\mathfrak{m} = \{0\}$ and restricted roots are the same as the roots of the complexifications $\mathfrak{sl}(n, \mathbb{C})$ and $\mathfrak{sp}(n, \mathbb{C})$ restricted to the normal real form.

One can define the set of **positive restricted roots** Λ^+ as a subset of Λ such that Λ is the disjoint union of Λ^+ and $-\Lambda^+$. We will soon see how to find such sets of positive roots in our favorite cases.

We need to use the positive roots to construct a certain nilpotent Lie subalgebra $\mathfrak{n}$ of $\mathfrak{g}$. By definition, a Lie algebra $\mathfrak{n}$ is said to be **nilpotent** if the lower central series $\mathfrak{n}^k$ defined by

$$\mathfrak{n}^0 = \mathfrak{n}, \ \mathfrak{n}^1 = [\mathfrak{n}, \mathfrak{n}], \ \mathfrak{n}^{k+1} = [\mathfrak{n}, \mathfrak{n}^k]$$

terminates; i.e., $\mathfrak{n}^k = \{0\}$, for some k.

Suppose that Λ^+ denotes the chosen set of positive roots of $\mathfrak{g}$. Define the **nilpotent Lie subalgebra $\mathfrak{n}$** *of* $\mathfrak{g}$ by:

$$\mathfrak{n} = \sum_{\alpha \in \Lambda^+}^{\oplus} \mathfrak{g}_\alpha.$$

We can also define the **opposite nilpotent subalgebra $\overline{\mathfrak{n}}$** of $\mathfrak{g}$ by:

$$\overline{\mathfrak{n}} = \sum_{\alpha \in \Lambda^+}^{\oplus} \mathfrak{g}_{-\alpha}.$$

To prove that $\mathfrak{n}$ is nilpotent, it suffices to know the following simple facts about roots.

Simple Facts About Roots

(1) Λ^+ is a finite set;
(2) $[\mathfrak{g}_\alpha, \mathfrak{g}_\beta] \subset \mathfrak{g}_{\alpha+\beta}$;
(3) $\alpha, \beta \in \Lambda^+$ implies that $\alpha + \beta$ is either a positive root or not a root at all.

Furthermore, if θ denotes the Cartan involution of $\mathfrak{g} = \mathfrak{k} \oplus \mathfrak{p}$, then θ interchanges the nilpotent algebra $\mathfrak{n}$ and its opposite $\overline{\mathfrak{n}}$; i.e., $\theta \mathfrak{n} = \overline{\mathfrak{n}}$. To see this, note that $\theta(X) = -X$ for all X in $\mathfrak{a} \subset \mathfrak{p}$ and θ preserves the Lie bracket.

From the preceding considerations, it is easy to obtain the **Iwasawa decomposition of the noncompact real semisimple Lie algebra**:

$$\mathfrak{g} = \mathfrak{k} \oplus \mathfrak{a} \oplus \mathfrak{n}.$$

For clearly, one has $\mathfrak{g} = \bar{\mathfrak{n}} \oplus \mathfrak{m} \oplus \mathfrak{a} \oplus \mathfrak{n}$. And $\mathfrak{k} \oplus \mathfrak{a} \oplus \mathfrak{n}$ is a direct sum, since

$$X + H + Y = 0, \quad \text{for } X \in \mathfrak{k}, \; H \in \mathfrak{a}, \; Y \in \mathfrak{n},$$

implies that

$$0 = \theta(X + H + Y) = X - H + \theta(Y).$$

Subtract the two equations to see that $2H + Y - \theta(Y) = 0$. This implies that $H = 0$ by the fact that the root space decomposition is a direct sum. Thus $Y = \theta(Y) = 0$ and $X = 0$.

To complete the proof of the Lie algebra Iwasawa decomposition, we need to only show that the dimensions are correct. It suffices to look at the following mapping:

$$\mathfrak{m} \oplus \bar{\mathfrak{n}} \to \mathfrak{k}, \qquad\qquad 1 - 1, \text{ onto}$$
$$X + Y \mapsto X + Y + \theta(Y), \quad \text{for } X \in \mathfrak{m}, \; Y \in \bar{\mathfrak{n}}.$$

Next we want to consider three examples: $\mathfrak{sl}(n, \mathbb{R})$, $\mathfrak{sp}(n, \mathbb{R})$, and $\mathfrak{su}(3, 1)$. The first two examples are **split** or **normal**, so that $\mathfrak{m} = \{0\}$, the restricted roots are restrictions of complex roots of the complexification, and all the roots spaces are one-dimensional real vector spaces.

Three Examples of Iwasawa Decompositions of Real Semisimple Lie Algebras

Example 2.1.5 ($\mathfrak{sl}(n, \mathbb{R})$).
Recall that the Cartan decomposition is $\mathfrak{sl}(n, \mathbb{R}) = \mathfrak{k} \oplus \mathfrak{p}_n$, where

$$\mathfrak{k} = \mathfrak{so}(n) = \left\{ X \in \mathbb{R}^{n \times n} \mid {}^t X = -X \right\},$$
$$\mathfrak{p}_n = \left\{ X \in \mathbb{R}^{n \times n} \mid {}^t X = X, \; \mathrm{Tr}\, X = 0 \right\}.$$

One can show that a maximal abelian subspace of $\mathfrak{p}_n$ is:

$$\mathfrak{a} = \left\{ H \in \mathbb{R}^{n \times n} \mid H \text{ is diagonal of trace } 0 \right\}.$$

Next let E_{ij} for $1 \leq i, j \leq n$ denote the matrix with 1 in the i, j place and 0's elsewhere. Then set $e_i(H) = h_i$ if H is a diagonal matrix with h_i as its ith diagonal entry. Let $\alpha_{ij} = e_i - e_j$. Then $[H, E_{ij}] = \alpha_{ij}(H)E_{ij}$ and we find the root space decomposition involves the

$$\mathfrak{g}_{\alpha_{ij}} = \mathbb{R}E_{ij} \text{ with } \Lambda^+ = \{\alpha_{ij} \mid 1 \le i < j \le n\}.$$

Thus

$$\mathfrak{n} = \sum_{1 \le i < j \le n}^{\oplus} \mathbb{R}E_{ij} = \text{the upper triangular real } n{\times}n \text{ matrices with 0 on the diagonal.}$$

It follows that the Lie algebra analogue of the Iwasawa decomposition of $SL(n, \mathbb{R})$ coming from (2.19) says:

$$\mathfrak{sl}(n, \mathbb{R}) = \mathfrak{so}(n) \oplus \mathfrak{a} \oplus \mathfrak{n},$$

with $\mathfrak{so}(n)$ denoting the skew symmetric $n \times n$ real matrices, $\mathfrak{a}$ equal to the $n \times n$ real diagonal trace zero matrices, and $\mathfrak{n}$ equal to the upper triangular $n \times n$ real matrices with zeros on the diagonal.

Example 2.1.6 ($\mathfrak{sp}(n, \mathbb{R})$).
   Recall that $\mathfrak{sp}(n, \mathbb{R})$ consists of matrices

$$(A, B, C) \doteq \begin{pmatrix} A & B \\ C & -{}^tA \end{pmatrix},$$

with $A, B, C \in \mathbb{R}^{n \times n}$ and B, C symmetric. We found the Cartan decomposition had:

$$\mathfrak{k} = \{(A, B, -B) \mid B \text{ symmetric, } A \text{ skew symmetric}\},$$
$$\mathfrak{p} = \{(A, B, B) \mid A, B \text{ symmetric}\}.$$

A calculation shows that a maximal abelian subspace of $\mathfrak{p}$ is:

$$\mathfrak{a} = \{(H, 0, 0) \mid H \text{ is real } n \times n \text{ diagonal}\}.$$

Suppose that the E_{ij}, $1 \le i < j \le n$, are as in Example 2.1.5. Set $G_{pq} = E_{pq} + E_{qp}$, for $1 \le p \le q \le n$. Then, if we abuse notation and write $H = (H, 0, 0)$, we have

$$[H, (E_{ij}, 0, 0)] = (e_i - e_j)(H)(E_{ij}, 0, 0),$$
$$[H, (0, G_{pq}, 0)] = (e_p + e_q)(H)(0, G_{pq}, 0),$$
$$[H, (0, 0, G_{pq})] = -(e_p + e_q)(H)(0, 0, G_{pq}).$$

It follows that we can take $\Lambda^+ = \{e_i - e_j \mid 1 \le i < j \le n\} \cup \{e_p + e_q \mid 1 \le p \le q \le n\}$.
Thus

$$\mathfrak{n} = \sum_{1 \le i < j \le n} {}^{\oplus} \mathbb{R}(E_{ij}, 0, 0) + \sum_{1 \le p \le q \le n} {}^{\oplus} \mathbb{R}(0, G_{pq}, 0)$$

$$= \{(A, B, 0) \mid A \text{ upper triangular, } 0 \text{ on diagonal, } B \text{ symmetric}\} .$$

So the **Iwasawa decomposition** is:

$$\mathfrak{sp}(n, \mathbb{R}) = \mathfrak{k} \oplus \mathfrak{a} \oplus \mathfrak{n},$$

where

$\mathfrak{k} = \{(A, B, -B) \mid B \text{ symmetric, } A \text{ skew-symmetric}\} ,$

$\mathfrak{a} = \{(H, 0, 0) \mid H \text{ diagonal}\} ,$

$\mathfrak{n} = \{(A, B, 0) \mid A \text{ upper triangular with } 0 \text{ on the diagonal, } B \text{ symmetric}\} .$

Example 2.1.7 ($\mathfrak{su}(3, 1)$).
First recall that

$$\mathfrak{su}(3, 1) = \left\{ X \in \mathfrak{sl}(4, \mathbb{C}) \mid -I_{3,1} \, {}^t\overline{X} \, I_{3,1} = X \right\}$$

where

$$I_{3,1} = \begin{pmatrix} 1 & & & 0 \\ & 1 & & \\ & & 1 & \\ 0 & & & -1 \end{pmatrix} .$$

The corresponding Lie group is $SU(3, 1) = \{g \in SL(4, \mathbb{C}) \mid {}^t\overline{g} I_{3,1} g = I_{3,1} \}$.
One sees easily that

$$\mathfrak{su}(3, 1) = \left\{ (A, b, c) \mid A \in \mathfrak{u}(3), \ b \in \mathbb{R}, \ c \in \mathbb{C}^{3 \times 1}, \ \mathrm{Tr}A + ib = 0 \right\} ,$$

where

$$(A, b, c) = \begin{pmatrix} A & c \\ {}^t\overline{c} & ib \end{pmatrix} .$$

We saw that the Cartan decomposition of $\mathfrak{su}(3, 1)$ involves

$$\mathfrak{k} = \{(A, b, 0) \mid A \in \mathfrak{u}(3), \ b \in \mathbb{R}, \ \mathrm{Tr}A + ib = 0\},$$
$$\mathfrak{p} = \left\{(0, 0, c) \mid c \in \mathbb{C}^{3 \times 1} \right\} .$$

A maximal abelian subspace of $\mathfrak{p}$ is $\mathfrak{a} = \mathbb{R}(0,0,e_1)$ where $e_1 = {}^t(1,0,0)$. Note that $(0,0,ie_1)$ does not commute with $(0,0,e_1)$. You need to multiply matrices to check these things (see Exercise 2.1.14 below). Similarly you can show that:

$$
\mathfrak{m} = \left\{ \begin{pmatrix} ib & 0 & 0 & 0 \\ 0 & u_1 & u_2 & 0 \\ 0 & u_3 & u_4 & 0 \\ 0 & 0 & 0 & ib \end{pmatrix} \middle| \; b \in \mathbb{R}, \; U = \begin{pmatrix} u_1 & u_2 \\ u_3 & u_4 \end{pmatrix} \in \mathfrak{u}(2), \; \mathrm{Tr}U + 2ib = 0 \right\}.
$$

To prove this, one must show that the matrices of $\mathfrak{m}$ centralize $\mathfrak{a}$ and that nothing else in $\mathfrak{k}$ does the same trick. Once again, this is checked by multiplying matrices. Thus we have come upon an example of a nonzero and rather fat $\mathfrak{m}$. The root space decomposition of $\mathfrak{su}(3, 1)$ is rather complicated. We find roots λ such that 2λ is also a root. Such things cannot happen for complex semisimple Lie algebras. And one finds root spaces $\mathfrak{g}_\lambda$ of dimension greater than one over $\mathbb{R}$.

The positive roots of $\mathfrak{su}(3, 1)$ are $\Lambda^+ = \{\lambda, 2\lambda\}$, where $\lambda((0, 0, e_1)) = 1$. And the root space decomposition of $\mathfrak{su}(3, 1)$ is:

$$
\mathfrak{su}(3, 1) = \mathfrak{a} \oplus \mathfrak{m} \oplus \mathfrak{g}_\lambda \oplus \mathfrak{g}_{-\lambda} \oplus \mathfrak{g}_{2\lambda} \oplus \mathfrak{g}_{-2\lambda}
$$
$$
\dim_{\mathbb{R}} \mathfrak{a} = 1, \quad \dim_{\mathbb{R}} \mathfrak{m} = 4, \quad \dim_{\mathbb{R}} \mathfrak{g}_\lambda = \dim_{\mathbb{R}} \mathfrak{g}_{-\lambda} = 4,
$$
$$
\dim_{\mathbb{R}} \mathfrak{g}_{2\lambda} = \dim_{\mathbb{R}} \mathfrak{g}_{-2\lambda} = 1.
$$

To see this, note that

$$
\left[\begin{pmatrix} 0 & 0 & 0 & 1 \\ 0 & 0 & 0 & 0 \\ 0 & 0 & 0 & 0 \\ 1 & 0 & 0 & 0 \end{pmatrix}, \begin{pmatrix} 0 & a & b & 0 \\ -\bar{a} & 0 & 0 & c \\ -\bar{b} & 0 & 0 & d \\ 0 & \bar{c} & \bar{d} & 0 \end{pmatrix} \right] = \begin{pmatrix} 0 & \bar{c} & \bar{d} & 0 \\ -c & 0 & 0 & \bar{a} \\ -d & 0 & 0 & \bar{b} \\ 0 & a & b & 0 \end{pmatrix} = k \begin{pmatrix} 0 & a & b & 0 \\ -\bar{a} & 0 & 0 & c \\ -\bar{b} & 0 & 0 & d \\ 0 & \bar{c} & \bar{d} & 0 \end{pmatrix}
$$

implies that $k = \pm 1$ and that $ka = \bar{c}$, $kb = \bar{d}$. Thus, if $k = 1$, we find that $\mathfrak{g}_\lambda$ is four-dimensional over $\mathbb{R}$:

$$
\mathfrak{g}_\lambda = \left\{ \begin{pmatrix} 0 & a & b & 0 \\ -\bar{a} & 0 & 0 & \bar{a} \\ -\bar{b} & 0 & 0 & \bar{b} \\ 0 & a & b & 0 \end{pmatrix} \middle| \; (a, b) \in \mathbb{C}^2 \right\}.
$$

If $k = -1$, we find that again $\mathfrak{g}_{-\lambda}$ is four-dimensional over $\mathbb{R}$:

$$
\mathfrak{g}_{-\lambda} = \left\{ \begin{pmatrix} 0 & a & b & 0 \\ -\bar{a} & 0 & 0 & -\bar{a} \\ -\bar{b} & 0 & 0 & -\bar{b} \\ 0 & -a & -b & 0 \end{pmatrix} \middle| \; (a, b) \in \mathbb{C}^2 \right\}.
$$

Recalling what it means to be in $\mathfrak{su}(3, 1)$, we see that it remains to deal with

$$
\mathfrak{g}_{2\lambda} = \mathbb{R} \begin{pmatrix} i & 0 & 0 & -i \\ 0 & 0 & 0 & 0 \\ 0 & 0 & 0 & 0 \\ i & 0 & 0 & -i \end{pmatrix} \quad \text{and} \quad \mathfrak{g}_{-2\lambda} = \mathbb{R} \begin{pmatrix} i & 0 & 0 & i \\ 0 & 0 & 0 & 0 \\ 0 & 0 & 0 & 0 \\ -i & 0 & 0 & -i \end{pmatrix}.
$$

The nilpotent Lie algebra $\mathfrak{n}$ is then:

$$
\mathfrak{n} = \mathfrak{g}_\lambda \oplus \mathfrak{g}_{2\lambda} = \left\{ \begin{pmatrix} ic & a & b & -ic \\ -\bar{a} & 0 & 0 & \bar{a} \\ -\bar{b} & 0 & 0 & \bar{b} \\ ic & a & b & -ic \end{pmatrix} \;\middle|\; a, b \in \mathbb{C}, \; c \in \mathbb{R} \right\}.
$$

Exercise 2.1.14. (a) Check the calculations in the preceding three examples.
(b) Perform the analogous calculation to that of part (a) in the case of the Lorentz algebra $\mathfrak{so}(3, 1)$.

Our next goal is to understand **the group level Iwasawa decomposition of a noncompact semisimple connected real Lie group G with finite center**:

$$
G = KAN,
$$

where K, A, N are connected Lie subgroups of G with Lie algebras $\mathfrak{k}, \mathfrak{a}, \mathfrak{n}$, respectively. G is actually diffeomorphic to the product $K \times A \times N$. The exponential maps $\mathfrak{k}$ onto the compact group K. And exp is a diffeomorphism which maps $\mathfrak{a}$ onto the abelian group A while taking addition to multiplication. The exponential is a diffeomorphism of $\mathfrak{n}$ onto the nilpotent group N. Recall that the exponential does not in general map $\mathfrak{g}$ onto G, nor is exp a *diffeomorphism* in general (see Exercise 2.1.6). For a proof that the exponential map is onto for abelian, nilpotent, and compact Lie groups, see Helgason [273, pp. 229, 56–58, 188–189].

In our discussion of the group level Iwasawa decomposition, we shall only consider the special case of the symplectic group. The proof of the global Iwasawa decomposition in the general case uses the Adjoint representation (see Helgason [273, 278] or Loos [408]).

Recall the following definition:

$$
Sp(n, \mathbb{R}) = \text{the symplectic group} = \left\{ g \in SL(2n, \mathbb{R}) \;\middle|\; {}^t g J_n g = J_n \right\},
$$

where

$$
J_n = \begin{pmatrix} 0 & I_n \\ -I_n & 0 \end{pmatrix}.
$$

It follows that

$$Sp(n, \mathbb{R}) = \left\{ \begin{pmatrix} A & B \\ C & D \end{pmatrix} \middle|\; {}^tAC = {}^tCA, \; {}^tBD = {}^tDB, \; {}^tAD - {}^tCB = I_n \right\}.$$

$$(2.21)$$

And **Lie subgroups** of $Sp(n, \mathbb{R})$ **which correspond to the Lie subalgebras** $\mathfrak{k}, \mathfrak{a}, \mathfrak{n}$ **in the Iwasawa decomposition** $\mathfrak{g} = \mathfrak{k} \oplus \mathfrak{a} \oplus \mathfrak{n}$ **are:**

$$\left.\begin{aligned}
K_n^* &= \left\{ \begin{pmatrix} A & B \\ -B & A \end{pmatrix} \middle|\; A + iB \in U(n) \right\}, \\
A_n^* &= \left\{ \begin{pmatrix} H & 0 \\ 0 & H^{-1} \end{pmatrix} \middle|\; H \text{ positive diagonal} \right\}, \\
N_n^* &= \left\{ \begin{pmatrix} A & B \\ 0 & {}^tA^{-1} \end{pmatrix} \middle|\; A \text{ upper triangular}; \; A\,{}^tB = B\,{}^tA \right\}.
\end{aligned}\right\}$$

$$(2.22)$$

We have seen that the **symmetric space** associated with $Sp(n, \mathbb{R})$ is:

$$Sp(n, \mathbb{R})/K_n^* \cong \mathcal{P}_n^* = \mathcal{P}_{2n} \cap Sp(n, \mathbb{R}) \tag{2.23}$$

(see Lemma 2.1.1). Now we wish to find (along with the Iwasawa decomposition) another realization of this symmetric space—**Siegel's upper half space**:

$$\mathcal{H}_n = \left\{ Z \in \mathbb{C}^{n \times n} \mid {}^tZ = Z, \; \mathrm{Im}\, Z \in \mathcal{P}_n \right\}. \tag{2.24}$$

Observe that we can define the following **actions** of $G = Sp(n, \mathbb{R})$ on the three versions of the symmetric space:

$$\left.\begin{aligned}
&\text{action of } g \in G \text{ on } G/K \text{ is } a_g(xK) = gxK \quad \text{for } x \in G; \\
&\text{action of } g \in G \text{ on } \mathcal{P}_n^* \text{ is } b_g(Y) = Y[g] = {}^tgYg \text{ for } Y \in \mathcal{P}_n^*; \\
&\text{action of } g \in G \text{ on } \mathcal{H}_n \text{ is } c_g(Z) = (AZ + B)(CZ + D)^{-1} \\
&\qquad\qquad \text{for } g = \begin{pmatrix} A & B \\ C & D \end{pmatrix} Z \in \mathcal{H}_n.
\end{aligned}\right\}$$

$$(2.25)$$

Exercise 2.1.15. (a) Prove formula (2.21).
(b) Check that $c_{hg}(Z) = c_h(c_g(Z))$ in formula (2.25).

The following lemma will allow us to identify all the versions of the symmetric space associated with the symplectic group. To see this, study the following diagram of mappings:

$$\left.\begin{aligned}
Sp(n, \mathbb{R})/K_n^* &\to \mathcal{P}_n^* \to \mathcal{H}_n \\
gK_n^* &\mapsto g\,{}^tg \mapsto X + iY.
\end{aligned}\right\}$$

$$(2.26)$$

Here X, Y come from the **partial Iwasawa decomposition** of $S \in \mathcal{P}_n^*$:

$$S = \begin{pmatrix} Y & 0 \\ 0 & Y^{-1} \end{pmatrix} \begin{bmatrix} I & 0 \\ X & I \end{bmatrix}, \quad \text{for } X = {}^t X \in \mathbb{R}^{n \times n}, \ Y \in \mathcal{P}_n. \tag{2.27}$$

Lemma 2.1.2 below gives the existence and uniqueness of this decomposition for every positive symplectic matrix. There is an equivalent Iwasawa decomposition obtained by applying matrix inverse to formula (2.27):

$$S = \begin{pmatrix} Y^{-1} & 0 \\ 0 & Y \end{pmatrix} \begin{bmatrix} I & -X \\ 0 & I \end{bmatrix}, \quad \text{for } X = {}^t X \in \mathbb{R}^{n \times n}, \ Y \in \mathcal{P}_n. \tag{2.28}$$

Exercise 2.1.16. (a) Show that the composition of the two maps in (2.26) takes gK_n^* with

$$g = \begin{pmatrix} A & B \\ C & D \end{pmatrix}$$

to $(Ai + B)(Ci + D)^{-1}$ in $\mathcal{H}_n$.

(b) Show also that the maps in (2.26) preserve the group actions in (2.25). More precisely, define $i_1 ({}^t g K_n) = I[g]$ for $g \in G$ and define $i_2(S) = X + iY$, for S with partial Iwasawa decomposition (2.27). Prove that

$$i_1 \circ a_g = b_{{}^t g} \circ i_1 \quad \text{and} \quad i_2 \circ b_{{}^t g} = c_g \circ i_2.$$

(c) Suppose that $Z^* = c_g(Z) = (AZ + B)(CZ + D)^{-1}$ for $Z \in \mathcal{H}_n$ and g as in part (a). Show that the imaginary part of Z^* is $Y^* = Y\{(CZ + D)^{-1}\}$, where Y is the imaginary part of Z and $Y\{W\} = {}^t \overline{W} Y W$. Here $\overline{W}$ is the matrix obtained from W by complex conjugation of all the entries of W. Then show that $Z^* \in \mathcal{H}_n$.

(d) Show that the Jacobian $|\partial Z^* / \partial Z| = |CZ + D|^{-n-1}$.

Hints. See Maass [426, p. 33].

(a) Note that

$$Z = (Ai + B)(Ci + D)^{-1}$$

$$= (Ai + B) \left(-{}^t Ci + {}^t D \right) \left(-{}^t Ci + {}^t D \right)^{-1} (Ci + D)^{-1}.$$

(c) Note that $c_g(W) - c_g(Z) = (W {}^t C + {}^t D)^{-1} (W - Z)(CZ + D)^{-1}$. To find Y^*, let $W = \overline{Z}$.

Lemma 2.1.2 (Iwasawa Decomposition for the Symplectic Group). *Here we use the notation (2.21)–(2.28).*

(a) *Every positive symplectic matrix has the partial Iwasawa decomposition given in (2.27) or (2.28). Thus the mappings in (2.26) are identifications of the three differentiable manifolds.*

(b) *The Iwasawa decomposition of $G = Sp(n, \mathbb{R})$ says that*

$$G = K_n^* A_n^* N_n^*, \quad \text{with} \quad K_n^*, A_n^*, N_n^* \text{ as in } (2.22).$$

Proof. (a) We know from (2.19) that we can write $S \in P_n^*$ as:

$$S = \begin{pmatrix} A & 0 \\ 0 & B \end{pmatrix} \begin{bmatrix} I & X \\ 0 & I \end{bmatrix} \quad \text{with } A, B \in P_n, \ X \in \mathbb{R}^{n \times n}.$$

Since S is symplectic, it follows that for J_n as defined in Exercise 2.1.7, we have $SJ_nS = J_n$. Thus $J_nSJ_n = -S^{-1}$.

The only way for J_nSJ_n to be equal to $-S^{-1}$ when S has the given partial Iwasawa decomposition is that

$$A = B^{-1} \quad \text{and} \quad X = {}^tX.$$

This is easily seen using again the fact that $J_n^2 = -I$. For

$$J_nSJ_n = J_n \begin{pmatrix} I & 0 \\ {}^tX & I \end{pmatrix} J_nJ_n \begin{pmatrix} A & 0 \\ 0 & B \end{pmatrix} J_nJ_n \begin{pmatrix} I & X \\ 0 & I \end{pmatrix} J_n = - \begin{pmatrix} B & 0 \\ 0 & A \end{pmatrix} \begin{bmatrix} I & 0 \\ -X & I \end{bmatrix}.$$

(b) Use part (a). This allows one to write $S \in P_n^*$ as

$$S = \begin{pmatrix} A & 0 \\ 0 & A^{-1} \end{pmatrix} \begin{bmatrix} I & B \\ 0 & I \end{bmatrix}, \quad \text{for } A \in P_n, \ B = {}^tB \in \mathbb{R}^{n \times n}.$$

Then express A as $A = H[Q]$ with H positive diagonal and Q upper triangular with 1's on the diagonal. This is possible by the Iwasawa decomposition for $GL(n, \mathbb{R})$. Thus

$$Y = \begin{pmatrix} H & 0 \\ 0 & H^{-1} \end{pmatrix} \begin{bmatrix} Q & QB \\ 0 & {}^tQ^{-1} \end{bmatrix},$$

which is the full Iwasawa decomposition of $Y \in P_n^*$. This translates to the Iwasawa decomposition for an element of the symplectic group using the Cartan decomposition (Lemma 2.1.1), completing the proof of Lemma 2.1.2. ∎

Exercise 2.1.17. Show that the Killing form on $\mathfrak{g} = \mathfrak{sp}(n, \mathbb{R})$ is

$$B(X, Y) = 4(n + 1)\mathrm{Tr}(XY).$$

Hint. Use the root space decomposition of $\mathfrak{g}$.

The Riemannian metric on $Sp(n, \mathbb{R})/K_n^* \cong \mathcal{P}_n^*$ is given by:

$$Q_Y(u, v) = \mathrm{Tr}(Y^{-1}uY^{-1}v),$$

for $Y \in \mathcal{P}_n^*$, $u, v \in T_Y(\mathcal{P}_n^*) = $ the tangent space to $\mathcal{P}_n^*$ at Y (see (2.14) and the analogous result for $SL(n, \mathbb{R})$). Here we have dropped the constant in the Killing form of Exercise 2.1.17. Using the notation of formula (1.11) in Section 1.1.3, if $dY = (dy_{ij}) \in T_Y(\mathcal{P}_n^*)$, $Y \in \mathcal{P}_n^*$, then the **arc length** on $\mathcal{P}_n^*$ is

$$ds^2 = \mathrm{Tr}(Y^{-1}dY\, Y^{-1}dY). \tag{2.29}$$

We want to show that the geodesics for this metric come from matrix exp and thus that $\mathcal{P}_n^*$ is a totally geodesic submanifold of $\mathcal{P}_{2n}$. We can use partial Iwasawa coordinates from Lemma 2.1.2 for this purpose. Now $W \in \mathcal{P}_n^*$ has partial Iwasawa decomposition

$$W = \begin{pmatrix} V & 0 \\ 0 & V^{-1} \end{pmatrix} \begin{bmatrix} I & X \\ 0 & I \end{bmatrix}, \quad V \in \mathcal{P}_n, \quad X = {}^t X \in \mathbb{R}^n. \tag{2.30}$$

Just as in Exercise 1.1.14 of Section 1.1.3, we obtain the following formula for the **arc length** on $\mathcal{P}_n^*$ **in partial Iwasawa coordinates** (2.30):

$$ds^2 = \mathrm{Tr}\left(\left(V^{-1}dV\right)^2 + \left(V\, d(V^{-1})\right)^2 + 2\left(V^{-1}\right)\left({}^t dX\right) V^{-1}\, dX \right). \tag{2.31}$$

Exercise 2.1.18. Prove formula (2.31). Then note that $d(V^{-1}) = -V^{-1}\, dV\, V^{-1}$ and use this to show that the arc length on $\mathcal{P}_n^*$ can be expressed as follows using partial Iwasawa coordinates (2.30):

$$ds^2 = 2\, \mathrm{Tr}\left(\left(V^{-1}dV\right)^2 + \left(V^{-1}\, dX\right)^2 \right).$$

Show that the action of $G = Sp(n, \mathbb{R})$ on $\mathcal{P}_n^*$ leaves the arc length invariant.

Using Exercise 2.1.18 we find that the **arc length on the Siegel upper half space** $\mathcal{H}_n$ is:

$$ds^2 = 2\, \mathrm{Tr}\left(V^{-1}dZ\, V^{-1}d\bar{Z}\right), \quad \text{if } Z = U + iV \in \mathcal{H}_n. \tag{2.32}$$

This is indeed the arc length considered by Siegel [565, Vol. II, p. 276].

Before proceeding to the study of geodesics in $\mathcal{P}_n^*$ or $\mathcal{H}_n$, we need to consider the analogue of polar coordinates in these spaces.

Lemma 2.1.3 (The Polar Decomposition of a Noncompact Semisimple Real Lie Group). *Let G be a noncompact connected real semisimple Lie group with connected Lie subgroups K and A, as in the Cartan and Iwasawa decompositions. Then G has the polar decomposition:*

$$G = KAK.$$

Proof. First we show that if the Lie algebra $\mathfrak{g}$ of G has Cartan decomposition $\mathfrak{g} = \mathfrak{k} \oplus \mathfrak{p}$ and Iwasawa decomposition $\mathfrak{g} = \mathfrak{k} \oplus \mathfrak{a} \oplus \mathfrak{n}$, then

$$\mathfrak{p} = \text{Ad}(K)\mathfrak{a}, \tag{2.33}$$

where the Adjoint representation Ad is defined in formula (2.11). To prove (2.33), choose H in $\mathfrak{a}$ so that its centralizer in $\mathfrak{p}$ is $\mathfrak{a}$; i.e., take $H \in \mathfrak{a}$ such that $\alpha(H) \neq 0$ for all roots $\alpha \in \Lambda$. Set K^* equal to $\text{Ad}_G K$ and suppose that X is in $\mathfrak{p}$. Now there is an element k_0 in K^* such that

$$B(H, \text{Ad}(k_0)X) = \text{Min}\{B(H, \text{Ad}(k)X) \mid k \in K^*\}.$$

Suppose that $T \in \mathfrak{k}$. Then the derivative at $t = 0$ of the following function $f(t)$ of the real variable t must be 0 by the first derivative test:

$$f(t) = B(H, \text{Ad}(\exp tT)\,\text{Ad}(k_0)X).$$

This implies using the fact that the derivative of Ad is ad:

$$B(H, (\text{ad}T)(\text{Ad}(k_0)X)) = 0 \text{ for all } T \text{ in } \mathfrak{k}.$$

Thus (by Exercise 2.1.3)

$$B(T, [H, \text{Ad}(k_0)X]) = 0 \text{ for all } T \text{ in } \mathfrak{k}.$$

Since $[\mathfrak{p}, \mathfrak{p}] \subset \mathfrak{k}$, and B is negative definite on $\mathfrak{k}$, it follows that $[H, \text{Ad}(k_0)X] = 0$ which says that $\text{Ad}(k_0)X \in \mathfrak{a}$, by the definition of H. The proof of Lemma 2.1.3 is completed by observing that (2.33) implies

$$\exp \mathfrak{p} = \exp(\text{Ad}(K))\mathfrak{a} = \text{Int}(K)(\exp \mathfrak{a}) = \bigcup_{k \in K} kAk^{-1}.$$

Lemma 2.1.3 follows from this equality and Lemma 2.1.1. ∎

Next we consider some examples.

Examples of the Polar Decomposition

Example 2.1.8 ($GL(n, \mathbb{R}) = O(n)A_n O(n)$, **Where** A_n **Consists of All Positive Diagonal Matrices**).
This is equivalent (via the Cartan decomposition) to saying that for any positive matrix Y in $\mathcal{P}_n$, there is an orthogonal matrix k in $O(n)$ and a positive diagonal

matrix a in A_n such that $Y = k^{-1}ak$. Thus the polar decomposition is just the **spectral theorem** for positive definite symmetric matrices, as we noted already in formula (1.22) of Section 1.1.4.

The next question is: **How unique are the a and k in the polar decomposition** of Y in $\mathcal{P}_n$? We saw in the paragraph after Exercise 1.1.24 of Section 1.1.4 that these coordinates give a $(2^n n!)$-fold covering of $\mathcal{P}_n$, since the entries of a are unique up to the action of the Weyl group of permutations of the diagonal entries and the matrices in $O(n)$ that commute with all the diagonal matrices must themselves be diagonal with entries ± 1.

Example 2.1.9 (Euler Angle Decomposition of the Compact Group $SO(3)$).
   A reference is Hermann [289, pp. 30–39]. Set $G = SO(3)$,

$$k = \begin{pmatrix} SO(2) & 0 \\ 0 & 1 \end{pmatrix}.$$

The Cartan decomposition of $\mathfrak{g} = \mathfrak{k} \oplus \mathfrak{p}$ is:

$$\mathfrak{so}(3) = \begin{pmatrix} \mathfrak{so}(2) & 0 \\ 0 & 0 \end{pmatrix} \oplus \left\{ \begin{pmatrix} 0 & c \\ -{}^t c & 0 \end{pmatrix} \,\middle|\, c \in \mathbb{R}^2 \right\}.$$

Then we can take the maximal abelian subspace of $\mathfrak{p}$ to be

$$\mathfrak{a} = \mathbb{R} \begin{pmatrix} 0 & 0 & 1 \\ 0 & 0 & 0 \\ -1 & 0 & 0 \end{pmatrix}.$$

And

$$\exp \left\{ t \begin{pmatrix} 0 & 0 & 1 \\ 0 & 0 & 0 \\ -1 & 0 & 0 \end{pmatrix} \right\} = \begin{pmatrix} \cos t & 0 & \sin t \\ 0 & 1 & 0 \\ -\sin t & 0 & \cos t \end{pmatrix}.$$

which is easily seen by writing out the series for the matrix exponential. Thus the **Euler angle decomposition** of g in $SO(3)$ is:

$$\begin{pmatrix} \cos u & \sin u & 0 \\ -\sin u & \cos u & 0 \\ 0 & 0 & 1 \end{pmatrix} \begin{pmatrix} \cos t & 0 & \sin t \\ 0 & 1 & 0 \\ -\sin t & 0 & \cos t \end{pmatrix} \begin{pmatrix} \cos v & \sin v & 0 \\ -\sin v & \cos v & 0 \\ 0 & 0 & 1 \end{pmatrix}.$$

The three Euler angles are u, t, v. Thus any rotation in 3-space is a product of three rotations about two axes.

Example 2.1.10 (The Dual Noncompact Group to $SO(3)$ Is the Lorentz Group $SO(2, 1)$).

The group $SO(2,1)$ again has an Euler angle decomposition that is well known to physicists. You need two angular variables and one real variable. One finds that the maximal abelian subalgebra $\mathfrak{a}$ of $\mathfrak{p}$ is:

$$\mathfrak{a} = \mathbb{R} \begin{pmatrix} 0 & 0 & 1 \\ 0 & 0 & 0 \\ 1 & 0 & 0 \end{pmatrix}.$$

Then

$$\exp\left\{ t \begin{pmatrix} 0 & 0 & 1 \\ 0 & 0 & 0 \\ 1 & 0 & 0 \end{pmatrix} \right\} = \begin{pmatrix} \cosh t & 0 & \sinh t \\ 0 & 1 & 0 \\ \sinh t & 0 & \cosh t \end{pmatrix}.$$

For $SO(3,1)$, these matrices are called "Lorentz boosts" (see Misner et al. [454, p. 67]). The A-part of this group does not get wound up like the A-part of the compact group in Example 2.1.9.

Example 2.1.11 (Euler Angles for $U(3,1)$).
   The physicist Wigner [668] considers this example, for which KAK is:

$$\begin{pmatrix} A & 0 \\ 0 & u \end{pmatrix} \begin{pmatrix} \cosh t & 0 & 0 & \sinh t \\ 0 & 0 & 0 & 0 \\ 0 & 0 & 0 & 0 \\ \sinh t & 0 & 0 & \cosh t \end{pmatrix} \begin{pmatrix} B & 0 \\ 0 & v \end{pmatrix},$$

for A, B in $U(3)$ and u, b in $i\mathbb{R}$.

Example 2.1.12 ($SU(2)$ Has Euler Angles: ($0 \le \theta \le \pi$, $0 \le \varphi \le 2\pi$, $0 \le \psi \le 4\pi$)).

$$\begin{pmatrix} \exp(i\varphi/2) & 0 \\ 0 & \exp(-i\varphi/2) \end{pmatrix} \begin{pmatrix} \cos(\theta/2) & \sin(\theta/2) \\ -\sin(\theta/2) & \cos(\theta/2) \end{pmatrix} \begin{pmatrix} \exp(i\psi/2) & 0 \\ 0 & \exp(-i\psi/2) \end{pmatrix}.$$

Exercise 2.1.19. Fill in the details of the derivations of polar decompositions in Exercises 2.1.9–2.1.12. How unique are these decompositions?

Example 2.1.13 (The Symplectic Group $Sp(n, \mathbb{R})$).
   The polar decomposition of $Sp(n, \mathbb{R})$ says:

$$Sp(n, \mathbb{R}) = K_n^* A_n^* K_n^*,$$

where

$$K_n^* = O(2n) \cap Sp(n, \mathbb{R}) \cong U(n),$$

$$A_n^* = A_{2n} \cap Sp(n, \mathbb{R}) = \{\text{positive diagonal symplectic matrices}\}.$$

How unique is this decomposition? This time it is not legal to permute all the $2n$ entries of the diagonal matrix in A_n^* because the matrix has to remain symplectic. The matrix looks therefore like

$$\begin{pmatrix} H & 0 \\ 0 & H^{-1} \end{pmatrix} \text{ with } H = \begin{pmatrix} a_1 & & 0 \\ & \ddots & \\ 0 & & a_n \end{pmatrix}, \quad a_j \text{ positive.}$$

Certainly it is legal to permute all the a_j. One can also send a_j to a_j^{-1}. The group generated by such transformations of the elements of A_n^* is the *Weyl group* of $Sp(n, \mathbb{R})$, which has order $n!2^n$. If we define A_n^{*+} to be the set of diagonal matrices of the form:

$$\begin{pmatrix} H & 0 \\ 0 & H^{-1} \end{pmatrix} \text{ with } H = \begin{pmatrix} a_1 & & 0 \\ & \ddots & \\ 0 & & a_n \end{pmatrix}$$

such that $1 \leq a_1 \leq a_2 \leq \cdots \leq a_n$, then the polar decomposition

$$P_n^* = A_n^{*+}[K_n^*]$$

is *unique*, up to the action of $M_n^* = $ the centralizer of A_n^* in K_n^*, which has order 2^n.

2.1.4   Geodesics and the Weyl Group

In order to discuss the uniqueness of the general polar decomposition, one needs to discuss the Weyl group for a general semisimple noncompact real Lie group. However, let us postpone this until we have obtained the geodesics in the symmetric space for the symplectic group.

Theorem 2.1.1 (Geodesics in the Symmetric Space of the Symplectic Group).

(a) *A geodesic segment in P_n^* of the form $T(t)$, for $0 \leq t \leq 1$, with $T(0) = I$ and $T(1) = Y \in P_n^*$ has the expression:*

$$T(t) = \exp\{tB[U]\}, \quad \text{for } 0 \leq t \leq 1,$$

provided that Y has polar decomposition from Lemma 2.1.3 with

$$Y = \exp B[U], \text{ for } U \in O(n) \text{ and}$$

$$B = \begin{pmatrix} H & 0 \\ 0 & -H \end{pmatrix} \text{ with } H = \begin{pmatrix} h_1 & & 0 \\ & \ddots & \\ 0 & & h_n \end{pmatrix}, \ h_j \in \mathbb{R}, \ 1 \le j \le n.$$

The length of the geodesic segment is:

$$\left(2 \sum_{j=1}^{n} h_j^2 \right)^{1/2}.$$

(b) Consider the geodesic through Z_0 and Z_1 in $\mathcal{H}_n$. Set

$$\rho(Z_1, Z_0) = (Z_1 - Z_0)(\overline{Z}_1 - Z_0)^{-1}(\overline{Z}_1 - \overline{Z}_0)(Z_1 - \overline{Z}_0)^{-1}.$$

A given pair of points Z_0, Z_1 in $\mathcal{H}_n$ can be transformed by the same matrix $M \in Sp(n, \mathbb{R})$ into another pair of points W_0, W_1 in $\mathcal{H}_n$ if and only if the matrices $\rho(Z_0, Z_1)$ and $\rho(W_0, W_1)$ have the same eigenvalues.

If $r_1, \ldots, r_n$ are the eigenvalues of the matrix $\rho(Z_0, Z_1)$, then the symplectic distance between Z_1 and Z_0 is:

$$s(Z_0, Z_1) = \sqrt{2} \left(\sum_{j=1}^{n} \log^2 \frac{1 + \sqrt{r_j}}{1 - \sqrt{r_j}} \right)^{1/2}.$$

Proof. See Maass [426, p. 39].

(a) The proof proceeds exactly as in the proof of Theorem 1.3.1 of Section 1.1.3. In the partial Iwasawa decomposition (2.30) of $T(t)$, we only decrease the arc length by taking X to be identically zero. Then by Exercise 2.1.18,

$$ds^2 = 2\text{Tr}\left(\left(V^{-1} \, dV \right)^2 \right),$$

and we know from Theorem 1.3.1 of Section 1.1.3 that this arc length is minimized by taking V to be diagonal. The rest of part (a) is immediate.

(b) Note that if $W_j = \left(A Z_j + B \right) \left(C Z_j + D \right)^{-1}$, for $j = 0, 1$,

$$\begin{pmatrix} A & B \\ C & D \end{pmatrix} \in Sp(n, \mathbb{R}),$$

then

$$\rho(W_1, W_0) = \left(Z_1 \, {}^t C + {}^t D\right)^{-1} \rho(Z_1, Z_0) \left(Z_1 \, {}^t C + {}^t D\right). \tag{2.34}$$

Using part (a), we need to only observe that with H as in part (a), we have

$$\rho(iH, iI) = (H - I)^2 (H + I)^{-2}.$$

The eigenvalues of $\rho(iH, iI)$ are $r_j = (h_j - 1)^2 (h_j + 1)^{-2}$. Thus

$$h_i = \frac{1 + \sqrt{r_j}}{1 - \sqrt{r_j}}.$$

This completes the proof of Theorem 2.1.1.

∎

It is possible to generalize Theorem 2.1.1 to all noncompact real symmetric spaces.

Exercise 2.1.20. Prove formula (2.34) which was used in the proof of part (b) of Theorem 2.1.1.
Hint. First show that

$$W_1 - W_0 = \left(Z_1 \, {}^t C + {}^t D\right)^{-1} (Z_1 - Z_0) (CZ_0 + D)^{-1}.$$

Theorem 2.1.2. *Suppose that G is a connected noncompact real semisimple Lie group.*

(a) A geodesic in G/K which passes through gK has the form:

$$\gamma_X(t) = g \exp(tX)K, \quad \text{for some } X \in \mathfrak{p}, \quad \text{with } t \in \mathbb{R}.$$

Here the Cartan decomposition of the Lie algebra $\mathfrak{g}$ of G is $\mathfrak{g} = \mathfrak{k} \oplus \mathfrak{p}$, and K is a connected Lie subgroup of G with Lie algebra $\mathfrak{k}$.
(b) Geodesics of G/K have the form $\gamma_X(t)$, for all $t \in \mathbb{R}$, using the notation of part (a). This means that G/K is a complete Riemannian manifold. Moreover, any two points of G/K can be joined by a geodesic segment of length equal to the Riemannian distance between the points.
(c) A geodesic through the origin K in G/K has the form

$$\gamma(t) = k \exp(tX)K \quad \text{for some } k \in K, \; X \in \mathfrak{a}, \quad \text{with } t \in \mathbb{R}.$$

Here the Iwasawa decomposition of $\mathfrak{g}$ is $\mathfrak{g} = \mathfrak{k} \oplus \mathfrak{a} \oplus \mathfrak{n}$.

Proof. (a) Let $g_0 K$ and $g_1 K$ be two cosets in G/K. Apply the transformation $a_{g_o^{-1}}$ from (2.25) to transform these cosets to K and $(g_0^{-1} g_1)K$. Now we have the polar decomposition (Lemma 2.1.3): $g_0^{-1}g_1 = k_1ak_2$, with $k_i \in K$ and $a \in A$. So the transformation $a_{k_1^{-1}}$ sends these cosets to K and aK. Thus we have reduced the proof to the case that $\gamma(t)$ is a geodesic with $\gamma(0) = K$ and $\gamma(1) = aK$, with $a \in A$. Write

$$\gamma(t) = a(t)n(t)K, \text{ using the Iwasawa decomposition.}$$

We want to show that $n(t) = e$, the identity in G. Then we would be reduced to the known result that straight lines in the Euclidean space $\mathfrak{a}$ are the geodesics.

The Riemannian structure on G/K comes from the Killing form B of formula (2.14). If $\pi : G \to G/K$ with $\pi(g) = gK$ and $\gamma(t) = w(t)K$, with $w(t) \in G$, $w(t) = a(t)n(t)$, then

$$Q_{\gamma(t)}(\gamma'(t), \gamma'(t)) = B\left(\left((d\pi)_{w(t)}(dL_{w(t)})_e\right)^{-1}\gamma'(t), \left((d\pi)_{w(t)}(dL_{w(t)})_e\right)^{-1}\gamma'(t)\right),$$

with $L_g(x) = gx$. Since $\gamma = \pi \circ w$, we have

$$Q_{\gamma(t)}(\gamma'(t), \gamma'(t)) = B\left(dL_{w(t)}^{-1}w'(t), dL_{w(t)}^{-1}w'(t)\right).$$

Now, we can calculate the **differential of multiplication** $w(t) = a(t)n(t)$ as follows—using Exercise 2.1.21 below:

$$w'(t) = (dL)_{w(t)}\left(\text{Ad}(n(t))^{-1}dL_{a(t)}^{-1}\left(a'(t)\right) + dL_{n(t)}^{-1}\left(n'(t)\right)\right). \tag{2.35}$$

Thus

$$Q_{\gamma(t)}(\gamma'(t), \gamma'(t)) = B\left(\text{Ad}(n(t))^{-1}dL_{a(t)}^{-1}(a'(t)) + dL_{n(t)}^{-1}(n'(t)), \text{ same}\right)$$

$$= B\left(dL_{a(t)}^{-1}(a'(t)) + \text{Ad}(n(t))dL_{n(t)}^{-1}(n'(t)), \text{ same}\right)$$

$$= Q_{a(t)}\left(a'(t), a'(t)\right) + Q_{n(t)}(n'(t), n'(t)),$$

since

$$0 = 2B\left(dL_{a(t)}^{-1}(a'(t)), \text{Ad}(n(t))dL_{n(t)}^{-1}(n'(t))\right),$$

because the first argument of the Killing form lies in $\mathfrak{a}$ and the second lies in $\mathfrak{n}$. See Exercise 2.1.22 below. Thus the distance is only made smaller by setting $n(t) = e$. It follows that we are reduced to the computation of the geodesics in the space $\mathfrak{a}$. The Killing form gives a metric on $\mathfrak{a}$ which is equivalent to the usual Euclidean metric. So the geodesics in $\mathfrak{a}$ are straight lines and the geodesics in $A = \exp\mathfrak{a}$ through e are of the form $\exp(tX)$, $t \in \mathbb{R}$, for some $X \in \mathfrak{a}$. This completes the proof of part (a) of Theorem 2.1.2.

(b) This is proved in Helgason [273, p. 56] using part (a).
(c) This follows from part (1) and the polar decomposition (Lemma 2.1.3).

∎

Exercise 2.1.21 (The Differential of Multiplication). Suppose that the Lie algebra $\mathfrak{g}$ can be decomposed into a direct sum of subalgebras

$$\mathfrak{g} = \mathfrak{m} \oplus \mathfrak{h}$$

and let $G \supset M, H$ be the corresponding connected Lie subgroups. If the map $\alpha : M \times H \to G$ is defined by $\alpha(m, h) = mh$, show that the differential is:

$$(d\alpha)_{(m,h)} (dL_m X, dL_h Y) = (dL_{mh}) \left(\mathrm{Ad}\left(h^{-1}\right) X + Y\right), \text{ for } X \in \mathfrak{m}, \ Y \in \mathfrak{h}.$$

Hint. Define $L_m \times L_h : M \times H \to M \times H$ by $(L_m \times L_h)(x, y) = (mx, hy)$, for $x \in M$, $y \in H$. Then

$$\alpha \circ (L_m \times L_h) = L_{mh} \circ \alpha \circ (\mathrm{Int}(h^{-1}) \times I).$$

Thus you can use formula (2.7) relating multiplication on G and Lie bracket on $\mathfrak{g}$ to show that:

$$(d\alpha)_{(e,e)}(X, Y)f = \frac{df}{dt}(\exp tX \exp tY)\Big|_{t=0} = (X + Y)f.$$

Exercise 2.1.22. Suppose that $\mathfrak{g}$ has the Iwasawa decomposition $\mathfrak{g} = \mathfrak{k} \oplus \mathfrak{a} \oplus \mathfrak{n}$ and Cartan involution θ. Consider the form $F(X, Y) = -B(X, \theta Y)$ for $X, Y \in \mathfrak{g}$ from formula (2.20) in Section 2.1.3. Then F is a positive definite bilinear form on $\mathfrak{g}$. Show that

$$X \in \mathfrak{k} \Rightarrow \mathrm{ad}\, X \text{ is skew symmetric;}$$

$$X \in \mathfrak{a} \Rightarrow \mathrm{ad}\, X \text{ is diagonal;}$$

$$X \in \mathfrak{n} \Rightarrow \mathrm{ad}\, X \text{ is upper triangular with 0 on the diagonal.}$$

Hint. (See Wallach [651, p. 166] or Helgason [273, p. 223].) You need to take an ordered set of positive roots: $\alpha_1, \alpha_2, \ldots, \alpha_m$. Then form an orthonormal basis of $\mathfrak{g}$ by taking orthonormal bases of

$$\mathfrak{g}_{\alpha_m}, \ldots, \mathfrak{g}_{\alpha_1}, \ \mathfrak{a} \oplus \mathfrak{m}, \ \theta(\mathfrak{g}_{\alpha_1}), \ldots, \theta(\mathfrak{g}_{\alpha_m}).$$

You need to use properties of the roots such as the fact that: $[\mathfrak{g}_\alpha, \mathfrak{g}_\beta] \subset \mathfrak{g}_{\alpha+\beta}$.

Next we consider the Weyl group of the symmetric space. See (1.296) of Section 1.5.3 for the definition in the case of $GL(n, \mathbb{R})$. As usual, suppose that G

is a noncompact real semisimple Lie group with the standard definitions of $K, \mathfrak{a}$, etc. Define the following subgroups of K:

$$M = \text{the \textbf{centralizer} of } \mathfrak{a} \text{ in } K = \{k \in K \mid \text{Ad}(k)|_{\mathfrak{a}} = \text{identity}\},$$
$$M' = \text{the \textbf{normalizer} of } \mathfrak{a} \text{ in } K = \{k \in K \mid \text{Ad}(k)\mathfrak{a} \subset \mathfrak{a}\}. \qquad (2.36)$$

Both M and M' are closed subgroups of K. The **Weyl group** of G/K is defined to be $W = M'/M$. Note that W is independent of the choice of $\mathfrak{a}$, by the conjugacy of all maximal abelian subspaces of $\mathfrak{p}$ (see the proof of Lemma 2.1.3). These definitions can also be made in the case that G is compact (see Helgason [273, p. 244]).

Theorem 2.1.3 (The Weyl Group).

(1) The Weyl group is a finite group contained in the orthogonal group in $GL(\mathfrak{a})$ with respect to the inner product on $\mathfrak{a}$ defined by the Killing form of $\mathfrak{g}$.

*(2) The Weyl group permutes the restricted roots. Define a **Weyl chamber** to be a connected component of*

$$\left(\mathfrak{a} - \bigcup_{\alpha \in \Lambda} \alpha^{-1}(0) \right), \quad \text{for } \alpha \in \Lambda.$$

Note that $\alpha^{-1}(0)$ is a hyperplane in $\mathfrak{a}$. The Weyl group also permutes the Weyl chambers. Moreover the action of the Weyl group on the Weyl chambers is simply transitive.

(3) For $\lambda \in \Lambda$, define $s_\lambda : \mathfrak{a} \to \mathfrak{a}$ by

$$s_\lambda(H) = H - 2(\lambda(H)/\lambda(H_\lambda))H_\lambda, \quad \text{where} \quad H_\lambda \in \mathfrak{a}$$

is defined by

$$B(H, H_\lambda) = \lambda(H), \text{ for all } H \in \mathfrak{a}.$$

Then s_λ is the reflection in the hyperplane $\lambda^{-1}(0)$. The Weyl group is generated by these reflections s_λ, for $\lambda \in \Lambda$.

Proof. We shall only prove part (1). For the other parts of the theorem, see Helgason [273, Ch. 7] or Wallach [651, pp. 77, 168]. By the Lie group/Lie algebra dictionary,

$$\text{the Lie algebra of } M = \text{Lie}(M) = \mathfrak{m} = \{X \in \mathfrak{k} \mid \text{ad}\,X|_{\mathfrak{a}} = 0\}.$$

If we can show that M' has the same Lie algebra as M, then it will follow that the quotient M'/M is both discrete and compact (thus finite). Suppose that T is in the Lie algebra of M'. Then write out the root space decomposition of T:

$$T = Y + \sum_{\lambda \in \Lambda} X_\lambda, \quad \text{for } Y \in \mathfrak{m} \oplus \mathfrak{a}, \ X_\lambda \in \mathfrak{g}_\lambda.$$

It follows that for all $H \in \mathfrak{a}$

$$[H, T] = \sum \lambda(H) X_\lambda \in \mathfrak{a} \text{ implies that } [H, T] = 0.$$

Here we have used the fact that the sum in the root space decomposition is direct.

To see that the group M'/M permutes the restricted roots is easy. To see that the reflections s_λ come from some $\mathrm{Ad}(k)$, $k \in K$, is harder. To see that the s_λ generate the Weyl group is even harder. Note that we cannot claim that the s_λ, with λ from a system of simple roots, generate the Weyl group. A system of simple roots has the property that any root is a linear combination of simple roots with integer coefficients that are either all positive or all negative (with $r = \dim \mathfrak{a}$ elements). Such simple root systems *do* give generators of the Weyl group in the case of *complex* semisimple Lie algebras. However, real Lie algebras are somewhat different, as we will see in the following examples. ∎

Exercise 2.1.23. (a) Why is it reasonable to call a Lie algebra "semisimple" if the Killing form is nondegenerate? What is the connection with the standard notion that an algebraic object is semisimple if it is a direct sum of simple objects?

(b) Why do we call a semisimple Lie algebra "compact" if its Killing form is negative definite? What is the connection with compact Lie groups? Can we drop the hypothesis that the Lie algebra be semisimple?

(c) Recall that we said a Lie algebra is "nilpotent" if all sufficiently long brackets must vanish. What is the connection with the usual idea of a nilpotent linear transformation (such as $\mathrm{ad}X$)?

Hints.

(a) See Helgason [273, pp. 121–122].
(b) See Helgason [273, pp. 120–122]. Think about $\mathbb{R}$ and $\mathbb{R}/\mathbb{Z}$.
(c) See Helgason [273, pp. 135–137].

Examples of Weyl Groups

Example 2.1.14 ($GL(n, \mathbb{R})$)**.**

Since $\mathrm{Ad} = \mathrm{Int}$ for matrix groups, we have:

$$M = \left\{ k \in O(n) \ \middle| \ kXk^{-1} = X, \text{ for any diagonal matrix } X \right\},$$
$$M' = \left\{ k \in O(n) \ \middle| \ kXk^{-1} \text{ is diagonal, for any diagonal matrix } X \right\}.$$

It follows that

$M = \{$diagonal matrices with entries $+1$ or $-1\}$,
$M' = \{$matrices with each row or column having exactly one non -0 entry of $\pm 1\}$.

Thus the Weyl group of $GL(n, \mathbb{R})$ is the group of all permutations of n objects as we also saw in Exercise 1.5.30 of Section 1.5.3.

Example 2.1.15 ($Sp(n, \mathbb{R})$). Here

$$K = \left\{ \begin{pmatrix} A & B \\ -B & A \end{pmatrix} \middle| A + iB \in U(n) \right\},$$

$$\mathfrak{a} = \left\{ \begin{pmatrix} H & 0 \\ 0 & -H \end{pmatrix} \middle| H \ n \times n \text{ real diagonal} \right\},$$

$$M = \left\{ \begin{pmatrix} A & 0 \\ 0 & A \end{pmatrix} \middle| A \text{ diagonal } n \times n, \text{ entries } \pm 1 \right\},$$

$$M' = \left\{ \begin{pmatrix} A & B \\ -B & A \end{pmatrix} \middle| A + B \text{ is in the } M' \text{ for } GL(n, \mathbb{R}) \right\}.$$

It follows that the Weyl group $W = M'/M$ contains all permutations of entries of H in

$$\begin{pmatrix} H & 0 \\ 0 & -H \end{pmatrix} \text{ in } \mathfrak{a},$$

as well as all possible changes of sign. So it has 2^n times $n!$ elements. For example, let $n = 3$ and

$$A = \begin{pmatrix} 0 & 1 & 0 \\ 1 & 0 & 0 \\ 0 & 0 & 0 \end{pmatrix}, B = \begin{pmatrix} 0 & 0 & 0 \\ 0 & 0 & 0 \\ 0 & 0 & 1 \end{pmatrix}, k = \begin{pmatrix} A & B \\ -B & A \end{pmatrix}, H = \begin{pmatrix} h_1 & 0 & 0 \\ 0 & h_2 & 0 \\ 0 & 0 & h_3 \end{pmatrix},$$

$$a = \begin{pmatrix} H & 0 \\ 0 & -H \end{pmatrix}, \text{ then } \text{Ad}(k)a = \begin{pmatrix} H' & 0 \\ 0 & -H' \end{pmatrix},$$

where

$$H' = \begin{pmatrix} h_2 & 0 & 0 \\ 0 & h_1 & 0 \\ 0 & 0 & -h_3 \end{pmatrix}.$$

Exercise 2.1.24. Check the results stated in Example 2.1.15 above for $Sp(n, \mathbb{R})$.

Example 2.1.16 ($SU(2, 1)$).
For this example,

$$K = \left\{ \begin{pmatrix} U & 0 \\ 0 & t \end{pmatrix} \middle| U \in U(2), \ t = (\det U)^{-1} \right\},$$

$$\mathfrak{a} = \mathbb{R} \begin{pmatrix} 0 & 0 & 1 \\ 0 & 0 & 0 \\ 1 & 0 & 0 \end{pmatrix},$$

$$M = \left\{ k = \begin{pmatrix} e^{i\alpha} & 0 & 0 \\ 0 & e^{i\beta} & 0 \\ 0 & 0 & e^{i\alpha} \end{pmatrix} \;\middle|\; \det k = 1 \right\},$$

$$M' = \left\{ k = \begin{pmatrix} e^{i\alpha} & 0 & 0 \\ 0 & e^{i\beta} & 0 \\ 0 & 0 & \pm e^{i\alpha} \end{pmatrix} \;\middle|\; \det k = 1 \right\}.$$

So the Weyl group has only two elements. The entries of the diagonal matrices in M' are supposed to be of complex norm 1.

Exercise 2.1.25. Verify the results stated in Example 2.1.16 for $SU(2, 1)$.

Now that we have described the Weyl group, it is possible to discuss the **nonuniqueness of the polar decomposition**. The precise result is that if we set $\mathfrak{a}' = \{H \in a \mid \lambda(H) \neq 0,$ for all $\lambda \in \Lambda\}$, $A' = \exp \mathfrak{a}'$, and define the map $f : (K/M) \times A' \to G/K$ by $f(kM, a) = (ka)K$, then the map f is $\#W$ to 1, regular, and onto an open submanifold of G/K whose complement in G/K has lower dimension (see Helgason [273, p. 381] or [278, p. 402] or Wallach [651]).

One should also consider the relation between the structure theory for $\mathfrak{g}$ a noncompact semisimple real Lie algebra and that for the complexification $\mathfrak{g}^c = \mathfrak{g} \otimes_{\mathbb{R}} \mathbb{C}$. The same question could be asked for the compact real form of $\mathfrak{g}^c$. As an example, consider $SU(2, 1)$ again. The **Cartan subalgebra** or maximal abelian subalgebra $\mathfrak{h}$ of $\mathfrak{su}(2, 1)$ containing $\mathfrak{a}$ is:

$$\mathfrak{h} = \left\{ \begin{pmatrix} a & 0 & b \\ 0 & c & 0 \\ b & 0 & a \end{pmatrix} \;\middle|\; a, c \in i\mathbb{R}, \; b \in \mathbb{R} \right\}.$$

Clearly the complexification of $\mathfrak{h}$ is a Cartan subalgebra of the complexification of $\mathfrak{g}$. This shows that much is missing from the complexification of $\mathfrak{a}$. One can show that the restricted roots are really restrictions of roots of the complexified Lie algebra (see Helgason [273, Ch. 6]). Again, some roots from the complexification may be missing in the real version of the Lie algebra.

2.1.5 Integral Formulas

Our next topic is integral formulas for noncompact semisimple real Lie groups. First perhaps we should discuss the Haar measures in $G, A, N,$ and K. See our earlier comments on this subject in Chapter 2 of Vol. I and Chapter 1 of this Volume. More details about Haar measures can be found in Helgason [273, Chapter 10]. Because Haar measure is unique up to a positive scalar multiple, we can define the **modular function** $\delta : G \to \mathbb{R}^+$ by the formula (assuming dg = left Haar measure):

$$\int f(gs^{-1})dg = \delta(s) \int f(g)dg. \tag{2.37}$$

For the left-hand side of the equality is a left G-invariant integral for fixed s. Thus it must be a positive constant times the Haar integral of f. It follows easily that δ is continuous, $\delta(st) = \delta(s)\delta(t)$, and $d(gs) = \delta(s)dg$. Thus the modular function relates right and left Haar measure. By definition, a **unimodular group** has $\delta = 1$ identically. Furthermore, it is easy to see that $d(g^{-1}) = \delta(g^{-1})dg$. If G is a Lie group, one also has $d(s^{-1}gs) = d(gs) = \delta(s)dg$. Thus $\det(\mathrm{Ad}(s^{-1})) = \delta(s)$, for all s in G.

We prove that compact, semisimple, and nilpotent Lie groups are all unimodular. Suppose first that K is compact. Then δ maps K onto a compact subgroup of $\mathbb{R}^+$ which must contain only one element, since otherwise powers would approach 0 or infinity. Suppose next that G is semisimple. Then $\mathrm{Ad}\,(s)$ leaves the Killing form invariant for s in G. But the Killing form of a semisimple group is nondegenerate and thus equivalent to

$$I_{p,q} = \begin{pmatrix} I_p & 0 \\ 0 & -I_q \end{pmatrix}, \quad \text{for some } p, q.$$

If $^t g I_{p,q} g = I_{p,q}$, the determinant of g must have absolute value 1. Finally suppose that N is nilpotent and connected. Then

$$\det(\mathrm{Ad}(n)) = \exp(\mathrm{Tr}(\mathrm{ad}(\log n))) = 1,$$

for $n \in N$, since ad $(\log n)$ is a nilpotent linear transformation.

Proposition 2.1.1 (The Integral Formula for the Iwasawa Decomposition).
Define $m_\lambda = \dim_\mathbb{R} \mathfrak{g}_\lambda$ and

$$J(a) = \prod_{0 < \lambda \in \Lambda^+} \exp(m_\lambda \lambda(\log a)),$$

for $a \in A$. Then

$$\int_A \int_N \int_K f(ank)\, da\, dn\, dk = \int_G f(g)\, dg,$$

where all the measures are left-invariant (and thus right-invariant) Haar measures on G, A, N, K. However, changing the order gives:

$$\int_K \int_A \int_N f(kan) J(a) \, dk \, da \, dn = \int_G f(g) \, dg.$$

Proof. In order to compute the Jacobian of the Iwasawa decomposition, we proceed as in Exercise 1.1.20 of Section 1.1.4. Thus we need the differential of $\mathrm{Int}(a)n = ana^{-1}$, for $n \in N$ and $a \in A$. We know that the differential of $\mathrm{Int}(a)$ is $\mathrm{Ad}(a)$, by definition. Thus if $a = \exp H$ for $H \in \mathfrak{a}$, we find that:

$$\det(\mathrm{Ad}(a)) = \det(\exp(\mathrm{ad}H)) = \exp\left(\mathrm{Tr}(\mathrm{ad}H)\right)$$

$$= \exp\left(\sum_{\lambda \in \Lambda^+} m_\lambda \lambda(H)\right) = \prod_{\lambda \in \Lambda^+} \exp\left(m_\lambda \lambda(H)\right),$$

which is simply $J(a)$, as defined in the proposition, since $H = \log a$. Here we have used the fact that:

$$\mathfrak{n} = \sum_{\lambda \in \Lambda^+}{}^{\oplus} \mathfrak{g}_\lambda, \qquad \mathfrak{g}_\lambda = \{X \in \mathfrak{g} \mid \mathrm{ad}H(X) = \lambda(H)X, \quad \text{for all } H \in \mathfrak{a}\}.$$

Note that $\mathrm{Int}(a) : N \to N$ for any $a \in A$.

The rest of the argument is really the same as that of Exercise 1.1.20 in Section 1.1.4, but we shall repeat it for completeness. First observe that the left Haar measures on G and K can be normalized so that if $d\bar{g}$ denotes the G-invariant measure on the symmetric space G/K, then the following equality prevails:

$$\int_G f(g) \, dg = \int_{\bar{g}=gK \in G/K} \int_{k \in K} f(gk) \, dk \, d\bar{g}.$$

Now G/K can be identified with AN. Thus we need to only show

$$\int_A \int_N f(an) \, dn \, da$$

gives a left AN-invariant integral on AN. Let $a_1 \in A$ and $n_1 \in N$. Then we have

$$\int_A \int_N f(a_1 n_1 an) \, dn \, da = \int_A \int_N f(a_1 a n_2 n) \, dn \, da, \quad \text{if } n_2 = a^{-1} n_1 a.$$

Since both da and dn are left invariant, the last integral is just

$$\int_A \int_N f(an) \, dn \, da.$$

This completes the proof of the first integral formula in the proposition.

Now we are ready to prove the second version of the integral formula for the Iwasawa decomposition. Using the differential of $\text{Int}(a)$ and the first integral formula, we get:

$$\int_G f(g)\, dg = \int_N \int_A \int_K f(nak) J(a)^{-1}\, dk\, da\, dn.$$

Now replace $f(g)$ by $f(g^{-1})$. This will reverse orders on the right-hand side and produce

$$\int_N \int_A \int_K f(k^{-1}a^{-1}n^{-1}) J(a)^{-1}\, dk\, da\, dn.$$

Finally the fact that G, N, A, K are all unimodular leads to the second integral formula for the Iwasawa decomposition. ■

Examples

(1) $G = GL(n,\mathbb{R})$.

$$J(a) = \prod_{1 \le i < j \le n} \frac{a_i}{a_j} = \prod_{i=1}^{n} a_i^{n-2i+1}.$$

(2) $G = Sp(n,\mathbb{R})$.

$$J(a) = \prod_{1 \le i < j \le n} \frac{a_i}{a_j} \prod_{1 \le i \le j \le n} a_i a_j = \prod_{i=1}^{n} a_i^{2(n+1-i)}.$$

In order to be more precise, we need to fix the invariant volumes on the symmetric spaces.

Invariant Volume Elements on the Symmetric Spaces of $GL(n,\mathbb{R})$ and $Sp(n,\mathbb{R})$

(1) We found in formula (1.16) of Section 1.1.4 that the $GL(n, \mathbb{R})$-invariant volume element on $\mathcal{P}_n$ is:

$$d\mu_n = |Y|^{-(n+1)/2} \prod_{1 \le i \le j \le n} dy_{ij}, \quad \text{if } Y = (y_{ij}) \in \mathcal{P}_n.$$

(2) Next we want to find the invariant volume element on the Siegel upper half space $\mathcal{H}_n$. The argument following (1.16) of Section 1.1.4 can be imitated to show that the $Sp(n, \mathbb{R})$-invariant volume on $\mathcal{H}_n$ is:

$$d\mu_n^*(Z) = |Y|^{-(n+1)} \prod_{1 \le i \le j \le n} dx_{ij}\, dy_{ij}, \quad \text{if } Z = X + iY \in \mathcal{H}_n,$$

with

$$X = (x_{ij}) \quad \text{and} \quad Y = (y_{ij}).$$

Let us prove this last formula. As in the case of $GL(n, \mathbb{R})$, it suffices to find the Jacobian of the action of a diagonal symplectic matrix on $\mathcal{H}_n$. So observe that the image of $X + iY \in \mathcal{H}_n$ under the matrix

$$\begin{pmatrix} a & 0 \\ 0 & a^{-1} \end{pmatrix} \in Sp(n, \mathbb{R}), \quad a = \begin{pmatrix} a_1 & & 0 \\ & \ddots & \\ 0 & & a_n \end{pmatrix},$$

is $aZa = Z[a] = X[a] + iY[a]$, according to (2.25). Thus the Jacobian of the transformation is:

$$\prod_{1 \le i \le j \le n} a_i a_j \prod_{1 \le i \le j \le n} a_i a_j = |a|^{2(n+1)}.$$

This shows that the measure $d\mu_n^*$ is $Sp(n, \mathbb{R})$-invariant.

Our next goal is to work out the integral formula for polar coordinates. First we need a Lemma.

Lemma 2.1.4 (The Integral Formula for Exp Restricted to p in the Cartan Decomposition). *There is a positive constant c such that:*

$$\int_{G/K} f(x)\, dx = c \int_{\mathfrak{p}} f(\exp Y) J(Y)\, dY,$$

where

$$J(X) = \det\left(\left. \frac{\sinh \mathrm{ad}X}{\mathrm{ad}X} \right|_{\mathfrak{p}} \right), \quad \text{for } X \in \mathfrak{p}.$$

Proof. First recall our calculation of the differential of exp in formula (2.8) or see Helgason [273, p. 95] for the general result. Observe that if $X, Y \in \mathfrak{p}$, then

$$(d\exp)_X Y = (dL_{\exp X})_e \circ \left\{ \sum_{n \geq 0} \frac{1}{(2n+1)!} (-\mathrm{ad}X)^{2n} \right\} Y.$$

For $[\mathfrak{k}, \mathfrak{p}] \subset \mathfrak{p}$ and $[\mathfrak{p}, \mathfrak{p}] \subset \mathfrak{k}$ follow from the properties of the Cartan involution. Therefore

$$(\mathrm{ad}X)^{2n+1} Y \in \mathfrak{k} \quad \text{and} \quad (\mathrm{ad}X)^{2n} Y \in \mathfrak{p}.$$

Since $\mathfrak{k} \cap \mathfrak{p} = 0$, we have the vanishing of the sum of the odd powers of $\mathrm{ad}X$ in the series expression for the differential of exp.

We can write $X \in \mathfrak{p}$ in the form

$$X = Y + \sum_{\lambda \in \Lambda^+} (X_\lambda - \theta X_\lambda), \quad Y \in \mathfrak{a},$$

for X_λ in a basis for $\mathfrak{g}_\lambda$. Note also that $H \in \mathfrak{a}$, $X_\lambda \in \mathfrak{g}_\lambda$ implies that

$$(\mathrm{ad}H)(X_\lambda - \theta X_\lambda) = \lambda(H)(X_\lambda + \theta X_\lambda), \quad \text{if } X_\lambda \in \mathfrak{g}_\lambda.$$

It follows that

$$(\mathrm{ad}H)^2 (X_\lambda - \theta X_\lambda) = (\lambda(H))^2 (X_\lambda - \theta X_\lambda).$$

Thus the differential at H has determinant

$$\left| (d\exp|_\mathfrak{p})_H \right| = \left| e^H \right| \prod_{0 < \lambda \in \Lambda} \frac{\sinh \lambda(H)}{\lambda(H)},$$

proving Lemma 2.1.4. This shows, in particular, that exp is a diffeomorphism when restricted to $\mathfrak{p}$. ∎

Proposition 2.1.2 (The Integral Formula for Polar Coordinates). *Suppose that the root space $\mathfrak{g}_\lambda$ has dimension m_λ for any restricted root λ. Then there is a positive constant c such that if dg denotes the Haar measure on G, then*

$$\int_G f(g) \, dg = c \int_K \int_A \int_K f(k_1 a k_2) D(a) \, dk_1 \, da \, dk_2,$$

where

$$D(a) = \prod_{0 < \lambda \in \Lambda} |\sinh(\lambda(\log a))|^{m_\lambda}, \quad \text{for } a \in A.$$

Proof. A reference for the proof is Helgason [273, Ch. 10]. The main step is the preceding lemma. Then one uses the fact that $\mathfrak{p} = \mathrm{Ad}(K)\mathfrak{a}$ from Lemma 2.1.3.

Observe first that $X \in \mathfrak{k}$ has the representation:

$$X = Y + \sum_{\lambda \in \Lambda^+} (X_\lambda + \theta X_\lambda), \quad \text{with} \quad Y \in \mathfrak{m}, \; X_\lambda \in \mathfrak{g}_\lambda,$$

where $\mathfrak{m}$ and $\mathfrak{g}_\lambda$ are from the root space decomposition of $\mathfrak{g}$ (see Helgason [273, p. 224]).

Define $f : K \times A \to P$ by $f(k, a) = kak^{-1} = p$. Suppose that $Y \in \mathfrak{k}$, $H \in \mathfrak{a}$. Then

$$\begin{aligned}
(df)_{(k,a)}(Y, H) &= \lim_{t \to 0} \frac{1}{t} \left\{ f(ke^{tY}, ae^{tH}) - f(k, a) \right\} \\
&= \lim_{t \to 0} \frac{1}{t} \left\{ ke^{tY} ae^{tH} e^{-tY} k^{-1} - kak^{-1} \right\} \\
&= k \left\{ \lim_{t \to 0} \frac{1}{t} \left(e^{tY} ae^{tH} e^{-tY} - a \right) \right\} k^{-1}.
\end{aligned}$$

Now suppose that $a = \exp(H_0)$ for $H_0 \in \mathfrak{a}$. Then the object inside the last limit is:

$$\begin{aligned}
e^{tY} e^{H_0 + tH} e^{-tY} - e^{H_0} &= \exp\left(H_0 + tH + t[Y, H_0 + tH] + o(t^2) \right) - \exp H_0 \\
&= \exp\left(H_0 + t(H + [Y, H_0]) + o(t^2) \right) - \exp H_0.
\end{aligned}$$

Use the chain rule to evaluate the derivative of the preceding quantity with respect to t at $t = 0$ and obtain:

$$(d\exp)_{H_0} \left(H + [Y, H_0] \right).$$

Take a basis of $\mathfrak{p}$ coming from $\mathfrak{a}$ and vectors $X_\lambda - \theta X_\lambda$ and a basis of $\mathfrak{k}$ coming from $\mathfrak{m}$ and vectors $X_\lambda + \theta X_\lambda$, with X_λ in the root spaces $\mathfrak{g}_\lambda$. One sees that for $Y = X_\lambda + \theta X_\lambda$, the preceding is:

$$(d\exp)_{H_0} \left(H - \lambda(H_0)(X_\lambda - \theta X_\lambda) \right).$$

Finally use the formula for the differential of $\exp$, along with the fact that the odd powers of $(\mathrm{ad}H_0)$ vanish, once again. This yields:

$$\begin{aligned}
&(d\exp)_{H_0} \left(H + [X_\lambda + \theta X_\lambda, H_0] \right) \\
&= e^{H_0} \left\{ H - \lambda(H_0) \sum_{n \geq 0} \frac{1}{(2n+1)!} \lambda(H_0)^{2n} (X_\lambda - \theta X_\lambda) \right\} \\
&= e^{H_0} \left\{ H - \sinh \lambda(H_0)(X_\lambda - \theta X_\lambda) \right\}.
\end{aligned}$$

This completes our discussion of Proposition 2.1.2. ∎

If G is a real noncompact semisimple Lie group, K a compact subgroup coming from the Cartan decomposition of G, the **boundary** of G/K can be defined as K/M. The group M was defined in (2.36). And we can identify this boundary with G/B, if B is the Borel or minimal parabolic subgroup $B = MAN$, as in (1.19) of Section 1.1.4. Furstenberg [194] and Moore [459] show that G/B is a "maximal boundary" in a certain probabilistic sense.

Example 2.1.17 $(G = SL(n, \mathbb{R}))$.

Here $B = MAN$ consists of all upper triangular matrices of determinant one. We can identify G/B as the **flag manifold**:

$$F_n = \{(V_1, \ldots, V_{n-1}) \mid V_i \text{ is a vector subspace of } \mathbb{R}^n, \ \dim_{\mathbb{R}} V_i = i, \ V_i \subset V_{i+1}\}.$$

The action of $g \in G$ on F_n is $g(V_1, \ldots, V_{n-1}) = (gV_1, \ldots, gV_{n-1})$. This action is easily seen to be transitive. To calculate the stability group of a point, let $e_i \in \mathbb{R}^n$ denote the column vector with ith coordinate one and the rest zero. Then set $V_i^0 = \mathbb{R}e_1 \oplus \cdots \oplus \mathbb{R}e_i$. Then g fixes V_i^0, for all i, means $g \in B$. See Exercise 1.3.11 of Section 1.3.6.

Exercise 2.1.26. Show that the Jacobian of the action of $g \in G$ on the boundary $G/MAN \cong K/M$ is given by the following integral formula:

$$\int\limits_{K/M} f(\bar{k}) \, d\bar{k} = \int\limits_{\bar{k}=kM \in K/M} f(g(\bar{k})) J(a(gk))^{-1} \, d\bar{k},$$

where $a(g)$ is the A-part of the KAN-Iwasawa decomposition of g, and $J(a)$ is the Jacobian of the Iwasawa decomposition in Proposition 2.1.1. Here $d\bar{k}$ is any K-invariant measure on K/M.

It is also possible to show (using the Bruhat decomposition of G described in Section 1.5.3 for $GL(n)$) that if $\overline{N}$ denotes the opposite nilpotent subgroup corresponding to the Lie subalgebra $\overline{\mathfrak{n}}$,

$$\overline{\mathfrak{n}} = \sum_{0 > \alpha \in \Lambda} \mathfrak{g}_\alpha,$$

then $\overline{N}MAN$ is an open subset of G with lower dimensional complement. See Lemma 1.3.2 of Section 1.3.3 for a proof of this result when $G = SL(n, \mathbb{R})$. This allows us to identify K/M with $\overline{N}$ as far as integration is concerned. In Section 1.3, we applied such a result to obtain the asymptotics of spherical functions. For more information on boundaries and compactifications of symmetric spaces, see Gérardin [216], Helgason [273–282], Koranyi [364], and the references mentioned when we defined the boundary of $\mathcal{P}_n$.

This concludes our discussion of the basic integral formulas for symmetric spaces. Next let us consider differential operators on the symmetric space G/K when G is a noncompact real semisimple Lie group.

2.1.6   Invariant Differential Operators

Let φ be a diffeomorphism of a manifold M. We say that a differential operator D on M is **invariant** under φ if D commutes with φ; i.e., if $D(f \circ \varphi) = (Df) \circ \varphi$, for all infinitely differentiable functions f on M. For each $g \in G$, we have a diffeomorphism a_g of the symmetric space G/K defined by $a_g(xK) = (gx)K$, for $g, x \in G$. Define $D(G/K)$ to be the set of all differential operators on G/K which are a_g-invariant for all $g \in G$. So $D(G/K)$ is the **algebra of invariant differential operators on G/K**. The Laplacian will, of course, be such an operator. In general, however, there will be invariant differential operators on G/K which are not polynomials in the Laplacian, just as for $\mathcal{P}_n$ (see Theorem 1.1.2 of Section 1.1.5). The following theorem is proved in Helgason [273, p. 432]). We will not prove it here.

Theorem 2.1.4 (Harish-Chandra and Chevalley). *Suppose that G is a noncompact real semisimple Lie group with* $\dim_{\mathbb{R}} \mathfrak{a} = r = $ ***rank of G/K****. Then the algebra $D(G/K)$ of all invariant differential operators on G/K is a commutative algebra. In fact, it is a polynomial ring with r algebraically independent generators.*

There is a close relation between $D(G/K)$ and $D(G) = $ the left-invariant differential operators on G or the **universal enveloping algebra** of G (see Helgason [273, Ch. 10]).

Question. Can one relate the G-invariant differential operators $D(G/K)$ for the following chain of inclusions of totally geodesic submanifolds?

$$\mathcal{P}_n \to \mathcal{H}_n \to \mathcal{P}_{2n}^* \qquad \subset \mathcal{P}_{2n},$$
$$Y \mapsto iY \mapsto \begin{pmatrix} Y & 0 \\ 0 & Y^{-1} \end{pmatrix}.$$

We know, for example, that the arc length on $\mathcal{H}_n$ is given by:

$$ds^2 = 2\mathrm{Tr}\left(\left(Y^{-1}dY\right)^2 + \left(Y^{-1}dX\right)^2 \right), \quad \text{for} \ \ Z = X + iY \in \mathcal{H}_n.$$

Therefore the Laplacian on $\mathcal{H}_n$ must be a sum of the Laplacian on $\mathcal{P}_n$ plus a term involving only differentiation with respect to X-variables. If follows that for functions of the Y-variable alone, the Laplacian on $\mathcal{H}_n$ coincides with that on $\mathcal{P}_n$, disregarding constants.

Let G be a noncompact real semisimple Lie group, as usual. A function $u : G/K \to \mathbb{C}$ is called **harmonic** if $Du = 0$ for any operator $D \in D(G/K)$

such that D annihilates constants. This definition was made by Godement [223]. Furstenberg [194] shows that, in fact, a bounded solution of $\Delta u = 0$ on G/K is automatically harmonic. Other references for potential theory (i.e., the study of harmonic functions) on symmetric spaces are Helgason [275] and Koranyi [364].

Theorem 2.1.5 (Godement's Mean Value Theorem). *Suppose $u : G/K \to \mathbb{C}$ is infinitely differentiable. Then u is harmonic if and only if*

$$\int_{k \in K} u(gkhK) \, dk = u(gK), \quad \text{for all} \quad g, h \in G.$$

Proof (Helgason [275, pp. 42–43]).
$\Rightarrow$ Let u be harmonic and

$$F(h) = \int_K u(gkhK) \, dk.$$

We want to show that $F(h) = F(e) = u(gK)$, $e = $ identity of G. Since F satisfies an elliptic partial differential equation with analytic coefficients, it follows, by a theorem of Bernstein, that F is analytic (see John [332, pp. 57, 142]). Now it suffices to show that

$$(DF)(e) = 0,$$

for every left invariant differential operator D on the Lie group G such that D annihilates the constants.

To show that DF vanishes at the identity, we must merely relate differential operators in $D(G/K)$ with those in $D(G)$. This is done in detail in Helgason [273, Ch. 10]. We merely sketch the process. Let us use the following notation for a diffeomorphism φ of G:

$$D^\varphi f = D(f \circ \varphi) \circ \varphi^{-1}, \quad \text{if} \quad D \in D(G).$$

Then for $D \in D(G)$ write

$$D^\# f = \int_K D^{R_k} f \, dk, \quad \text{if} \quad R_k x = xk \quad \text{for} \quad x \in G.$$

Now it can be shown that $D^\#$ is a differential operator which is invariant under all the R_k, $k \in K$ and thus gives rise to an operator $\tilde{D}$ in $D(G/K)$. And we find that by hypothesis:

$$(DF)(e) = (D^\# F)(e) = \int_K (\tilde{D}u)(gkK) \, dk = 0.$$

⇐ Assume that u has the mean value property stated in the theorem and that $D \in D(G/K)$ annihilates constants. As usual, set $a_g(hK) = ghK$, for g, h in G. Thus

$$\int_{k \in K} u(a_{gk}(x)) \, dk = u(gK), \quad \text{if} \quad x \in G/K.$$

Apply D to both sides of this equation considered as functions of $x \in G/K$ to obtain

$$\int_{k \in K} (Du)(a_{gk}(x)) dk = 0,$$

since D and a_g commute. Take x to be the coset K in G/K to see that $Du = 0$. ∎

Theorem 2.1.6 ((Furstenberg) (Poisson Integral Formula)). *Let us suppose* $u \colon G/K \to \mathbb{C}$ *is a bounded harmonic function. Then there is a bounded measurable function* $\widehat{u} \colon K/M \to \mathbb{C}$ *such that*

$$u(gK) = \int_{\bar{k} \in K/M} \widehat{u}(g(\bar{k})) \, d\bar{k}, \quad \text{for all} \ \ g \in G. \tag{2.38}$$

Here $d\bar{k}$ *is the unique K-invariant measure on the boundary* K/M *such that*

$$\int_{K/M} d\bar{k} = 1.$$

And conversely, given $\widehat{u}$, *as above, the function* u *on* G/K *defined by (2.38) is harmonic. We can rewrite formula (2.38) as:*

$$u(x) = \int_{\bar{k} \in K/M} \widehat{u}(\bar{k}) P(x, \bar{k}) \, d\bar{k},$$

*where **Poisson's kernel** $P(x, \bar{k})$ is:*

$$P(gK, kM) = d\left(g^{-1}\left(\bar{k}\right)\right) / d\bar{k} = J^{-1}\left(a\left(g^{-1}k\right)\right). \tag{2.39}$$

Here J denotes the Jacobian of the Iwasawa decomposition from Proposition 2.1.1 and $a(g)$ *is the A-part of the KAN Iwasawa decomposition of* $g \in G$.

Proof. Sketch (See Helgason [275, pp. 42–52].)
⇒ We need to know that the Borel subgroup $B = MAN$ has the following **fixed point property**. Suppose that B acts continuously on a locally convex topological vector space by linear transformations leaving a nonempty compact convex set invariant. Then B has a fixed point in the convex set. Assuming this result, suppose $u \colon G/K \to \mathbb{C}$ is bounded and harmonic. Define the set

$$Q_u = \left\{ w \in L^\infty(G) \;\middle|\; \begin{array}{l} ||w||_\infty = \text{l.u.b.} \{|w(h)| \mid h \in G\} \le ||u||_\infty \\[2mm] u(gK) = \int_K w(gkh)\,dk, \quad \text{for all } g, h \in G \end{array} \right\}.$$

By Godement's Mean Value Theorem, $u \circ \pi = \tilde{u} \in Q_u$, where $\pi : G \to G/K$ is defined by $\pi(g) = gK$.

Suppose that MAN leaves u_1 in Q_u fixed. Set $\widehat{u}(gMAN) = u_1(g)$, for all $g \in G$. Then $\widehat{u}$ has the required property.

$\Leftarrow$ If $\widehat{u}$ is as described in the theorem, then u defined by (2.38) is easily shown to have the mean value property. ∎

Exercise 2.1.27. Prove this last statement; i.e., the $\Leftarrow$ of Theorem 2.1.6.

Another standard result in potential theory generalizes as follows.

Theorem 2.1.7. *Suppose that F is continuous on the boundary of G/K and that $P(x, b)$ =Poisson's kernel from (2.39). Set*

$$u(gK) = \int_{K/M} P(gK, \bar{k}) F(\bar{k}) \, d\bar{k}, \quad \text{for } g \in G.$$

Then u has boundary values given by F; i.e.,

$$\lim_{t \to \infty} u\left((k \exp tH)K\right) = F(kM), \quad \text{for } k \in K, \ H \in \mathfrak{a}^+,$$

where $\mathfrak{a}^+$ is a fixed Weyl chamber in $\mathfrak{a}$ (from the Iwasawa decomposition).

Proof (Helgason [275, pp. 47–48]). First one must identify the boundary K/M (up to set of measure 0) with $\bar{N}$ the Lie subgroup of G corresponding to the Lie subalgebra

$$\bar{\mathfrak{n}} = \sum_{\alpha \in \Lambda^-} {}^\oplus \mathfrak{g}_\alpha,$$

where the Weyl chamber is

$$\mathfrak{a}^+ = \{H \in \mathfrak{a} \mid \alpha(H) < 0 \text{ if } \alpha \in \Lambda^-\}.$$

Set $a_t = \exp(tH)$, for $t \in \mathbb{R}$. Write $k(g) = $ the K-part in the KAN Iwasawa decomposition of $g \in G$ and obtain:

$$\int_{K/M} F(a_t(\bar{k})) \, d\bar{k} = \int_{\bar{N}} F(k(\text{Int}(a_t)\bar{n})M) \, \frac{d\bar{k}}{d\bar{n}} \, d\bar{n},$$

since $a_t \bar{n} MAN = a_t \bar{n} a_t^{-1} MAN$. Set $\bar{n} = \exp \sum_{\alpha<0} X_\alpha$, for $X_\alpha \in \mathfrak{g}_\alpha$. Then

$$\text{Int}\,(\exp tH)\,\bar{n} = \exp\left(\text{Ad}(\exp tH)\sum_{\alpha<0} X_\alpha\right) = \exp\left(e^{\text{ad}tH}\sum_{\alpha<0} X_\alpha\right)$$

$$= \exp\left(\sum_{\alpha<0} e^{t\alpha(H)}\right) X_\alpha \to e, \quad \text{as} \quad t \to \infty,$$

where e denotes the identity, because $\alpha(H) < 0$ if $H \in \mathfrak{a}^+$. ∎

2.1.7 Special Functions and Harmonic Analysis on Symmetric Spaces

It is now possible to discuss various types of special functions on the symmetric space $K\backslash G$ of a noncompact real semisimple Lie group G. We shall view G as acting on the right in order to remain close to the notation that we used in Section 1.2. The basic eigenfunction of the invariant differential operators on $K\backslash G$ is the **power function** $p(Kg)$ defined as follows. Let $\lambda : \mathfrak{a} \to \mathbb{C}$ be a linear functional over $\mathbb{R}$; i.e., $\lambda \in \mathfrak{a}^*$. For $g \in G$, with Iwasawa decomposition $g = kan$, write $H(g) = \log a \in \mathfrak{a}$. Then define the **power function**

$$p_\lambda(Kg) = \exp(\lambda(H(g))). \tag{2.40}$$

The power function is indeed an eigenfunction for all the G-invariant differential operators $D \in D(K\backslash G)$. The proof is the same as that for Proposition 1.2.1 of Section 1.2.1. We know that if $t = a_1 n_1$, for $a_1 \in A$, $n_1 \in N$, we have

$$p_\lambda((Kx)t) = p_\lambda(Kx)p_\lambda(Kt),$$

since $x = kan$ implies that $xa_1 = kana_1 = kaa_1(a_1^{-1}na_1)$ with $a_1^{-1}na_1 \in N$. Then, if $D \in D(K\backslash G)$, $t \in AN$, and $Ky = (Kx)t$,

$$Dp_\lambda(Ky) = (Dp_\lambda)(Kxt) = (Dp_\lambda(Kx))p_\lambda(Kt).$$

Set $x = e =$ the identity, to complete the proof that the power function is indeed an eigenfunction for all the invariant differential operators on $K\backslash G$.

Define a **spherical function** of $K\backslash G$ to be a function $f : K\backslash G \to \mathbb{C}$ such that $f(K) = 1$ and such that f is a K-invariant eigenfunction of all the invariant differential operators in $D(K\backslash G)$.

Spherical functions can be built up out of power functions as in Theorem 1.2.3 of Section 1.2.3. The following theorem is proved in Helgason [273, Ch. 10]. In fact, the proof that we gave in Section 1.2.3 generalizes. It is also possible to extend the rest of the Theorem 1.2.3 of Section 1.2.3 to $K\backslash G$.

Theorem 2.1.8 (Harish-Chandra). *A spherical function has the form*

$$h_\lambda(Kg) = \int_K p_\lambda(Kgk) \, dk,$$

where $\lambda = i\mu - \rho$, $\rho = \frac{1}{2}\sum_{\alpha>0}\alpha$. *Moreover spherical functions* $h_{i\mu-\rho}$ *are invariant under the Weyl group acting on the* μ*-variable. Here* $\mu \in \mathfrak{a}^* = $ *the dual vector space of* $\mathfrak{a}$.

Harish-Chandra [263] obtained the asymptotics of the spherical function:

$$h_{i\mu-\rho}(\exp H) \sim e^{-\rho(H)} \sum_{s\in W} c(s\mu)e^{is\mu(H)}, \quad \text{as} \quad H \to \infty, \quad H \in \mathfrak{a}^+,$$

$$c(\mu) = \int_{\overline{N}} \exp\{(-i\mu - \rho)(H(\overline{n}))\} \, d\overline{n}.$$

$$(2.41)$$

Gindikin and Karpelevic [220] obtained the explicit formula for the c-function:

$$c(\mu) = I(i\mu)/I(\rho), \quad \text{where} \quad I(\nu) = \prod_{\alpha>0} B\left(\frac{1}{2}m_\alpha, \frac{1}{4}m_{\alpha/2} + \frac{(\nu,\alpha)}{(\alpha,\alpha)}\right), \quad \nu \in \mathfrak{a}^*,$$

$$(2.42)$$

where B is the beta function (not to be confused with the Killing form), $m_\alpha = \dim_{\mathbb{R}} \mathfrak{g}_\alpha$, $\rho = \frac{1}{2}\sum_{\alpha>0}\alpha$. Here (ν,α) denotes the inner product on the dual space $\mathfrak{a}^*$ induced by the Killing form of $\mathfrak{g}$ restricted to $\mathfrak{a}$ (a form which is automatically positive definite).

This concludes what we have to say about spherical functions. It would be interesting to look at analogues of Bessel and Whittaker functions for general symmetric spaces, but we will not do this here.

The asymptotics and functional equations of the spherical functions h_λ are sufficient to study the **Helgason–Fourier transform** of $f : K\backslash G \to \mathbb{C}$ defined by:

$$\mathcal{H}f(\lambda,\overline{k}) = \int_{x\in K\backslash G} f(x)\overline{p_\lambda(xk)} \, dx. \qquad (2.43)$$

Here $\lambda \in (\mathfrak{a}^*)^c$, the complexification of the dual vector space to $\mathfrak{a}$, $\overline{k} = kM \in K/M$, xg, for $x \in K\backslash G$ and $g \in G$, denotes the right action given by $(Kh)g = K(hg)$, and dx denotes the G-invariant volume on the symmetric space $K\backslash G$.

The **inversion formula** for this transform is due to Harish-Chandra and Helgason (see Helgason [273–282]) and says that

$$f(x) = \int\limits_{\mu \in \mathfrak{a}^*} \int\limits_{B=K/M} \mathcal{H}f(i\mu + \rho, \bar{k}) \, p_{i\mu+\rho}(xk) \, |c(\mu)|^{-2} \, d\mu \, d\bar{k}, \qquad (2.44)$$

with a suitable normalization of the Euclidean measure on the real vector space $\mathfrak{a}^*$, which is the dual space to $\mathfrak{a}$. The proof of (2.44) is analogous to that of Theorem 1.3.1 in Section 1.3. Helgason [282, Ch. IV] gives a detailed account of the transform for K bi-invariant functions on G. Information on the history of the subject can be found in the same place.

We leave it to the reader to note the remaining properties of the Helgason transform, analogous to those listed in Theorem 1.3.1 in Section 1.3.

Example 2.1.18 ($G = Sp(n, \mathbb{R})$).

Recall our identifications in formula (2.26) in Section 2.1.3 of $K \backslash G$ and $\mathcal{H}_n$ with $\mathcal{P}_n^*$ via:

$$W = \begin{pmatrix} Y & 0 \\ 0 & Y^{-1} \end{pmatrix} \begin{bmatrix} I & X \\ 0 & I \end{bmatrix}, \quad {}^tX = X \in \mathbb{R}^{n \times n}, \ Y \in \mathcal{P}_n.$$

Thus, the power function is:

$$p_s(W) = \prod_{j=1}^{n} |Y_j|^{s_j}, \quad \text{for } W, Y \text{ as above, } s \in \mathbb{C}^n.$$

Viewing $\mathcal{P}_n$ as the subset of $\mathcal{P}_n^*$ consisting of the W with $X = 0$ in the partial Iwasawa decomposition above, it follows that the power function on $\mathcal{P}_n^*$ restricts to the power function on $\mathcal{P}_n$ which was defined in Equation (1.41) of Section 1.2.1.

A possible analogue of the **gamma function** for $\mathcal{P}_n^*$ is the Helgason–Fourier transform of $\exp(-\text{Tr}(W))$:

$$\int\limits_{\mathcal{P}_n^*} \exp(-\text{Tr}(W)) p_s(W) \, d\mu_n^*(W)$$

$$= \int\limits_{Y \in \mathcal{P}_n} \int\limits_{\substack{X \in \mathbb{R}^{n \times n} \\ X = {}^tX}} \exp\left\{ -\text{Tr}\left(Y + Y^{-1} + Y^{-1}[X] \right) \right\} p_s(Y) |Y|^{-(n+1)/2} \, d\mu_n(Y) \, dX$$

$$= \pi^{n(n+1)/4} K_n(s^\# | I, I), \quad s^\# = s - \left(0, \dots, 0, \tfrac{1}{2} \right),$$

where K_n denotes the K-Bessel function for $\mathcal{P}_n$ defined by formula (1.61) of Section 1.2.2. We saw the case $n = 1$ of this result in formula (3.141) in Section 3.7 of Volume I. The present formula should allow one to generalize (3.142) of Section 3.7, Vol. I, to $Sp(n, \mathbb{Z})$ using the spectral resolution of the G-invariant differential operators on $L^2(\mathcal{H}_n / Sp(n, \mathbb{Z}))$.

Example 2.1.19 ((The Heat Equation on $K \setminus G$) (Gangolli [197, pp. 108–109])).
We want to find $u(Kx, t)$ such that

$$\begin{cases} \Delta u = u_t, & \text{where } \Delta = \text{ the Laplacian for } K \setminus G, \\ u(Kx, 0) = f(Kx), & \text{for some given } K\text{-invariant function } f \text{ on } K \setminus G. \end{cases}$$

Now, it can be shown that

$$\Delta p_{i\mu-\rho} = -\{(\mu, \mu) + (\rho, \rho)\} p_{i\mu-\rho}.$$

Thus the same sort of argument that worked in § 1.3.4 shows that **the heat kernel is**

$$G_t(Kx) = \int_{\mu \in \mathfrak{a}^*} \exp\left(-\{(\mu, \mu) + (\rho, \rho)\}t\right) h_{i\mu-\rho}(Kx) \, |c(\mu)|^{-2} \, d\mu$$

and

$$u(t, Kx) = G_t * f, \quad \text{where the convolution is over } G.$$

Recall that convolution was defined in formula (1.24). Gangolli [197] shows that $G_t(Kx)$ has the standard properties of the fundamental solution of the heat equation, just as we saw for $\mathcal{P}_n$ in Exercise 1.3.8 of Section 1.3.4.

As we noted earlier, Ólafsson and Schlichtkrull [479] consider the holomorphic extension of the heat transform $G_t * f$, for f an L^2 function on a general symmetric space. The extension is to the complex crown of the symmetric space. Helgason's conjecture on eigenfunctions of the invariant differential operators being reconstructible from their hyperfunction boundary values can be considered using the crown of the symmetric space. See Gindikin [219].

Helgason [275, pp. 67–68] solves the wave equation on a symmetric space using the Radon transform on $K \setminus G$. He also discusses Huyghen's principle for a symmetric space. It is also shown by Helgason that eigenfunctions of $D(G/K)$ can be expressed as a Poisson integral over the boundary of the symmetric space.

2.1.8 An Example of a Symmetric Space of Type IV: The Quaternionic Upper Half 3-Space

References for this example include Belinfante and Kolman [41], Bougerol [73], Elstrodt et al. [165–168], Jauch [331], Kubota [373–375], Maass [426, Ch 1], Mennicke [442, 443], Sarnak [526], and Marie-France Vignéras [633].

First we need a brief review of **quaternions**. See also Volume I, p. 218. The quaternions, denoted $\mathbb{H}$ for Hamilton, form a division ring or noncommutative field:

$$\mathbb{H} = \mathbb{R} \oplus \mathbb{R}i \oplus \mathbb{R}j \oplus \mathbb{R}k,$$

where $ij = k = -ji$, $jk = i = -kj$, $ki = j = -ik$, $i^2 = j^2 = k^2 = -1$. The **reduced norm** of a quaternion $q = a + bi + cj + dk$, with a, b, c, d real is $\mathrm{Nrd}\,(q) = qq^c$, with the **conjugate** $q^c = a - bi - cj - dk$. The Euclidean length of the quaternion thought of as a vector in $\mathbb{R}^4$ is:

$$\|q\| = \sqrt{a^2 + b^2 + c^2 + d^2} = \sqrt{\mathrm{Nrd}\,(q)},$$

All goes very much as with the complex numbers except that things do not commute.

It is possible to represent quaternions by complex 2×2 matrices via:

$$1 \mapsto \begin{pmatrix} 1 & 0 \\ 0 & 1 \end{pmatrix}$$

$$i \mapsto \begin{pmatrix} 0 & -i \\ -i & 0 \end{pmatrix} = -i\sigma_1$$

These matrices are $-i$ times the Pauli matrices σ_1, σ_2, σ_3 from quantum mechanics.

$$j \mapsto \begin{pmatrix} 0 & 1 \\ -1 & 0 \end{pmatrix} = -i\sigma_2$$

$$k \mapsto \begin{pmatrix} i & 0 \\ 0 & -i \end{pmatrix} = -i\sigma_3.$$

One can view $SU(2)$ as the unit quaternions via such an identification. Thus $SU(2)$ is simply connected. Call the preceding map from quaternions to matrices f. Then we claim $SU(2) \cong \{f(q) \mid \|q\| = 1\}$.

The group $K = SU(2)$ can be mapped onto $SO(3, \mathbb{R})$ via a homomorphism of fundamental importance in the Dirac theory of electron spin. The map is given by taking Q in $SU(2)$ to $A = (a_{ij})$ in $\mathbb{R}^{3 \times 3}$ via $a_{ij} = \mathrm{Tr}(Q\sigma_i{}^t\overline{Q}\sigma_j)/2$. The map is onto with kernel the center of $SU(2)$.

After this brief discussion of quaternions, we can give various descriptions of a symmetric space that has been of interest to number theorists and physicists. The space is

$$SL(2, \mathbb{C})/SU(2).$$

It fits into type IV of Cartan's classification of symmetric spaces. We can identify this space as the space of positive Hermitian matrices of determinant one:

$$\mathcal{SP}_2^c = \left\{ Y \in \mathbb{C}^{2 \times 2} \mid Y = {}^t\overline{Y},\ Y \text{ positive},\ |Y| = 1 \right\}.$$

The identification is:

$$
\begin{aligned}
SL(2, \mathbb{C})/SU(2) &\to \mathcal{SP}_2^c, \\
gSU(2) &\mapsto g\,{}^t\overline{g}.
\end{aligned}
\tag{2.45}
$$

Here, we define a **positive Hermitian matrix** Y to be a Hermitian matrix $Y \in \mathbb{C}^{n \times n}$ such that $Y\{x\} = {}^t\bar{x}Yx > 0$ for all $x \in \mathbb{C}^n - 0$. These matrices are quite analogous to ordinary positive matrices. We could rewrite Chapter 3 of Vol. I and Chapter 1 of this Volume in the Hermitian case, if we had the time. Mercifully Elstrodt et al. [168] have done this and more.

By generalizing the Iwasawa decomposition, one sees that the coset representatives $g \in SL(2, \mathbb{C})$ for $SL(2, \mathbb{C})/SU(2)$ can be chosen to have the form:

$$g = \begin{pmatrix} \sqrt{t} & z/\sqrt{t} \\ 0 & 1/\sqrt{t} \end{pmatrix}, \ z \in \mathbb{C}, \ t > 0, \ z = x + iy. \tag{2.46}$$

This allows us to identify $SL(2, \mathbb{C})/SU(2)$ with **the quaternionic upper half space**:

$$\mathcal{H}^c = \{z + kt = x + iy + kt \mid x, y \in \mathbb{R}, \ t > 0\}. \tag{2.47}$$

Thus the elements of $\mathcal{H}^c$ are quaternions with j-coordinate equal to zero and positive k-coordinate. The mapping from $SL(2, \mathbb{C})/SU(2)$ to $\mathcal{H}^c$ sends $gSU(2)$ with g given by (2.46) to $z + kt$.

The **action** of a matrix

$$g = \begin{pmatrix} a & b \\ c & d \end{pmatrix} \in SL(2, \mathbb{C})$$

on an element q of the quaternionic upper half plane is:

$$g(q) = (aq + b)(cq + d)^{-1} = q^* \ \text{with} \ t^* = t\|cq + d\|^{-2}. \tag{2.48}$$

Recall that it is all right to divide by quaternions (on one side or the other), but it is not all right to interchange the order of multiplication.

The **action** of $g \in SL(2, \mathbb{C})$ on $Y \in S\mathcal{P}_2^c$ is

$$Y \mapsto Y\{g\} = {}^t\bar{g}Yg. \tag{2.49}$$

Using this action we identify our symmetric space as $SU(2)\backslash SL(2, \mathbb{C})$; i.e., we consider left rather than right cosets.

Exercise 2.1.28. Check that the three group actions are preserved in our identifications of $SL(2, \mathbb{C})/SU(2)$ with $\mathcal{H}^c$ and $S\mathcal{P}_2^c$.

Exercise 2.1.29. Show that the invariant arc length, volume element, and Laplacian on $\mathcal{H}^c$ are given by:

$$ds^2 = t^{-2}(dx^2 + dy^2 + dt^2);$$
$$d\mu = t^{-3} \, dx \, dy \, dt;$$
$$\Delta = t^2(\partial^2/\partial x^2 + \partial^2/\partial y^2 + \partial^2/\partial t^2) - t\partial/\partial t.$$

Exercise 2.1.30. (a) Show that a spherical function on $\mathcal{SP}_2^c$ has the form

$$h_\lambda(Y) = \frac{2\sin(\lambda r/2)}{\lambda \sinh r}, \quad \text{if } Y = a_r[k],$$

$$a_r = \begin{pmatrix} \exp(r/2) & 0 \\ 0 & \exp(-r/2) \end{pmatrix}, \quad k \in SU(2) = K.$$

Here r is the geodesic radial coordinate in the polar coordinate decomposition of Y in $\mathcal{SP}_2^c$.

(b) Show that if f is in $L^1(\mathcal{SP}_2^c/K)$, then

$$\int\limits_{\mathcal{SP}_2^c} f(Y)d\mu = \int\limits_{\mathbb{R}} f(a_r)\sinh^2 r\, dr.$$

(c) Use part (b) to show that the Helgason–Fourier transform for K-invariant functions on $\mathcal{SP}_2^c/K$ has the form:

$$\widehat{f}(\lambda) = \frac{2}{\lambda i}\int\limits_{\mathbb{R}} \exp(i\lambda t/2)f(a_t)\sinh t\, dt.$$

(d) Use the inversion formula for the ordinary Euclidean Fourier transform from Section 1.2 of Volume I to show that the spectral measure for Fourier inversion on $\mathcal{SP}_2^c$ is:

$$\frac{|\lambda|^2}{16}d\lambda, \quad \text{where} \quad d\lambda = \text{Lebesgue measure on } \mathbb{R}.$$

(e) Find the fundamental solution for the heat equation on $\mathcal{SP}_2^c$.
Hint. See Bougerol [73], Karpelevich et al. [340], or Burridge and Papanicolaou [92].

Exercise 2.1.30 shows that harmonic analysis on $SL(2,\mathbb{C})$ is far simpler than that on $SL(2,\mathbb{R})$. This is an example of a general phenomenon (see Helgason [276, p. 31] for the generalization of part (a) of Exercise 2.1.30). It would also be nice to consider analogues of Bessel and Whittaker functions for $\mathcal{H}^c$.

This completes our brief sketch of the theory of harmonic analysis on general symmetric spaces. There are many applications, other than those mentioned so far. For example, Resnikoff [506] considers the consequences of using the geometries of the spaces $(\mathbb{R}^+)^3$ or $(\mathbb{R}^+ \times SL(2,\mathbb{R})/SO(2))$ as models for color perception. An experiment is posed for distinguishing which geometry gives a more accurate model. Other references for the general theory are Gurarie [254] and Wawrzyńczyk [657].

2.2 Geometry and Analysis on $\Gamma \backslash G/K$

"Say what you know, do what you must, and whatever will be, will be."

Sofya Kovalevskaya's maxim from her paper [368] quoted in Pelageya Kochina
[358, p. 168]

2.2.1 Fundamental Domains

Our goal for the remainder of this volume is to give a very brief sketch of parts of the story of harmonic analysis on $\Gamma \backslash G/K$, and automorphic forms for certain subgroups Γ of G **acting discontinuously** on the symmetric space $X = G/K$. This means that for each $x \in X$, the set of images of x under Γ has no limit point in X. We will concentrate on two specific discontinuous groups: $GL(n, \mathfrak{O}_K)$, where $\mathfrak{O}_K$ is the ring of integers in an algebraic number field K,[2] and the **Siegel modular group** $Sp(n, \mathbb{Z})$. Here $GL(n, \mathfrak{O}_K)$ is the **modular group over an algebraic number field** which consists of $n \times n$ matrices γ such that both γ and γ^{-1} have entries in $\mathfrak{O}_K$. See Section 1.4 of Volume I for the necessary definitions from algebraic number theory. The group $Sp(n, \mathbb{Z})$ consists of all symplectic $2n \times 2n$ integral matrices.

There are many reasons to study such discontinuous groups. Of course knowledge of $GL(n, \mathfrak{O}_K)$ and related groups leads to greater understanding of the arithmetic of K itself. For example, there are applications to explicit class field theory, distribution of Gauss sums, values of Dedekind zeta functions and L-functions, asymptotics of units, elliptic curves, quadratic forms, and abelian varieties. References for these subjects include: Andrianov [13, 14], Bruinier et al. [82], Borel and Casselman [66], Borel and Mostow [68], Elstrodt et al. [165–168], Freitag [185, 186], Gelbart [208, 209], Goldfeld et al. [231], Heath-Brown and Patterson [267], Hecke [268, pp. 21–114], Jacquet and Langlands [324], Klingen [355], Kubota [373–376], Langlands [394], Maass [426], Mennicke [442, 443], Saito [525], Sarnak [527], Séminaire Cartan [547], Shimura [555], Shintani [558–560], Siegel [563, 565], Tunnell [620, 621], and Weil [660].

The Siegel modular group $Sp(n, \mathbb{Z})$ and kindred groups appear in many diverse areas of physics, often via the connections with abelian integrals and Riemann theta functions which arise in many theories from boson fields to solitons. See Cartier's article in Borel and Mostow [68, pp. 361–386], Cooke [124], Dubrovin et al. [143], Linda Keen [345], Pelageya Kochina [358], Sofya Kovalevskaya [368], Gérard Lion and Michèle Vergne [406], Lonngren and Scott [407], McKean and Trubowitz [440], Monastyrsky and Perelomov [457], Mumford [471], Novikov [474], Perelomov [484], Shale [551], and Wallach [652].

---

[2]Hopefully the beleaguered reader will not be too confused by our use of K for the maximal compact subgroup of G as well as an algebraic number field.

Theta functions also play a major role in the analytic theory of quadratic forms. See Siegel [565, Vol. I, pp. 326–405, 410–443, 469–548] and Weil [662, Vol. 2, pp. 1–157].

Here we seek to outline the foundations of a building which would ultimately encompass the generalization of Sections 3.3–3.7 of Volume I to these new discontinuous groups Γ. Our achievements will be pitiful compared with what is required. In particular, we will not say much about extensions of Section 3.7 of Volume I; i.e., the analogues of the non-Euclidean Poisson summation formula and the Selberg trace formula. Such results have already found various arithmetic and geometric applications, e.g., in computing dimensions of spaces of holomorphic automorphic forms. There are also results on units in number fields over imaginary quadratic fields and elliptic curves over imaginary quadratic fields. References for such work include: Christian [109], Eie [160], Efrat [152], Elstrodt et al. [165–168], Hashimoto's article in Hejhal et al. [272, pp. 253–276], Hashimoto [264], Langlands [389–395], Mennicke [442, 443], Morita [463], Müller [469], Petra Ploch [489], Sarnak [527], Tanigawa [591], Marie-France Vignéras [630, 631], Yamazaki [673], and Zograf [677].

General references for this section include: Andrianov [9–14], Baily [32], Hel Braun [74–76], Bruinier et al. [82], Christian [108], Elstrodt et al. [168], Freitag [185, 186], Gelfand, Graev, and Piatetski-Shapiro [214], Hecke [268], Hirzebruch and Van der Geer [299], Klingen [355], Maass [426], Mennicke [442], Séminaire Cartan [547], Shimura [554], Siegel [563–565], and Weil [658, 660, 662].

The following quote is from Van der Geer in [82, p. 182]:

> The general theory of automorphic representations provides a generalization of the theory of elliptic modular forms. But despite the obvious merits of this approach some of the attractive explicit features of the $g = 1$ [i.e., $SL(2, \mathbb{R})$] theory are lost in the generalization.

Our first topic is fundamental domains $K \backslash G / \Gamma$, for our favorite examples.

Example 2.2.1 (The Picard Modular Group).
Let K be the number field $\mathbb{Q}(i)$ with ring of Gaussian integers

$$\mathfrak{O}_K = \mathbb{Z}[i] = \{x + iy \mid x, y \in \mathbb{Z}\}.$$

Here $i = \sqrt{-1}$. The **Picard modular group** is defined to be

$$\Gamma = SL(2, \mathfrak{O}_K) = \left\{ \gamma = \begin{pmatrix} a & b \\ c & d \end{pmatrix} \;\middle|\; a, b, c, d \in \mathfrak{O}_K, \; \det \gamma = 1 \right\}.$$

The group $SL(2, \mathfrak{O}_K)$ acts discontinuously on the quaternionic upper half space $\mathcal{H}^c$ by fractional linear transformation defined by formula (2.48) in Section 2.1.8. An equivalent version of this action from formula (2.49) in the same subsection gives the action of $\gamma \in SL(2, \mathfrak{O}_K)$ on a positive determinant one Hermitian matrix $Y \in \mathcal{SP}_2^c$ via:

$$Y \mapsto Y\{\gamma\} = {}^t\overline{\gamma} Y \gamma.$$

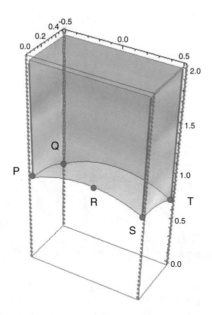

Fig. 2.1 A fundamental domain for $SL(2, \mathbb{Z}[i])$ in the quaternionic upper half plane:

$$D = \left\{ x+iy+kt \; \middle| \; \begin{array}{c} |x| \leq .5, \; t > 0 \\ 0 \leq y \leq .5 \\ x^2+y^2+t^2 \geq 1 \end{array} \right\}.$$

The labeled points are:

$$P = \left(-\frac{1}{2}, 0, \sqrt{3}/2\right), Q = \left(-\frac{1}{2}, \frac{1}{2}, 1/\sqrt{2}\right),$$

$$R = (0,0,1), S = \left(\frac{1}{2}, 0, \sqrt{3}/2\right), T = \left(\frac{1}{2}, \frac{1}{2}, 1/\sqrt{2}\right)$$

A fundamental domain for the action of $SL(2, \mathfrak{O}_K)$ on $\mathcal{H}^c$ was determined by Picard [488] and is pictured in Figure 2.1. Various views of a tessellation of hyperbolic 3-space $\mathcal{H}^c$ obtained by transforming this fundamental domain by elements of $\Gamma = SL(2, \mathfrak{O}_K)$ are shown in Figures 2.2, 2.3, 2.4, 2.5, and 2.6. Figures 2.7 and 2.8 show Cayley transforms of this tessellation which are inside of the unit sphere in 3-space. The figures are shown in stereo. If you stare at the two versions of the picture, one for each eye, you should be able to see a 3D tessellation. All of the tessellations were created by Mark Eggert using one of UCSD's VAX computers in the 1980s. Part of a tessellation obtained by taking a union of the fundamental domain in Figure 2.1 with its image under $y \mapsto -y$ can be found just before the index of this volume.

As for $SL(2, \mathbb{Z})$ (see Exercise 3.3.1, Volume I), the sides of the fundamental domain are mapped to each other by generators of Γ, which are in this case:

Fig. 2.2 Tessellation of the quaternionic upper halfplane from $SL(2, \mathbb{Z}[i])$ in stereo. It may help to put a division between the two halves of this figure and those that follow, in order to produce the 3D effect. The figure was created by Mark Eggert using the UCSD VAX computer

Fig. 2.3 Tessellation of the quaternionic upper half plane from $SL(2, \mathbb{Z}[i])$ in stereo. The figure was created by Mark Eggert using the UCSD VAX computer

$$\begin{pmatrix} 1 & 1 \\ 0 & 1 \end{pmatrix}, \begin{pmatrix} 1 & i \\ 0 & 1 \end{pmatrix}, \begin{pmatrix} i & 0 \\ 0 & -i \end{pmatrix}, \begin{pmatrix} 0 & -1 \\ 1 & 0 \end{pmatrix}.$$

Given any imaginary quadratic number field $K = Q(\sqrt{D})$, of discriminant $D > 0$, one can consider $SL(2, \mathfrak{O}_K)$, $\mathfrak{O}_K = $ the ring of algebraic integers in K,

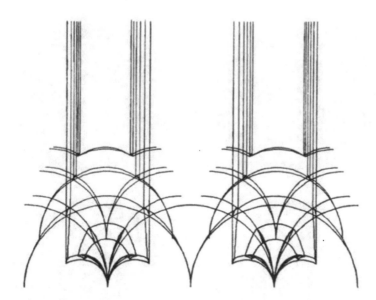

Fig. 2.4 Tessellation of the quaternionic upper half plane from $SL(2, \mathbb{Z}[i])$ in stereo. The figure was created by Mark Eggert using the UCSD VAX computer

Fig. 2.5 Tessellation of the quaternionic upper half plane from $SL(2, \mathbb{Z}[i])$ in stereo. The figure was created by Mark Eggert using the UCSD VAX computer

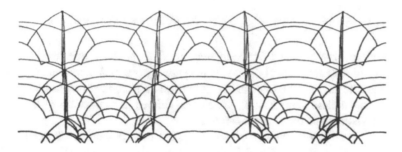

Fig. 2.6 Tessellation of the quaternionic upper half plane from $SL(2, \mathbb{Z}[i])$ in stereo. The figure was created by Mark Eggert using the UCSD VAX computer

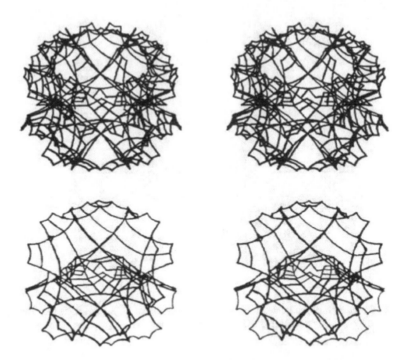

Fig. 2.7 Stereo tessellation of the unit sphere obtained by mapping the preceding tessellations of the quaternionic upper half plane into the unit sphere by a Cayley transform. The mapping for this figure and Figure 2.8 is

$$q \mapsto (q - k)(-kq + 1)^{-1}.$$

The figure was created by Mark Eggert using the UCSD VAX computer

$$SL(2, \mathfrak{O}_K) = \left\{ \gamma = \begin{pmatrix} a\, b \\ c\, d \end{pmatrix} \,\middle|\, a, b, c, d \in \mathfrak{O}_K, \; \det \gamma = 1 \right\}.$$

This investigation was begun by Bianchi [50]. Humbert [310] showed that the **volume of the fundamental domain** $SL(2, \mathfrak{O}_K) \backslash \mathcal{H}^c$ is

$$\frac{|D_K|^{3/2} \zeta_K(2)}{4\pi^2}, \quad \text{where } \zeta_K(s) = \quad \text{the Dedekind zeta function of } K$$

(see Section 1.4 of Volume I for the definitions of ζ_K and the discriminant D_K). The geometry of the fundamental domain is thus closely associated with the arithmetic of the number field K. In particular, the number of cusps of the fundamental domain is equal to the class number of K, which was defined in Section 1.4 of Vol. I. We will demonstrate this fact in Proposition 2.1.1.

Siegel gives two methods to prove formulas for the volume of fundamental domains of this sort. See Siegel [565, Vol. I, pp. 464–465, Vol. II, pp. 330–331, and Vol. III, pp. 39–46, 328–333]. One of Siegel's methods is the one we used in Theorem 1.4.4 of Section 1.4.4 to find the volume of the fundamental domain

Fig. 2.8 Stereo tessellation of the unit sphere obtained by mapping the preceding tessellations of the quaternionic upper half plane into the unit sphere by a Cayley transform. The figure was created by Mark Eggert using the UCSD VAX computer

for $GL(n, \mathbb{Z})$ via Siegel's integral formula. The other method involves finding the residues of Eisenstein series like (1.174) in Section 1.4.1, using the method of theta functions.

References for fundamental domains in quaternionic upper half space include: Ahlfors [2], Elstrodt et al. [165–168], Humbert [309, 310], Kubota [373–376], Mennicke [442, 443], Milnor [450], Sarnak [527], and Stark [575].

The next example to be considered is the analogue of Example 2.2.1 for real quadratic fields K.

Example 2.2.2 (The Hilbert Modular Group).

Suppose that K is a real quadratic number field and $K = \mathbb{Q}(\sqrt{D})$ has positive discriminant D; e.g., $K = \mathbb{Q}(\sqrt{2})$. Such a field has, as mentioned in Section 1.4 of Volume I, two conjugations mapping K into the field of real numbers and denoted $x^{(1)}, x^{(2)}$, for $x \in K$. If $x = a + b\sqrt{d} \in K$, with $a, b \in \mathbb{Q}$, then $x^{(1)} = x$ and $x^{(2)} = a - b\sqrt{D}$. Form the group

$$\Gamma = SL(2, \mathfrak{O}_K) = \left\{ \gamma = \begin{pmatrix} a & b \\ c & d \end{pmatrix} \,\middle|\, a, b, c, d \in \mathfrak{O}_K, \det \gamma = 1 \right\}.$$

This is called the **Hilbert modular group** for the field K. Then Γ acts discontinuously on the product $\mathcal{H}^2$ of two ordinary upper half planes via

$$\gamma(z^{(1)}, z^{(2)}) = (\gamma^{(1)} z^{(1)}, \gamma^{(2)} z^{(2)}), \quad \text{for} \ \ z^{(j)} \in \mathcal{H}, \tag{2.50}$$

and $\gamma^{(j)}$ denoting the matrix each of whose entries is obtained by taking the jth conjugate of the corresponding entry of $\gamma \in \Gamma$, $j = 1, 2$. Here $\gamma^{(j)}$ acts on $z^{(j)}$ by fractional linear transformation as in Chapter 3 of Volume I.

Exercise 2.2.1. (a) Show that the action of $\Gamma = SL(2, \mathfrak{O}_K)$, for a real quadratic field K, on the ordinary upper half plane H via $z \mapsto \gamma z$, for $z \in H$, is not discontinuous.
(b) Show that the action of $\Gamma = SL(2, \mathfrak{O}_K)$, for a real quadratic field K, on $\mathcal{H}^2$ defined by (2.50) is discontinuous.
Hint. (a) Make use of the units in K; that is $x \in \mathfrak{O}_K$ such that $x^{-1} \in \mathfrak{O}_K$.

It is an easy matter to generalize to $SL(2, \mathfrak{O}_K)$ where K is any **totally real algebraic number field** K; i.e., a number field K such that every conjugation is real-valued. Then the Hilbert modular group $SL(2, \mathfrak{O}_K)$ acts discontinuously on $\mathcal{H}^m$, where m is the degree of K over $\mathbb{Q}$.

The fundamental domains $\Gamma \backslash \mathcal{H}^m$ have been rather intensively studied. They are $2m$-dimensional and complicated by the existence of units of infinite order in $\mathfrak{O}_K$. The formula for the **volume of the fundamental domain** is

$$2(-2\pi)^m \zeta_K(-1) = 2\pi^{-m} D_K^{3/2} \zeta_K(2),$$

where D_K is the absolute value of the discriminant of K and $\zeta_K(s)$ is Dedekind's zeta function of K. See Klingen [351] for a proof of a much more general result.

References for these results include: Blumenthal [52], Harvey Cohn [116], Freitag [186], Giraud [221], Gundlach [253], Hammond [259], Hirzebruch [296], Hirzebruch and Van der Geer [299], Hirzebruch and Zagier [300, 301], Humbert [309], Klingen [350–352], Maass [415, 416], Resnikoff [505], Shimizu [553], Shimura [554], Siegel [563, 565], Thomas and Vasquez [613], and Weisser [665]. See our earlier comments on volumes of fundamental domains for the Picard type groups.

Example 2.2.3 (The Modular Group over any Number Field).

Suppose that K is any number field, $\mathfrak{O}_K$ its ring of integers, and m its degree over $\mathbb{Q}$. Then, as in Section 1.4 of Volume I, K must be equal to $\mathbb{Q}(a)$ for some complex number a with minimal polynomial $f(x)$ and K is isomorphic to the quotient $\mathbb{Q}[x] / ((f(x)))$. So

$$\left. \begin{array}{l} K \otimes_{\mathbb{Q}} \mathbb{R} \cong \mathbb{R}[x]/(f(x)) \cong \sum_{j=1}^{r_1+r_2} \oplus \, E_j, \\ \\ E_j \cong \begin{cases} \mathbb{R}, j = 1, \ldots, r_1; \\ \mathbb{C}, j = r_1 + 1, \ldots, r_1 + r_2. \end{cases} \end{array} \right\} \tag{2.51}$$

Therefore we have m conjugations sending K into E_j by mapping x to $x^{(j)}$, for $j = 1, \ldots, r_1$, and mapping x to $x^{(j)}$ or $\overline{x^{(j)}}$, for $j = r_1 + 1, \ldots, r_1 + r_2$.

What is a positive matrix over a number field? We are actually seeking the "infinite prime part" of an adelic symmetric space (see Cassels and Fröhlich [101], Gelbart [208], Gelfand et al. [214], and Weil [658, 660, 662]).

Define a **positive quadratic form** Y **over the number field** K to be a vector

$$Y = \left(Y^{(1)}, \ldots, Y^{(r_1 + r_2)} \right),$$

with $Y^{(j)} \in \mathcal{P}_n$, for $j = 1, \ldots, r_1$ and $Y^{(j)} \in \mathcal{P}_n^c$, $j = r_1 + 1, \ldots, r_1 + r_2$. Here $\mathcal{P}_n$ is the symmetric space of positive real $n \times n$ matrices studied in Chapter 1 of this volume, while $\mathcal{P}_n^c$ is the symmetric space of positive $n \times n$ Hermitian complex matrices; i.e.,

$$\mathcal{P}_n^c = \left\{ Y \in \mathbb{C}^{n \times n} \mid {}^t\overline{Y} = Y, \ Y \text{ positive} \right\} \cong U(n) \backslash GL(n, \mathbb{C}).$$

A complex Hermitian matrix Y is called **positive** if $Y\{x\} = {}^t\overline{x} Y x > 0$ for every column vector $x \in \mathbb{C}^n - 0$. Set $\mathcal{P}_n^K =$ the **space of positive quadratic forms** over K. Clearly this symmetric space will generalize the two preceding examples, if we restrict to the **determinant one subspace**

$$\mathcal{SP}_n^K = \left\{ Y \in \mathcal{P}_n^K \mid |Y^{(j)}| = 1, \ j = 1, \ldots, r_1 + r_2 \right\}.$$

Set

$$\Gamma = GL(n, \mathfrak{O}_K) = \left\{ \gamma \in \mathfrak{O}_K^{n \times n} \mid \gamma^{-1} \in \mathfrak{O}_K^{n \times n} \right\}.$$

The **action of the modular group** $GL(n, \mathfrak{O}_K)$ on $Y \in \mathcal{P}_n^K$ is given by:

$$Y \mapsto Y\{A\}, \quad \text{with} \ (Y\{A\})^{(j)} = \left(Y^{(j)} \right) \{A^{(j)}\} = {}^t\overline{A^{(j)}} Y^{(j)} A^{(j)},$$

for $j = 1, \ldots, r_1 + r_2$. Here $A^{(j)}$ denotes the matrix all of whose entries are the jth conjugate of the corresponding entries in A.

There is a long history of looking only at the "infinite prime" part of the symmetric space rather than the adelic version which includes an infinite number of p-adic components, one for each finite prime p—a much more recent construct. Some references are Hecke [268, pp. 21–55], Humbert [309, 310], Klingen [350–352], Ramanathan [497, 498], Siegel [563], Weil [658], and Weyl [666, Vol. IV, pp. 232–264]. Much of our discussion here was inspired by working with John Hunter who considered number-theoretic applications of analogues of Siegel's integral formula (Proposition 2.1.2 of Section 1.4.4) for $SL(n, \mathfrak{O}_K)$, K imaginary quadratic. I regret that John's death prevents publication of his thesis work (see Hunter [311]).

Exercise 2.2.2. Show that $GL(n, \mathfrak{O}_K)$ consists of all matrices in $\mathfrak{O}_K^{n \times n}$ whose determinant is a unit in $\mathfrak{O}_K$.

Fundamental domains for $\mathcal{P}_n^K / GL(n, \mathfrak{O}_K)$ were discussed by Humbert [309], who generalized many of the results that we presented in Section 1.4 for the case that K is the field of rational numbers. Siegel [565] obtains analogues of many of the results of Section 1.4 in various places. For example, Siegel [565, Vol. I, p. 475] gives an analogue of Lemma 1.4.2 of Section 1.4.2. And Siegel [565, pp. 464–465] obtains a formula for the volume of the fundamental domain, in a paper which was to be corrected later [565, Vol. III, pp. 328–333].

We choose not to rewrite all of Section 1.4 in this case. Instead we take up some aspects of the theory when $\Gamma = SL(2, \mathfrak{O}_K)$. In particular, we discuss a result of Maass [416] correcting an error of Blumenthal [52]. This error also appears in Hecke's first paper (see Hecke [268, pp. 21–55 and the notes at the end of the volume]). We want to show that the cusps of the fundamental domain $S\mathcal{P}_2^K / SL(2, \mathfrak{O}_K)$ are in one-to-one correspondence with the ideal classes of K. The cusps are the points of the fundamental domain which are equivalent to infinity under the action of $SL(2, K)$. Thus they are elements of $\widehat{K} = K \cup \{\infty\}$.

Proposition 2.2.1 (The Cusp-Ideal Class Correspondence). *The cusps of the fundamental domain for $S\mathcal{P}_2^K / SL(2, \mathfrak{O}_K)$, K any number field, are in one-to-one correspondence with the ideal class group I_K of K.*

Proof. See Siegel [563, p. 242]. Let h denote the class number of K. Choose fixed integral ideals $\mathfrak{a}_1, \ldots, \mathfrak{a}_h$ representing the ideal class group I_K. We want to show that the elements of $\widehat{K} = K \cup \{\infty\}$ are divided into h equivalence classes by the action of

$$\gamma = \begin{pmatrix} a & b \\ c & d \end{pmatrix} \in SL(2, \mathfrak{O}_K) \text{ on } x \in \widehat{K} \text{ defined by } \gamma(x) = \frac{ax + b}{cx + d}, \ \gamma(\infty) = a/c.$$

Suppose that $x = p/s$ with $p, s \in \mathfrak{O}_K$. Here we write $\infty = 1/0$. Then define $f(x)$ to be the integral ideal (p, s) which is generated by p and s. Note that the ideal class of $f(x)$ is well defined. For if $x = p_1/s_1 = p_2/s_2$, then $\mathfrak{a}_2 = k\mathfrak{a}_1$, for $k = p_2/p_1$, where $\mathfrak{a}_i = f(p_i/s_i)$.

So there is an induced map $\overline{f} : \widehat{K}/SL(2, \mathfrak{O}_K) \to I_K$, since if $\gamma \in SL(2, \mathfrak{O}_K)$ and

$$\gamma = \begin{pmatrix} a & b \\ c & d \end{pmatrix}, \text{ then } f(\gamma(p/s)) = (ap + bs, cp + ds) \subset (p, s).$$

The reverse inclusion must hold as well because the determinant of γ is one.

The map $\overline{f}$ is onto, since every ideal in $\mathfrak{O}_K$ has at most two generators (see Pollard [490]).

In order to show that $\overline{f}$ is one-to-one, you will probably first think of the following argument. Suppose that $f(p_1/s_1) = kf(p_2/s_2)$ for some $k \in K$. If $k = \omega/\tau$, for $\omega, \tau \in \mathfrak{O}_K$, then we see that

$$\tau(ap_1 + bs_1) = \omega p_2, \text{ and } \tau(cp_1 + ds_1) = \omega s_2,$$

for some

$$\gamma = \begin{pmatrix} a & b \\ c & d \end{pmatrix} \in GL(2, \mathfrak{O}_K).$$

It follows that

$$\frac{ap_1 + bs_1}{cp_1 + ds_1} = \frac{p_2}{s_2}.$$

This says that p_1/s_1 and p_2/s_2 are indeed equivalent modulo $GL(2, \mathfrak{O}_K)$. But unfortunately we need to know that they are equivalent modulo $SL(2, \mathfrak{O}_K)$. The difference between special and general linear groups over $\mathfrak{O}_K$ can be rather large, thanks to the presence of lots of units.

Since our boat seems to have stopped moving, we take a different tack. This time we follow Siegel's argument. Suppose that $\mathfrak{a}^{-1}$ denotes the inverse ideal to $\mathfrak{a}$. Then $\mathfrak{a}\mathfrak{a}^{-1} = \mathfrak{O}_K = (1)$ and thus

$$p_1 v_1 - s_1 u_1 = 1 \quad \text{and} \quad p_2 v_2 - s_2 u_2 = 1 \quad \text{for some } u_i, v_i \in \mathfrak{a}_i^{-1}.$$

Set

$$A_i = \begin{pmatrix} p_i & u_i \\ s_i & v_i \end{pmatrix}, \quad i = 1, 2,$$

and note that although A_i only has entries in K, the product $A_2 A_1^{-1}$ is actually in $SL(2, \mathfrak{O}_K)$. For we have assumed that $\bar{f}(p_i/s_i)$ both equal $\mathfrak{a}_1$, say. Thus if you compute the first entry of $A_2 A_1^{-1}$, for example, you find it is $p_2 v_1 - u_2 s_1$, which is in the ideal $\mathfrak{a}_2 \mathfrak{a}_1^{-1} + \mathfrak{a}_1 \mathfrak{a}_2^{-1} = \mathfrak{O}_K$, since $\mathfrak{a}_1 = \mathfrak{a}_2$. It is important that we have chosen a fixed set of representatives of our ideal classes. But note here that $f(p_1/s_1) = kf(p_2/s_2)$, for $k \in K$, implies that we can assume $k = 1$ by replacing p_2/s_2 by $(kp_2)/(ks_2)$. To see that $A_2 A_1^{-1}$ does indeed take p_1/s_1 to p_2/s_2, note that A_i maps ∞ to p_i/s_i (acting by fractional linear transformation).

This completes the proof of Proposition 2.2.1. ∎

Lemma 2.2.1. *The stabilizer in $SL(2, \mathfrak{O}_K)$ of a cusp $x_i = p_i/s_i$ for the fundamental domain $S\mathcal{P}_2^K/SL(2, \mathfrak{O}_K)$ is defined by:*

$$\Gamma_{x_i} = \{\gamma \in SL(2, \mathfrak{O}_K) \mid \gamma x_i = x_i\}.$$

Suppose that $x_i = A_i \infty$ and that

$$A_i = \begin{pmatrix} p_i & u_i \\ s_i & v_i \end{pmatrix} \in SL(2, K), \quad \text{with } \mathfrak{a}_i = (p_i, s_i), \; u_i, v_i \in \mathfrak{a}_i^{-1},$$

for $i = 1, 2, \ldots, h$. *Here the ideals* $\mathfrak{a}_1, \ldots, \mathfrak{a}_h$ *represent the ideal class group* I_K. *And* $\mathfrak{a}^{-1}$ *is the inverse ideal to* $\mathfrak{a}$. *Let* U_K *denote the group of units in* $\mathfrak{O}_K$. *Then the stabilizer of a cusp has the form:*

$$\Gamma_{x_i} = \left\{ A_i \begin{pmatrix} w & z \\ 0 & w^{-1} \end{pmatrix} A_i^{-1} \,\middle|\, z \in \mathfrak{a}_i^{-2}, \ w \in U_K \right\}.$$

Proof. See Exercise 2.2.3 below. The result is clearly true for the infinite cusp. ∎

Exercise 2.2.3. Prove the formula for the stabilizer of the infinite cusp in Lemma 2.2.1 above. Then deduce the result for an arbitrary cusp.
Hint. You can find the details in Siegel [563, p. 245]. If γ stabilizes x_i, then $A_i^{-1}\gamma A_i$ stabilizes infinity and therefore

$$A_i^{-1}\gamma A_i = \begin{pmatrix} w & z \\ 0 & w^{-1} \end{pmatrix}.$$

To see that z must lie in $\mathfrak{a}_i^{-2}$, just multiply out the matrices. And note that

$$(ap_i + bs_i)s_i = (cp_i + ds_i)p_i \ \text{ if } \ \gamma = \begin{pmatrix} a & b \\ c & d \end{pmatrix}.$$

But then it follows that w must be a unit, since, after division by $\mathfrak{a}_i^2$, we see that:

$$\frac{(ap_i + bs_i)}{\mathfrak{a}_i} = \frac{(p_i)}{\mathfrak{a}_i}, \quad \frac{(cp_i + ds_i)}{\mathfrak{a}_i} = \frac{(s_i)}{\mathfrak{a}_i}.$$

You also have to multiply out the following matrices:

$$\begin{pmatrix} p_i & u_i \\ s_i & v_i \end{pmatrix} \begin{pmatrix} w & z \\ 0 & w^{-1} \end{pmatrix} = \begin{pmatrix} a & b \\ c & d \end{pmatrix} \begin{pmatrix} p_i & u_i \\ s_i & v_i \end{pmatrix},$$

to see that w is a unit.

Example 2.2.4 (The Siegel Modular Group).
   The Siegel modular group $Sp(n, \mathbb{Z})$ is the group of all symplectic matrices with integer entries. It acts discontinuously on Siegel's upper half space $\mathcal{H}_n$ (or on the space of positive symplectic matrices $\mathcal{P}_n^*$) considered in Section 2.1.
   Siegel [565, Vol. II, pp. 300–301] shows that a fundamental domain for $Sp(n, \mathbb{Z}) \backslash \mathcal{H}_n$ can be obtained by generalizing the method of perpendicular bisectors from Exercise 3.3.6 of Volume I. That is, taking $\Gamma = Sp(n, \mathbb{Z})$, we consider the domain:

$$\mathcal{D}_n = \{ Z \in \mathcal{H}_n \mid d(Z, W) \le d(Z, \gamma W) \ \text{for all} \ \gamma \in \Gamma \}.$$

Here $W \in \mathcal{H}_n$ is chosen to be a point not fixed by Γ.

As we quoted at the end of Section 1.4.3, Siegel [565, Vol. II, p. 309] said: "The application of the general method [stated above]... would lead to a rather complicated shape of the frontier [boundary] of F." So Siegel goes on to consider another fundamental domain for $Sp(n, \mathbb{Z}) \backslash \mathcal{H}_n$. Before we can say what this new domain is, we need to make a definition. Call $C, D \in \mathbb{Z}^{n \times n}$ a **coprime symmetric pair** if $C\,^t D = D\,^t C$ and the matrices C and D are relatively prime in the sense that: if for any matrix $G \in \mathbb{Q}^{n \times n}$, the matrices GC and GD are both integral, then G must be integral; i.e., in $\mathbb{Z}^{n \times n}$.

Exercise 2.2.4. Show that $C, D \in \mathbb{Z}^{n \times n}$ are a coprime symmetric pair if and only if $(C \; D)$ can be completed to a matrix

$$\begin{pmatrix} A & B \\ C & D \end{pmatrix} \in Sp(n, \mathbb{Z}).$$

Now we can obtain Siegel's fundamental domain $\mathcal{D}_n^S$ for $Sp(n, \mathbb{Z}) \backslash \mathcal{H}_n$, as in Siegel [565, Vol. II, p. 108], by finding an analogue of the highest point method from Exercise 3.3.1 of Vol. I. For this, we need to recall from Exercise 2.1.16 of Section 2.1 that if $W = (AZ + B)(CZ + D)^{-1}$, then the imaginary part of W is $Y \left\{ (CZ + D)^{-1} \right\}$ if Y is the imaginary part of Z and $Y\{A\} = {}^t\overline{A} Y A$. We will take the **height** of $Z = X + iY \in \mathcal{H}_n$ to be the determinant $|Y|$. So we see that $|W| = |Y| \, \|CZ + D\|^{-2}$. This concept of height leads to the following construction for a fundamental domain by a highest point method.

The **Siegel fundamental domain** $\mathcal{D}_n^S$ for $Sp(n, \mathbb{Z}) \backslash \mathcal{H}_n$ is the set of $Z \in \mathcal{H}_n$ such that the following three statements hold (if we ignore boundary identifications):

(1) $\|CZ + D\| \geq 1$, for all coprime symmetric pairs C, D with $C \neq 0$;
(2) $Y = \mathrm{Im}\, Z \in \mathcal{M}_n = $ Minkowski's fundamental domain for $\mathcal{P}_n \backslash GL(n, \mathbb{Z})$;
(3) $X = \mathrm{Re}\, Z$, $X = \left(x_{ij} \right)$, with $\left| x_{ij} \right| \leq 1/2$, $1 \leq i, j \leq n$.

$$(2.52)$$

Again there is a certain relation between Siegel's fundamental domain and matrices in $Sp(n, \mathbb{Z})$:

(1) $\begin{pmatrix} A & B \\ C & D \end{pmatrix} \in Sp(n, \mathbb{Z})$ with $C \neq 0$;

(2) $\begin{pmatrix} U & 0 \\ 0 & {}^t U^{-1} \end{pmatrix}$, $U \in GL(n, \mathbb{Z})$;

(3) $\begin{pmatrix} I & N \\ 0 & I \end{pmatrix}$, $N = {}^t N \in \mathbb{Z}^{n \times n}$.

In fact, Maass [426, § 11] shows that $Sp(n, \mathbb{Z})$ can be generated by matrices of the form

$$\begin{pmatrix} I & N \\ 0 & I \end{pmatrix}, \ N = {}^t N \in \mathbb{Z}^{n \times n} \ \text{ and } \ J = \begin{pmatrix} 0 & -I \\ I & 0 \end{pmatrix}.$$

The Siegel fundamental domain $\mathcal{D}_n^S$ can be shown to be closed, connected, and bounded by finitely many algebraic hypersurfaces. Compactifications have been studied and their singularities resolved (see Ash et al. [30], Chai [103], Van der Geer in Bruinier et al. [82] and Namikawa [472]). Gottschling [238] found the explicit list of (28) inequalities defining the fundamental domain $\mathcal{D}_2^S$ for $Sp(2, \mathbb{Z})$. As far as I know, no one has written down the explicit inequalities for $\mathcal{D}_n^S$, when n is larger than 2. Other references for related facts are Christian [108], Freitag [185], Maass [426], Séminaire H. Cartan [547], and Siegel [564, 565]. Van der Geer shows that there is a canonical 1–1 correspondence between the set of isomorphism classes of principally polarized abelian varieties of dimension n and the orbit space $Sp(n, \mathbb{Z}) \backslash \mathcal{H}_n$ (see [82], p. 202]).

The symplectic **volume of the fundamental domain** $Sp(n, \mathbb{Z}) \backslash \mathcal{H}_n$ was computed by Siegel [565, Vol. II, p. 279] to be:

$$2 \prod_{k=1}^{n} \Lambda(k), \ \text{ if } \ \Lambda(s) = \pi^{-s} \Gamma(s) \zeta(2s).$$

Setting $V_n = \mathrm{Vol}(\mathcal{D}_n^S)$, it follows that

$$V_1 = \frac{\pi}{3}, \quad V_2 = \frac{\pi^3}{270}, \quad V_3 = \frac{\pi^6}{127575}, \quad V_4 = \frac{\pi^{10}}{200930625}.$$

Klingen [351] generalized this result to the **Hilbert–Siegel modular group** which is defined for any totally real algebraic number field K by:

$$Sp(n, \mathfrak{O}_K) = \left\{ \gamma \in \mathfrak{O}_K^{n \times n} \mid {}^t \gamma J \gamma = J \right\}, \quad J = \begin{pmatrix} 0 & -I_n \\ I_n & 0 \end{pmatrix}.$$

It is possible to connect the volume of the fundamental domain for $Sp(n, \mathbb{Z})$ with the Euler characteristic of the fundamental domain via the Gauss–Bonnet theorem (see Siegel [565, Vol. II, p. 277, 331], Harder [261] and Klingen [351]). Harder's result is very general. However, the symmetric spaces involved do not include those without complex structure like $\mathcal{SP}_n$ for $n > 2$ or the quaternionic upper half plane.

Mathematicians have studied much more general arithmetic groups Γ and their fundamental domains, including that for $GL(n)$ over a simple associative algebra. Some references are: Borel [65], Hel Braun [75, 76], L. Cohn [120], Feit [176], Krieg [370], Ramanathan [497, 498], Resnikoff and Tai [509], Siegel [565, Vol. III, pp. 143–153; Vol. II, pp. 390–405], Weyl [666, Vol. IV, pp. 232–264], and Weil [658]. Margulis [433] has characterized arithmetic subgroups of connected noncompact Lie groups G of the sort we consider.

Before we leave the subject of fundamental domains, let us give an example of a discrete subgroup Γ of isometries of G/K such that $\Gamma\backslash G/K$ is compact. This example comes from notes of D. Sullivan. Take

$$G = \{g \in GL(n+1, \mathbb{R}) \mid {}^t g \varphi g = \varphi\},$$

where

$$\varphi = \begin{pmatrix} 1 & & & 0 \\ & \ddots & & \\ 0 & & 1 & \\ & & & -\sqrt{2} \end{pmatrix}.$$

Let $\Gamma = G \cap GL(n+1, \mathfrak{O}_L)$ for $L = \mathbb{Q}\left(\sqrt{2}\right)$. Now Γ is a discrete group of isometries of G/K. We can identify G with the Lorentz-type group $O(n, 1)$ and G/K with one sheet of the hyperboloid consisting of the set of points $x \in \mathbb{R}^{n+1}$ such that $\varphi(x) = -1$.

It can be shown that the quotient $\Gamma\backslash G/K$ is compact. We sketch the proof. Note that when $L = \mathbb{Q}\left(\sqrt{2}\right)$, $\mathfrak{O}_L = \mathbb{Z}\left[\sqrt{2}\right]$ which is not discrete in $\mathbb{R}$. Thus we must work harder than usual to show that Γ is actually a discrete subgroup of G. Note the conjugations map:

$$\begin{array}{ccc} \mathfrak{O}_L & \to & \mathbb{R}^2, \\ m + n\sqrt{2} & \mapsto & \left(m + n\sqrt{2}, m - n\sqrt{2}\right) \quad \text{for } m, n \in \mathbb{Z}. \end{array}$$

This induces a mapping which sends Γ discretely into $GL(n+1, \mathbb{R})^2$. Now $\gamma \in \Gamma$ leaves φ invariant and thus γ' leaves φ' invariant where γ' denotes the matrix formed by conjugating all entries of γ; i.e., sending $m + n\sqrt{2}$ to $m - n\sqrt{2}$. Now

$$\varphi' = \begin{pmatrix} 1 & & & 0 \\ & \ddots & & \\ 0 & & 1 & \\ & & & \sqrt{2} \end{pmatrix}$$

and thus the matrices leaving φ' invariant form a compact group. It follows that Γ is a discrete subgroup of G. Why?

To see that $K\backslash G/\Gamma$ is compact, one need to use some version of the Hermite–Mahler compactness theorem in Exercise 1.4.14, Section 1.4.2.

A similar example is $G = \{g \in GL(4, \mathbb{R}) \mid {}^t g \chi g = \chi\}$, where

$$\chi = \begin{pmatrix} 1 & 0 & 0 & 0 \\ 0 & 1 & 0 & 0 \\ 0 & 0 & 1 & 0 \\ 0 & 0 & 0 & -7 \end{pmatrix}.$$

Let $\Gamma = G \cap GL(4, \mathbb{Z})$. It is clear that Γ is discrete. To see that the fundamental domain $K \backslash G / \Gamma$ is compact, identify $K \backslash G / \Gamma$ with a subset S of $\mathcal{P}_4 / GL(4, \mathbb{Z})$ by mapping Kg to $I[g]$, as usual. Now to see S has compact closure, we need to only show that $Y \in S$ implies $|Y|$ bounded above and $m_Y = \min \{Y[a] \mid a \in \mathbb{Z}^4 - 0\}$ bounded below. The fact that m_Y is bounded from below comes from the fact that $\chi[x] = 0$ has no solution $x \in \mathbb{Z}^4 - 0$. For the neighborhood U of 0 defined by

$$U = \{x \in \mathbb{R}^4 \mid \chi[x] \le 1/2\}$$

contains no point of $g(\mathbb{Z}^4 - 0)$, for $g \in G$.

To see that $\chi[x] \ne 0$ for $x \in \mathbb{Z}^4 - 0$, one looks at

$$x_1^2 + x_2^2 + x_3^2 - 7t^2 \equiv 0 \pmod 8.$$

Since any integer can be written as a sum of 4 squares, the quadratic form

$$\varphi = x_1^2 + \cdots + x_n^2 - 7t^2$$

does vanish on $\mathbb{Z}^n - 0$ when n is larger than 3. That's why we took the form φ for general n.

There are other references for examples in which $K \backslash G / \Gamma$ is compact; e.g., Borel [65, p. 57]. Borel notes that if G is the orthogonal group of a quadratic form over $\mathbb{Q}$ in n variables and Γ is a group of units of a lattice in $\mathbb{Q}^n$, then $G_{\mathbb{R}} / \Gamma$ is compact iff the form F does not represent 0 over $\mathbb{Q}$. Here Borel uses another sort of compactness criterion. See also Borel [64], Mostow [466], and Mostow and Tamagawa [467]. The last reference proves that if G is a semisimple algebraic matrix group defined over the field $\mathbb{Q}$ and having no unipotent or parabolic elements other than the identity, then $G / G_{\mathbb{Z}}$ is compact. Elstrodt et al. [168] give many examples of $\Gamma \subset G = SL(2, \mathbb{C})$.

2.2.2   Automorphic Forms

Having considered the analogue of Section 1.4 for the types of arithmetic groups in Examples 2.2.1–2.2.4 above, we next begin the study of automorphic forms for these arithmetic groups. The theory of holomorphic forms has received the most attention.

This restricts us here to consideration of automorphic forms for the Hilbert modular group $SL(2, \mathfrak{O}_K)$, for a totally real field K, or the Siegel modular group $Sp(n, \mathbb{Z})$, or the Hilbert–Siegel modular group. It is also possible to discuss non-holomorphic automorphic forms on these symmetric spaces. Such forms satisfy some sort of differential equation involving the invariant differential operators on the symmetric space. Harish-Chandra made a very general definition of automorphic form on a Lie group (see Borel's article in Borel and Mostow [68, p. 199]). Let us begin with a brief sketch of the holomorphic theory which models itself on that from Sections 3.4 and 3.6 of Volume I.

Example 2.2.5 (Holomorphic Hilbert Modular Forms).

Some references for holomorphic Hilbert modular forms are Blumenthal [52], Bruinier's article in [82], Freitag [186], Gundlach [253], Herrmann [292], Hirzebruch [296, 298], Maass [415, 416], Resnikoff [505], Shimura [554], Siegel [563], Marie-France Vignéras [630–632], Van der Geer and Zagier [206], and Zagier [674, 675].

Suppose that K is a totally real algebraic number field of degree m with ring of integers $\mathfrak{O}_K$. We say that a function $f : \mathcal{H}^m \to \mathbb{C}$ is an (entire) **Hilbert modular form of weight k** belonging to $\Gamma = SL(2, \mathfrak{O}_K)$ if f has the following two properties, **assuming that m is greater than one**:

(1) $f(z)$ is holomorphic on $\mathcal{H}^m$;
(2) $f(\gamma z) = \mathrm{N}(cz + d)^k f(z)$, for all

$$\gamma = \begin{pmatrix} a & b \\ c & d \end{pmatrix} \in SL(2, \mathfrak{O}_K), \ z \in \mathcal{H}^m.$$

When f satisfies (1) and (2) we say that f is in $\mathcal{M}(SL(2, \mathfrak{O}_K), k)$. The notation in (2) means that for $z = (z^{(1)}, \ldots, z^{(m)})$, we have

$$\gamma(z) = \left(\gamma^{(1)} z^{(1)}, \ldots, \gamma^{(m)} z^{(m)} \right),$$

where $\gamma^{(j)}$ denotes the matrix obtained from γ by conjugating each entry of γ by the jth conjugation of the number field K. Here $\gamma^{(j)}$ acts on $z^{(j)}$ by fractional linear transformation. And the **norm** Nz is defined by:

$$\mathrm{N}(z) = \prod_{j=1}^{m} z^{(j)}, \quad \text{for } z \in \mathcal{H}^m. \tag{2.53}$$

One can also look at forms with different weights for each conjugate. This is considered in some of the references listed above but we will not go there.

In analogy to our definition of the norm of $z \in \mathcal{H}^m$, we also define the **trace** as follows:

$$\mathrm{Tr}\,(z) = \sum_{j=1}^{m} z^{(j)}, \quad \text{for } z \in \mathcal{H}^m. \tag{2.54}$$

Note that if $z \in K$, instead of $\mathcal{H}^m$, the norm and trace are the usual ones for the field extension $K/\mathbb{Q}$.

Of course, one might expect that, as in the case that $m = [K : \mathbb{Q}] = 1$, we should also require a Hilbert modular form to satisfy certain growth conditions at the cusps of the fundamental domain. For the cusp at infinity, one would just require that $f(z)$ be holomorphic at infinity. It was proved by Götzky [239] that this growth condition is unnecessary when $K = \mathbb{Q}(\sqrt{5})$. Gundlach generalized Götzky's result to any totally real number field (see Siegel [563]).

Siegel [563] is a good reference for the basic facts about Hilbert modular forms. For example, Siegel [563, p. 215] shows that a Hilbert modular form of weight $k < 0$ is identically zero, while a Hilbert modular form of weight 0 must be a constant. The argument is analogous to that of Hecke for $SL(2, \mathbb{Z})$ given in Volume I.

Exercise 2.2.5. Show that mk must be an even integer if $\mathcal{M}(SL(2, \mathfrak{O}_K), k)$ is nonzero.

Now to discuss Fourier expansions of Hilbert modular forms, we need to recall the concept of different $\mathfrak{d}_K$ of a number field K (see Section 1.4 of Volume I). The inverse different is the dual lattice to the lattice $\mathfrak{O}_K$ of the number field (cf. pages 80–81 of Volume I). With the trace defined by (2.54), the duality is with respect to the form $\mathrm{Tr}(\alpha\beta)$, and $\mathfrak{d}^{-1} = \{\beta \in K \mid \mathrm{Tr}(\alpha\beta) \in \mathbb{Z} \text{ for all } \alpha \in \mathfrak{O}_K\}$. Fourier expansions are sums running over this dual lattice. Suppose now that f is in $\mathcal{M}\,(SL(2, \mathfrak{O}_K), k)$. Then $f(z)$ is periodic under translations from elements of the lattice $\mathfrak{O}_K$, since matrices

$$\begin{pmatrix} 1 & a \\ 0 & 1 \end{pmatrix} \quad \text{are in } SL(2, \mathfrak{O}_K) \text{ when } a \in \mathfrak{O}_K.$$

It follows that $f(z)$ has a **Fourier expansion at the infinite cusp** of the form:

$$f(z) = c(0) + \sum_{0 \ll b \in \mathfrak{d}_K^{-1}} c(b)\, \exp\{2\pi i \mathrm{Tr}\,(zb)\},$$

where, for $z \in \mathcal{H}^m$ and b a totally positive element of K, we define zb to be the element of $\mathcal{H}^m$ with jth coordinate $z^{(j)}b^{(j)}$. The sum is over b in the inverse different and such that $0 \ll b$, which means that b is **totally positive**; i.e., all conjugates $b^{(j)}$ are positive for $j = 1, \ldots, m$.

Fourier expansions at other cusps x_j can be described using the matrices A_j such that $A_j \infty = x_j$. One must also make use of the formula for the stabilizer of a cusp in Lemma 2.2.1.

Exercise 2.2.6. Show how to obtain the Fourier expansion at infinity of a Hilbert modular form $f(z)$, making use of the Cauchy–Riemann equations to show that the coefficients have the form $c(b) \exp\{-2\pi \operatorname{Tr}(yb)\}$, if $z = x + iy \in \mathcal{H}^m$.

One example of a Hilbert modular form is the **Eisenstein series** corresponding to an integral ideal $\mathfrak{a}$ in K defined by:

$$E_k(\mathfrak{a}, z) = \sum_{\substack{c, d/U_K \\ (c,d) = \mathfrak{a}}} N(cz + d)^{-k}, \quad k > 2.$$

The sum is over a complete system of representatives for pairs c, d which generate the ideal $\mathfrak{a} = (c, d)$ under the equivalence relation:

$$(c, d) \sim (uc, ud) \text{ for a unit } u \in U_K, \text{ with } Nu = +1.$$

The norm of $cz + d$ is defined by (2.53).

In fact, the Eisenstein series E_k will vanish identically if k is odd and K has a unit of norm -1. It is possible to use an integral test to obtain the convergence of E_k for $k > 2$ (see Siegel [563]). Moreover, Siegel [563, p. 292] proves the vanishing of the lead coefficient or constant term of the Fourier expansion of $E_k(\mathfrak{a}, z)$ with respect to the cusp corresponding to an ideal $\mathfrak{b}$ if $\mathfrak{b}$ is not in the same ideal class as $\mathfrak{a}$. This is to be expected when we recall that Proposition 2.2.1 gave the cusp-ideal class correspondence. Thus one demonstrates the linear independence of the Eisenstein series corresponding to ideals $\mathfrak{a}_1, \ldots, \mathfrak{a}_h$ representing the ideal classes in the ideal class group I_K.

Define a **cusp form** to be a Hilbert modular form f such that $f(z)$ approaches zero as z approaches any cusp of the fundamental domain. Let $\mathcal{S}(SL(2, \mathfrak{O}_K), k)$ be the **vector space of cusp forms of weight k** for $SL(2, \mathfrak{O}_K)$. Suppose that the ideals $\mathfrak{a}_1, \ldots, \mathfrak{a}_h$ represent the ideal classes in the ideal class group of K. Then we have the direct sum decomposition:

$$\mathcal{M}(SL(2, \mathfrak{O}_K), k) = \left(\sum_{i=1}^{h} {}^{\oplus} \mathbb{C} E_k(\mathfrak{a}_i, z) \right) \oplus \mathcal{S}(SL(2, \mathfrak{O}_K), k)$$

(see Siegel [563, p. 294]).

It is also possible to define Poincaré series as in Vol. I, Section 3.4.7. Look at the cusp p/s corresponding to the ideal $\mathfrak{a}$ via Proposition 2.2.1 and let $A \in SL(2, K)$ have the property that $A(\infty) = p/s$. Define a **Poincaré series** by:

$$f_k(\mathfrak{a}, \lambda, z) = \sum_{(c,d)=\mathfrak{a}} N(cz + d)^{-k} \exp\{2\pi i \operatorname{Tr}(\lambda \gamma z)\}.$$

Here λ is a totally positive element of the ideal $\mathfrak{a}^2 \mathfrak{d}_K^{-1}$, where $\mathfrak{d}_K$ is the different of K. The sum is over pairs of generators c, d of the ideal $\mathfrak{a}$ such that

$$\gamma = \begin{pmatrix} a & b \\ c & d \end{pmatrix} = A^{-1}\tau, \quad \text{for some } \tau \in SL(2, \mathfrak{D}_K), \quad \text{with } A(\infty) = p/q.$$

See Siegel [563, p. 230]). It can be shown that Poincaré series are cusp forms of weight k. Maass has proved that the Poincaré series and Eisenstein series generate the space $\mathcal{M}(SL(2, \mathfrak{D}_K), k)$, for $k \geq 2$. The Poincaré series can vanish identically, but not for large enough weights. See Siegel [563] for more information on the subject, also Bruinier's article in [82] and Freitag [186].

When $m = [K : \mathbb{Q}]$ is larger than one, a function $f(z)$ which is meromorphic on $\mathcal{H}^m$ is called a **Hilbert modular function** if $f(\gamma z) = f(z)$ for all $\gamma \in SL(2, \mathfrak{D}_K)$ and $z \in \mathcal{H}^m$. In the case that $m = 1$ and $K = \mathbb{Q}$, we would add a further requirement that $f(z)$ have at most a pole at the infinite cusp. This need not be assumed when $m \geq 2$. A Hilbert modular function which is holomorphic in $\mathcal{H}^m$ is automatically holomorphic at the cusps and thus must be a modular form of weight zero and therefore a constant. For there are no isolated singularities in several complex variables. Thus, when $m \geq 2$ there does not exist an analogue of the elliptic modular invariant $J(z)$ from Section 3.4.3 of Volume I. This fact was first noted by Götzky [239]. There are errors in Hecke's early papers due to the lack of knowledge of Götzky's result. These early Hecke papers seek to solve Hilbert's 12th problem which asks for an explicit construction of class fields (extension fields having abelian Galois groups) over arbitrary algebraic number fields using automorphic forms (see Hecke [268, p. 942]). Siegel [563] gives a proof that the Hilbert modular functions form an algebraic function field of n variables.

Herrmann [292] investigated the theory of Hecke operators for Hilbert modular forms. This theory has been extended to Picard modular groups by Styer [584]. Shimura [554] describes a very general theory of Hecke operators. Adelic versions of Hecke theory also exist for quite general groups (see Jacquet and Langlands [324], Gelbart [208], and Weil [660]). We leave it to the beleaguered reader to define the Petersson inner product of two cuspidal Hilbert modular forms of weight k and to obtain an analogue of Theorem 3.6.3 in Volume I for Hilbert modular forms.

The correspondence between Hilbert modular forms and Dirichlet series has been much studied (see the preceding references and Stark [575]). The situation is much like that investigated by Weil [662, Vol. III, pp. 165–172] for congruence subgroups of $SL(2, \mathbb{Z})$. One must have functional equations for L-functions that have been "twisted" by Hecke grossencharacters for K, in order to know that the corresponding function $f(z), z \in \mathcal{H}^m$, is a Hilbert modular form for $SL(2, \mathfrak{D}_K)$.

Let us just consider the simplest example of Hecke theory over number fields. Define the **theta function** corresponding to an ideal $\mathfrak{a}$ of a totally real number field K by:

$$\theta(\mathfrak{a}, z) = \sum_{a \in \mathfrak{a}} \exp\left\{-\pi \operatorname{Tr}\left(za^2\right)\right\}.$$

By slightly altering the proof of Theorem 1.4.2 in Volume I, we can view the ideal class zeta functions (which occur as partial sums of the Dedekind zeta function) of a totally real number field as Mellin transforms of this theta function (see Hecke [268, p. 227]). Similarly Hecke obtained the analytic continuation of his L-functions by showing them to be Mellin transforms of theta functions. Generalizations of this theta function are considered by Eichler [159], Kloosterman [356], and Schoeneberg [537]. These authors also look at the effect of Hecke operators on such theta functions.

If Γ is a subgroup of $SL(2, \mathfrak{O}_K)$ without elliptic fixed points, either the Selberg trace formula or the Hirzebruch–Riemann–Roch theorem can be used to compute the dimension of the space of Hilbert modular forms (see Ash et al. [30], Hirzebruch [296–298], Langlands [389], and Shimizu [553], as well as Section 2.2.3). If, for example, K is real quadratic, Γ of index a in $SL(2, \mathfrak{O}_K)/\pm I$, and Γ acts freely on $\mathcal{H}^2$, then, for $k \geq 3$, we have:

$$\dim \mathcal{S}(\Gamma, k) = \frac{k(k-2)}{2} \zeta_K(-1)a + \chi, \quad \chi = 1 + \dim \mathcal{S}(\Gamma, 2).$$

We will say more about the use of trace formulas to compute such dimensions at the end of this chapter. See also Freitag [186].

Example 2.2.6 (Holomorphic Siegel Modular Forms).

Some references for this section are Andrianov [9–14], Baily [32], Böcherer [54, 55], Hel Braun [74], Christian [108, 110], Eichler [157–159], Feit [175, 176], Freitag [185], Garrett [202], Hoobler and Resnikoff [304], Igusa [315, 316], Kaori Imai (Ota) [317], Kalinin [338], Karel [339], Klingen [350, 353–355], Maass [414, 419, 426], Morita [463], Resnikoff [507, 508], Shimura [556], Siegel [564, 565], Tsao [616], Van der Geer's article in Bruinier et al. [82], Weissauer [663, 664], and Yamazaki [673].

We will say that a function f on Siegel's upper half space $\mathcal{H}_n$, $n > 1$, is a holomorphic **Siegel modular form of weight** k and write $f \in \mathcal{M}(Sp(n, \mathbb{Z}), k)$ if it satisfies the following two conditions:

(1) f is holomorphic on $\mathcal{H}_n$;
(2) $f(\gamma Z) = |CZ + D|^k f(Z)$, for all

$$\gamma = \begin{pmatrix} A & B \\ C & D \end{pmatrix} \text{ in } Sp(n, \mathbb{Z}), \quad Z \in \mathcal{H}_n.$$

Some people say that $f \in \mathcal{M}(Sp(n, \mathbb{Z}), k)$ is a Siegel modular form of **genus** n, while others say f is a Siegel modular form of **degree** n. I have decided not to take sides, probably because of a memory of the time that I looked up the German word for genus in my dictionary.

There are generalizations to vector valued Siegel modular forms transforming according to a finite dimensional representation of $GL(n, \mathbb{R})$. See van der Geer's article in Bruinier et al. [82, p. 187].

One might expect to add a third condition that f must be bounded in the region $\operatorname{Im}Z = Y \geq Y_0 > 0$, where $Y \geq Y_0$ means that $Y - Y_0$ lies in the closure of $\mathcal{P}_n$. Koecher shows that this third condition is unnecessary when n is bigger than one (see Proposition 2.2.2 below and Maass [426, Section 13]).

Thus we have another analogue of the space of ordinary modular forms which was studied in Section 3.4 of Volume I. It can be shown that when k is larger than one, $\mathcal{M}(Sp(n, \mathbb{Z}), k) \neq 0$ implies $k \in \mathbb{Z}$ and $nk \in 2\mathbb{Z}$. It can also be proved that Siegel modular forms of negative weight must vanish while those of weight zero must be constant.

Exercise 2.2.7. Prove the last statements.
Hint. See Klingen [355].

Next we want to consider Fourier expansions of Siegel modular forms. First note that if X lies in the lattice of integral $n \times n$ symmetric matrices, then the dual lattice with respect to the form $\operatorname{Tr}(TX)$ consists of the $n \times n$ semi-integral symmetric matrices $T = (t_{ij})$. Here "semi-integral" means that $t_{jj} \in \mathbb{Z}$ and $t_{ij} \in \frac{1}{2}\mathbb{Z}$, when $i \neq j$.

If f is a Siegel modular form of weight k, then $f(X + iY)$ has period one in each entry of the symmetric matrix X. This implies that f has a **Fourier expansion**:

$$f(Z) = \sum_{\substack{0 \leq T = {}^t T \\ T \text{ semi-integral}}} a(T) \exp\{2\pi i \operatorname{Tr}(TZ)\}. \tag{2.55}$$

Here $T \geq 0$ means that $T[x] \geq 0$ for all $x \in \mathbb{R}^n$.

To see that the Fourier coefficient has the form $a(T) \exp\{2\pi i \operatorname{Tr}(TZ)\}$, one must use the fact $f(Z)$ satisfies the Cauchy–Riemann equations in each variable. The sum is over semi-integral matrices because that is the dual lattice to $\mathbb{Z}^{n \times n}$ with respect to the form $\operatorname{Tr}(TX)$.

There is also an expansion known as the Fourier–Jacobi expansion of Piatetski-Shapiro which has proved to be useful. See van der Geer's article in Bruinier et al. [82, p. 196] and Freitag [184, p. 101].

Exercise 2.2.8. Prove what we just said about (2.55)
Hint. The sum is over nonnegative matrices T by Lemma 2.2.2 below.

Lemma 2.2.2. *Let $f \in \mathcal{M}(Sp(n, \mathbb{Z}), k)$ have Fourier coefficients $a(T)$ as in (2.55) above. Then $f(Z)$ is bounded in every domain $Y \geq Y_0 > 0$ (for fixed Y_0) if and only if the Fourier coefficient $a(T) \neq 0$ implies that $T \geq 0$.*

Proof. See Maass [426, pp. 183–184].
$\Rightarrow$ Suppose that $|f(Z)| \leq C(Y_0)$ in $Y \geq Y_0 > 0$. Then

$$a(T) \exp\{-2\pi \operatorname{Tr}(TY)\} = \int_{{}^tX = X \in [0,1]^{n \times n}} f(X + iY) \exp\{-2\pi i \operatorname{Tr}(TX)\} \, dX.$$

In order to show that $a(T) \neq 0$ implies that $T \geq 0$, use the above equality to obtain the bound:

$$|a(T)| \leq C(Y_0) \exp\{2\pi \operatorname{Tr}(TY)\}.$$

If T is not ≥ 0, then one can show that the right-hand side of this inequality can be made arbitrarily small. Thus $a(T)$ must vanish when T is not ≥ 0.
$\Leftarrow$ Suppose the Fourier expansion of $f(Z)$ has the form

$$f(Z) = \sum_{\substack{{}^tT = T \geq 0 \\ \text{semi-integral}}} a(T) \exp\{2\pi i \operatorname{Tr}(TZ)\}.$$

The convergence of the Fourier series implies that if $Y \geq Y_0 \geq aI > 0$

$$|f(Z)| \leq C\left(\frac{1}{2}Y_0\right) \sum_{T \geq 0} \exp\{-\pi a \operatorname{Tr}(T)\}. \tag{2.56}$$

The series on the right in (2.56) is easily seen to converge. ∎

Exercise 2.2.9. Fill in the details in the proof of Lemma 2.2.2 above.

Proposition 2.2.2 (Koecher). *If $f \in \mathcal{M}(Sp(n, \mathbb{Z}), k)$, then f is bounded in any domain $Y \geq Y_0 > 0$. In particular, f is bounded in the fundamental domain $\mathcal{D}_n^S$ for $Sp(n, \mathbb{Z})$, which is defined in formula (2.52).*

Proof. See Maass [426, pp. 185–187]. For $n = 1$, the result is part of the definition of modular form. When n is larger than one, consider the Fourier expansion (2.55) of f. Note that $a(T[U]) = a(T)$ for all $U \in GL(n, \mathbb{Z})$. The main step in the proof of Proposition 2.2.2 is the proof of the claim that when T is not ≥ 0, then the number

$$c(T, -m) = \#\{T[U] \mid U \in SL(n, \mathbb{Z}), \operatorname{Tr}(T[U]) = -m\}$$

is greater than or equal to one for infinitely many $m \geq 1$. Therefore $a(T)$ must vanish in this case.

To prove this claim, observe that, upon setting $U = I + b\,{}^td$, with $b, d \in \mathbb{Z}^n$, such that ${}^tbd = 0$, we have

$$|U| = 1 \quad \text{and} \quad \operatorname{Tr}(T[U]) = \operatorname{Tr}(T) + 2\operatorname{Tr}\left(T\,b\,{}^td\right) + T[b]\left({}^tdd\right).$$

If T were not ≥ 0, then we could choose $b \in \mathbb{Z}^n$ so that $T[b] < 0$ and $\,^t bd = 0$, for any $d \in \mathbb{Z}^n$. And then $c(T, -m)$ must indeed be ≥ 1 for infinitely many m. The proof of Proposition 2.2.2 is completed using Lemma 2.2.2. ∎

An example of a Siegel modular form of even weight k in $\mathcal{M}(Sp(n, \mathbb{Z}), k)$ is given by the **Eisenstein series**:

$$E_k(Z) = \sum_{C,D} |CZ + D|^{-k}, \text{ for even } k > n + 1,$$

where the sum is over coprime symmetric pairs C, D of matrices in $\mathbb{Z}^{n \times n}$ modulo the equivalence relation

$$(C \; D) \sim (UC \; UD), \text{ for } U \in GL(n, \mathbb{Z}).$$

Hel Braun [74] proved the convergence of the Eisenstein series in the stated region. Hel Braun was a student of Siegel. She was one of the few women mathematics professors in Germany during her life. I had the opportunity to meet her at an Oberwohlfach meeting. She had much to say about Siegel. It would be interesting to know more of her life. She only discusses the beginnings in her book [77].

See Freitag [185, pp. 66–67] for a convergence proof using a sort of integral test. Freitag [185, p. 67] and Maass [426, Section 14] consider more general Eisenstein series involving modular forms for $Sp(r, \mathbb{Z})$, $r < n$, which were introduced by Klingen [353]. See Van der Geer's article in Bruinier et al. [82, p. 194] for a theorem of Hel Braun on the Klingen Eisenstein series and their behavior under the Siegel Φ-operator defined below in equation (2.57). Another reference is Klingen [355, p. 68], who notes [355, p. 63] that "There is no reasonable way to introduce Eisenstein series for odd weights."

Siegel [565, Vol. II, p. 133] gives the Fourier expansion of the Eisenstein series E_k (k even and $> n + 1$) for $Sp(n, \mathbb{Z})$ (see also Baily [32]). Let

$$T[U] = \begin{pmatrix} T_1 & 0 \\ 0 & 0 \end{pmatrix} \text{ for } U \in GL(n, \mathbb{Z}), \; T_1 \in \mathbb{Z}^{r \times r},$$

and set $D(T) = |T_1|$. The term corresponding to T in the Fourier expansion of $E_k(Z)$ is:

$$(-1)^{rk/2} 2^{r(k-(r-1)/2)} \prod_{j=0}^{r-1} \frac{\pi^{k-j/2}}{\Gamma(k - j/2)} D(T)^{k-(r+1)/2} \sum_{\substack{^t R = R \in (\mathbb{Q}/\mathbb{Z})^{r \times r}}} e^{2\pi i \mathrm{Tr}(T_1 R)} \nu(R)^{-k}.$$

Here $\nu(R)$ is the product of the reduced denominators of the elementary divisors of R. An analogous result for Eisenstein series for $GL(n, \mathbb{Z})$ was discussed in Exercise 1.5.36 of Section 1.5.3 (see also Terras [596]). Siegel used his main theorem on quadratic forms to deduce the rationality of the coefficients of the

Eisenstein series E_k for $Sp(n, \mathbb{Z})$. Baily [32, Ch. 12] gives another derivation. Similar and more general results of this type are obtained by Karel [339] and Tsao [616]. Kaufhold [343] finds even more explicit results for Fourier coefficients of Eisenstein series for $Sp(2, \mathbb{Z})$. When the Eisenstein series have rational coefficients and the Eisenstein series generate the full field of automorphic or Siegel modular functions (i.e., meromorphic functions satisfying condition (2) in the definition of Siegel modular form and having weight $k = 0$) for $Sp(n, \mathbb{Z})$, then the algebraic variety which is the Satake compactification of the fundamental domain $Sp(n, \mathbb{Z}) \backslash \mathcal{H}_n$ is a variety defined over $\mathbb{Q}$ (see Baily [32, p. 238]). More information on Eisenstein series can be found in the references mentioned at the beginning of this discussion of holomorphic Siegel modular forms. Some of the references consider non-holomorphic Eisenstein series as well.

An example of a modular form for a congruence subgroup of $Sp(n, \mathbb{Z})$ is the **theta function** defined for $Q \in \mathcal{P}_n \cap \mathbb{Z}^{n \times n}$, $Z \in \mathcal{H}_n$, by:

$$\theta(Z) = \sum_{A \in \mathbb{Z}^{k \times k}} \exp \{\pi i \mathrm{Tr} \, (ZQ\,[A])\} .$$

The theta function in formula (1.172) of Section 1.4.1. is simply a restriction of this symplectic theta function to $Z = iY$, $Y \in \mathcal{P}_n$. Eichler [156] shows that in the special case $n = 1$ the theta function is a modular form for a congruence subgroup of $SL(2, \mathbb{Z})$ by considering it as a special value of a theta function on $\mathcal{H}_k$. Andrianov and Maloletkin [15] generalize this result to any n—evaluating the 8th root of unity involved when k is even. Stark [576] extends these results further by evaluating the 8th root of unity in a case that can be reached by theorems on matrix primes in progressions (see also Styer [584]). Other references for theta functions are Freitag [185] and Igusa [316].

Theta functions and zeta functions for indefinite quadratic forms have been studied by many authors (see Koecher [360], Maass [418, 423, 425], Siegel [565, Vol. I, pp. 410–443; Vol. II, pp. 41–96, 421–466; Vol. III, pp. 85–91, 105–142, 154–177], Andrianov and Maloletkin [16], and Friedberg [190]).

It is possible to obtain examples of modular forms of various types by integrating against theta functions of indefinite quadratic forms. For example, one can obtain holomorphic Hilbert modular forms for a real quadratic field by integrating ordinary holomorphic modular forms for $SL(2, \mathbb{Z})$ multiplied by an appropriate theta function. One can similarly lift Maass wave forms for $SL(2, \mathbb{Z})$. And one can replace real by imaginary quadratic fields, etc. References for such constructions, often referred to as "base change," include Friedberg [190], Goldfeld [230], Kudla [378, 379], Stark [579], Tsuyumine [618], Marie-France Vignéras [634], and Waldspurger [641].

Theta functions have many other applications. There are, in fact, entire books devoted to them (e.g., Igusa [316] and Mumford [471]). Siegel's main theorem on quadratic forms can be viewed as an equality between linear combinations of theta functions and generalized Eisenstein series (see Siegel [565, Vol. I, pp. 326–405, 410–443, 469–548]). Theta functions on the Siegel upper half space can be used to obtain an expression for the generalization of elliptic integrals known as abelian integrals (see Siegel [564]). This leads to many applications in physics. We have already mentioned the work of Sofya Kovalevskaya [368] and Dubrovin et al. [143]. Other references for applications to the Korteweg–deVries equation are McKean and Trubowitz [440] and Novikov [474]. Siegel's work on quadratic forms has been connected with quantum mechanics and representation theory via the Segal–Shale–Weil representation (see Cartier's talk in Borel and Mostow [68, pp. 361–368], Gérard Lion and Michèle Vergne [406], Shale [551], Weil [662, Vol. III, pp. 1–157], and Wallach [652]). The role of theta functions in algebraic geometry and purely algebraic constructions of these functions are discussed in Mumford [471] and Shafarevitch [549]. Connections with Jordan algebras are pursued by Resnikoff [508]. Connections with knot theory and physics are to be found in Gelca [211].

In order to discuss cusp forms, Siegel defined the **Φ-operator** taking $f \in \mathcal{M}(Sp(n, \mathbb{Z}), k)$ to $f | \Phi \in \mathcal{M}(Sp(n-1, \mathbb{Z}), k)$ by:

$$f | \Phi(W) = \lim_{t \to \infty} f \begin{pmatrix} W & 0 \\ 0 & it \end{pmatrix}, \text{ for } W \in \mathcal{H}_{n-1}. \tag{2.57}$$

A Siegel modular form $f \in \mathcal{M}(Sp(n, \mathbb{Z}), k)$ is said to be a **cusp form** if $f | \Phi$ vanishes identically. Let $\mathcal{S}(Sp(n, \mathbb{Z}), k)$ denote the **space of Siegel cusp forms of weight k**. The Fourier expansion (2.55) of $f \in \mathcal{S}(Sp(n, \mathbb{Z}), k)$ can have no terms corresponding to singular symmetric semi-integral matrices T. Here a singular T is one with determinant equal to zero.

Exercise 2.2.10. Prove the last statement about cusp forms.
Hint. See Klingen [355].

There is an opposite concept to that of cusp form—the concept of **singular form** which is a form in $\mathcal{M}(Sp(n, \mathbb{Z}), k)$ whose Fourier coefficients $a(T)$ in (2.55) vanish unless the T are singular matrices. Certain theta functions give examples. An integer $k \geq 0$ is said to be a **singular weight** for $Sp(n, \mathbb{Z})$ if $k < n/2$. Maass [426] showed singular forms have singular weights. Resnikoff [507] and Freitag [184] proved the converse. See Klingen [355, Chapter 8].

Let us say a bit about dimensions of spaces of Siegel modular forms for $Sp(n, \mathbb{Z})$. Christian showed that for nonvanishing forms to exist, the weights must be integers. See Klingen [355, p. 43]. Moreover the weights must be nonnegative. See Klingen [355, p. 47]. Finally any form of weight 0 is constant. See Klingen [355, p. 49].

It can be proved that there is a positive constant c_n depending only on n such that

$$\dim \mathcal{M}(Sp(n, \mathbb{Z}), k) \leq c_n k^{n(n+1)/2}.$$

See Eicher [157, 158], Freitag [185, p. 52], Klingen [355, p. 51], Maass [426, p. 194], and Siegel [565, Vol. II, pp. 97–137]. The main principle needed to prove such an inequality is that which gave us Theorem 3.5.4 in Volume I. This principle says: The vanishing of sufficiently many terms in the Fourier series of $f \in \mathcal{M}(Sp(n, \mathbb{Z}), k)$ implies the vanishing of f itself. Freitag [185, p. 50] shows that $\mathcal{S}(Sp(n, \mathbb{Z}), k)$ vanishes in the following situations:

$n = 1$	$k < 12$
$n = 2$	$k < 9$
$n = 3$	$k < 8$
$n = 4$	$k < 5$

See the very last pages of this volume for more information.

It is possible to use the Selberg trace formula or the Hirzebruch–Riemann–Roch theorem to give formulas for dimensions of spaces of Siegel cusp forms for congruence subgroups of $Sp(n, \mathbb{Z})$ acting without elliptic fixed points. See Arakawa [19], Christian [109], Hirzebruch [297, Appendix], Langlands [389], Morita [463], Petra Ploch [489], and Yamazaki [673]. See also Eie [160] and Hashimoto [264] for the case of $Sp(n, \mathbb{Z})$ for small n. Let us examine one such computation—that of Arakawa [19] using the Selberg trace formula. One begins with the dimension formula of Godement which writes the dimension of the space of cusp forms as an integral of a sum over Γ. The identity and parabolic elements of Γ produce the only nonzero contributions to the trace formula. Work of Shintani [559] is needed to compute special values of zeta functions arising in the calculations. See the last section of this book for a few more details and references.

The spaces $\mathcal{M}(Sp(2, \mathbb{Z}), k)$ were completely determined by Igusa [315]. See also Freitag [183].

The **Petersson inner product** for weight k Siegel modular forms f, g for $\Gamma = Sp(n, \mathbb{Z})$, at least one of which is a cusp form, is defined by:

$$\langle f, g \rangle = \int_{\Gamma \backslash \mathcal{H}_n} f(Z)\overline{g(Z)} \det\left(\mathrm{Im}\,(Z)\right)^k d\mu_n^*,$$

where $d\mu_n^*$ is the invariant volume element on $\mathcal{H}_n$.

Hecke operators for $Sp(n, \mathbb{Z})$ were first systematically investigated by Maass [414]. Let m be a positive integer and J_n as in the definition of $Sp(n, \mathbb{Z})$, define

$$M_n = \left\{ g \in \mathbb{Z}^{2n \times 2n} \mid {}^t g J_n g = m J_n \right\}.$$

Then we will characterize the **mth Hecke operator** $T(m)$ by what it does to a form $f \in \mathcal{M}(Sp(n, \mathbb{Z}), k)$, namely:

$$T(m)f(Z) = m^{nk - n(n+1)/2} \sum_{\gamma = \begin{pmatrix} * & * \\ C & D \end{pmatrix} \in Sp(n, \mathbb{Z}) \backslash M_n} |CZ + D|^{-k} f(\gamma Z).$$

It is possible to show that symplectic Hecke operators have similar properties to those of Hecke operators for $GL(n, \mathbb{Z})$ which were obtained in Theorem 1.5.2 of Section 1.5.2. The Euler products involved are more complicated though. Some other references for these Hecke operators are Andrianov [9–14], Freitag [185], and Shimura [554].

There are many sorts of **Dirichlet series or L-functions associated with Siegel modular forms** $f \in \mathcal{M}(Sp(n, \mathbb{Z}), k)$. For simplicity, let us assume that f is a cusp form. Then, given a Maass form v for $GL(n, \mathbb{Z})$ in $\mathcal{A}(GL(n, \mathbb{Z}), \lambda)$ as in Section 1.5.1, we can consider the following Mellin transform:

$$M(f, v) = \int_{Y \in \mathcal{P}_n / GL(n, \mathbb{Z})} f(iY) v(Y) \, d\mu_n(Y).$$

Suppose that $f(Z)$ has the Fourier expansion (2.55) above and that

$$v(Y) = |Y|^s u(Y^o), \quad \text{for } Y^o = |Y|^{-1/n} Y \in \mathcal{SP}_n, \quad s \in \mathbb{C}.$$

Set

$$u^*(W) = u(W^{-1}).$$

As in the proof of part (5) of Theorem 1.5.2 of Section 1.5.2, we have:

$$M(f, v) = \sum_{T>0} \int_{Y \in \mathcal{P}_n / GL(n, \mathbb{Z})} a(T) \exp\{-2\pi \operatorname{Tr}(TY)\} v(Y) \, d\mu_n(Y)$$

$$= (2\pi)^{-s} \Gamma_n(r(u, s)) \sum_{0 < T / GL(n, \mathbb{Z})} a(T) |T|^{-s} u^*(T),$$

for some $r(u, s) \in \mathbb{C}^n$. Thus it is natural to associate to $f \in \mathcal{S}(Sp(n, \mathbb{Z}), k)$ with Fourier expansion (2.55) having Fourier coefficients $a(T)$, the Dirichlet series:

$$L(f, v) = \sum_{\substack{0 < T / GL(n, \mathbb{Z}) \\ T \text{ symmetric, semi-integral}}} a(T) v\left(T^{-1}\right).$$

Maass [426, Section 15] obtains the analytic continuation and functional equation of $L(f, v)$, even when f is not a cusp form. See the proof of part (5) of Theorem 1.5.2 of Section 1.5.2 for a discussion of a similar analytic continuation.

Harmonic analysis on $\mathcal{P}_n / GL(n, \mathbb{Z})$ gives a converse to this Hecke correspondence between f and $L(f, v)$, as Kaori Imai (Ota) [317] shows when $n = 2$. Weissauer [663] obtains a converse result for congruence subgroups of $Sp(n, \mathbb{Z})$, for general n. It would be nice not to have to know that there are Dirichlet series with functional equations for all the v in $\mathcal{A}(GL(n, \mathbb{Z}), \lambda)$. Just how many such v are necessary is a very interesting question. Similar questions exist for Weil's theory of the Hecke correspondence for congruence subgroups of $SL(2,\mathbb{Z})$ (see Section 3.6 of Volume I).

Andrianov [10–14] investigates various sorts of Dirichlet series associated with eigenfunctions of Hecke operators. The language of adelic representation theory leads to the same sort of results. See Piatetski-Shapiro's talk in Borel and Casselman [66, Vol. 1, pp. 185–188]. See also Piatetski-Shapiro [486]. Tamara Veenstra [626] investigates L-functions corresponding to Siegel eigenforms of Hecke operators and shows that the p-factors for almost all primes determine the L-function.

Bounds on Fourier coefficients of Siegel cusp forms are investigated by Kathrin Bringmann [80]. The explicit action of the standard generators of the Hecke algebra on Fourier coefficients of Siegel modular forms of half-integral weight is studied by Lynne Walling [654].

Very general Poincaré series for $\Gamma_n = Sp(n, \mathbb{Z})$ are considered by Klingen [355, Ch. 6]. One example which was introduced by Maass is:

$$g_n^k(Z, t) = \sum_{\Upsilon_n \backslash \Gamma_n} |CZ + D|^{-k} \exp\left(t\,(Az + B)\,(Ca + D)^{-1}\right), \qquad (2.58)$$

for $k > 2n$, $kn \equiv 0 (\mathrm{mod}\,2)$, $Z \in \mathcal{H}_n$, and half integral positive t. Here Υ_n is the subgroup of matrices of the form $\begin{pmatrix} \pm I_n & * \\ 0 & \pm I_n \end{pmatrix}$. Klingen [355, p. 90] notes the main properties of the Maass Poincaré series; the most important being

(1) The Petersson inner product $\langle f, g_n^k(*, t) \rangle$ pulls out the tth Fourier coefficient of a Siegel cusp form f of weight k multiplied by a constant and a power of t.
(2) The Maass Poincaré series (2.58) span the space $\mathcal{S}\,(Sp(n, \mathbb{Z}), k)$ for a finite set of values of t.

Example 2.2.7 (Eisenstein Series for $GL(2, \mathfrak{O}_K)$).

References for this subject include: Asai [27], Efrat and Sarnak [154], Elstrodt et al. [168], Fueter [191], Gelbart [208], Grosswald [250], Hecke [268], Hoffstein [303], Hunter [311], Jacquet and Langlands [324], Mennicke [442], Mordell [460], Ramanathan [497, 498], Sarnak [527], Siegel [565, Vol. I, pp. 173–179], Stark [579], Tamagawa [587], Terras [597–600], and Weil [660].

Let K be an algebraic number field of degree m over $\mathbb{Q}$. We use the notation set up earlier during the discussion of the fundamental domain for $GL(n, \mathfrak{O}_K)$. The **Epstein zeta function** for $GL(n, \mathfrak{O}_K)$ is defined for $Y \in \mathcal{P}_2^K$, $\mathfrak{a}$ an ideal of K, and s a complex number with $\mathrm{Re}\,s > 1$, by:

$$Z(\mathfrak{a}, Y, s) = \sum_{0 \neq b \in \mathfrak{a}^2 / U_K} N\,(Y\,\{b\})^{-s}. \qquad (2.59)$$

Here the **norm** $N\,(Y\,\{b\})$ is defined by

$$N\left(Y\{b\}\right) = \prod_{j=1}^{r_1+r_2} \left(\overline{{}^tb^{(j)}}Y^{(j)}b^{(j)}\right)^{e_j}, \text{ for } e_j = \begin{cases} 1, j = 1, \ldots, r_1, \\ 2, j = r_1 + 1, \ldots, r_1 + r_2. \end{cases}$$

(2.60)

The sum in (2.59) is over a complete system of nonzero column vectors in $\mathfrak{a}^2$ inequivalent under the equivalence relation

$${}^tb = (b_1, b_2) \sim (b_1u, b_2u), \text{ for } u \in U_K = \text{the group of units of } \mathfrak{O}_K.$$

In order to prove the convergence of Epstein's zeta function (2.59), one could devise an integral test similar to that used in the case that K is the field of rational numbers (see Corollary 1.4.4 in Section 1.4.4). Related methods are used by Siegel [563, p. 290] and Godement in Borel and Mostow [68, p. 207]. It is also possible to deduce the convergence from bounds on theta functions as in Ramanathan [498, p. 54].

Exercise 2.2.11. Obtain the analytic continuation and functional equation of Epstein's zeta function for $GL(2, \mathfrak{O}_K)$ in (2.59) by imitating Riemann's proof of the analytic continuation of $\zeta(s)$ given in Section 1.4 of Volume I. See also Hecke's proof of the analytic continuation of the Dedekind zeta function in Lang [386, pp. 255–258]. You will find that $Z(\mathfrak{a}, Y, s)$ has a simple pole at $s = 1$ and a functional equation relating it to $Z(\mathfrak{a}', Y^{-1}, 1 - s)$, where $\mathfrak{a}' = (\mathfrak{a}\mathfrak{d}_K)^{-1}$, if $\mathfrak{d}_K$ is the different of K.

We have a **simultaneous Iwasawa decomposition** of the vector $Y \in \mathcal{P}_2^K$

$$Y = \begin{pmatrix} v & 0 \\ 0 & w \end{pmatrix} \begin{bmatrix} 1 & q \\ 0 & 1 \end{bmatrix}, \text{ with } v, w \in Y \in \mathcal{P}_1^K, \ q \in \mathbb{R} \otimes_{\mathbb{Q}} K. \tag{2.61}$$

This means that for $1 \le j \le r_1 + r_2$,

$$Y^{(j)} = \begin{pmatrix} v^{(j)} & 0 \\ 0 & w^{(j)} \end{pmatrix} \begin{bmatrix} 1 & q^{(j)} \\ 0 & 1 \end{bmatrix}, \text{ with } v^{(j)}, w^{(j)} > 0,$$

$$q^{(j)} \in E_j, \quad E_j = \begin{cases} \mathbb{R}, j = 1, \ldots r_1, \\ \mathbb{C}, j = r_1 + 1, \ldots, r_1 + r_2. \end{cases}$$

Clearly the Epstein zeta function $Z(\mathfrak{a}, Y, s)$ has the **invariance property**:

$$Z(\mathfrak{a}, Y, s) = Z(\mathfrak{a}, Y\{\gamma\}, s) \text{ for all } \gamma \in GL(2, \mathfrak{O}_K), Y \in \mathcal{P}_2^K.$$

It follows that if we view $Z(\mathfrak{a}, Y, s)$ as a function of the q-variable in the Iwasawa decomposition (2.61) of Y, we are looking at a function that is periodic modulo $\mathfrak{O}_K$. Thus we can obtain a Fourier expansion of $Z(\mathfrak{a}, Y, s)$ in the q-variable.

Let us eliminate the dependence on the ideal $\mathfrak{a}$ by defining a new zeta summed over classes C in the ideal class group I_K:

$$Z^*(Y, s) = \sum_{C \in I_k} N\mathfrak{b}^{2s} Z(\mathfrak{b}, Y, s), \quad \text{for} \quad \mathfrak{b} \in C. \tag{2.62}$$

Here $N\mathfrak{b}$ denotes the norm of the ideal $\mathfrak{b}$, which is $\#(\mathfrak{O}_K/\mathfrak{b})$. Note that it does not matter what ideal $\mathfrak{b} \in C$ is chosen. Because the ideal class group I_K is finite, so is the sum in (2.62) provided that $s \neq 1$.
    Set

$$A = 2^{-r_2} \pi^{-m/2} D_K^{1/2}, \quad D_K = \text{the absolute value of the discriminant of } K,$$

and define **Dedekind's zeta function** by

$$\zeta_K(s) = \sum_{\mathfrak{b} \subset \mathfrak{O}_K} N\mathfrak{b}^{-s}, \quad \text{for} \quad \text{Re } s > 1,$$

where the sum is over ideals $\mathfrak{b}$ of $\mathfrak{O}_K$. As proved in Section 1.4 of Volume I, the functional equation of Dedekind's zeta function is:

$$\Lambda_K(s) = A^s \Gamma\left(\frac{s}{2}\right)^{r_1} \Gamma(s)^{r_2} \zeta_K(s) = \Lambda_K(1-s). \tag{2.63}$$

Motivated by this functional equation, we define:

$$\Lambda^*(Y, s) = A^{2s} \Gamma(s)^{r_1} \Gamma(2s)^{r_2} Z^*(Y, s).$$

Theorem 2.2.1 (Fourier Expansion of Epstein's Zeta Function for K). *Using the notation of formulas (2.59)–(2.63), we have the Fourier expansion:*

$$\Lambda^*(Y, s) = Nv^{-s}\Lambda_K(2s) + Nv^{-\frac{1}{2}} Nw^{\frac{1}{2}-s}\Lambda_K(2s - 1)$$

$$+ \frac{2^{r_1+r_2} D_K^{s-\frac{1}{2}}}{Nv^{\frac{1}{2}} N|Y|^{-\frac{1}{4}+s/2}} \sum_{0 \neq u \in \mathfrak{d}_K^{-1}} |Nu|^{s-\frac{1}{2}} \sigma_{1-2s}(u\mathfrak{d}_K) e^{2\pi i \text{Tr}(qu)}$$

$$\times \prod_{j=1}^{r_1+r_2} K_{e_j(s-\frac{1}{2})}\left(2\pi e_j \sqrt{\left(\frac{wu^2}{v}\right)^{(j)}}\right),$$

*where $K_s(y)$ is the ordinary K-Bessel function and for any ideal $\mathfrak{b} \subset \mathfrak{O}_K$ the function $\sigma_s(\mathfrak{b})$ is the **divisor function**:*

$$\sigma_s(\mathfrak{b}) = \sum_{\mathfrak{c}|\mathfrak{b}} N\mathfrak{c}^s.$$

Here $\mathfrak{c}|\mathfrak{b}$ is equivalent to $\mathfrak{c} \supset \mathfrak{b}$.

Proof. See Terras [598, 599]. The idea is to generalize Exercise 3.5.4 in Volume I. Set

$$\Lambda(\mathfrak{a}, Y, s) = A^{2s}\Gamma(s)^{r_1}\Gamma(2s)^{r_2}Z(\mathfrak{a}, Y, s).$$

Then, if Y has the Iwasawa decomposition (2.61), it follows that:

$$Y\left\{\begin{matrix} a \\ b \end{matrix}\right\} = v\{a + qb\} + w\{b\}.$$

Thus

$$\Lambda(\mathfrak{a}, Y, s) = Nv^{-s}\Lambda_K(\mathfrak{a}, 2s) + A^{2s}\Gamma(s)^{r_1}\Gamma(2s)^{r_2}\sum_{\substack{0\neq b\in\mathfrak{a}/U_K \\ a\in\mathfrak{a}}} N\left(v\{a+qb\} + w\{b\}\right)^{-s},$$

if $\mathrm{Re}\,s > 1$ and we define

$$\Lambda_K(\mathfrak{a}, 2s) = A^{2s}\Gamma(s)^{r_1}\Gamma(2s)^{r_2}\sum_{0\neq b\in\mathfrak{a}/U_K} |Nb|^{-2s}.$$

The Poisson summation formula from Section 1.3 of Volume I or Weil [661, p. 106] shows that the sum over $a \in \mathfrak{a}$ equals the sum of Fourier transforms over $c \in \mathfrak{a}' = (\mathfrak{a}\partial_K)^{-1}$ which is the dual ideal to $\mathfrak{a}$. The Fourier transforms here are:

$$\widehat{f}(b, c) = \int_{x\in K\otimes_\mathbb{Q}\mathbb{R}} N\left(v\{a + qb\} + w\{b\}\right)^{-s}\exp\left(-2\pi i\mathrm{Tr}\left({}^t cx\right)\right)\,d\mu(x),$$

where the measure $d\mu(x)$ is chosen so that

$$\int_{x\in K\otimes_\mathbb{Q}\mathbb{R}/\mathfrak{a}} d\mu(x) = 1.$$

Now the ideal $\mathfrak{a}$ has an **integral basis**, i.e.,

$$\mathfrak{a} = \sum_{j=1}^{m}{}^{\oplus} \mathbb{Z}w_j \quad\text{and}\quad K\otimes_\mathbb{Q}\mathbb{R} = \sum_{j=1}^{m}{}^{\oplus}\mathbb{R}w_j.$$

So if

$$x = \sum_{j=1}^{m}x_jw_j, \quad x_j \in \mathbb{R},$$

we can take

$$d\mu(x) = \prod_{j=1}^{m} dx_j \quad \text{with} \quad dx_j = \text{Lebesgue measure on } \mathbb{R}.$$

We can also see that

$$K \otimes_{\mathbb{Q}} \mathbb{R} \cong \sum_{j=1}^{r_1+r_2} \oplus\, E_j, \quad \text{where} \quad E_j = \begin{cases} \mathbb{R}, j = 1, \ldots r_1, \\ \mathbb{C}, j = r_1 + 1, \ldots, r_1 + r_2. \end{cases}$$

Therefore we can define the mapping:

$$T : K \otimes_{\mathbb{Q}} \mathbb{R} \to \sum_{j=1}^{r_1+r_2} \oplus\, E_j \text{ by } \quad T\left(\sum_{j=1}^{m} x_j w_j\right) = y = \left(y^{(1)}, \ldots, y^{(r_1+r_2)}\right),$$

$$\text{where} \quad y^{(i)} = \sum_{j=1}^{m} x_j w_j^{(i)}.$$

(2.64)

The Jacobian of the map T is (**Exercise**)

$$\left|\frac{\partial y}{\partial x}\right| = \left|\frac{\partial\left(y_1, \ldots, y_{r_1}, \text{Re}\, y_{r_1+1}, \text{Im}\, y_{r_1+1}, \ldots, \text{Re}\, y_{r_1+r_2}, \text{Im}\, y_{r_1+r_2}\right)}{\partial\left(x_1, \ldots, x_m\right)}\right| = 2^{-r_2} D_K^{\frac{1}{2}} N\mathfrak{a}.$$

Therefore

$$\widehat{f}(b, c) = D_K^{-\frac{1}{2}} N\mathfrak{a}^{-1} 2^{r_2} \prod_{j=1}^{r_1+r_2} \int_{E_j} N_{E_j/\mathbb{R}}\left(\left(v\,\{a + qb\} + w\,\{b\}\right)^{(j)}\right)^{-s}$$

$$\times \exp\left(-2\pi i\, \text{Tr}_{E_j/\mathbb{R}}\left((cy)^{(j)}\right)\right)\, dy^{(j)}.$$

Make the change of variables $x_j = \left((w\,\{b\})^{-\frac{1}{2}} t\,(y + qb)\right)^{(j)}$, where $v = t^2$, $t^{(j)} > 0$, to see that:

$$\widehat{f}(b, c) = D_K^{-\frac{1}{2}} N\mathfrak{a}^{-1} 2^{r_2}\, Nv^{-\frac{1}{2}} N\,(w\,\{b\})^{\frac{1}{2}-s}\, \exp\left(2\pi i\, \text{Tr}\,(cqb)\right)$$

$$\times \prod_{j=1}^{r_1+r_2} \int_{E_j} \left(1 + \overline{x}_j x_j\right)^{-se_j} \exp\left(-2\pi i\, \text{Tr}_{E_j/\mathbb{R}}\left(\left(\frac{c}{t}\,(w\,\{b\})^{\frac{1}{2}}\right)^{(j)} x_j\right)\right)\, dx_j.$$

Define

$$I_j(a, s) = \int_{E_j} \left(1 + \overline{x}x\right)^{-se_j} \exp\left(-2\pi i\, \text{Tr}_{E_j/\mathbb{R}}\,(ax)\right)\, dx.$$

By part (a) of Exercise 3.2.1 in Volume I, we find that for $j = 1, \ldots, r_1$:

$$I_j(a, s) = \begin{cases} 2\pi^{\frac{1}{2}} \Gamma(s)^{-1} |\pi a|^{s-\frac{1}{2}} K_{s-\frac{1}{2}} (2\pi |a|), & a \neq 0, \\ \Gamma\left(\frac{1}{2}\right) \Gamma\left(s - \frac{1}{2}\right) \Gamma(s)^{-1}, & a = 0. \end{cases}$$

For $j = r_1 + 1, \ldots, r_1 + r_2$, we must compute:

$$I_j(a_1 + ia_2, s) = \int\limits_{x_1 + ix_2 \in \mathbb{C}} \left(1 + x_1^2 + x_2^2\right)^{-2s} \exp\left(-4\pi i \left(a_1 x_1 - a_2 x_2\right)\right) dx$$

$$= k_{1,2} \left(2s | I, 2\pi (a_1, a_2)\right),$$

where $k_{1,2}$ is the function defined in formula (1.60) in Section 1.2.2. We can use part (2) of Theorem 1.2.2 in Section 1.2.2 to see that in terms of the K-Bessel function defined by (1.61) in Section 1.2.2, we have, for $j = r_1 + 1, \ldots, r_1 + r_2$,

$$I_j(a_1 + ia_2, s) = \pi \Gamma(2s)^{-1} K_1 \left(1 - 2s | 4\pi^2 \left(a_1^2 + a_2^2\right), 1\right).$$

It follows then that for $j = r_1 + 1, \ldots, r_1 + r_2$, we have the following formula when $a = a_1 + ia_2 \in \mathbb{C} = E_j$,

$$I_j(a, s) = \begin{cases} 2^{2s} \Gamma(2s)^{-1} \left(\pi^2 \left(a_1^2 + a_2^2\right)\right)^{s-\frac{1}{2}} K_{2s-1} \left(4\pi \sqrt{a_1^2 + a_2^2}\right), & a = a_1 + ia_2 \neq 0, \\ \pi \Gamma(2s-1) \Gamma(2s)^{-1}, & a = 0. \end{cases}$$

Substituting these results into the original Poisson sum leads to:

$$\Lambda(a, Y, s) = Nv^{-s} \Lambda_K(a, 2s) + Nv^{-\frac{1}{2}} Nw^{\frac{1}{2}-s} Na^{-1} \Lambda_K(a, 2s - 1)$$

$$+ N|Y|^{\frac{1}{4}-\frac{s}{2}} Nv^{-\frac{1}{2}} A^{2s-1} Na^{-1} 2^{r_1 + 2s r_2}$$

$$\times \sum_{\substack{0 \neq b \in a/U_K \\ 0 \neq c \in (a\partial_K)^{-1}}} \left|\frac{Nc}{Nb}\right|^{s-\frac{1}{2}} e^{2\pi i \mathrm{Tr}(qbc)} \prod_{j=1}^{r_1 + r_2} K_{e_j\left(s - \frac{1}{2}\right)} \left(2\pi e_j \sqrt{\frac{w^{(j)}}{v^{(j)}}} \left|b^{(j)} c^{(j)}\right|\right).$$

To complete the proof of Theorem 2.2.1, suppose that C is an ideal class of K and $b \in C$ is as in formula (2.62). Note that the equation

$$ab = b\partial_K$$

defines a one-to-one mapping from ideals $a \in C^{-1}$ onto elements $b \in b \mod U_K$. Set $u = bc$, for $c \in (b\partial_K)^{-1}$. Then define the map

$$L : (b/U_K) \times (b\partial_K)^{-1} \to \left(C^{-1}\right) \times (\partial_K)^{-1} \quad \text{by} \quad L(b, c) = (a, u = bc).$$

The map L is easily seen to be one-to-one. It is not onto, since the image consists of $(\mathfrak{a}, u)$ such that $\mathfrak{a}$ divides $\partial_K u$.

Finally observe that

$$N\partial_K^{2s-1} \left| \frac{Nc}{Nb} \right|^{s-\frac{1}{2}} = N\mathfrak{a}^{1-2s} |Nu|^{s-\frac{1}{2}}.$$

This completes the proof of Theorem 2.2.1. ∎

We have corrected the following Corollaries after reading the observation of Elstrodt et al. [168, p. 395]. We had made the mistake of thinking that $e_1 = 1$. But recall that Nv involves $v_1^{e_1}$ Of course in the case considered by Elstrodt, Grunewald, and Mennicke $e_1 = 2$ as the field K is imaginary quadratic.

Corollary 2.2.1 (Relations Between $\zeta_K(s)$ and $\zeta_K(s-1)$). *Set*

$$M_s(z) = K_s(z) + \frac{2}{e_1} z \frac{d}{dz} K_s(z)$$

and

$$T(s, u) = M_{e_1 s}\left(2\pi e_1 \left| u^{(1)} \right|\right) \prod_{j=2}^{r_1 + r_2} K_{e_j s}\left(2\pi e_j \left| u^{(j)} \right|\right).$$

Then

$$(1 - s) \Lambda_K (2s - 1) + s\Lambda_K(2s)$$

$$= -e_1 2^{r_1 + r_2 - 1} D_K^{s-\frac{1}{2}} \sum_{0 \neq u \in (\partial_K)^{-1}} |Nu|^{s-\frac{1}{2}} \sigma_{1-2s} (u\partial_K) T\left(s - \frac{1}{2}, u\right).$$

Proof. This is the analogue of a generalization of part (b) of Exercise 3.5.7 of Volume I.

Substitute

$$Y = \begin{pmatrix} v & 0 \\ 0 & 1 \end{pmatrix} \quad \text{in} \quad Z^* (Y, s)$$

and use the following functional equation:

$$Z^* \left(\begin{pmatrix} v & 0 \\ 0 & 1 \end{pmatrix}, s \right) = Z^* \left(\begin{pmatrix} 1 & 0 \\ 0 & v \end{pmatrix}, s \right)$$

plus Theorem 2.2.1 to deduce that

$$Nv^{-s}\Lambda_K(2s) + Nv^{-\frac{1}{2}}\Lambda_K(2s-1)$$
$$+\frac{2^{r_1+r_2}D_K^{s-\frac{1}{2}}}{Nv^{\frac{1}{4}+s/2}} \sum_{0\neq u\in\partial_K^{-1}} |Nu|^{s-\frac{1}{2}}\sigma_{1-2s}(u\partial_K) \prod_{j=1}^{r_1+r_2} K_{e_j(s-\frac{1}{2})}\left(2\pi e_j\frac{|u^{(j)}|}{\sqrt{v^{(j)}}}\right)$$
$$= \Lambda_K(2s) + Nv^{\frac{1}{2}-s}\Lambda_K(2s-1)$$
$$+\frac{2^{r_1+r_2}D_K^{s-\frac{1}{2}}}{Nv^{-\frac{1}{4}+s/2}} \sum_{0\neq u\in\partial_K^{-1}} |Nu|^{s-\frac{1}{2}}\sigma_{1-2s}(u\partial_K) \prod_{j=1}^{r_1+r_2} K_{e_j(s-\frac{1}{2})}\left(2\pi e_j\sqrt{v^{(j)}}\,|u^{(j)}|\right).$$

Differentiate this equation with respect to v_1 and set all $v_j = 1$, $j = 1,\ldots,r_1 + r_2$ to finish the proof of Corollary 2.2.1. ∎

The following corollary gives upper bounds for the product of the class number and the regulator, which should be compared with that obtained by Lang [386, p. 261]. A lower bound is more difficult to obtain. See p. 74 of Volume I for the definition of the regulator.

Corollary 2.2.2 (A Formula for the Product of the Class Number and the Regulator).
Let K be any algebraic number field of degree m, with w_K = the number of roots of unity in K, R_K = the regulator of K, h_K = the class number of K, D_K = the absolute value of the discriminant, $e_1 = 1$ if the field K has any real conjugate fields and 2 otherwise, ∂_K = the different, r_2 = the number of complex conjugate fields, ζ_K = the Dedekind zeta function, $T(s,u)$ as defined in Corollary 2.2.1. Then

$$h_K R_K/w_K = 2(2\pi)^{-m} D_K \zeta_K(2) + e_1 2^{r_2} D_K^{\frac{1}{2}} \sum_{0\neq u\in\partial_K^{-1}} |Nu|^{\frac{1}{2}} \sigma_{-1}(u\partial_K) T\left(\frac{1}{2},u\right).$$

Proof. Let s approach 1 in Corollary 2.2.1. ∎

Exercise 2.2.12. Compute the Jacobian of the mapping T in formula (2.64).

Exercise 2.2.13. Complete the proof of Corollary 2.2.2.

Corollary 2.2.3. *If K is a totally real algebraic number field, then, using the notation of Corollary 2.2.2, we have:*

$$h_K R_K = 4(2\pi)^{-m} D_K \zeta_K(2) - \pi 2^{3-m} D_K^{\frac{1}{2}} \sum_{0\neq u\in(\partial_K)^{-1}} |u^{(1)}|\sigma_{-1}(u\partial_K) e^{-2\pi(|u^{(1)}|+\cdots+|u^{(m)}|)}.$$

Proof. Since $K_{\frac{1}{2}}(z) = (2z/\pi)^{-\frac{1}{2}} e^{-z}$, we see that $M_{1/2}(z) = -(2\pi z)^{\frac{1}{2}} e^{-z}$. The result follows easily then from Corollary 2.2.2. ∎

When $K = \mathbb{Q}$, Corollary 2.2.1 gives formulas relating $\zeta(2n)$ and $\zeta(2n+1)$ (see Exercise 3.5.7 of Vol. I). Formulas of this sort have been studied by many authors, without, however, leading to information on the rationality, irrationality, algebraicity, or transcendence of $\zeta(2n+1)$, $n = 1, 2, \ldots$. See Hunter [311].

Siegel [565, Vol. I, pp. 173–179] used the Fourier expansion of Eisenstein series for $GL(2, \mathcal{O}_K)$ to obtain the analytic continuation and functional equation of the Dedekind zeta function. Mordell [460, pp. 518 ff.] also derives the Fourier expansion of the Eisenstein series for $GL(2, \mathcal{O}_K)$.

Hoffstein [303] has used Fourier expansions of Eisenstein series for $GL(2, \mathcal{O}_K)$, K a real quadratic field, to study the real zeros of these series. Asai [27] uses such Fourier expansions to generalize Kronecker's limit formula (see Exercise 3.5.6 in Vol. I). See also Zagier [674].

Grosswald [250] obtains results related to that in Corollary 2.2.1—formulas for the Dedekind zeta function involving the Meijer's G-function.

Our final goal in this section is to describe the relation between the Epstein zeta function (2.59) and the Eisenstein series for $SL(2, \mathcal{O}_K)$. First recall that Proposition 2.2.1 gave a correspondence between the cusps of $S\mathcal{P}_2^K/SL(2, \mathcal{O}_K)$ and ideal classes in the ideal class group I_K:

$$
\begin{array}{ccc}
\widehat{K}/SL(2, \mathcal{O}_K) & \leftrightarrow & I_K \\
\text{represented by cusps} & & \text{represented by ideals} \\
x_1, \ldots, x_h & & \mathfrak{a}_1, \ldots, \mathfrak{a}_h.
\end{array}
$$

The map was obtained by setting $x_i = p_i/s_i$ with $p_i, s_i \in \mathcal{O}_K$,

$$
A_i = \begin{pmatrix} p_i & u_i \\ s_i & v_i \end{pmatrix} \in SL(2, K), \quad (p_i, s_i) = \begin{pmatrix} \text{the ideal generated} \\ \text{by } p_i \text{ and } s_i \end{pmatrix} = \mathfrak{a}_i,
$$

$$
x_i = A_i \infty, \qquad u_i, v_i \in \mathfrak{a}_i^{-1}.
$$

We showed in Lemma 2.2.1 that

$$
\Gamma_{x_i} = \{\gamma \in SL(2, \mathcal{O}_K) \mid \gamma x_i = x_i\}
$$

$$
= \left\{ A_i \begin{pmatrix} w & z \\ 0 & w^{-1} \end{pmatrix} A_i^{-1} \;\middle|\; z \in \mathfrak{a}_i^{-2}, \; w \in U_K = \text{units of } \mathcal{O}_K \right\}.
$$

We can now define an **Eisenstein series corresponding to the cusp** x_i (cf. Kubota [377]):

$$
E_i(Y, s) = N\mathfrak{a}_i^{2s} \sum_{\gamma \in SL(2, \mathcal{O}_K)/\Gamma_{x_i}} N\left((v(Y\{\gamma A_i\}))\right)^{-s}, \quad \text{for } \operatorname{Re} s > 1. \tag{2.65}
$$

Here we use the notation that if $Y \in \mathcal{P}_2^K$ has Iwasawa decomposition (2.61), we write $v(Y)$ for the v-coordinate of Y. It is also the upper left entry of Y. Now taking the v-part of $Y\{\gamma A_i\}$ amounts to taking $Y\{g\}$, where g is the first column of γA_i. Since such a g must generate $\mathfrak{a}_i$, we find that

$$E_i(Y,s) = N\mathfrak{a}_i^{2s} \sum_{\substack{g \in \mathfrak{a}_i^2/U_K \\ \text{entries of } g \text{ generate } \mathfrak{a}_i}} N(Y\{g\})^{-s}, \text{ for } \operatorname{Re} s > 1. \tag{2.66}$$

Exercise 2.2.14. Prove formula (2.66) for all $Y \in \mathcal{P}_2^K$. Then show that the Eisenstein series E_i is dependent only on the ideal class containing the ideal $\mathfrak{a}_i$, and not on the choice of $\mathfrak{a}_i$ in that ideal class. Finally, prove that, if we define for an ideal class $C \in I_K$ the **ideal class zeta function**:

$$\zeta(C,s) = \sum_{\mathfrak{c} \in C} N\mathfrak{c}^{-s}, \text{ for } \operatorname{Re} s > 1,$$

then we have a relation between Epstein's zeta function (2.59) and the Eisenstein series:

$$Z(\mathfrak{O}_K, Y, s) = \sum_{i=1}^{h} \zeta(C_i, 2s)E_i(Y,s),$$

using the notation C_i for the ideal class containing $\mathfrak{a}_i$, $i = 1, \ldots, h$.

In order to generalize the relation obtained in Exercise 2.2.14 between $Z(\mathfrak{O}_K, Y, s)$ and the vector of Eisenstein series E_i to $Z(\mathfrak{a}_i, Y, s)$, we need a **matrix of ideal class zeta functions**:

$$M_K(s) = (\zeta(C(i,j), 2s))_{1 \leq i,j \leq h}, \tag{2.67}$$

where $C(i,j)$ is the ideal class containing the ideal $\mathfrak{a}_j \mathfrak{a}_i^{-1}$. Define also the **column vector of Epstein zeta functions**:

$$\vec{Z}(Y,s) = {}^t(Z(\mathfrak{a}_1, Y, s), \ldots, Z(\mathfrak{a}_h, Y, s)), \tag{2.68}$$

and the **column vector of Eisenstein series**:

$$\vec{E}(Y,s) = {}^t(E_1(Y,s), \ldots, E_h(Y,s)). \tag{2.69}$$

Proposition 2.2.3 (The Relation Between Epstein's Zeta Function and the Eisenstein Series for $SL(2, \mathfrak{O}_K)$). *Using the notation (2.59), (2.65), (2.67)–(2.69), we have the following equality for $\operatorname{Re} s > 1$:*

$$\vec{Z}(Y,s) = M_K(s)\vec{E}(Y,s).$$

Proof. We have the following chain of equalities:

$$Z(\mathfrak{a}_i, Y, s) = N\mathfrak{a}_i^{2s} \sum_{0 \neq g \in \mathfrak{a}_i^2/U_K} N(Y\{g\})^{-s} = N\mathfrak{a}_i^{2s} \sum_{\substack{\mathfrak{a}_i | \mathfrak{b}}} \sum_{\substack{g \in \mathfrak{b}^2/U_K \\ (g_1,g_2)=\mathfrak{b}}} N(Y\{g\})^{-s}$$

$$= \sum_{\mathfrak{a}_i | \mathfrak{b}} \left(\frac{N\mathfrak{b}}{N\mathfrak{a}_i} \right)^{-2s} N\mathfrak{b}^{2s} \sum_{\substack{g \in \mathfrak{b}^2/U_K \\ (g_1,g_2)=\mathfrak{b}}} N(Y\{g\})^{-s}.$$

Here the inner sum is over 2-vectors g with entries in the ideal $\mathfrak{b}$ modulo the unit group U_K such that the entries of g generate the ideal $\mathfrak{b}$. Then set $\mathfrak{b}/\mathfrak{a}_i = \mathfrak{c}$ and observe that $\mathfrak{c}$ runs through all integral ideals in the ideal class $C(i,j)$ containing the ideal $\mathfrak{a}_j \mathfrak{a}_i^{-1}$, to complete the proof. ∎

The formula in Proposition 2.2.3 raises certain questions.

Questions Arising from Proposition 2.2.3

(1) Is it possible to diagonalize the matrix $M_K(s)$ using characters of the ideal class group?
(2) What does this have to do with Hecke operators for $GL(2, \mathfrak{O}_K)$?
(3) What does this have to do with the analogue of Siegel's integral formula for $GL(n, \mathfrak{O}_K)$? See Proposition 1.4.2 of Section 1.4.4 for the integral formula when $K = \mathbb{Q}$.

Here we shall discuss only question 1. Hecke operators for these groups are treated by Herrmann [292], Shimura [554], and Styer [584]. Siegel's integral formula for $SL(n, \mathfrak{O}_K)$, K imaginary quadratic, is considered by Hunter [311]. See also Elstrodt et al. [168] as well as Efrat and Sarnak [154].

The ideal class group I_K is a finite abelian group of order h. Let $\widehat{I}_K$ denote the **dual group** of characters $\chi : I_K \to \mathbb{T}$, where $\mathbb{T}$ is the circle group of complex numbers of norm 1; i.e.,

$$\mathbb{T} = \{z \in \mathbb{C} \mid |z| = 1\}.$$

That is, χ is a homomorphism of multiplicative groups. Then the dual group is:

$$\widehat{I}_K = \{\chi_1, \ldots, \chi_h\}.$$

We can diagonalize the matrix M_K using Fourier transforms on I_K. Suppose that C_i is the ideal class containing the ideal $\mathfrak{a}_i$, for $i = 1, \ldots, h$. Define

$$Z(C_i) = Z(\mathfrak{a}_i, Y, s), \quad E(C_i) = E(\mathfrak{a}_i, Y, s), \quad \zeta(C) = \zeta(C, 2s). \tag{2.70}$$

Our proof of Proposition 2.2.3 rested on the equation:

$$Z(C_i) = \sum_{j=1}^{h} \zeta(C_j/C_i)E(C_j). \tag{2.71}$$

Now define **convolution** of functions $f : I_K \to \mathbb{C}$ by:

$$(f * g)(C_i) = h^{-1} \sum_{j=1}^{h} f\left(C_i/C_j\right) g(C_j). \tag{2.72}$$

Thus formula (2.71) says that

$$Z(C) = h\zeta(C^{-1}) * E(C). \tag{2.73}$$

As for the group of real numbers (see part (4) of Theorem 1.2.1 in Volume I), the Fourier transform can be used to simplify this convolution equation. We define the **Fourier transform** of a function

$$f : I_K \to \mathbb{C}$$

at the character $\chi \in \widehat{I}_K$ by:

$$\widehat{f}(\chi) = h^{-1} \sum_{y \in I_K} f(y)\overline{\chi(y)}. \tag{2.74}$$

Since I_K is a finite abelian group, there are no convergence problems. In fact, the theory of Fourier transforms on finite abelian groups has many applications, since it is just what is needed for the fast Fourier transform, an idea which has speeded computation of such transforms immensely. See Terras [608, 609] for more information on finite and fast Fourier transforms.

Proposition 2.2.4 (Some Properties of the Fourier Transform on I_K).

(1) Convolution.

$$\widehat{f * g}(\chi) = \widehat{f}(\chi) \cdot \widehat{g}(\chi).$$

(2) Inversion.

$$f(x) = \sum_{\chi \in \widehat{I}_K} \widehat{f}(\chi)\chi(x), \quad \text{for all } x \in I_K.$$

Proof. (1) Note that

$$\widehat{f * g}(\chi) = h^{-2} \sum_{z \in I_K} \sum_{y \in I_K} f(zy^{-1}) g(y) \overline{\chi(z)}$$

$$= h^{-2} \sum_{y \in I_K} g(y) \sum_{w=zy^{-1} \in I_K} f(w) \overline{\chi(wy)} = \widehat{f}(\chi) \cdot \widehat{g}(\chi).$$

(2) Observe that

$$h^{-1} \sum_{\chi \in \widehat{I}_K} \chi(x) \sum_{y \in I_K} f(y) \overline{\chi(y)} = \sum_{y \in I_K} f(y) h^{-1} \sum_{\chi \in \widehat{I}_K} \chi(xy^{-1}) = f(x),$$

since we have:

$$h^{-1} \sum_{\chi \in \widehat{I}_K} \chi(xy^{-1}) = \begin{cases} 0, & x \neq y, \\ 1, & x = y; \end{cases} \text{ and } \overline{\chi(y)} = \chi(y)^{-1}. \tag{2.75}$$

∎

Exercise 2.2.15. Prove formula (2.75) above.

For

$$x \in \widehat{I}_K, \quad Y \in \mathcal{P}_2^K, \quad s \in \mathbb{C} \text{ with } \mathrm{Re}\, s > 1,$$

define the **zeta function**:

$$Z(\chi, Y, s) = \sum_{\mathfrak{a}} N\mathfrak{a}^{2s} \chi(\mathfrak{a}) \sum_{0 \neq g \in \mathfrak{a}^2/U_K} N(Y\{g\})^{-s}, \tag{2.76}$$

where the outer sum is over all ideals $\mathfrak{a}$ of $\mathfrak{O}_K$ and the character χ of the ideal class group is regarded in the obvious way as a function of ideals. Then $Z(\chi, Y, s)$ is the Fourier transform of $Z(C)$ (times h), where $Z(C)$ is defined in formula (2.70).

Similarly, define the **Eisenstein series** associated with $\chi \in \widehat{I}_K$, $Y \in \mathcal{P}_2^K$, and $s \in \mathbb{C}$ with $\mathrm{Re}\, s > 1$ by:

$$E(\chi, Y, s) = \sum_{\mathfrak{a}} N\mathfrak{a}^{2s} \chi(\mathfrak{a}) \sum_{\substack{g \in \mathfrak{a}^2/U_K \\ (g_1, g_2) = \mathfrak{a}}} N(Y\{g\})^{-s}, \tag{2.77}$$

where the outer sum is over all ideals $\mathfrak{a}$ of $\mathfrak{O}_K$ and the inner sum is over column vectors $g = {}^t(g_1, g_2)$ such that the ideal $\mathfrak{a}$ is generated by g_1 and g_2 and the vectors g form a complete set of representatives for the equivalence relation obtained from multiplication by units. Then $E(\chi, Y, s)$ is h times the Fourier transform of $E(C)$ defined by (2.70).

Proposition 2.2.5 (The Diagonalization of the Relation Between Epstein's Zeta Function and the Eisenstein Series for $SL(2, \mathfrak{O}_K)$). *Using the definitions (2.76), (2.77) and setting*

$$L(\chi, s) = \sum_{\mathfrak{a}} \chi(\mathfrak{a}) N\mathfrak{a}^{-s}, \quad \text{for } \mathrm{Re}\, s > 1,$$

with the sum running over all ideals $\mathfrak{a}$ of $\mathfrak{O}_K$, we have

$$Z(\overline{\chi}, Y, s) = L(\chi, s)E(\overline{\chi}, Y, s).$$

Proof. This is just the convolution property in Proposition 2.2.4 for the special case of the functions from Proposition 2.2.3. ∎

Our discussion of nonholomorphic automorphic forms for $GL(2, \mathfrak{O}_K)$ is now at an end, although there still remains much to do if we wish to extend all of Chapter 3 of Volume I and Chapter 1 of this volume to $GL(n, \mathfrak{O}_K)$. For we have not even begun the theory of Hecke operators, the Hecke correspondence, the Selberg trace formula. See Arthur [21–25], Bernstein and Gelbart [47], Frenkel [187], Gelbart [208], Goldfeld and Hundley [232], Jacquet and Langlands [324], and Weil [660] for a general adelic version of the subject. Many papers on automorphic forms are reviewed in *Math. Reviews*. See also the collections of math. reviews in LeVeque [401] and Guy [255] as well as a third volume compiled by the *Math. Reviews* staff [436]. In the next section we seek to address some examples of higher rank trace formulas.

2.2.3 Trace Formulas

In this final section we discuss special cases of the Selberg trace formula which can be applied to the special cases of automorphic forms just discussed.

Trace Formula for Discrete Γ Acting on the Quaternionic Upper Half Plane

Let us give a brief sketch of the Selberg trace formula for cocompact discrete subgroups Γ of $G = SL(2, \mathbb{C})$ or $PSL(2, \mathbb{C})$, following Elstrodt et al. [168], Define $\gamma \in \Gamma$ to be

> **parabolic**   if $\mathrm{Tr}\,(\gamma) \in \mathbb{R}$ and $|\mathrm{Tr}\,(\gamma)| = 2$;
> **hyperbolic** if $\mathrm{Tr}\,(\gamma) \in \mathbb{R}$ and $|\mathrm{Tr}\,(\gamma)| > 2$;
> **elliptic**   if $\mathrm{Tr}\,(\gamma) \in \mathbb{R}$ and $|\mathrm{Tr}\,(\gamma)| < 2$;
> **loxodromic** otherwise.

If $\gamma \in \Gamma$ is hyperbolic or loxodromic, it is conjugate in G to a diagonal matrix with diagonal entries $a(\gamma)$ and its reciprocal. We may assume $|a(\gamma)| > 1$. Define the **norm** of hyperbolic or loxodromic γ to be $N(\gamma) = |a(\gamma)|^2$. Define $N(\gamma) = 1$ is γ is elliptic.

Recall that **the quaternionic upper half plane** is

$$\mathcal{H}^c = \left\{ z + kt = x + iy + kt \,|\, x, y \in \mathbb{R}, \, t > 0 \right\},$$

with Laplacian $\Delta = t^2(\partial^2/\partial x^2 + \partial^2/\partial y^2 + \partial^2/\partial t^2) - t\partial/\partial t$. Note that $\Delta t^{1+s} = (s^2 - 1) t^{1+s}$. Thus it is natural to write the eigenvalue of $-\Delta$ acting on t^{1+s} in the form $\lambda = 1-s^2$. Similarly the eigenvalue of $-\Delta$ acting on t^{1+r} can be written in the form $\mu = 1-r^2$. Let $R_\lambda = (-\Delta - \lambda I)^{-1}$ denote the **resolvent** operator. This is not trace class but $R_\lambda R_\mu$ is trace class. Recall the **resolvent equation** $(\lambda - \mu) R_\lambda R_\mu = R_\lambda - R_\mu$.

Suppose that $\{e_n\}_{n \geq 0}$ is a complete orthonormal set of eigenfunctions of Δ on the (compact) fundamental domain of Γ. Write $-\Delta e_n = \lambda_n e_n$ with $\lambda_n = 1 - s_n^2$. Here we may assume that $s_n = it_n$, with $t_n \geq 0$, except for a finite number of n with $s_n \in [-1, +1]$. Why? Recall the fact that $\lambda_n \geq 0$.

Elstrodt et al. [168] obtain a special case of the **Selberg trace formula** says (assuming $\Gamma\backslash\mathcal{H}^c$ is compact):

$$(\lambda - \mu) \operatorname{Tr}\left(R_\lambda R_\mu \right) = \sum_{n \geq 0} \left(\frac{1}{s^2 - s_n^2} - \frac{1}{r^2 - r_n^2} \right) = \frac{-\mathrm{vol}(\Gamma\backslash\mathcal{H}^c)}{4\pi}(s - r)$$

$$+ \frac{1}{2s} \sum_{\{\gamma\}} \frac{\log(N(\gamma_0))}{|\varepsilon(\gamma)| \left|\operatorname{Tr}(\gamma)^2 - 4\right|} N(\gamma)^{-s} - \frac{1}{2r} \sum_{\{\gamma\}} \frac{\log(N(\gamma_0))}{|\varepsilon(\gamma)| \left|\operatorname{Tr}(\gamma)^2 - 4\right|} N(\gamma)^{-r}.$$

The sums over $\{\gamma\}$ range over the noncentral conjugacy classes of Γ. The element γ_0 is a hyperbolic or loxodromic element of the centralizer, $Z(\gamma)$, of γ in Γ having minimal norm. The maximal finite subgroup in the centralizer of γ is denoted $\varepsilon(\gamma)$. See Elstrodt et al. [168, p. 199]. They use the result to study the Selberg zeta function for Γ and to prove the **Weyl law** for the asymptotic behavior of the counting function

$$\#\left\{ n \mid \lambda_n = 1 - s_n^2 = 1 + t_n^2, \, t_n < T \right\} \sim \frac{\mathrm{vol}(\Gamma\backslash\mathcal{H}^c)}{6\pi^2} T^3, \quad \text{as } T \to \infty.$$

Error terms are also obtained. See [168, p. 211]. Moreover Elstrodt, Grunewald, and Mennicke consider the case that $\Gamma\backslash\mathcal{H}^c$ is finite volume but not compact. They give examples of groups Γ such that $\Gamma\backslash\mathcal{H}^c$ is compact as well as noncompact finite volume examples (see [168, Chapter 10]).

Sarnak [527] applies the Selberg trace formula for $SL(2, \mathfrak{O}_K)$, K imaginary quadratic of class number one, to extend his results on the asymptotics of units in number fields.

Özlem Imamoğlu and Nicole Raulf [319] use the Selberg trace formula for $\Gamma = SL(2, \mathfrak{O}_K)$, K imaginary quadratic of class number one, to study the distribution of Hecke eigenvalues for Γ. They assume that $\{e_n\}_{n \geq 0}$ forms a complete orthonormal set of eigenfunctions of Δ in $L^2(\Gamma \backslash \mathcal{H}^c)$ and that, in addition $T_{\mathfrak{p}} e_n = \rho_n(\mathfrak{p}) e_n$, for the Hecke operator $T_{\mathfrak{p}}$ associated with the prime ideal $\mathfrak{p}$ of $\mathfrak{O}_K$. Since the class number is one, $\mathfrak{p} = v \mathfrak{O}_K$ for some element $v \in \mathfrak{O}_K$. The element v is unique up to multiplication by a unit in $\mathfrak{O}_K$. The **Hecke operator** T_v associated with any nonzero element $v \in \mathfrak{O}_K$ is defined as a sum over $SL(2, \mathfrak{O}_K) \backslash M_v$, where

$$M_v = \left\{ A \in \mathfrak{O}_K^{2 \times 2} \,\middle|\, \det A = v \right\}.$$

More explicitly

$$(T_v f)(z) = \frac{1}{Nv} \sum_{A \in M_v / SL(2, \mathfrak{O}_K)} f(Az).$$

The main result proved by Imamoğlu and Raulf [319] is that the sequence of Hecke eigenvalues $\{\rho_n(\mathfrak{p})\}_{n \geq 1}$ is equidistributed according to the measure

$$d\mu_{\mathfrak{p}}(x) = \begin{cases} \dfrac{1}{2\pi} \left(1 + \dfrac{1}{N\mathfrak{p}} \right) \dfrac{\sqrt{4 - x^2}}{\left(1 + \frac{1}{N\mathfrak{p}} \right)^2 - \frac{x^2}{N\mathfrak{p}}}, & \text{if } |x| < 2, \\[4mm] 0, & \text{otherwise.} \end{cases}$$

This measure approaches an analogue of the Sato–Tate or semi-circle measure as $N\mathfrak{p} \to \infty$. To prove the result, they first use the trace formula to obtain a Weyl law for powers of the Hecke eigenvalues. Then they use the method of moments to obtain the main result.

Trace Formula for Discrete Γ Acting on $\mathcal{H}^m$

If one wants to study groups such as the Hilbert modular group $\Gamma = SL(2, \mathfrak{O}_K)$, K a totally real number field, it helps to have a trace formula for $\Gamma \backslash \mathcal{H}^m$, where m is the degree of K over $\mathbb{Q}$. See Efrat [152], Freitag [186], Müller [469], Shimizu [553], and Zograf [677]. We consider here a special case of the trace formula which gives rise to a formula for the dimension of the space of Hilbert cusp forms of weight 2 when K is real quadratic following Freitag [186].

For $\gamma \in SL(2, \mathbb{R})^m$, write

$$\gamma = \left(\underbrace{\gamma^{(1)}, \ldots, \gamma^{(k)}}_{\text{hyperbolic}}, \underbrace{\gamma^{(k+1)}, \ldots, \gamma^{(l)}}_{\text{parabolic}}, \underbrace{\gamma^{(l+1)}, \ldots \gamma^{(m)}}_{\text{elliptic}} \right).$$

This means that after conjugation

$$\gamma^{(i)} = \begin{pmatrix} \varepsilon_i & 0 \\ 0 & 1/\varepsilon_i \end{pmatrix}, \quad \text{for} \quad i = 1, \ldots, k;$$

$$\gamma^{(i)} = \begin{pmatrix} 1 & a_i \\ 0 & 1 \end{pmatrix}, \quad \text{for} \quad i = k+1, \ldots, l;$$

$$\gamma^{(i)} \in SO(2, \mathbb{R}), \quad \text{for} \quad i = l+1, \ldots, m.$$

One creates a self-reproducing kernel starting with (see Freitag [184, p. 74])

$$k(z, w) = N\left(\frac{z - \overline{w}}{2i}\right) = \prod_{j=1}^{m} \left(\frac{z_j - \overline{w_j}}{2i}\right)^{-2} \quad \text{for} \quad z, w \in \mathcal{H}^m.$$

Then define for $\gamma \in SL(2, \mathbb{R})^m$

$$k(\gamma, z) = \left[\frac{k(\gamma z, z)}{k(z, z)}\right]^r j(\gamma, z),$$

$$\text{where} \quad j(\gamma, z) = N(cz + d)^{-2} = \prod_{j=1}^{m} \left(c^{(j)} z_j + d^{(j)}\right)^{-2},$$

$$\text{for} \quad \gamma = \begin{pmatrix} a & b \\ c & d \end{pmatrix} \in SL(2, \mathbb{R})^m.$$

Let ℓ denote the order of the kernel of the natural projection of Γ into $(SL(2, \mathbb{R})/\{\pm I\})^m$. Define for an integer $r \geq 2$,

$$K(z, w) = \frac{1}{\ell} \sum_{\gamma \in \Gamma} k(\gamma w, z)^r j(\gamma, w)^r.$$

Freitag [186, p. 79] proves that

$$\dim \mathcal{S}(\Gamma, 2r) = \left(\frac{2r - 1}{4\pi}\right) \int_{\Gamma \backslash \mathcal{H}^m} \frac{K(z, z)}{k(z, z)^r} N y^{-2} dx dy.$$

One then writes the sum over Γ giving $K(z, z)$ as a sum over conjugacy classes and finds that the only contribution comes from the central terms and the elliptic terms, assuming $\Gamma \backslash \mathcal{H}^m$ compact. Assuming $\Gamma \backslash \mathcal{H}^m$ compact and Γ irreducible (meaning that the restriction of each of the m projections of $SL(2, \mathbb{R})^m$ into $SL(2, \mathbb{R})$ is 1–1), the result is (cf. Freitag [184, p. 89]:

$$\dim \mathcal{S}(\Gamma, 2r) = \frac{\text{vol}(\Gamma \backslash \mathcal{H}^m)}{(2r - 1)^m} + \sum_{a} E_r(\Gamma, a),$$

where the sum over a is over a set of representatives of Γ-classes of elliptic fixed points with

$$E_r(\Gamma, a) = \frac{1}{|\Gamma_a|} \sum_{\substack{\gamma \in \Gamma_a \\ \gamma \neq \text{identity}}} N \frac{\zeta^r}{1 - \zeta},$$

Here the **stabilizer** of a is $\Gamma_a = \{\gamma \in \Gamma \mid \gamma a = a\}$. An elliptic γ is conjugate in $SL(2, \mathbb{C})$ with a matrix $\begin{pmatrix} \zeta & 0 \\ 0 & \zeta^{-1} \end{pmatrix}$, $\zeta^h = 1$, for some h. If $\Gamma \backslash \mathcal{H}^m$ has cusps as is the case for the Hilbert modular group, then there will also be terms corresponding to the cusps κ of the form $L(\Gamma, \kappa)$, a Shimizu L-series. See [186, p. 110].

Putting this together for the case of the Hilbert modular group (for which the fundamental domain does have cusps), one obtains the formula for the **arithmetic genus** $= g = 1 + (-1)^m \dim \mathcal{S}(\Gamma, 2)$, $m > 1$. Freitag [186, p. 130] gives the result for the Hilbert modular group for $K = \mathbb{Q}\left(\sqrt{p}\right)$, where p is a prime. For example, $g = 1$, if $p = 2, 3, 5$, and, if the prime p is greater than 5 and $p \equiv 1 \pmod 4$,

$$g = 1 + \dim \mathcal{S}(SL(2, \mathfrak{O}_{\mathbb{Q}(\sqrt{d})}), 2) = \frac{\zeta_K(-1)}{2} + \frac{h(-4p)}{8} + \frac{h(-3p)}{6},$$

where $h(d)$ is the class number of $\mathbb{Q}\left(\sqrt{d}\right)$. Another reference is Hirzebruch [296]. Helen Grundman and Lisa E. Lippencott [252] have computed the arithmetic genus for many examples of totally real degree 4 fields over $\mathbb{Q}$.

It is also possible to compute dimensions of spaces of holomorphic cusp forms using generalized Riemann–Roch theorems.

Trace Formula for Γ Acting on the Siegel Upper Half Space

Many people have used the trace formula to compute the dimension of the space of holomorphic cusp forms of weight k for a subgroup Γ of $Sp(n, \mathbb{R})$. We mention only a few: Arakawa [19], Christian [109], Eie [160], Hashimoto [264], Morita [463], Tsushima [617], and Wakatsuki [640]. Yamazaki [673] obtained similar results using the Riemann–Roch theorem. See also the review of Christian's paper by Resnikoff in *Math. Reviews*, 53 #2841. The method goes back to Selberg as well as Godement's Séminaire Cartan [547] lectures in which one uses a self-reproducing or Bergman kernel in the trace formula. This is the same method used for Hilbert modular forms. See Klingen [355, p. 76] for a discussion of the reproducing kernel.

Of course, one must also compute all the conjugacy classes of Γ as well as the orbital integrals corresponding to each class. For example when $n = 3$ there are 300 conjugacy classes. It is not surprising that mistakes might be made. Anyway if you read [*Math.*] *Reviews in Number Theory, Vol. 2B*, for the period 1984–1996, especially pages 576 and 586, you will find much heated discussion. Here we will definitely not delve into the details of these computations, nor take sides in this war.

What is the kernel used in the trace formula to obtain the dimension of $S(Sp(n, \mathbb{Z}), k)$, the space of Siegel cusp forms of weight k? Here we follow Arakawa [19] and Hashimoto's discussion in [272, pp. 253–276]. It starts out for $Z, W \in \mathcal{H}_n$ as $k(Z, W) = \det\left(\frac{Z-\overline{W}}{2i}\right)$. See Klingen [355, p. 76] who uses the Cayley transformation to transform the integrals from $\mathcal{H}_n$ to the generalized unit disc:

$$\mathcal{D}_n = \left\{ W \in \mathbb{C}^{n \times n} \mid {}^t W = W, \; I - \overline{W}W \in \mathcal{P}_n \right\}.$$

which is a bounded domain. The Cayley transform is:

$$\begin{aligned} \mathcal{H}_n &\to \mathcal{D}_n \\ Z &\mapsto (Z - iI)(Z + iI)^{-1}. \end{aligned}$$

The self-reproducing formula on $\mathcal{D}_n$ involves the Bergman kernel and was studied by Hua [308].

Then, in the usual way of trace formulas, following Godement [547], one gets a self-reproducing kernel on the Hilbert space $L^2(\Gamma\backslash\mathcal{H}_n, \det(\mathrm{Im}(W))^k d\mu^*)$, using the measure for the Petersson inner product. The self-reproducing kernel for the weight k cusp forms has the following form, with $j(\gamma, Z) = \det(CZ + D)$, if $\gamma = \begin{pmatrix} A & B \\ C & D \end{pmatrix} \in \Gamma$,

$$K(Z, W) = a_n(k) \sum_{\gamma \in \Gamma} \det\left(\frac{Z - \gamma \overline{W}}{2i}\right)^{-k} j(\gamma, \overline{W})^{-k},$$

where $a_n(k)$ is a constant. Note that the kernel is the symplectic analogue of that for the Hilbert modular group. Moreover, the trace of the kernel should give the dimension of the space of Siegel cusp forms of weight k.

Next one must split Γ into conjugacy classes and evaluate orbital integrals for each conjugacy class. See Wakatsuki [640, pp. 203–204]. There are seven basic types: central, elliptic, hyperbolic, elliptic-hyperbolic, unipotent (or parabolic), quasi-unipotent, hyperbolic-unipotent. What happens next is similar to what happened in the Hilbert modular case. One finds that the orbital integrals vanish unless they correspond to $\gamma \in \Gamma$ which are central, elliptic, unipotent, or quasi-unipotent. Dumping factors are needed for the last two types of terms. The elliptic terms were evaluated by Langlands in [389].

Let us just mention some results on dimensions of spaces of Siegel modular forms of small weight for $Sp(n, \mathbb{Z})$, where $n = 2$. Some references are William Duke and Özlem Imamoḡlu [147], Gerard van der Geer's article in Bruinier et al. [82], Jun-ichi Igusa [315], Helmut Klingen [355], and Martin Raum et al. [500], and Tsuymine [619]. If $\Gamma = Sp(2, \mathbb{Z})$, the ring of Siegel modular forms of even weights is generated by the four Eisenstein series E_4, E_6, E_{10}, E_{12}. See Klingen [355, p. 123]. Moreover the generators are algebraically independent. Thus, for even k, the dimension of the space of Siegel modular forms $\mathcal{M}(Sp(2, \mathbb{Z}), k)$ is the number of nonnegative integer solutions (a, b, c, d) of $k = 4a + 6b + 10c + 12d$. The first cusp form occurs at weight 10. Breeding [78], Van der Geer in [82, p. 233] and Wakatsuki [640, p. 249] give the following list of dimensions

$$d_k(2) = \dim \mathcal{S}(Sp(2, \mathbb{Z}), k)$$

of the space of Siegel cusp forms of even weight for $Sp(2, \mathbb{Z})$:

k	10	12	14	16	18	20	22
$d_k(2)$	1	1	1	2	2	3	4

These authors also give more general tables for modular forms transforming according to a representation of $GL(2, \mathbb{C})$ and for more general $\Gamma \subset Sp(2, \mathbb{R})$ such as congruence subgroups and quaternion groups. Wakatsuki notes that one can use such dimension formulas to aid in understanding the Jacquet–Langlands–Ihara correspondence for $Sp(2, \mathbb{R})$.

The dimensions of the spaces of Siegel cusp forms for $Sp(3, \mathbb{Z})$ were computed by Eie and Lin [161] as well as Tsuyumine [619]. One finds for example that the first time (for even k) that $\dim \mathcal{M}(Sp(3, \mathbb{Z}), k) \geq 2$ is $k = 10$.

Tsuymine [619] gives a long table for

$$md_k(3) = \dim \mathcal{M}(Sp(3, \mathbb{Z}), k)$$

as well as 34 generators of the ring of even weight modular forms for $Sp(3, \mathbb{Z})$. He notes that the ring of Siegel modular forms for $Sp(3, \mathbb{Z})$ cannot be generated by Eisenstein series. Instead he makes use of theta series known as theta constants. We reproduce a bit of Tsuymine's table here:

k	0	2	4	6	8	10	12	14	16	18	20
$md_k(3)$	1	0	1	1	1	2	4	3	7	8	11

See Poor and Yuen [492] for dimensions of spaces of Siegel modular forms of low weight for $Sp(4, \mathbb{Z})$. They find for example that $\dim \mathcal{S}(Sp(4, \mathbb{Z}), 8) = 1$, $\dim \mathcal{S}(Sp(4, \mathbb{Z}), 12) = 2$, and that the dimensions in lower weights are 0. William Duke and Özlem Imamoḡlu [147] find a multitude of results of this sort for small weights and small n in $Sp(n, \mathbb{Z})$. There is a conjectural formula for dimensions of spaces $\mathcal{M}(Sp(n, \mathbb{Z}), k)$ in general. See T. Ibukiyama and H. Saito [314]. Wikipedia has a long table of $\dim \mathcal{M}(Sp(n, \mathbb{Z}), k)$ going out to $n = 9$. For more information on Siegel modular forms for $Sp(n, \mathbb{Z})$, $n \leq 4$, see the website: www.lmfdb.org.

Applications of Siegel modular forms are discussed by A. Ghitza in [217]. Duke [145] gives applications to coding theory. Applications to cryptography can be found in Kirsten Eisenträger and Kristin Lauter [162] as well as Kristin Lauter and Tonghai Yang [397].

References

1. M. Abramowitz, I. Stegun, *Handbook of Mathematical Functions* (Dover Publications, New York, 1965)
2. L. Ahlfors, *Möbius Transformations in Several Dimensions*. Lecture Notes (University of Minnesota, Minneapolis, 1981)
3. G. Akemann, J. Baik, P. Di Francesco, *The Oxford Handbook of Random Matrix Theory* (Oxford University Press, Oxford, 2011)
4. F. Alayont, Meromorphic continuation of spherical cuspidal data Eisenstein series. Can. J. Math. **59**, 1121–1134 (2007)
5. F. Alayont, Residues of Eisenstein series via Maass-Selberg relations, Ph.D. thesis, University of Minnesota, 2003
6. N.T. Allen, On the spectra of certain graphs arising from finite fields. Finite Fields Appl. **4**, 393–440 (1998)
7. G.W. Anderson, A. Guionnet, O. Zeitouni, *An Introduction to Random Matrices* (Cambridge University Press, Cambridge, 2010)
8. T.W. Anderson, *An Introduction to Multivariate Statistical Analysis* (Wiley, New York, 1958)
9. A.N. Andrianov, Spherical functions for $GL(n)$ over local fields and summation of Hecke series. Math. U.S.S.R. Sbornik **12**, 429–452 (1970)
10. A.N. Andrianov, Dirichlet product in the theory of Siegel modular forms of genus 2. Proc. Steklov Inst. Math. **112**, 70–93 (1971)
11. A.N. Andrianov, Euler products corresponding to Siegel modular forms of genus two. Russ. Math. Surv. **29**, 43–110 (1974)
12. A.N. Andrianov, On zeta functions of Rankin type associated with Siegel modular forms, in *Modular Functions of One Variable VI*. Lecture Notes in Mathematics, vol. 627 (Springer, New York, 1977), pp. 325–338
13. A.N. Andrianov, *Quadratic Forms and Hecke Operators* (Springer, New York, 1987)
14. A.N. Andrianov, *Introduction to Siegel Modular Forms and Dirichlet Series* (Springer, New York, 2009)
15. A.N. Andrianov, G.N. Maloletkin, Behavior of theta series of degree N under modular substitutions. Izv. Akad. Nauk. S.S.S.R. **39**, 243–258 (1975)
16. A.N. Andrianov, G.N. Maloletkin, Behavior of theta-series of genus n of indeterminate quadratic forms under modular substitution (Russian). Trudy Mat. Inst. Steklov **148**, 5–15, 271 (1978)
17. T. Apostol, *Calculus*, vols. I, II (Blaisdell, Waltham, MA, 1967)
18. T. Arakawa, Dirichlet series corresponding to Siegel's modular forms. Math. Ann. **238**, 157–174 (1978)

© Springer Science+Business Media New York 2016

A. Terras, *Harmonic Analysis on Symmetric Spaces—Higher Rank Spaces, Positive Definite Matrix Space and Generalizations*, DOI 10.1007/978-1-4939-3408-9

19. T. Arakawa, The dimension of the space of cusp forms on the Siegel upper half plane of degree 2 related to a quaternion unitary group. J. Math. Soc. Jpn. **33**, 125–145 (1981)
20. G. Arfken, *Mathematical Methods for Physicists* (Academic, New York, 1970)
21. J. Arthur, A trace formula for reductive groups I. Duke Math. J. **45**, 911–952 (1978)
22. J. Arthur, A trace formula for reductive groups II. Compos. Math. **40**(1), 87–121 (1980)
23. J. Arthur, The trace formula in invariant form. Ann. Math. **114**, 1–74 (1981)
24. J. Arthur, Automorphic representations and number theory, in *Canadian Mathematical Society Conference Proceedings*, vol. 1 (American Mathematical Society, Providence, RI, 1981), pp. 3–54
25. J. Arthur, The trace formula for noncompact quotient, in *Proceedings of International Congress of Mathematicians, Warsaw, 1983*
26. E. Artin, *Collected Papers* (Addison-Wesley, Reading, MA, 1965)
27. T. Asai, On a certain function analogous to $\log |\eta(z)|$. Nagoya Math. J. **40**, 193–211 (1970)
28. A. Ash, Cohomology of congruence subgroups of $SL_n(\mathbb{Z})$. Math. Ann. **249**, 55–73 (1980)
29. A. Ash, D. Grayson, P. Green, Computations of cuspidal cohomology of congruence subgroups of $SL(3, \mathbb{Z})$. J. Number Theory **19**, 412–436 (1984)
30. A. Ash, D. Mumford, M. Rapoport, Y. Tai, *Smooth Compactifications of Locally Symmetric Spaces* (Mathematical Science Press, Brookline, MA, 1975)
31. K.E. Aubert, E. Bombieri, D. Goldfeld, in *Number Theory, Trace Formulas and Discrete Groups, Symposium in honor of Atle Selberg University of Oslo*, June 14–20, 1987 (Academic, Boston, MA, 1989)
32. W.L. Baily, *Introductory Lectures on Automorphic Forms* (Princeton University Press, Princeton, NJ, 1973)
33. W.L. Baily, A. Borel, Compactification of arithmetic quotients of bounded symmetric domains. Ann. Math. **84**, 442–528 (1966)
34. K.M. Ball, The lower bound for the optimal density of lattice packings. Int. Math. Res. Not. **10**, 217–221 (1992)
35. C.M. Ballantine, Ramanujan type buildings. Can. J. Math. **52**(6), 1121–1148 (2000)
36. E.S. Barnes, The complete enumeration of extreme senary forms. Philos. Trans. R. Soc. Lond. **249**, 461–506 (1957)
37. P. Barrucand, H. Williams, L. Baniuk, A computational technique for determining the class number of a pure cubic field. Math. Comput. **30**, 312–323 (1976)
38. H.J. Bartels, Nichteuklidische Gitterpunktprobleme und Gleichverteilung in linear algebraischen Gruppen. Comment. Math. Helvetici **57**, 158–172 (1982)
39. A.O. Barut, R. Rączka, *Theory of Group Representations and Applications* (Polish Scientific Publishers, Warsaw, 1977)
40. H. Bass, Algebraic K-theory, in *Proceedings of the International Congress of Mathematicians*, Vancouver, 1974, vol. I, pp. 277–283
41. J.G.F. Belinfante, B. Kolman, *A Survey of Lie Groups and Lie Algebras and Computational Methods* (SIAM, Philadelphia, PA, 1972)
42. T. Bengtson, Bessel functions on $\mathcal{P}_n$. Pac. J. Math. **108**, 19–30 (1983)
43. F.A. Berezin, Laplace operators on a semisimple Lie group. Am. Math. Soc. Transl. (2) **21**, 239–339 (1962)
44. F.A. Berezin, I.M. Gelfand, Some remarks on the theory of spherical functions on symmetric Riemannian manifolds. Am. Math. Soc. Transl. (2) **21**, 193–238 (1962)
45. M. Berger, *Geometry Revealed: A Jacob's Ladder to Modern Higher Geometry*, Translated by M. Senechal, L. J. Jointly published by Springer and Cassini Éditeurs, Paris, 2010
46. A. Berman, R.J. Plemmons, *Nonnegative Matrices in the Mathematical Sciences* (Academic, New York, 1979)
47. J. Bernstein, S. Gelbart, *An Introduction to the Langlands Program* (Birkhäuser/Springer, Boston, 2004)
48. T.S. Bhanu-Murthy, Plancherel's measure for the factor space $SL(n, \mathbb{R})/SO(n)$. Dokl. Akad. Nauk., S.S.S.R. **133**, 503–506 (1960)
49. C. Bian, Computing $GL(3)$ automorphic forms. Bull. Lond. Math. Soc. **42**(5), 827–842 (2010)

50. L. Bianchi, Geometrische Darstellung der Gruppen linearer Substitutionen mit ganzen complexen Coefficienten nebst Anwendungen auf die Zahlentheorie. Math. Ann. **38**, 313–333 (1891)

51. G.L. Blankenship, Perturbation theory for stochastic ordinary differential equations with applications to optical waveguide analysis, in *Applications of Lie Group Theory to Nonlinear Network Problems* (Western Periodicals Co., N. Hollywood, CA, 1974), pp. 51–77

52. O. Blumenthal, Über Modulfunktionen von mehreren Veranderlichen. Math. Ann. **56**, 509–548 (1903); **58**, 497–527 (1904)

53. S. Böcherer, Über die Fourierkoeffizienten der Siegelschen Eisensteinreihen. Manuscripta Math. **45**, 273–288 (1984)

54. S. Böcherer, Über die Funktionalgleichung automorpher *L*-Funktionen zur Siegelschen Modulgruppe. J. Reine Angew. Math. **362**, 46–168 (1985)

55. S. Bochner, Bessel functions and modular relations of higher type and hyperbolic differential equations, *Comm. Sem. Math. U. Lund*, Tome suppl., 12–20 (1952)

56. S. Bochner, W.T. Martin, *Several Complex Variables* (Princeton University Press, Princeton, NJ, 1948)

57. O. Bohigas, Quantum chaos. Nucl. Phys. A **751**, 343c–372c (2005)

58. O. Bohigas, M.-J. Giannoni, Chaotic motion and random matrix theories, in *Mathematical and Computational Methods in Nuclear Physics*. Lecture Notes in Physics, vol. 209 (Springer, Berlin, 1984), pp. 1–99

59. O. Bohigas, R.U. Haq, A. Pandey, Fluctuation properties of nuclear energy levels and widths: comparison of theory with experiment, in *Nuclear Data for Science and Technology*, ed. by K.H. Böckhoff (Reidel, Dordrecht, 1983), pp. 809–813

60. A.R. Booker, Uncovering a new *L*-function. Not. Am. Math. Soc. **55**(9), 1088–1094 (2008)

61. A. Borel, Les fonctions automorphes de plusieurs variables complexes. Bull. Soc. Math. Fr. **80**, 167–182 (1952)

62. A. Borel, Les espaces hermitiens symétriques, Séminaire Bourbaki, Paris, 1952

63. A. Borel, Arithmetic properties of algebraic groups, in *Proceedings of the International Congress of Mathematicians*, Stockholm, 1962

64. A. Borel, Compact Clifford-Klein forms of symmetric spaces. Topology **2**, 111–121 (1963)

65. A. Borel, *Introduction aux Groupes Arithmétiques* (Hermann, Paris, 1969)

66. A. Borel, W. Casselman, *Automorphic Forms, Representations, and L-Functions, Proceedings of Symposia in Pure Mathematics*, vol. 33 (American Mathematical Society, Providence, RI, 1979)

67. A. Borel, L. Ji, *Compactifications of Symmetric and Locally Symmetric Spaces* (Birkhäuser, Boston, MA, 2006)

68. A. Borel, G. Mostow, *Algebraic Groups and Discontinuous Subgroups, Proceedings of Symposia in Pure Mathematics*, vol. 9 (American Mathematical Society, Providence, RI, 1966)

69. A. Borel, J.-P. Serre, Corners and arithmetic groups. Comm. Math. Helv. **48**, 436–491 (1973)

70. A. Borel, N. Wallach, *Continuous Cohomology, Discrete Subgroups and Representations of Reductive Groups* (Princeton University Press, Princeton, NJ, 1980)

71. F. Bornemann, On the numerical evaluation of Fredholm determinants. Math. Comput. **79**, 871–915 (2010)

72. P. Bougerol, Comportement asymptotique des puissances de convolution d'une probabilité sur un espace symétrique. *Astérisque*, 74, 24–95, (1980)

73. P. Bougerol, Un Mini-cours sur les Couples de Gelfand. Pub. du Lab. de Statistiques et Probabilités de U. Paul Sabatier, N° 01–83, 1983

74. H. Braun, Konvergenz verallgemeinerter Eisensteinscher Reihen. Math. Z. **44**, 387–397 (1939)

75. H. Braun, Hermitian modular functions, I, II. Ann. Math. **50, 51**, 827–855/92–104 (1949, 1950)

76. H. Braun, Der Basissatz für Hermitesche Modulformen. Abh. aus dem Math. Sem. d. U. Hamburg **19**, 134–148 (1955)

77. H. Braun, in *Eine Frau und die Mathematik 1933–1940. Der Beginn einer wissenschaftlichen Laufbahn*, ed. by M. Koecher (Springer, Berlin, 1990)

78. J. Breeding II, Dimensions of spaces of Siegel cusp forms of degree 2. arXiv:1209.3088v2 [math.RT] 17 Sep 2012

79. A.J. Brentjes, Multi-dimensional continued fraction algorithms, in *Computational Methods in Number Theory*, ed. by H.W. Lenstra, R. Tijdeman (Mathematica Centrum, Amsterdam, 1982), pp. 287–320

80. K. Bringmann, Estimates of Fourier coefficients of Siegel cusp forms for subgroups and in the case of small weight. J. Lond. Math. Soc. **73**, 31–47 (2006)

81. T. Broecker, T. tom Dieck, *Representations of Compact Lie Groups* (Springer, New York, 1985)

82. J.H. Bruinier, G. van der Geer, G. Harder, D. Zagier, *The 1-2-3 of Modular Forms* (Springer, Berlin, 2008)

83. D. Bump, *Automorphic Forms on GL(3)*. Lecture Notes in Mathematics, vol. 1083 (Springer, New York, 1984)

84. D. Bump, *Automorphic Forms and Representations* (Cambridge University Press, Cambridge, 1998)

85. D. Bump, S. Friedberg, On Mellin transforms of unramified Whittaker functions on $GL(3, \mathbb{C})$. J. Math. Anal. Appl. **139**(1), 205–216 (1989)

86. D. Bump, S. Friedberg, The exterior square automorphic L-functions on $GL(n)$, in *Festschrift in Honor of I. I. Piatetski-Shapiro on the Occasion of His Sixtieth Birthday, Part II (Ramat Aviv, 1989) Israel Mathematical Conference Proceedings*, vol. 3 (Weizmann, Jerusalem, 1990), pp. 47–65

87. D. Bump, S. Friedberg, D. Goldfeld, Poincaré series and Kloosterman sums, in *The Selberg Trace Formula and Related Topics*. Contemporary Mathematics, vol. 53 (American Mathematical Society, Providence, RI, 1986), pp. 39–49

88. D. Bump, S. Friedberg, D. Goldfeld, *Multiple Dirichlet Series, L-Functions and Automorphic Forms* (Birkhäuser/Springer, New York, 2010)

89. D. Bump, D. Goldfeld, A Kronecker limit formula for cubic fields, in *Modular Forms*, ed. by R.A. Rankin (Horwood, Chichester (distrib. Wiley), 1984), pp. 43–49

90. D. Bump, J. Hoffstein, Cubic metaplectic forms on $GL(3)$. Invent. Math. **84**, 481–505 (1986)

91. M. Burger, P. Sarnak, Ramanujan duals II. Invent. Math. **106**, 1–11 (1991)

92. R. Burridge, G. Papanicolaou, The geometry of coupled mode propagation in one-dimensional random media. Commun. Pure Appl. Math. **25**, 715–757 (1972)

93. C.J. Bushnell, I. Reiner, L-functions of arithmetic orders and asymptotic distribution of ideals. J. für die reine und angew. Math. **327**, 156–183 (1981)

94. E. Cartan, Sur une classe remarquable d'espaces de Riemann. Bull. Soc. Math. Fr. **54**, 214–264 (1926)

95. E. Cartan, Sur une classe remarquable d'espaces de Riemann. Bull. Soc. Math. Fr. **55**, 114–134 (1927)

96. E. Cartan, Sur la détermination d'un système orthogonal complet dans un espace de Riemann symétrique clos, Rend. Circ. Mat. Palermo. **53**, 217–252 (1929)

97. R.W. Carter, *Simple Groups of Lie Type* (Wiley, New York, 1972)

98. W. Casselman, $GL(n)$, in *Algebraic Number Fields*, ed. by A. Frohlich (Academic, London, 1977), pp. 663–704

99. J.W.S. Cassels, *An Introduction to the Geometry of Numbers* (Springer, Berlin, 1959)

100. J.W.S. Cassels, *Rational Quadratic Forms* (Academic, New York, 1978)

101. J.W.S. Cassels, A. Fröhlich, *Algebraic Number Theory* (Thompson, Washington, DC, 1967)

102. S. Catto, J. Huntley, N.-J. Moha, D. Tepper, Spectral theory of automorphic forms and analysis of invariant differential operators on $SL_3(\mathbb{Z})$ with applications. arXiv, Apr 2003

103. C.-L. Chai, Siegel moduli schemes and their compactifications over $\mathbb{C}$, in *Arithmetic Geometry*, ed. by G. Cornell, J. Silverman (Springer, New York, 1986)

104. C. Chevalley, *Theory of Lie Groups* (Princeton University Press, Princeton, NJ, 1946)

105. P. Chiu, Covering with Hecke points. J. Number Theory **53**, 25–44 (1995)

106. Y. Choquet-Bruhat, C. DeWitt-Morette, M. Dillard-Bleick, *Analysis, Manifolds, and Physics* (North-Holland, New York, 1977)
107. S. Chowla, A. Selberg, On Epstein's zeta function. J. Reine Angew. Math. **227**, 86–110 (1967)
108. U. Christian, *Siegelsche Modulfunktionen*. Lectures U. Göttingen, 1974–1975
109. U. Christian, Berechnung des Ranges der Schar der Spitzenformen zur Modulgruppe zweiten Grades und Stufe $q > 2$. J. Reine Angew. Math. **277**, 130–154 (1975); **296**, 108–118 (1977)
110. U. Christian, *Selberg's Zeta-, L-, and Eisensteinseries*. Lecture Notes in Mathematics, vol. 1030 (Springer, New York, 1983)
111. U. Christian, Maassche *L*-Reihen und eine Identität für Gaussche Summen. Abh. Math. Sem. U. Hamburg, **54**, 29–32 (1984)
112. B. Cipra, *What's Happening in the Mathematical Sciences, 1998–1999* (American Mathematical Society, Providence, RI, 1999)
113. L. Clozel, H. Oh, E. Ullmo, Hecke operators and equidistribution of Hecke points. Invent. Math. **144**, 327–351 (2001)
114. J.E. Cohen, Ergodic theorems in demography. Bull. Am. Math. Soc. **1**, 275–295 (1979)
115. J.E. Cohen, H. Kesten, C.M. Newman (eds.), *Random Matrices and Their Applications, Contemporary Mathematics*, vol. 50 (American Mathematical Society, Providence, RI, 1986)
116. H. Cohn, On the shape of the fundamental domain of the Hilbert modular group, in *Proceedings of Symposia in Pure Mathematics*, vol. VIII (American Mathematical Society, Providence, RI, 1965), pp. 190–202
117. H. Cohn, New upper bounds on sphere packings II. Geom. Topol. **6**, 329–353 (2002)
118. H. Cohn, N. Elkies, New upper bounds on sphere packings I. Ann. Math. **157**, 689–714 (2003)
119. H. Cohn, A. Kumar, Optimality and uniqueness of the Leech lattice among lattices. Ann. Math. **170**, 1003–1050 (2009)
120. L. Cohn, *The Dimension of Spaces of Automorphic Forms on a Certain Two-Dimensional Complex Domain, Memoirs American Mathematical Society*, vol. 158 (American Mathematical Society, Providence, RI, 1975)
121. B. Conrey, A guide to random matrix theory for number theorists, manuscript
122. J.H. Conway, N.J.A. Sloane, *Sphere Packings, Lattices, and Groups* (Springer, New York, 1993)
123. J.H. Conway, C. Goodman-Strauss, N.J.A. Sloane, Recent progress in sphere packing,, in *Current Developments in Mathematics, 1999 (Cambridge, MA)* (International Press, Somerville, MA, 1999), pp. 37–76
124. R. Cooke, *The Mathematics of Sonya Kovalevskaya* (Springer, New York, 1984)
125. R. Courant, D. Hilbert, *Methods of Mathematical Physics*, vol. I (Wiley-Interscience, New York, 1961)
126. H. Cramér, *Mathematical Methods of Statistics* (Princeton University Press, Princeton, NJ, 1946)
127. C.W. Curtis, Representations of finite groups of Lie type. Bull. Am. Math. Soc. **1**, 721–757 (1979)
128. T.W. Cusick, L. Schoenfeld, A table of fundamental pairs of units in totally real cubic fields. Math. Comput. **48**, 147–158 (1987)
129. H. Davenport, *Selected Topics in the Geometry of Numbers*. Lecture Notes (Stanford University Press, Stanford, 1950)
130. H. Davenport, *Multiplicative Number Theory* (Springer, New York, 1981)
131. A. Deitmar, J.W. Hoffman, The Ihara-Selberg zeta function for PGL_3 and Hecke operators. Int. J. Math. **17**(2), 13–155 (2006)
132. A. Deitmar, M. Pavey, A prime geodesic theorem for SL_4. Ann. Glob. Anal. Geom. **33**, 161–205 (2008)
133. B.N. Delone, D.K. Faddeev, *The Theory of Irrationalities of the Third Degree*. Translations of Mathematical Monographs, vol. 10 (American Mathemtical Society, Providence, RI, 1964)
134. B.N. Delone, S.S. Ryskov, Extremal problems in the theory of positive quadratic forms. Proc. Steklov Inst. Math. **112**, 211–231 (1971)
135. M. Deuring, *Algebren* (Springer, New York, 1968)

136. B. Diehl, Die analytische Fortsetzung der Eisensteinreihe zur Siegelschen Modulgruppe. J. for die Reine und Angew. Math. **317**, 40–73 (1980)

137. J. Dieudonné, *Treatise on Analysis*, vols. I–VI (Academic, New York, 1969–1978)

138. H. Donnelly, On the cuspidal spectrum for finite volume symmetric spaces. J. Differ. Geom. **17**, 239–253 (1982)

139. V.G. Drinfeld, Two-dimensional ℓ-adic representations of the fundamental group of a curve over a finite field and automorphic forms on $GL(2)$. Am. J. Math. **105**, 85–114 (1983)

140. V.G. Drinfeld, Langlands conjecture for $GL(2)$ over function field, in *Proceedings of International Congress of Mathematicians*, Helsinki, 1978, pp. 565–574

141. V.G. Drinfeld, Moduli varieties of F-sheaves. Funct. Anal. Appl. **21**, 107–122 (1987)

142. V.G. Drinfeld, The proof of Petersson's conjecture for GL(2) over a global field of characteristic p. Funct. Anal. Appl. **22**, 28–43 (1988)

143. B.A. Dubrovin, V.B. Matveev, S.P. Novikov, Non-linear equations of Korteweg-deVries type, finite-zone linear operators, and abelian varieties. Russ. Math. Surv. **31**, 59–146 (1976)

144. W. Duke, Hecke's representation for L-functions for GL_n, preprint

145. W. Duke, On codes and Siegel modular forms. Int. Math. Res. Not. **5**, 125–136 (1993)

146. W. Duke, Ö. Imamoḡlu, A converse theorem and the Saito-Kurokawa lift. Int. Math. Res. Not. **7**, 347–355 (1996)

147. W. Duke, Ö. Imamoḡlu, Siegel modular forms of small weight. Math. Ann. **310**, 73–82 (1998)

148. W. Duke, Ö. Imamoḡlu, Special values of multiple gamma functions. J. de Théorie des Nombres de Bordeaux **18**, 113–123 (2006)

149. H. Dym, H.P. McKean, *Fourier Series and Integrals* (Academic, New York, 1972)

150. F.J. Dyson, Unfashionable pursuits. Math. Intell. **5**, 47–54 (1983)

151. A. Edelman, N.R. Rao, Random matrix theory. Acta Numerica , 1–65 (2005)

152. I. Efrat, *The Selberg Trace Formula for $PSL_2(\mathbb{R})^n$*. Memoirs of the American Mathematical Society, vol. 65, Number 359 (American Mathematical Society, Providence, RI, 1987)

153. I. Efrat, On a $GL(3)$-analog of $|\eta(z)|$. J. Number Theory **40**(2), 174–186 (1992)

154. I. Efrat, P. Sarnak, The determinant of the Eisenstein matrix and Hilbert class fields. Trans. Am. Math. Soc. **290**, 815–824 (1985)

155. L. Ehrenpreis, F. Mautner, Some properties of the Fourier transform on semisimple Lie groups, I–III. Ann. Math. **61**, 406–439 (1955); Trans. Am. Math. Soc. **84, 90**, 1–55, 431–483 (1957, 1959)

156. M. Eichler, *Introduction to the Theory of Algebraic Numbers and Functions* (Academic, New York, 1966)

157. M. Eichler, Zur Begründung der Theorie der automorphen Funktionen in mehreren Variablen. *Aeq. Math.* **3**, 93–111 (1969)

158. M. Eichler, Über die Anzahl der linear unabhängigen Siegelschen Modulformen von gegebenem Gewicht. Math. Ann. **213**, 281–291 (1975)

159. M. Eichler, On theta functions of real algebraic number fields. Acta Arith. **33**, 269–292 (1977)

160. M. Eie [W.-C. Yu], *Dimensions of Spaces of Siegel Cusp Forms of Degree 2 and 3*. Memoirs of the American Mathematical Society, vol. 50 (American Mathematical Society, Providence, RI, 1984)

161. M. Eie [W.-C. Yu], T.Y. Lin, Dimension formulae for the vector spaces of Siegel cusp forms of degree three. Am. J. Math. **108**, 1059–1087 (1986)

162. K. Eisenträger, K. Lauter, A CRT algorithm for constructing genus 2 curves over finite fields, in *Arithmetics, Geometry and Coding Theory (AGCT 2005), Sémin. Cong.*, vol. 21 (Soc. Math. France, Paris, 2010), pp. 161–176

163. N. Elkies, Lattices, linear codes, and invariants. Not. Am. Math. Soc. **47**, 1238–1245/1382–1391 (2000)

164. P.D.T.A. Elliott, C. Moreno, F. Shahidi, On the absolute value of Ramanujan's τ-function. Math. Ann. **266**, 507–511 (1984)

165. J. Elstrodt, F. Grunewald, J. Mennicke, On the group $PSL_2(\mathbb{Z}[i])$, in *Journées Arithmétiques, 1980, Exeter*, ed. by J.V. Armitage. LMS Lecture Notes (Cambridge University Press, Cambridge, 1982)

166. J. Elstrodt, F. Grunewald, J. Mennicke, Discontinuous groups on 3-dimensional hyperbolic space: analytical theory and arithmetic applications. Russ. Math. Surv. **38**, 137–168 (1983)

167. J. Elstrodt, F. Grunewald, J. Mennicke, $PSL(2)$ over imaginary quadratic integers. Astérisque **94**, 43–60 (1983)

168. J. Elstrodt, F. Grunewald, J. Mennicke, *Groups Acting on Hyperbolic Space* (Springer, Berlin, 1998)

169. P. Epstein, Zur Theorie allgemeiner Zetafunktionen, I, II. Math. Ann. **56**, **63**, 614–644, 205–216 (1903, 1907)

170. A. Erdélyi et al., *Higher Transcendental Functions*, vols. I, II, III (McGraw-Hill, New York, 1953–1955)

171. A. Eskin, H. Oh, Ergodic theoretic proof of equidistribution of Hecke points. Ergod. Th. & Dynam. Sys. **26**, 163–167 (2006)

172. D.W. Farmer, S. Koutsoliotas, S. Lemurell, Maass forms on $GL(3)$ and $GL(4)$. Int. Math. Res. Not. **22**, 6276–6301 (2014)

173. R.D. Farrell, *Techniques of Multivariate Calculus*. Lecture Notes in Mathematics, vol. 520 (Springer, New York, 1976)

174. R.D. Farrell, *Multivariate Calculation, Use of the Continuous Groups* (Springer, New York, 1985)

175. P. Feit, Locating the poles of Eisenstein series of level 1 on SL_n, SP_n and $SU(n,n)$, preprint

176. P. Feit, *Poles and Residues of Eisenstein Series for Symplectic and Unitary Groups*. Memoirs of the American Mathematical Society, vol. 346 (American Mathematical Society, Providence, RI, 1986)

177. L. Fejes Tóth, *Regular Figures* (MacMillan, New York, 1964)

178. W. Feller, *An Introduction to Probability Theory and Its Applications*, vols. I, II (Wiley, New York, 1950, 1966)

179. H.R.P. Ferguson, R.W. Forcade, Generalization of the Euclidean algorithm for real numbers to all dimensions higher than 2. J. fur die reine und angew Math. **334**, 171–181 (1984)

180. R.A. Fisher, The sampling distribution of some statistics obtained from nonlinear equations. Ann. Eugen. **9**, 238–249 (1939)

181. M. Flensted-Jensen, Spherical functions on a real semisimple Lie group. A method of reduction to the complex case. J. Funct. Anal. **30**, 106–146 (1978)

182. Y. Z. Flicker, *The Trace Formula and Base Change for GL(3)*. Lecture Notes in Mathematics, vol. 927 (Springer, New York, 1982)

183. E. Freitag, Zur Theorie der Modulformen zweiten Grades. Nachr. Akad. Wiss. Göttingen, II, Math.-Phys. Kl., 151–157 (1965)

184. E. Freitag, Holomorphe Differentialformen zu Kongruenzgruppen der Siegelschen Modulgruppe. Inv. Math. **30**, 181–196 (1975)

185. E. Freitag, *Siegelsche Modulfunktionen* (Springer, New York, 1983)

186. E. Freitag, *Hilbert Modular Forms* (Springer, Berlin, 1990)

187. E. Frenkel, Lectures on the Langlands program and conformal field theory, in *Frontiers in Number Theory, Physics, and Geometry, II* (Springer, Berlin, 2007), pp. 387–533

188. E. Frenkel, *Love and Math. The Heart of Hidden Reality* (Basic Books, New York, 2013)

189. S. Friedberg, *Lectures on Modular Forms and Theta Series Correspondences* (Middle East Technical U. Foundation, Ankara, 1985)

190. S. Friedberg, A global approach to the Rankin-Selberg convolution for $GL(3, \mathbb{Z})$. Trans. Am. Math. Soc. **300**, 159–174 (1987)

191. R. Fueter, Über automorphe Funktionen der Picard'schen Gruppe I. Comm. Math. Helv. **3**, 42–68 (1931)

192. H. Funk, Beiträge zur Theorie der Kugelfunktionen. Math. Ann. **77**, 136–152 (1916)

193. H. Furstenberg, Noncommuting random products. Trans. Am. Math. Soc. **108**, 377–428 (1963)

194. H. Furstenberg, A Poisson formula for semi-simple Lie groups. Ann. Math. **77**, 335–386 (1963)

195. R. Gangolli, Isotropic infinitely divisible measures on symmetric spaces. Acta Math. **111**, 213–246 (1964)

196. R. Gangolli, On the Plancherel formula and the Paley-Wiener theorem for spherical functions on semisimple Lie groups. Ann. Math. **93**, 150–165 (1971)

197. R. Gangolli, Spectra of discrete uniform subgroups, in *Geometry and Analysis on Symmetric Spaces*, ed. by W. Boothby, G. Weiss (Dekker, New York, 1972), pp. 93–117

198. R. Gangolli, Spherical functions on semisimple Lie groups, in *Geometry and Analysis on Symmetric Spaces*, ed. by W. Boothby, G. Weiss (Dekker, New York, 1972), pp. 41–92

199. R. Gangolli, G. Warner, On Selberg's trace formula. J. Math. Soc. Jpn. **27**, 328–343 (1975)

200. R. Gangolli, G. Warner, Zeta functions of Selberg's type for some noncompact quotients of symmetric spaces of rank one. Nagoya Math. J. **78**, 1–44 (1980)

201. P.R. Garabedian, *Partial Differential Equations* (Wiley, New York, 1964)

202. P. Garrett, Arithmetic properties of Fourier-Jacobi expansions of automorphic forms in several variables. Am. J. Math. **103**, 1103–1134 (1981)

203. P. Garrett, Decomposition of Eisenstein series: Rankin triple products. Ann. Math. **125**, 209–237 (1987)

204. P. Garrett, Truncation and Maaß-Selberg relations. http://www.math.umn.edu/~garrett/, 19 Feb 2005

205. C.F. Gauss, *Werke* (Königlichen Gesellshaft der Wissenshaften, Göttingen, 1870–1927)

206. G. van der Geer, D. Zagier, The Hilbert modular group for the field $\mathbb{Q}\left(\sqrt{3}\right)$. Invent. Math. **42**, 93–133 (1977)

207. S. Gelbart, Bessel functions, representation theory, and automorphic functions, in *Proceedings of Symposia in Pure Mathematics*, vol. 26 (American Mathematical Society, Providence, RI, 1973), pp. 343–345

208. S. Gelbart, *Automorphic Forms on Adele Groups* (Princeton University Press, Princeton, NJ, 1975)

209. S. Gelbart, An elementary introduction to the Langlands program. Bull. Am. Math. Soc. **10**, 177–220 (1984)

210. S. Gelbart, H. Jacquet, A relation between automorphic representations of $GL(2)$ and $GL(3)$, Ann. Sci. Ecole Norm. Sup. **11**, 471–552 (1978)

211. R. Gelca, *Theta Functions and Knots* (World Scientific, Hackensack, NJ, 2014)

212. I.M. Gelfand, Spherical functions on symmetric spaces. Dokl. Akad. Nauk. S.S.S.R. **70**, 5–8 (1950)

213. I.M. Gelfand, M.I. Graev, Analogue of the Plancherel formula for the classical groups. Trudy Moscow Mat. Obšč, 375–404 (1955) (also in Gelfand & Naimark, 1957)

214. I.M. Gelfand, M.I. Graev, I.I. Piatetski-Shapiro, *Representation Theory and Automorphic Functions* (Saunders, Philadelphia, PA, 1966)

215. I.M. Gelfand, M.A. Naimark, *Unitary Representations of the Classical Groups* (German translation) (Akademie-Verlag, Berlin, 1957)

216. P. Gérardin, On harmonic functions on symmetric spaces and buildings, in *Canadian Mathematical Society Conference Proceedings*, vol. 1 (American Mathematical Society, Providence, RI, 1981), pp. 79–92

217. A. Ghitza, *An Elementary Introduction to Siegel Modular Forms*. Lecture Notes from a Talk Given at the Analytic Number Theory Seminar at the U. of Illinois at Urbana-Champaign, 2004

218. S.G. Gindikin, Analysis in homogeneous domains. Russ. Math. Surv. **19**, 1–90 (1964)

219. S.G. Gindikin, Helgason's conjecture in complex analytical interior, in *Representation Theory, Complex Analysis, and Integral Geometry* (Birkhauser/Springer, New York, 2012), pp. 87–95

220. S. Gindikin, F. Karpelevic, Plancherel measures of Riemannian symmetric spaces of nonpositive curvature. Sov. Math. Dokl. **3**, 962–965 (1962)

221. J. Giraud, *Surfaces d'Hilbert-Blumenthal, I, II, III*. Springer Lecture Notes in Mathematics, vol. 868 (Springer, New York, 1981), pp. 1–18, 19–34, 35–37

222. R. Godement, A theory of spherical functions. Trans. Am. Math. Soc. **73**, 496–556 (1952)
223. R. Godement, Une généralization du théorème de la moyenne pour les fonctions harmoniques. C. R. Acad. Sci. Paris **234**, 2137–2139 (1952)
224. R. Godement, Introduction aux travaux de A. Selberg, Séminaire Bourbaki, exp. 144, Paris, 1957
225. R. Godement, Introduction à la théorie de Langlands, Séminaire Bourbaki, exp. 244, Paris, 1962
226. R. Godement, La formule des traces de Selberg considerée comme source de problèmes mathématiques, Séminaire Bourbaki, exp. 244, Paris, 1962
227. R. Godement, *Notes on Jacquet-Langlands Theory* (Institute for Advanced Study, Princeton, NJ, 1970)
228. R. Godement, H. Jacquet, *Zeta Functions of Simple Algebras*. Lecture Notes in Mathematics, vol. 260 (Springer, New York, 1972)
229. B. Goldfarb, On integral assembly maps for lattices in SL_3, preprint
230. D. Goldfeld, *Automorphic Forms and L-Functions for the Group $GL(n, \mathbb{R})$* (Cambridge U. Press, Cambridge, 2006) [with an appendix by Kevin A. Broughan]
231. D. Goldfeld, J. Hoffstein, S.J. Patterson, On automorphic functions of half-integral weight with applications to elliptic curves, in *Number Theory Related to Fermat's Last Theorem*, ed. by N. Koblitz (Birkhäuser, Boston, MA, 1982), pp. 153–193
232. D. Goldfeld, J. Hundley, *Automorphic Representations and L-Functions for the General Linear Group*, vols. I, II (Cambridge University Press, Cambridge, 2011)
233. D. Goldfeld, J. Jorgenson, P. Jones, D. Ramakrishnan, K.A. Ribet, J.Tate (eds.), *Number Theory, Analysis, and Geometry: In Memory of Serge Lang* (Springer, New York, 2012)
234. R. Goodman, Horospherical functions on symmetric spaces, in *Canadian Mathematical Society Conference Proceedings*, vol. 1 (American Mathematical Society, Providence, 1981), pp. 125–133
235. R. Goodman, N. Wallach, Conical vectors and Whittaker vectors. J. Funct. Anal. **39**, 199–279 (1980)
236. M. Goresky, Compactifications and cohomology of modular varieties, in *Harmonic Analysis, the Trace Formula, and Shimura Varieties, Clay Mathematics Proceedings*, vol. 4 (American Mathematical Society, Providence, RI, 2005), pp. 551–582
237. D. Gordon, D. Grenier, A. Terras, Hecke operators and the fundamental domain for $SL(3, \mathbb{Z})$. Math. Comput. **48**, 159–178 (1987)
238. E. Gottschling, Explizite Bestimmung der Randflächen des Fundamentalbereiches der Modulgruppe zweiten Grades. Math. Ann. **138**, 103–124 (1959)
239. F. Götzky, Über eine zahlentheoretische Anwendung von Modulfunktionen zweier Veränderlicher. Math. Ann. **100**, 411–437 (1928)
240. P. Graczyk, A central limit theorem on the space of positive definite symmetric matrices. Annales de l'institut Fourier **42**(4), 857–874 (1992)
241. D. Grenier, Fundamental domains for $\mathcal{P}_n/GL(n, \mathbb{Z})$ and applications in number theory, Ph.D. thesis, U.C.S.D., 1986
242. D. Grenier, Fundamental domains for the general linear group. Pac. J. Math. **132**(2), 293–317 (1988)
243. D. Grenier, An analogue of Siegel's ϕ-Operator for automorphic forms for $GL_n(\mathbb{Z})$. Trans. Am. Math. Soc. **333**(1), 463–477 (1992)
244. W. Greub, S. Halperin, R. Vanstone, *Connections, Curvature, and Cohomology*, vol. II (Academic, New York, 1973)
245. C. Grosche, Selberg supertrace formula for super Riemann surfaces, analytic properties of Selberg super zeta-functions and multiloop contributions for the fermionic string. Commun. Math. Phys. **133**(3), 433–485 (1990)
246. K.I. Gross, W.J. Holman, R.A. Kunze, A new class of Bessel functions and applications in harmonic analysis, in *Proceedings of Symposia in Pure Mathematics*, vol. 33 (American Mathematical Society, Providence, RI, 1979), pp. 407–415

247. K.I. Gross, R.A. Kunze, Bessel functions and representation theory, I, II. J. Funct. Anal. **22**, 73–105 (1976); **25**, 1–49 (1976)
248. K.I. Gross, D. St. P. Richards, Special functions of matrix argument, I: algebraic induction, zonal polynomials and hypergeometric functions. Trans. Am. Math. Soc. **301**, 781–811 (1987)
249. E. Grosswald, *Topics from the Theory of Numbers* (MacMillan, New York, 1966)
250. E. Grosswald, Relations between values at integer arguments of Dirichlet series satisfying functional equations, in *Proceedings of the Symposium of Pure Mathematics*, vol. 24 (American Mathematical Society, Providence, RI, 1973)
251. P.M. Gruber, C.G. Lekkerkerker, *Geometry of Numbers* (North-Holland, Amsterdam, 1987)
252. H.G. Grundman, L.E. Lippincott, Computing the arithmetic genus of Hilbert modular fourfolds. Math. Comput. **75**(255), 1553–1560 (2006)
253. K.-B. Gundlach, Die Bestimmung der Funktionen zu einigen Hilbertschen Modulgruppen. J. für die Reine und Angew. Math. **220**, 109–153 (1965)
254. D. Gurarie, *Symmetries and Laplacians. Introduction to Harmonic Analysis, Group Representations and Applications* (Elsevier, Amsterdam, 1992)
255. R.K. Guy, *Reviews in Number Theory, 1973–1983* (American Mathematical Society, Providence, RI, 1984)
256. L.R. Haff, P.T. Kim, J.-Y. Koo, D. St. P. Richards, Minimax estimation for mixtures of Wishart distributions. Ann. Stat. **39**(6), 3417–3440 (2011)
257. T. C. Hales, S. P. Ferguson, (J.C. Lagarias, Ed.), *The Kepler Conjecture: The Hales-Ferguson Proof by Thomas Hales and Samuel Ferguson* (Springer, New York, 2011)
258. R.W. Hamming, *Coding and Information Theory* (Prentice-Hall, Englewood Cliffs, NJ, 1980)
259. W. Hammond, The modular groups of Hilbert and Siegel. Am. J. Math. **88**, 497–516 (1966)
260. H. Hancock, *Development of the Minkowski Geometry of Numbers*, vols. I, II (Dover Publications, New York, 1939)
261. G. Harder, A Gauss-Bonnet formula for discrete arithmetically defined groups. Ann. Sci. Éc. Norm. Sup. **4**, 409–455 (1971)
262. Harish-Chandra, *Automorphic Forms on Semi-simple Lie Groups*. Lecture Notes in Mathematics, vol. 62 (Springer, New York, 1968)
263. Harish-Chandra, *Collected Papers, I–IV* (Springer, New York, 1984)
264. K. Hashimoto, The dimension of the spaces of cusp forms on Siegel upper half plane of degree 2, I. J. Fac. Sci. U. Tokyo **30**, 403–488 (1983); II, Math. Ann. **266**, 539–559 (1984)
265. M. Hashizume, Whittaker models for real reductive groups. Jpn. J. Math. **5**, 349–401 (1979)
266. D. Healy, A relationship between harmonic analysis on $SU(2)$ and on $SL(2, \mathbb{C})/SU(2)$, Ph.D. thesis, U.C.S.D., 1986
267. D.R. Heath-Brown, S.J. Patterson, The distribution of Kummer sums at prime arguments. J. Reine und Angew. Math. **310**, 111–130 (1979)
268. E. Hecke, *Mathematische Werke* (Vandenhoeck und Ruprecht, Göttingen, 1970)
269. D. Hejhal, *The Selberg Trace Formula for PSL(2, $\mathbb{R}$), I, II*. Lecture Notes in Mathematics, vols. 548, 1001 (Springer, New York, 1976, 1983)
270. D. Hejhal, The Selberg trace formula and the Riemann zeta function. Duke Math. J. **43**, 441–482 (1976)
271. D. Hejhal, Roots of quadratic congruences and eigenvalues of the non-Euclidean Laplacian, in *The Selberg Trace Formula and Related Topics*. Contemporary Mathematics, vol. 53 (American Mathematical Society, Providence, RI, 1986), pp. 277–339
272. D. Hejhal, P. Sarnak, A. Terras (eds.), *The Selberg Trace Formula and Related Topics*. Contemporary Mathematics, vol. 53 (American Mathematical Society, Providence, RI, 1986)
273. S. Helgason, *Differential Geometry and Symmetric Spaces* (Academic, New York, 1962)
274. S. Helgason, An analogue of the Paley-Wiener theorem for the Fourier transform on certain symmetric spaces. Math. Ann. **165**, 297–308 (1966)
275. S. Helgason, Lie groups and symmetric spaces, in *Battelle Rencontres* ed. by C.M. DeWitt, J.A. Wheeler (Benjamin, New York, 1968), pp. 1–71

276. S. Helgason, *Analysis on Lie Groups and Homogeneous Spaces, American Mathematical Society Regional Conference*, vol. 14 (American Mathematical Society, Providence, RI, 1971) [corrected, 1977]
277. S. Helgason, Functions on symmetric spaces, in *Proceedings of Symposia in Pure Mathematics*, vol. 26 (American Mathematical Society, Providence, RI, 1973), pp. 101–146
278. S. Helgason, *Differential Geometry, Lie Groups and Symmetric Spaces* (Academic, New York, 1978)
279. S. Helgason, A duality for symmetric spaces with applications to group representations, III, Tangent space analysis. Adv. Math. **36**, 297–323 (1980)
280. S. Helgason, *The Radon Transform* (Birkhäuser, Boston, MA, 1980)
281. S. Helgason, *Topics in Harmonic Analysis on Homogeneous Spaces* (Birkhäuser, Boston, 1981)
282. S. Helgason, *Groups and Geometric Analysis* (Academic, New York, 1984)
283. S. Helgason, K. Johnson, The bounded spherical functions on symmetric spaces. Adv. Math. **3**, 586–593 (1969)
284. J.W. Helton, A simple test to determine gain bandwidth limitations, in *Proceedings of I.E.E.E. International Conference on Circuits and Systems*, 1977
285. J.W. Helton, The distance of a function to H^∞ in the Poincaré metric. Electrical power transfer. J. Funct. Anal. **38**, 273–314 (1980)
286. J.W. Helton, Non-Euclidean functional analysis and electronics. Bull. Am. Math. Soc. **7**, 1–64 (1982)
287. R.A. Herb, P. Sally, The Plancherel formula, the Plancherel theorem, and the Fourier transform of orbital integrals, in *Representation Theory and Mathematical Physics*. Contemporary Mathematics, vol. 557 (American Mathematical Society, Providence, RI, 2011), pp. 3–22
288. R.A. Herb, J.A. Wolf, The Plancherel theorem for general semisimple Lie groups. Compos. Math. **57**, 271–355 (1986)
289. R. Hermann, *Lie Groups for Physicists* (Benjamin, New York, 1966)
290. R. Hermann, *Fourier Analysis on Groups and Partial Wave Analysis* (Benjamin, New York, 1969)
291. C. Hermite, *Oeuvres, Vols. I–IV* (Gauthiers-Villars, Paris, 1905–1917)
292. O. Herrmann, Über Hilbertsche Modulfunktionen und die Dirichletschen Reihen mit Eulerscher Produktentwicklung. Math. Ann. **127**, 357–400 (1954)
293. C. Herz, Bessel functions of matrix argument. Ann. Math. **61**, 474–523 (1955)
294. K. Hey, *Analytische Zahlentheorie in Systemen Hyperkomplexer Zahlen* (Inaug.-Diss., Hamburg, 1929)
295. H. Heyer (ed.), *Probability Measures on Groups IX*. Lecture Notes in Mathematics, vol. 1379 (Springer, Berlin, 1989)
296. F. Hirzebruch, Hilbert modular surfaces. *L'Enseignement Math., 21* (1973), U. Geneve
297. F. Hirzebruch, *Topological Methods in Algebraic Geometry* (Springer, New York, 1966)
298. F. Hirzebruch, *Modular Functions of One Variable VI*, The ring of Hilbert modular forms for real quadratic fields of small discriminant. Lecture Notes in Mathematics, vol. 627 (Springer, New York, 1977), pp. 288–323
299. F. Hirzebruch, G. van der Geer, *Lectures on Hilbert Modular Surfaces* (Presses U. de Montreal, Montreal, QC, 1981)
300. F. Hirzebruch, D. Zagier, Intersection numbers of curves on Hilbert modular surfaces and modular forms of Nebentypus. Invent. Math. **36**, 57–113 (1976)
301. F. Hirzebruch, D. Zagier, Classification of Hilbert modular surfaces, in *Complex Analysis and Algebraic Geometry* (Iwanami Shoten, Tokyo, 1977), pp. 43–77
302. H. Hochstadt, *Integral Equations* (Wiley-Interscience, New York, 1973)
303. J. Hoffstein, Real zeros of Eisenstein series. Math. Z. **181**, 179–190 (1982)
304. R.T. Hoobler, H.L. Resnikoff, Normal connections for automorphic embeddings, preprint
305. C. Hooley, On the distribution of the roots of polynomial congruences. Mathematika **11**, 39–49 (1964)

306. L. Hörmander, *An Introduction to Complex Analysis in Several Variables* (Van Nostrand, Princeton, NJ, 1966)
307. P.L. Hsu, On the distribution of the roots of certain determinantal equations. Ann. Eugen. **9**, 250–258 (1939)
308. L.K. Hua, *Harmonic Analysis of Functions of Several Complex Variables in the Classical Domains*, Transl. of Math. Monographs, vol. 6 (American Mathematical Society, Providence, RI, 1963)
309. G. Humbert, Théorie de la réduction des formes quadratiques définis positives dans un corps algébrique K fini. Comm. Math. Helv. **12**, 263–306 (1939/1940)
310. G. Humbert, Sur la mesure des classes d'Hermite de discriminant donné dans un corps quadratique imaginaire, et sur certains volumes non euclidiens. C. R. Paris **169**, 448–454 (1919)
311. J. Hunter, Harmonic analysis over imaginary quadratic number fields, Ph.D. thesis, U.C.S.D., 1982.
312. N. Hurt, *Geometric Quantization in Action* (D. Reidel, Amsterdam, 1983)
313. N. Hurt, Propagators in quantum mechanics on multiply connected spaces, in *Group Theoretical Methods in Physics*. Lecture Notes in Physics, vol. 50 (Springer, New York, 1976), pp. 182–192
314. T. Ibukiyama, H. Saito, On zeta functions associated to symmetric matrices and an explicit conjecture on dimensions of Siegel modular forms of general degree. Int. Math. Res. Not. **8**, 161–169 (1992)
315. J.-I. Igusa, On Siegel modular forms of genus 2, I, II. Am. J. Math. **84**, 175–200 (1962); **86**, 392–412 (1964)
316. J.-I. Igusa, *Theta Functions* (Springer, New York, 1964)
317. K. Imai (Ota, Ohta), Generalization of Hecke's correspondence to Siegel modular forms. Am. J. Math. **102**, 903–936 (1980)
318. K. Imai (Ota), A. Terras, Fourier expansions of Eisenstein series for $GL(3, \mathbb{Z})$. Trans. Am. Math. Soc. **273**, 679–694 (1982)
319. Ö. Imamoḡlu. N. Raulf, On the behavior of eigenvalues of Hecke operators. Math. Res. Lett. **17**, 51–67 (2010)
320. A.E. Ingham, An integral which occurs in statistics, Proc. Camb. Philos. Soc. **29**, 271–276 (1933)
321. H. Iwaniec, P. Sarnak, Perspectives on the analytic theory of L-functions. Geom. Funct. Anal. (Special Vol., Part II): GAFA 2000, Tel Aviv, 705–741 (1999)
322. K. Iwasawa, On some types of topological groups. Ann. Math. **50**, 507–558 (1949)
323. H. Jacquet, Les fonctions de Whittaker associées aux groupes de Chevalley. Bull. Soc. Math. Fr. **95**, 243–309 (1967)
324. H. Jacquet, R.P. Langlands, *Automorphic Forms on GL(2)*. Lecture Notes in Mathematics, vol. 114 (Springer, New York, 1970)
325. H. Jacquet, I.I. Piatetski-Shapiro, J. Shalika, Automorphic forms on $GL(3)$, I, II. Ann. Math. **109**, 169–212, 213–258 (1979)
326. H. Jacquet, J. Shalika, On Euler products and the classification of automorphic representations, I, II. Am. J. Math. **103**, 499–558, 777–815 (1981)
327. H. Jacquet, J. Shalika, A non-vanishing theorem for zeta functions of $GL(n)$. Invent. Math. **38**, 1–16 (1976)
328. A.J. James, Zonal polynomials of the real positive definite symmetric matrices. Ann. Math. **75**, 456–469 (1961)
329. A.J. James, Distributions of matrix variates and latent roots derived from normal samples. Ann. Math. Stat. **35**, 475–501 (1964)
330. A.J. James, Special functions of matrix and single argument in statistics, in *Theory and Applications of Special Functions*, ed. by R. Askey (Academic, New York, 1975), pp. 497–520
331. J.M. Jauch, Projective representations of the Poincaré group, in *Group Theory and Its Applications*, ed. by E.M. Loebl (Academic, New York, 1968), pp. 131–182

332. F. John, *Plane Waves and Spherical Means Applied to Partial Differential Equations* (Wiley-Interscience, New York, 1955)
333. J. Jorgenson, S. Lang, *Spherical Inversion on $SL_n(\mathbb{R})$* (Springer, New York, 2001)
334. J. Jorgenson, S. Lang, $Pos_n(\mathbb{R})$ *and Eisenstein Series*. Lecture Notes in Mathematics, vol. 1868 (Springer, Berlin, 2005)
335. J. Jorgenson, S. Lang, *Heat Eisenstein Series on $SL_n(\mathbb{C})$*. Memoirs of the American Mathematical Society, vol. 946 (American Mathematical Society, Providence, RI, 2009)
336. A. Joux, J. Stern, Lattice reduction: a toolbox for the cryptanalyst. J. Cryptol. **11**, 161–185 (1998)
337. G.A. Kabatiansky, V.I. Levenshtein, Bounds for packings on the sphere and in space. Probl. Inf. Transm. **14**, 1–17 (1978)
338. V.I. Kalinin, Eisenstein series on the symplectic group. Math. U.S.S.R. Sbornik **32**, 449–476 (1977)
339. M. Karel, Eisenstein series and fields of definition, Compos. Math. **32**, 225–291 (1976)
340. F.I. Karpelevich, V.N. Tutubalin, M.G. Shur, Limit theorems for the compositions of distributions in the Lobachevsky plane and space. Theory Probab. Appl. **4**, 399–402 (1959)
341. N. Katz, P. Sarnak, *Random Matrices, Frobenius Eigenvalues and Monodromy* (American Mathematical Society, Providence, RI, 1999)
342. N. Katz, P. Sarnak, Zeroes of zeta functions and symmetry. Bull. Am. Math. Soc. **36**(1), 1–26 (1999)
343. G. Kaufhold, Dirichletsche Reihen mit Funktionalgleichung in der Theorie der Modulfunktion 2. Grades. Math. Ann. **137**, 454–476 (1959)
344. D.A. Kazhdan, S.J. Patterson, *Metaplectic Forms, Inst. des Hautes Etudes Scientifiques, Publ. Math.*, vol. 59, Paris (1984)
345. L. Keen (ed.), *The Legacy of Sonya Kovalevskaya, Contemporary Mathematics*, vol. 64 (American Mathematical Society, Providence, RI, 1987)
346. O.-H. Keller, Geometrie der Zahlen, Enzyklop. der Math. Wissenschaften, I.2.2. Aufl. Heft. 11, III
347. T. Kemp, *Introduction to Random Matrix Theory*. U.C.S.D. Lecture Notes for Math 247A (2013)
348. A.B. Kirillov, Unitary representations of nilpotent Lie groups. Russ. Math. Surv. **17**, 53–104 (1962)
349. A.B. Kirillov, *Elements of the Theory of Representations* (Springer, New York, 1976)
350. H. Klingen, Eisensteinreihen zur Hilbertschen Modulgruppe n-ten Grades. Nachr. Akad. Wiss. Göttingen, 87–104 (1960)
351. H. Klingen, Volumbestimmung des Fundamentalbereichs der Hilbertschen Modulgruppe n-ten Grades. J. für die Reine und Angew. Math. **206**, 9–19 (1961)
352. H. Klingen, Über die Werte der Dedekindschen Zetafunktion. Math. Ann. **145**, 265–272 (1962)
353. H. Klingen, Zum Darstellungssatz fiir Siegelsche Modulformen. Math. Z. **102**, 30–43 (1967); **105**, 399–400 (1968)
354. H. Klingen, On Eisenstein series and some applications, in *Automorphic Forms of Several Variables, Katata Conference, 1983* (Birkhäuser, Boston, MA, 1984)
355. H. Klingen, *Introductory Lectures on Siegel Modular Forms* (Cambridge University Press, Cambridge, 1990)
356. H.D. Kloosterman, Thetareihen in total reellen algebraischen Zahlkörpern. Math. Ann. **103**, 279–299 (1930)
357. A. Hibner Koblitz, *A Convergence of Lives. Sofia Kovalevskaia: Scientist, Writer, Revolutionary* (Birkhäuser, Boston, MA, 1983)
358. P. Kochina, *Love and Mathematics: Sofya Kovalevskaya* (Mir Publishers, Moscow, 1985)
359. M. Koecher, Über Dirichlet-Reihen mit Funktionalgleichung. J. Reine und Angew. Math. **192**, 1–23 (1953)
360. M. Koecher, Über Thetareihen indefiniter quadratischer Formen. Math. Nachr. **9**, 51–85 (1953)

361. M. Koecher, Gruppen und Lie Algebren rationaler Funktionen. Math. Z. **109**, 349–392 (1969)
362. A.N. Kolmogorov, S.V. Fomin, Introductory Real Analysis (Dover Publications, New York, 1975)
363. T. Koornwinder, Jacobi functions and analysis on noncompact semisimple Lie groups, in *Special Functions, Group Theoretical Aspects and Applications*, ed. by R. Askey (D. Reidel, Boston, MA, 1984), pp. 1–85
364. A. Koranyi, A survey of harmonic functions on symmetric spaces, in *Proceedings of Symposia in Pure Mathematics*, vol. 35 (American Mathematical Society, Providence, RI, 1979), pp. 323–344
365. A. Korkine, G. Zolotareff, Sur les formes quadratiques. Math. Ann. **6**, 366–389 (1869)
366. A. Korkine, G. Zolotareff, Sur les formes quadratiques positives. Math. Ann. **11**, 242–292 (1877)
367. B. Kostant, On Whittaker vectors and representation theory. Invent. Math. **48**, 101–184 (1978)
368. S. Kovalevskaya, Sur le probleme de la rotation d'un corps solide autour d'un point fixe. Acta Math. **12**, 177–232 (1889)
369. E. Kowalski, G. Ricotta, Fourier coefficients of $GL(N)$ automorphic forms in arithmetic progressions (2014, preprint)
370. A. Krieg, *Modular Forms on Half-Spaces of Quaternions*. Lecture Notes in Mathematics, vol. 1143 (Springer, New York, 1985)
371. A. Krieg, *Hecke Algebras*. Memoirs of the American Mathematical Society, vol. 87, no. 435 (American Mathematical Society, Providence, RI, 1990)
372. B. Krötz, E. Opdam, Analysis on the crown domain. Geom. Funct. Anal. **18**(4), 1326–1421 (2008)
373. T. Kubota, Über diskontinuierliche Gruppen Picardschen Typus und zugehörige Eisenstein-sche Reihen. Nagoya Math. J. **32**, 259–271 (1968)
374. T. Kubota, On a special kind of Dirichlet series. J. Math. Soc. Jpn. **20**, 193–207 (1968)
375. T. Kubota, *On Automorphic Functions and the Reciprocity Law in a Number Field*. Lectures in Mathematics (Kyoto U., Kinokuniya Book Store Co., Ltd., Tokyo, 1969)
376. T. Kubota, Some results concerning reciprocity law and real analytic automorphic functions, in *Proceedings of Symposia in Pure Mathematics*, vol. 20 (American Mathematical Society, Providence, RI, 1971), pp. 382–395
377. T. Kubota, *Elementary Theory of Eisenstein Series* (Wiley, New York, 1973)
378. S. Kudla, Relations between automorphic forms produced by theta functions, in *Modular Functions of One Variable VI*. Lecture Notes in Mathematics, vol. 627 (Springer, New York, 1977), pp. 277–285.
379. S. Kudla, Theta functions and Hilbert modular forms. Nagoya Math. J. **69**, 97–106 (1978)
380. H.B. Kushner, The linearization of the product of two zonal polynomials. SIAM J. Math. Anal. **19**(3), 687–717 (1988)
381. L. Lafforgue, Chtoucas de Drinfeld et correspondance de Langlands. Invent. Math. **147**, 1–241 (2002)
382. J.C. Lagarias, H.W. Lenstra, C.P. Schnorr, Korkin-Zolotarev bases and successive minima of a lattice and its reciprocal lattice. Combinatorica **10** (4), 333–348 (1990)
383. J.C. Lagarias, A. Odlyzko, Solving low-density subset sum problems, in *Proceedings of the 24th Annual IEEE Symposium on Foundations of Computer Science*, 1983, pp. 1–10
384. J.-L. Lagrange, *Oeuvres*, vols. I–XIV (Gauthier-Villars, Paris) MDCCCXCII–MDCCCLVII
385. S. Lang, *Analysis I* (Addison-Wesley, Reading, MA, 1968); new edition, *Undergraduate Analysis* (Springer, New York, 1997)
386. S. Lang, *Algebraic Number Theory* (Addison-Wesley, Reading, MA, 1968)
387. S. Lang, $SL_2(\mathbb{R})$ (Addison-Wesley, Reading, MA, 1975)
388. S. Lang, *Real Analysis* (Addison-Wesley, Reading, MA, 1983)
389. R. Langlands, The dimension of spaces of holomorphic forms. Am. J. Math. **85**, 99–125 (1963)
390. R. Langlands, *Euler Products* (Yale University Press, New Haven, 1967)

391. R. Langlands, *Problems in the Theory of Automorphic Forms*. Lecture Notes in Mathematics, vol. 170 (Springer, New York, 1970), pp. 18–61
392. R. Langlands, *Eisenstein Series*. Lecture Notes in Mathematics, vol. 544 (Springer, New York, 1976)
393. R. Langlands, *L*-functions and automorphic representations, in *Proceedings of International Congress of Mathematics*, Helsinki, 1978
394. R. Langlands, *Base Change for GL*(2) (Princeton University Press, Princeton, NJ, 1980)
395. R. Langlands, Review of Osborne and Warner, Eisenstein systems. Bull. Am. Math. Soc. **9**, 351–361 (1983)
396. E. Lapid, W.Müller, Spectral asymptotics for arithmetic quotients of $SL(n, \mathbb{R})/SO(n)$. Duke Math. J. **149**, (1), 117–156 (2009)
397. K. Lauter, T. Yang, Computing genus 2 curves from invariants on the Hilbert moduli space. J. Number Theory **131**, 936–958 (2011)
398. N.N. Lebedev, *Special Functions and Their Applications* (Dover, New York, 1972)
399. R. Lee, J. Schwermer, Cohomology of arithmetic subgroups of SL_3 at infinity. J. für die Reine und Angew. Math. **330**, 100–131 (1982)
400. R. Lee, R.H. Szczarba, Homology and cohomology of congruence subgroups. Proc. Natl. Acad. Sci. U.S.A. **72**, 651–653 (1975).
401. W.J. LeVeque, *Reviews in Number Theory* (American Mathematical Society, Providence, RI, 1974)
402. W.-C.W. Li, Ramanujan graphs and Ramanujan hypergraphs, in *IAS/Park City Math. Series*, vol. 12 (2002), pp. 401–427
403. W.-C.W. Li, Ramanujan hypergraphs. Geom. Funct. Anal. **14**(2), 380–399 (2004)
404. W.-C.W. Li, P. Solé, Spectra of regular graphs and hypergraphs and orthogonal polynomials. Eur. J. Comb. **17**, 461–477 (1996)
405. E. Lindenstrauss, A. Venkatesh, Existence and Weyl's law for spherical cusp forms. Geom. Funct. Anal. **17**, 220–251 (2007)
406. G. Lion, M. Vergne, *The Weil Representation, Maslov Index and Theta Series* (Birkhäuser, Boston, MA, 1980)
407. K. Lonngren, A. Scott (eds.), *Solitons in Action* (Academic, New York, 1978)
408. O. Loos, *Symmetric Spaces, I, II* (Benjamin, New York, 1969)
409. A. Lubotzky, R. Phillips, P. Sarnak, Hecke operators and distributing points on the sphere. I, Frontiers of the mathematical sciences, 1985 (New York, 1985). Commun. Pure Appl. Math. **39**, S149–S186 (1986)
410. A. Lubotzky, R. Phillips, P. Sarnak, Hecke operators and distributing points on S_2, II. Commun. Pure Appl. Math. **40**(4), 401–420 (1987)
411. A. Lubotzky, B. Samuels, U. Vishne, Explicit constructions of Ramanujan complexes of type $\widetilde{A}_d$. Eur. J. Comb. **26**(6), 965–993 (2005)
412. W. Luo, Z. Rudnick, P. Sarnak, On Selberg's eigenvalue conjecture. Geom. Funct. Anal. **5**(2), 387–401 (1995)
413. W. Luo, Z. Rudnick, P. Sarnak, On the generalized Ramanujan conjecture for GL(n), in *Proceedings of Symposia in Pure Mathematics*, vol. 66, Pt. II (American Mathematical Society, Providence, RI, 1999), pp. 301–310
414. H. Maass, Die Primzahlen in der theorie der Siegelschen Modulfunktionen. Math. Ann. **117**, 538–578 (1940)
415. H. Maass, Zur Theorie der automorphen Funktionen von *n* Veränderlichen. Math. Ann. **117**, 538–578 (1940)
416. H. Maass, Über Gruppen von hyperabelschen Transformationen. Sitz.-Ber. der Heidelberg Akad. Wiss., Math.-Nat., Kl. 2 Abh. (1940)
417. H. Maass, Über eine neue Art von nichtanalytischen automorphen Funktionen und die Bestimmung Dirichletscher Reihen durch Funktionalgleichung. Math. Ann. **121**, 141–183 (1949)
418. H. Maass, Automorphe Funktionen und indefinite quadratische Formen. Sitz.-Ber. der Heidelberg Akad. Wiss., Math.-Nat. Kl., 1 Abh. (1949)

419. H. Maass, Modulformen zweiten Grades und Dirichletreihen. Math. Ann. **122**, 90–108 (1950)
420. H. Maass, Spherical functions and quadratic forms. J. Indian Math. Soc. **20**, 117–162 (1956)
421. H. Maass, Zetafunktionen mit Grössencharakteren und Kugelfunktionen. Math. Ann. **134**, 1–32 (1957)
422. H. Maass, Zur Theorie der Kugelfunktionen einer Matrix-variablen. Math. Ann. **135**, 391–416 (1958)
423. H. Maass, Über die raumliche Verteilung der Punkte in Gittem mit indefiniter Metrik. Math. Ann. **138**, 287–315 (1959)
424. H. Maass, *Lectures on Modular Forms of One Complex Variable* (Tata Institute of Fundamental Research, Bombay, 1964)
425. H. Maass, Modulformen zu indefiniten quadratischen Formen. Math. Scand. **17**, 41–55 (1965)
426. H. Maass, *Siegel's Modular Forms and Dirichlet Series*. Lecture Notes in Mathematics, vol. 216 (Springer, New York, 1971)
427. I. Macdonald, Some conjectures for root systems. SIAM J. Math. Anal. **13**, 988–1007 (1982)
428. I. Macdonald, *Symmetric Functions and Hall Polynomials* (Clarenden Press, Oxford, 1979, 1995)
429. G. Mackey, *The Theory of Group Representations* (University of Chicago Press, Chicago, Illinois, 1976)
430. G. Mackey, *Unitary Group Representations in Physics, Probability and Number Theory* (Benjamin/Cummings, Reading, MA, 1978)
431. G. Mackey, Harmonic analysis as the exploitation of symmetry—a historical survey. Rice U. Stud. Houston, TX **64** (1978)
432. K. Mahler, On Minkowski's theory of reduction of positive quadratic forms. Q. J. Math. Oxford Ser. **9**, 259–262 (1938)
433. G. Margulis, *Discrete Subgroups of Semisimple Lie Groups* (Springer, Berlin, 1991)
434. M. G. Martínez, The finite upper half space and related hypergraphs. J. Number Theory **84**, 342–360 (2000)
435. M.G. Martínez, H.M. Stark, A.A. Terras, Some Ramanujan hypergraphs associated to $GL(n, \mathbb{F}_q)$. Proc. Am. Math. Soc. **129**(6), 1623–1629 (2001)
436. Mathematical Reviews Staff, *Reviews in Number Theory 1984–1996* (American Mathematical Society, Providence, RI, 1997)
437. K. Maurin, *General Eigenfunction Expansions and Unitary Representations of Topological Groups* (Polish Scientific Publishers, Warsaw, 1968)
438. F. Mautner, Geodesic flows on symmetric Riemannian spaces. Ann. Math. **65**, 416–431 (1957)
439. F. Mautner, Spherical functions and Hecke operators, in *Lie Groups and Their Representations, Proceedings of Summer School Bolya Janos Math. Soc.*, Budapest, vol. 1971 (Halsted, New York, 1975), pp. 555–576
440. H.P. McKean, E. Trubowitz, Hill's operator and hyperelliptic function theory in the presence of infinitely many branch points. Commun. Pure Appl. Math. **29**, 143–226 (1976)
441. M.L. Mehta, *Random Matrices and the Statistical Theory of Energy Levels* (Elsevier/Academic, San Diego, CA, 1967, 1991, 2004)
442. J. Mennicke, *Vorträge über Selbergs Spurformel I* (University of Bielefeld, Bielefeld)
443. J. Mennicke, Lectures on discontinuous groups on 3-dimensional hyperbolic space, in *Modular Forms Conference*, Durham, 1983
444. P. Menotti, E. Onofri, The action of $SU(N)$ lattice gauge theory in terms of the heat kernel on the group manifold, in *Lattice Gauge Theories and Monte Carlo Simulation*, ed. by C. Rebbi (World Scientific Publication, Singapore, 1983), pp. 447–459
445. B. Mezhericher, Evaluating Jacquet's Whittaker functions and Maass forms for $SL(3, \mathbb{Z})$. Math. Comput. **80**, 2299–2312 (2011)
446. S.D. Miller, Cusp forms on $SL_3(\mathbb{Z}) \backslash SL_3(\mathbb{R})/SO_3(\mathbb{R})$, Ph.D. thesis, Princeton University, 1997
447. S.D. Miller, On the existence and temperedness of cusp forms for $SL_3(\mathbb{Z})$. J. Reine Angew. Math. **533**, 127–169 (2001)

448. S.J. Miller, R. Takloo-Bighash, *An Invitation to Modern Number Theory* (Princeton University Press, Princeton, NJ, 2006)
449. J. Milnor, Hilbert's problem 18, in *Proceedings of Symposia in Pure Mathematics*, vol. 28 (American Mathematical Society, Providence, RI, 1976), pp. 491–506
450. J. Milnor, Hyperbolic geometry: the first 150 years. Bull. Am. Math. Soc. **6**, 9–24 (1982)
451. J. Milnor, D. Husemoller, *Symmetric Bilinear Forms* (Springer, New York, 1973)
452. H. Minc, *Nonnegative Matrices* (Wiley, New York, 1988)
453. H. Minkowski, *Gesammelte Abhandlungen* (Chelsea, New York, 1911) [reprinted 1967]
454. C.W. Misner, K.S. Thorne, J.A. Wheeler, *Gravitation* (Freeman, San Francisco, CA, 1973)
455. C. Mœglin, J.-L. Waldspurger, Le spectre résiduel de $GL(n)$. Ann. scient. Éc. Norm. Sup., 4e série, t. **22**, 605–674 (1989)
456. C. Mœglin, J.-L. Waldspurger, *Spectral Decomposition and Eisenstein Series: Une Paraphrase de l' Ecriture*. Cambridge Tracts in Mathematics, No. 113 (Cambridge University Press, 1995)
457. M.I. Monastyrsky, A.M. Perelomov, Coherent states and bounded homogeneous domains. Rep. Math. Phys. **6**, 1–14 (1974)
458. H. Montgomery, The pair correlation of zeros of the zeta function, in *Proceedings of Symposia in Pure Mathematics*, vol. 24 (American Mathematical Society, Providence, RI, 1973), pp. 181–193
459. C. Moore, Compactifications of symmetric spaces. Am. J. Math. **86**, 201–218 (1964)
460. L.J. Mordell, On Hecke's modular functions, zeta functions, and some other analytic functions in the theory of numbers, Proc. Lond. Math. Soc. **32**, 501–556 (1931)
461. C. Moreno, F. Shahidi, The L-functions $L(s, Sym^m(r), \pi)$. Can. Math. Bull. **28**, 405–410 (1985)
462. C. Moreno, F. Shahidi, The fourth moment of the Ramanujan tau function. Math. Ann. **266**, 233–239 (1983)
463. Y. Morita, An explicit formula for the dimension of spaces of Siegel modular forms of degree 2. J. Fac. Sci. Univ. Tokyo **21**, 167–248 (1974)
464. D.F. Morrison, *Multivariate Statistical Methods* (McGraw-Hill, New York, 1976)
465. G. Mostow, Some new decomposition theorems for semisimple Lie groups. Mem. Am. Math. Soc. **14**, 31–54 (1955)
466. G. Mostow, Discrete subgroups of Lie groups. Adv. Math. **15**, 112–123 (1975)
467. G. Mostow, T. Tamagawa, On the compactness of arithmetically defined homogeneous spaces. Ann. Math. **76**, 440–463 (1962)
468. R.J. Muirhead, *Aspects of Multivariate Statistical Theory* (Wiley, New York, 1978)
469. W. Müller, Signature defects of cusps of Hilbert modular varieties and values of L-series at $s = 1$, Report Math., Akad. der Wiss. der D.D.R., Inst. für Math., Berlin, 1983; J. Diffl. Geom. **20**, 55–119 (1984)
470. W. Müller, B. Speh, with appendix by E. Lapid, Absolute convergence of the spectral side of the Arthur trace formula for GL_n. Geom. Funct. Anal. **14**, 58–93(2004)
471. D. Mumford, *Tata Lectures on Theta, I, II* (Birkhäuser, Boston, 1982, 1984)
472. Y. Namikawa, *Toroidal Compactification of Siegel Spaces*. Lecture Notes in Mathematics, vol. 812 (Springer, New York, 1980)
473. D. Newland, Kernels in the Selberg trace formula on the k-regular tree and Zeros of the Ihara Zeta function, Ph.D. thesis, U.C.S.D., 2005
474. S.P. Novikov, A method for solving the periodic problem for the KdV equation and its generalizations. Rocky Mt. J. Math. **8**, 83–93 (1978)
475. N.V. Novikova (N.V. Zaharova), Korkin-Zolotarev reduction domains of positive quadratic forms in $n \leq 8$ variables and a reduction algorithm for these domains. Soviet Math. Dokl. **27**, 557–560 (1983)
476. M.E. Novodvorsky, I.I. Piatetski-Shapiro, Rankin-Selberg method in the theory of automorphic forms, in *Proceedings of Symposia in Pure Mathematics*, vol. 30 (American Mathematical Society, Providence, RI, 1976), pp. 297–301

477. A. Odlyzko, On the distribution of spacings between zeros of the zeta function. Math. Comput. **48**, 273–308 (1987)

478. H. Oh, Harmonic analysis and Hecke operators, in *Rigidity in Dynamics and Geometry (Cambridge, 2000)* (Springer, Berlin, 2002), pp. 363–378

479. G. Ólafsson, H. Schlichtkrull, Representation theory, Radon transform and the heat equation on a Riemannian symmetric space, in *Group Representations, Ergodic Theory, and Mathematical Physics: A Tribute to George W. Mackey*. Contemporary Mathematics, vol. 449 (American Mathematical Society, Providence, RI, 2008), pp. 315–344

480. H. Ooguri, Interview with Edward Witten. Not. Am. Math. Soc. **62**(5), 492–506 (2015)

481. F.W. Olver, D.W. Lozier, R.F. Boisvert, C.W. Clark (eds.), *NIST Handbook of Mathematical Functions* (Cambridge University Press, Cambridge, 2010)

482. M.S. Osborne, G. Warner, *The Theory of Eisenstein Systems* (Academic, New York, 1981)

483. R.A. Pendavingh, S.H.M. Van Zwam, New Korkin-Zolotarev inequalities. SIAM J. Optim. **18** (1), 364–378 (2007)

484. A. Perelomov, *Generalized Coherent States and Their Applications* (Springer, New York, 1986)

485. I.I. Piatetski-Shapiro, *Automorphic Functions and the Geometry of the Classical Domains* (Gordon and Breach, New York, 1969)

486. I.I. Piatetski-Shapiro, Euler subgroups, in *Lie Groups and Their Representations, Summer School Bolyai Janos Math. Soc.*, ed. by I.M. Gelfand (Halsted, New York, 1975), pp. 597–620

487. I.I. Piatetski-Shapiro, Cuspidal automorphic representations associated to parabolic subgroups and Ramanujan conjecture, in *Number Theory Related to Fermat's Last Theorem*, ed. by N. Koblitz (Birkhäuser, Boston, MA, 1982)

488. E. Picard, Sur un groupe de transformations des points de l'espace situés du même coté d'un plan. Bull. Soc. Math. de France **12** , 43–47 (1844)

489. P. Ploch, *Bestimmung von Konjugiertenklassen und Beweis von Verschwindungssätzen, die bei der Berechnung des Ranges der Schar der Spitzenformen zur Siegelschen Modulgruppe vierten Grades und Stufe q $\geq$ 3 Auftreten*, Dissertation, Göttingen, 1985

490. H. Pollard, *The Theory of Algebraic Numbers* (Mathematical Association of America, Washington, DC, 1961)

491. A.M. Polyakov, Quantum geometry of bosonic strings. Phys. Lett. **103B**, 207 (1981); Quantum geometry of fermionic strings. Phys. Lett. **103B**, 211 (1981)

492. C. Poor, D.S. Yuen, Dimensions of spaces of Siegel modular forms of low weight in degree 4, Bull. Austral. Math. Soc. **54**, 309–315 (1996)

493. S.J. Press, *Applied Multivariate Analysis* (Holt, Rinehart and Winston, New York, 1972)

494. N.V. Proskurin, Expansions of automorphic functions. J. Sov. Math. **26**, 1908–1921 (1984)

495. N.J. Pullman, *Matrix Theory and Its Applications* (Dekker, New York, 1976)

496. M.S. Raghunathan, *Discrete Subgroups of Lie Groups* (Springer, New York, 1972)

497. K.G. Ramanathan, Quadratic forms over involutorial division algebras II. Math. Ann. **143**, 293–332 (1961)

498. K.G. Ramanathan, Zeta functions of quadratic forms. Acta Arith. **7**, 39–69 (1961)

499. M. Ratner, On Raghunathan's conjecture. Ann. Math. **134**, 545–607 (1991)

500. M. Raum, N.C. Ryan, N.-P. Skoruppa, G. Tornaría, Explicit computations of Siegel modular forms of degree 2. arXiv:1205.6255v3 [math.NT] 4 Jun 2012

501. M. Reed, B. Simon, *Methods of Modern Mathematical Physics, Vol. I, Functional Analysis* (Academic, New York, 1972)

502. A. Regev, letter of 8/30/79, Math. Dept., Weizmann Inst. Science, Rehovot, Israel (1979)

503. R. Remak, Über die Minkowskische Reduktion der definiten quadratischen Formen. Compos. Math. **5**, 368–391 (1938)

504. P. Ren, T. Aleksić, R.C. Wilson, E.R. Hancock, Hypergraphs, Characteristic polynomials and the Ihara Zeta function, in *CAIP 2009*. Lecture Notes in Computer Science, vol. 5702 (Springer, New York, 2009), pp. 369–376

505. H. Resnikoff, On the graded ring of Hilbert modular forms associated with $\mathbb{Q}(\sqrt{5})$. Math. Ann. **203**, 161–170 (1974)
506. H. Resnikoff, Differential geometry and color perception. J. Math. Biol. **1**, 97–131 (1974)
507. H. Resnikoff, Automorphic forms of singular weight are singular forms. Math. Ann. **215**, 175–193 (1975)
508. H. Resnikoff, Theta functions for Jordan algebras. Invent. Math. **31**, 87–104 (1975)
509. H. Resnikoff, Y.-S. Tai, On the structure of a graded ring of automorphic forms on the 2-dimensional complex ball, I, II. Math. Ann. **238**, 97–117 (1978); **258**, 367–382 (1982)
510. J. Rhodes, Modular forms on p-adic planes, Ph.D. thesis, M.I.T., 1986
511. D.S.P. Richards, The central limit theorem on spaces of positive definite matrices. J. Multivar. Anal. **29**, 326–332 (1989)
512. W. Roelcke, Über die Wellengleichung bei Grenzkreisgruppen erster Art, *Sitzber. Akad. Heidelberg, Math.-naturwiss. Kl* (1953/1955)
513. A. Rogers, *Supermanifolds: Theory and Applications* (World Scientific, Hackensack, NJ, 2007)
514. C.A. Rogers, *Packing and Covering* (Cambridge University Press, Cambridge, 1964)
515. M. Rosen, *Number Theory in Function Fields* (Springer, New York, 2002)
516. J. Rosenberg, A quick proof of Harish-Chandra's Plancherel theorem for spherical functions on a semisimple Lie groups. Proc. Am. Math. Soc. **63**, 143–149 (1977)
517. S.N. Roy, P-statistics or some generalizations in analysis of variance appropriate to multivariate problems. Sankyha **4**, 381–396 (1939)
518. M. Rubinstein, Computational methods and experiments in analytic number theory, in *Recent Perspectives in Random Matrix Theory and Number Theory*. London Mathematical Society Lecture Note Series, vol. 322 (Cambridge University Press, Cambridge, 2005), pp. 425–506
519. J. A. Rush, On Hilbert's 18th problem: packing, preprint
520. J.A. Rush, N. Sloane, An improvement to the Minkowski-Hlawka bound for packing superballs. Mathematika **34**, 8–18 (1987)
521. S.S. Ryskov, The theory of Hermite-Minkowski reduction of positive quadratic forms. J. Sov. Math. **6**, 651–676 (1976)
522. S.S. Ryskov, The geometry of positive quadratic forms. Am. Math. Soc. Transl. **109**(2), 27–32 (1977)
523. S.S. Ryskov, E.P. Baranovskii, Classical methods in the theory of lattice packings. Russ. Math. Surv. **34**, 1–68 (1979)
524. A.A. Sagle, R.E. Walde, *Introduction to Lie Groups and Lie Algebras* (Academic, New York, 1973)
525. H. Saito, *Automorphic Forms and Algebraic Extensions of Number Fields*. Lecture Notes in Mathematics, vol. 8 (Kinokuniya Book Store, Tokyo, 1975)
526. P. Sarnak, The arithmetic and geometry of some hyperbolic three-manifolds. Acta Math. **151**, 253–295 (1983)
527. P. Sarnak, Arithmetic quantum chaos. Israel Math. Conf. Proc. **8**, 183–236 (1995) (published by the American Mathematical Society)
528. P. Sarnak, Notes on the generalized Ramanujan conjectures, in *Harmonic Analysis, the Trace Tormula, and Shimura Varieties, Clay Mathematics Proceedings*, vol. 4 (American Mathematical Society, Providence, RI, 2005), pp. 659–685
529. P. Sarnak, F. Shahidi (eds.), *Automorphic Forms and Applications, IAS/Park City Math. Series*, vol. 12 (American Mathematical Society, Providence, RI, 2007)
530. P. Sarnak, A. Strömbergsson, Minima of Epstein's zeta function and heights of flat tori. Invent. Math. **165**, 115–151 (2006)
531. I. Satake, On the compactification of the Siegel space. J. Indian Math. Soc. **20**, 259–281 (1956)
532. I. Satake, Theory of spherical functions on reductive algebraic groups over p-adic fields. Publ. Math. **18** (Inst. des Hautes Études, Paris (1963)
533. I. Satake, Review of Ash, Mumford, Rapoport, and Tai in Math. Reviews **56**, #15642

534. G. Schiffman, Integrales d'entrelacement et fonctions de Whittaker. Bull. Soc. Math. France **99**, 3–72 (1971)
535. W. Schmidt, The distribution of sublattices of $\mathbb{Z}^m$. Monat. fur Math. **125**, 37–81 (1998)
536. C. Schmit, Quantum and classical properties of some billiards on the hyperbolic plane, in *Chaos and Quantum Physics*, ed. by M.-J. Giannoni et al. (Elsevier, New York, 1991), pp. 333–369
537. B. Schoeneberg, Das Verhalten von mehrfachen Thetareihen bei Modulsubstitutionen. Math. Ann. **116**, 511–523 (1939)
538. R.L.E. Schwarzenberger, *N-Dimensional Crystallography* (Pitman, San Francisco, CA, 1980)
539. J. Schwermer, *Eisensteinreihen und die Kohomologie von Kongruenzuntergruppen von SL(n, $\mathbb{Z}$)*. Bonner Math. Schriften, **99** (1977)
540. J. Schwermer, *Kohomologie arithmetisch definierter Gruppen und Eisensteinreihen*. Lecture Notes in Mathematics, vol. 988 (Springer, New York, 1983)
541. L.A. Seeber, Untersuchungen über die Eigenschaften der Positive Ternaren Quadratischen Formen, Freiburg, 1831
542. A. Selberg, Remarks on a multiple integral. Norsk Mat. Tidsskr. **26**, 71–78 (1944) [in Norwegian]
543. A. Selberg, Harmonic analysis and discontinuous groups in weakly symmetric Riemannian spaces with applications to Dirichlet series, J. Indian Math. Soc. **20**, 47–87 (1956)
544. A. Selberg, A new type of zeta function connected with quadratic forms. Report of the Institute in the Theory of Numbers, University of Colorado, Boulder, CO, 1959, pp. 207–210
545. A. Selberg, Discontinuous groups and harmonic analysis, in *Proceedings of International Congress of Mathematics*, Stockholm, 1962, pp. 177–189
546. A. Selberg, *Collected Papers*, vols. I, II (Springer, Berlin, 1989, 1991)
547. Séminaire H. Cartan, 1957/1958, *Fonctions Automorphes* (Benjamin, New York, 1967)
548. J.-P. Serre, Cohomologie des groupes discrets, in *Prospects in Mathematics* (Princeton University Press, Princeton, NJ, 1971), pp. 77–169
549. I.R. Shafarevitch, *Basic Algebraic Geometry* (Springer, New York, 1974)
550. F. Shahidi, On the Ramanujan conjecture and finiteness of poles for certain *L*-functions Ann. Math. **127**, 547–584 (1988)
551. D. Shale, Linear symmetries of free boson fields. Trans. Am. Math. Soc. **103**, 149–167 (1962)
552. I.A. Shalika, The multiplicity one theorem for $GL(n)$. Ann. Math. **100**, 171–193 (1974)
553. H. Shimizu, On discontinuous groups operating on the product of the upper half planes. Ann. Math. **77**, 33–71 (1963)
554. G. Shimura, *Introduction to the Arithmetic Theory of Automorphic Functions* (Princeton University Press, Princeton, NJ, 1971)
555. G. Shimura, Confluent hypergeometric functions on tube domains. Math. Ann. **260**, 269–302 (1982)
556. G. Shimura, On Eisenstein series. Duke Math. J. **50**, 417–476 (1983)
557. G. Shimura, *The Map of My Life* (Springer, New York, 2008)
558. T. Shintani, On "liftings" of holomorphic automorphic forms (a representation-theoretic interpretation of the recent work of Saito), U.S.-Japan Sem., Ann Arbor, MI, 1975
559. T. Shintani, On zeta-functions associated with the vector space of quadratic forms. J. Fac. Sci. Univ. Tokyo **22**, 25–65 (1975)
560. T. Shintani, On an explicit formula for class 1 "Whittaker functions" over $\mathcal{P}$-adic fields. Proc. Jpn. Acad. **52**, 180–182 (1976)
561. C. L. Siegel, *Lectures on the Geometry of Numbers* (New York University, New York, 1945–1946)
562. C.L. Siegel, *Lectures on Quadratic Forms* (Tata Institute of Fundamental Research, Bombay, 1957).
563. C.L. Siegel, *Lectures on Advanced Analytic Number Theory* (Tata Institute of Fundamental Research, Bombay, 1957)
564. C.L. Siegel, *Topics in Complex Function Theory* (Wiley-Interscience, New York, 1969–1973)
565. C.L. Siegel, *Gesammelte Abhandlungen*, vols. I–IV (Springer, New York, 1966, 1979)

566. C.L. Siegel, *Lectures on the Geometry of Numbers*, rewritten by K. Chandrasekharan, Springer, Berlin, 1989

567. F. Sigrist, Sphere packing. Math. Intelligencer **5**, 34–38 (1983)

568. N.J.A. Sloane, The packing of spheres. Sci. Am. **250**, 116–125 (1984)

569. N.J.A. Sloane, Binary codes, lattices, and sphere-packings, in *Combinatorial Surveys, Proceedings of 6th British Combinat. Conf.* (Academic, London, 1977), pp. 117–164

570. L. Solomon, Partially ordered sets with colors, in *Proceedings of Symposia in Pure Mathematics*, vol. 34 (American Mathematical Society, Providence, RI, 1979), pp. 309–329

571. C. Soulé, Cohomology of $SL_3(\mathbb{Z})$. C. R. Acad. Sci. Paris **280**, 251–254 (1975)

572. C. Soulé, The cohomology of $SL(3, \mathbb{Z})$. Topology **17**, 1–22 (1978)

573. B. Speh, The unitary $GL(3, \mathbb{R})$ and $GL(4, \mathbb{R})$. Math. Ann. **258**, 113–133 (1981)

574. E. Stade, D. Wallace, Weyl's law for $SL(3, \mathbb{Z})\backslash SL(3, \mathbb{R})/SO(3, \mathbb{R})$. Pac. J. Math. **173**(1) (1996), 241–261

575. H.M. Stark, M.I.T. and U.C.S.D number theory course lecture notes, unpublished

576. H.M. Stark, On the transformation formula for the symplectic theta function and applications. J. Fac. Sci. Univ. Tokyo **29**, 1–12 (1982)

577. H. M. Stark, Fourier coefficients of Maass wave forms, in *Modular Forms*, ed. by R.A. Rankin (Horwood, Chichester, 1984), pp. 263–269 [distributed by Wiley]

578. H.M. Stark, Modular forms and related objects. Can. Math. Soc. Conf. Proc. **7**, 421–455 (1987)

579. H.M. Stark, On modular forms from L-functions in number theory, I, II, preprint

580. C. Storm, The zeta function of a hypergraph. Electron. J. Comb. **13**, #R84 1 (2006)

581. G. Strang, *Linear Algebra and Its Applications* (Academic, New York, 1976)

582. G. Strang, G.J. Fix, *An Analysis of the Finite Element Method* (Prentice-Hall, Englewood Cliffs, NJ, 1973)

583. H. Strassberg, L functions for $GL(n)$. Math. Ann. **245**, 23–36 (1979)

584. R. Styer, Hecke theory over complex quadratic fields, Ph.D. thesis, M.I.T., 1981

585. A. Takemura, *Zonal Polynomials, Inst. of Math. Stat. Lecture Notes*, vol. 4 (Institute of Mathematical Statistics, Hayward, CA, 1984)

586. L.A. Takhtadzhyan, A.I, Vinogradov, Theory of Eisenstein series for the group $SL(3, \mathbb{R})$ and its application to a binary problem. J. Soviet Math. **18**, 293–324 (1982)

587. T. Tamagawa, On some extensions of Epstein's Z-series, in *Proc. Internatl. Symp. on Alg. No. Theory*, Tokyo-Nikko, 1955, pp. 259–261

588. T. Tamagawa, On Selberg's trace formula. J. Fac. Sci. Univ. Tokyo, Sec. I **8**, 363–386 (1960)

589. T. Tamagawa, On the zeta-functions of a division algebra. Ann. Math. **77**, 387–405 (1963)

590. P.O. Tammela, The Minkowski reduction domain for positive quadratic forms of seven variables. Proc. Steklov Inst. Math. **67**, 108–143 (1977) [translation, J. Soviet Math. **16**, 836–857 (1981)]

591. Y. Tanigawa, Selberg trace formula for Picard groups, in *Proceedings of the International Symposium on Algebraic Number Theory*, Tokyo, 1977, pp. 229–242

592. T. Tao, V. Vu, From the Littlewood-Offord problem to the circular law: universality of the spectral distributions of random matrices. Bull. Am. Math. Soc. **46**, 377–396 (2009)

593. A. Terras, A generalization of Epstein's zeta function. Nagoya J. Math. **42**, 173–188 (1971)

594. A. Terras, Functional equations of generalized Epstein zeta functions in several variables. Nagoya J. Math. **44**, 89–95 (1971)

595. A. Terras, Bessel series expansions of the Epstein zeta function and the functional equation. Trans. Am. Math. Soc. **183**, 477–486 (1973)

596. A. Terras, Fourier coefficients of Eisenstein series of one complex variable for the special linear group. Trans. Am. Math. Soc. **205**, 97–114 (1975)

597. A. Terras, The Fourier expansion of Epstein's zeta function for totally real algebraic number fields and some consequences for Dedekind's zeta function. Acta Arith. **30**, 187–197 (1976)

598. A. Terras, The Fourier expansion of Epstein's zeta function over an algebraic number field and its consequences for algebraic number theory. Acta Arith. **32**, 37–53 (1977)

599. A. Terras, A relation between $\zeta_K(s)$ and $\zeta_K(s-1)$ for any algebraic number field K, in *Algebraic Number Fields*, ed. by A. Fröhlich (Academic, New York, 1977), pp. 475–483

600. A. Terras, Applications of special functions for the general linear group to number theory, Sem. Delange-Pisot-Poitou, 1976–1977, exp. 23

601. A. Terras, Real zeroes of Epstein's zeta function for ternary positive quadratic forms. Ill. J. Math. **23**, 1–14 (1979)

602. A. Terras, Integral formulas and integral tests for series of positive matrices. Pac. J. Math. **89**, 471–490 (1980)

603. A. Terras, The minima of quadratic forms and the behavior of Epstein and Dedekind zeta functions. J. Number Theory **12**, 258–272 (1980)

604. A. Terras, On automorphic forms for the general linear group. Rocky Mt. J. Math. **12**, 123–143 (1982)

605. A. Terras, Special functions for the symmetric space of positive matrices. SIAM J. Math. Anal. **16**, 620–640 (1985)

606. A. Terras, The Chowla-Selberg method for Fourier expansion of higher rank Eisenstein series. Can. Math. Bull. **28**, 280–294 (1985)

607. A. Terras, Some simple aspects of the theory of automorphic forms for $GL(n, \mathbb{Z})$. J. Contemp. Math. **53**, 409–447 (1986)

608. A. Terras, *An Introduction to Number Theory with the Aid of a Computer*, U. C. S. D. Lecture Notes

609. A. Terras, *Fourier Analysis on Finite Groups and Applications* (Cambridge University Press, Cambridge, 1999)

610. A. Terras, Arithmetical quantum chaos, in *IAS/Park City Mathematical Series*, vol. 12 (2002), pp. 333–375

611. A. Terras, *Zeta Functions of Graphs: A Stroll Through the Garden* (Cambridge University Press, Cambridge, 2011)

612. A. Terras, *Harmonic Analysis on Symmetric Spaces—Euclidean Space, the Sphere, and the Poincaré Upper Half Plane* (Springer, New York, 2013)

613. E. Thomas, A.T. Vasquez, On the resolution of cusp singularities and the Shintani decomposition in totally real cubic number fields. Math. Ann. **247**, 1–20 (1980)

614. T.M. Thompson, *From Error-Correcting Codes Through Sphere Packings to Simple Groups* (Mathematical Association of America, Washington, DC, 1983)

615. C.A. Tracy, H. Widom, Introduction to random matrices, in *Geometric and Quantum Aspects of Integrable Systems*. Lecture Notes in Physics, vol. 424 (Springer, Berlin, 1993), pp. 103–130

616. L.-C. Tsao, The rationality of the Fourier coefficients of certain Eisenstein series on tube domains. Compos. Math. **32**, 225–291 (1976)

617. R. Tsushima, On dimension formula for Siegel modular forms, in *Automorphic Forms and Geometry of Arithmetic Varieties*. Advanced Studies in Pure Mathematics, vol. 15 (Academic, Boston, MA, 1989), pp. 41–64

618. S. Tsuyumine, Construction of modular forms by means of transformation formulas for theta series. Tsukuba J. Math. **3**, 59–80 (1979)

619. S. Tsuyumine, On Siegel modular forms of degree three. Am. J. Math. **108**, 755–862, 1001–1003 (1986)

620. J. Tunnell, On the local Langlands conjecture for $GL(2)$. Invent. Math. **46**, 179–200 (1978)

621. J. Tunnell, Artin's conjecture for representations of octahedral type. Bull. Am. Math. Soc. **5**, 173–175 (1981)

622. S. Vance, Improved sphere packing lower bounds from Hurwitz lattices. Adv. Math. **227**, 2144–2156 (2011)

623. V.S. Varadarajan, *Lie Groups, Lie Algebras, and Their Representations* (Prentice-Hall, Englewood Cliffs, NJ, 1974)

624. V.S. Varadarajan, *Harmonic Analysis on Real Reductive Groups*. Lecture Notes in Mathematics, vol. 576 (Springer, New York, 1977)

625. V.S. Varadarajan, Eigenfunction expansions on semisimple Lie groups, in *Harmonic Analysis and Group Representations*, ed. by A. Figa-Talamanca (C.I.M.E. Inti. Math. Summer Center, Liguori editoro, Napoli, 1982), pp. 351–422

626. T. Veenstra, Siegel modular forms, L-functions and Satake parameters. J. Number Theory **87**, 15–30 (2001)

627. A.B. Venkov, On the trace formula for $SL(3, \mathbb{Z})$. J. Soviet Math. **12**, 384–424 (1979)

628. B.A. Venkov, Über die Reduction positiver quadratischer Formen. Izv. Akad. Nauk. S.S.S.R. **4** (1940), 37–52

629. M. Vergne, Representations of Lie groups and the orbit method, in *Emmy Noether in Bryn Mawr*, ed. by B. Srinivasan, J. Sally (Springer, New York, 1983), pp. 59–101

630. M.-F. Vignéras, Invariants numériques des groups de Hilbert. Math. Ann. **224**, 189–215 (1976)

631. M.-F. Vignéras, Genre arithmétique des groupes modulaires de Hilbert et nombres de classes de quaternions, Sém. de Théorie de Nombres, 1975–1976, U. Bordeaux, Exp. 28, Talence, 1976

632. M.-F. Vignéras, Invariants des groupes modulaires de Hilbert sur un corps quadratique, Sém. Delange-Pisot-Poitou, 1975–1976, Exp. 69, Paris, 1977

633. M.-F. Vignéras, Séries théta des formes quadratiques indéfinies, *Modular Functions of One Variable VI*. Lecture Notes in Mathematics, vol. 627 (Springer, New York, 1977), pp. 227–240

634. M.-F. Vignéras, *Arithmétique des Algèbres de Quaternions*. Lecture Notes in Mathematics, vol. 800 (Springer, New York, 1980)

635. N.J. Vilenkin, *Special Functions and the Theory of Group Representations, Translations of Mathematical Monographs*, vol. 22 (American Mathematical Society, Providence, RI, 1968)

636. G.F. Voronoi, Propriétées des formes quadratiques positives parfaites. J. Reine Angew. Math. **133**, 97–178 (1908)

637. G.F. Voronoi, Nouvelles applications des paramètres continus à la théorie des formes quadratiques. J. Reine Angew. Math. **134**, 198–287 (1908); **136**, 67–178 (1909)

638. B.L. van der Waerden, Die Reduktionstheorie der positiven quadratischen Formen. Acta Math. **96**, 265–309 (1956)

639. B.L. van der Waerden, Punktverteilungen auf der Kugel und Informationstheorie. Die Naturwissenschaften **48**, 189–192 (1961)

640. S. Wakatsuki, Dimension formulas for spaces of vector-valued Siegel cusp forms of degree 2. J. Number Theory **132**, 200–253 (2012)

641. J.-L. Waldspurger, Formes quadratiques à 4 variables et relèvement. Acta Arith. **36**, 377–405 (1980)

642. D. Wallace (Andreoli), Selberg's trace formula and units in higher degree number fields, Ph.D. thesis, U.C.S.D., 1982

643. D. Wallace, Conjugacy classes of hyperbolic matrices in $SL(n, \mathbb{Z})$ and ideal classes in an order, Trans. Am. Math. Soc. **283**, 177–184 (1984)

644. D. Wallace, Explicit form of the hyperbolic term in the Selberg trace formula for $SL(3, \mathbb{R})$ and Pell's equation for hyperbolics in $SL(3, \mathbb{R})$, J. Number Theory **29**, 127–133 (1986)

645. D. Wallace, A preliminary version of the Selberg trace formula for $SL(3, \mathbb{Z}) \backslash SL(3, \mathbb{R})/SO(3, \mathbb{R})$, in *The Selberg Trace Formula and Related Topics*. Contemporary Mathematics, vol. 53 (American Mathematical Society, Providence, RI, 1986), pp. 11–15

646. D. Wallace, Maximal parabolic terms in the Selberg trace formula for $SL(3, \mathbb{Z}) \backslash SL(3, \mathbb{R})/SO(3, \mathbb{R})$. J. Number Theory **29**(2), 101–117 (1988)

647. D. Wallace, Minimal parabolic terms in the Selberg trace formula for $SL(3, \mathbb{Z}) \backslash SL(3, \mathbb{R})/SO(3, \mathbb{R})$. J. Number Theory **32**(1), 1–13 (1989)

648. D. Wallace, Terms in the Selberg trace formula for $SL(3, \mathbb{Z}) \backslash SL(3, \mathbb{R})/SO(3, \mathbb{R})$ associated to Eisenstein series coming from a minimal parabolic subgroup. Trans. Am. Math. Soc. **327**(2), 781–793 (1991)

649. D. Wallace, The loxodromic term of the Selberg trace formula for $SL(3, \mathbb{Z}) \backslash SL(3, \mathbb{R})/SO(3, \mathbb{R})$. Proc. Am. Math. Soc. **113**(1), 5–9 (1991)

650. D. Wallace, The Selberg trace formula for $SL(3, \mathbb{Z}) \backslash SL(3, \mathbb{R})/SO(3, \mathbb{R})$. Trans. Am. Math. Soc. **345**(1), 1–36 (1994)
651. N. Wallach, *Harmonic Analysis on Homogeneous Spaces* (Dekker, New York, 1973)
652. N. Wallach, *Symplectic Geometry and Fourier Analysis* (Mathematical Science Press, Brookline, MA, 1977)
653. N. Wallach, Lecture Notes, 1981 NSF-CBMS Regional Conf. on Representations of Semisimple Lie Groups and Applications to Analysis, Geometry and Number Theory, unpublished
654. L.H. Walling, A formula for the action of Hecke operators on half-integral weight Siegel modular forms and applications. J. Number Theory **133**(5), 1608–1644 (2013)
655. G. Warner, *Harmonic Analysis on Semi-simple Lie Groups, I, II* (Springer, New York, 1972)
656. G. Warner, Selberg's trace formula for non-uniform lattices: the rank one case, in *Studies in Algebra and Number Theory*. Advances in Math., Suppl. Studies, vol. 6 (Academic, New York, 1979), pp. 1–142
657. A. Wawrzyńczyk, *Group Representations and Special Functions* (Reidel, Boston, MA, 1984)
658. A. Weil, *Discontinuous Subgroups of Classical Groups* (University of Chicago, Chicago, IL, 1958)
659. A. Weil, *L'intégration dans les Groupes Topologiques* (Hermann, Paris, 1965)
660. A. Weil, *Dirichlet Series and Automorphic Forms*. Lecture Notes in Mathematics, vol. 189 (Springer, New York, 1971)
661. A. Weil, *Basic Number Theory* (Springer, New York, 1974)
662. A. Weil, *Oeuvres Scientifiques, Collected Papers (1926–1978)*, vols. I–III (Springer, New York, 1979)
663. R. Weissauer, Siegel modular forms and Dirichlet series, preprint
664. R. Weissauer, Eisensteinreihen vom Gewicht $n + 1$ sur Siegelschen Modulgruppe n-ten Grades. Math. Ann. **268**, 357–377 (1984)
665. D. Weisser, The arithmetic genus of the Hilbert modular variety and the elliptic fixed points of the Hilbert modular group. Math. Ann. **257**, 9–22 (1981)
666. H. Weyl, *Gesammelte Abhandlungen* (Springer, New York, 1968)
667. E.P. Wigner, Random matrices in physics. SIAM Rev. **9**(1), 1–23 (1967)
668. E.P. Wigner, On a generalization of Euler's angles, in *Group Theory and Its Applications*, ed. by E.M. Loebl (Academic, New York, 1968), pp. 119–129
669. H. Williams, J. Broere, A computational technique for evaluating $L(1, \chi)$ and the class number of a real quadratic field. Math. Comput. **30**, 887–893 (1976)
670. J. Wishart, The generalized product moment distribution in samples from a normal multivariate population. Biometrika **20A**, 32–43 (1928)
671. S.-T. Wong, *The Meromorphic Continuation and Functional Equations of Cuspidal Eisenstein Series for Maximal Cuspidal Subgroups*. Memoirs of the American Mathematical Society, vol. 83, No. 423 (American Mathematical Society, Providence, RI, 1990)
672. B.G. Wybourne, *Classical Groups for Physicists* (Wiley, New York, 1974)
673. T. Yamazaki, On Siegel modular forms of degree 2. Am. J. Math. **98**, 39–53 (1973)
674. D. Zagier, A Kronecker limit formula for real quadratic fields. Math. Ann. **213**, 153–184 (1975)
675. D. Zagier, Modular forms associated to real quadratic fields. Invent. Math. **30**, 1–46 (1975)
676. R.I. Zimmer, *Ergodic Theory and Semisimple Groups* (Birkhäuser, Boston, MA, 1984)
677. P.G. Zograf, The Selberg trace formula for the Hilbert modular group of a real quadratic algebraic number field (Russian). Zap. Nauchn. Sem. Leningrad, Otdel/Mat. Inst. Steklov (LOMI) **100**, 26–47, 173 (1980)

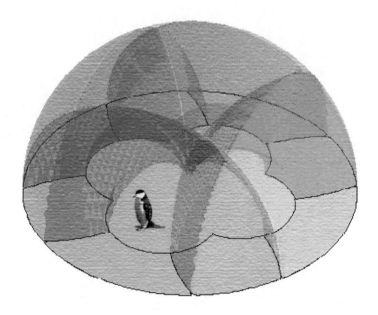

"Mathematics tries to replace reality with a dream of order. It is perhaps for this reason that mathematicians are often such strange and socially inept people. To devote oneself to mathematics is to turn away from the physical world and meditate about an ideal world of thoughts. The striking thing is that these pure mathematical meditations can in fact make fairly good predictions about messy matter. Eclipses are predicted; bridges are built; computers function .... "

From Rudy Rucker, *Mind Tools*, Houghton Mifflin, Boston, 1987, p. 156.

Index

A

A, $\mathfrak{a}$, abelian Lie group and its Lie algebra, 6, 84, 133, 360, 366

$\mathfrak{a}_n$, $\mathfrak{sl}(n+1, \mathbb{C})$, Lie algebra of the special linear group, 346

abelian integral, 162, 401, 426

$\mathcal{AC}(\Gamma, \lambda)$, space of Maass cusp forms, 238

$\mathcal{A}^0(\Gamma, \lambda)$, space of Maass forms on determinant one surface, 240

action of

   G on G/K, 9, 10, 12, 29, 75, 118, 122, 349, 367, 399

   G on boundary, 25, 126, 154, 389

adelic theory, 4, 5, 56, 70, 158, 216, 237, 245, 259, 272, 304, 319, 332, 409, 420, 429, 442

adjoint of a differential operator, 50, 110, 269

adjoint representation, 154, 340, 345, 358, 366

$\mathcal{A}(\Gamma, \lambda)$, space of Maass forms or automorphic forms, 236, 238–240, 242, 244, 246, 258, 259, 266, 270, 272, 275–277, 282–284, 303, 306, 308–310, 312–315, 428

algebraic integer, 404

algebraic number field, 159, 171, 273, 330, 338, 401, 408, 417, 429, 436

analytic continuation, 55, 156, 231, 245, 252, 259, 271, 304, 318, 421, 428, 437

arc length, 8, 16, 27, 40, 242, 370, 375, 390, 399

arithmetic quantum chaos, 136

Arthur trace formula, 239, 319, 442

Artin conjecture, L-functions, and reciprocity, 5, 158, 237, 314, 332

Arzelà-Ascoli theorem, 323

associated parabolic subgroups, 275, 320, 327

asymptotics

   Eisenstein series, 277, 280

   K-Bessel functions, 60

   number-theoretical quantities, 6, 139, 164, 232, 239, 322, 401, 443

   spherical functions, 86, 93, 98, 111, 123, 128, 134, 395

asymptotics/functional equations principle, 108, 122, 277, 327, 395

automorphic form, 4, 159, 162, 204, 236, 266, 332, 355, 416, 420, 428, 442

automorphic function, 425

B

Babylonian reduction, 15

base change, 159, 332, 425

Bergman kernel, 446

Bessel function

   I-Bessel function, 284

   J-Bessel function, 56, 66, 96

   K-Bessel function, 56, 65, 71, 283, 286, 287, 396, 431

   k-Bessel function, 58, 61, 70, 130, 295, 306

beta function, 60, 110, 128, 301, 395

$\mathfrak{b}_n$, $\mathfrak{so}(2n+1, \mathbb{C})$, the Lie algebra of an orthogonal group, 346

Borel or minimal parabolic subgroup, 279, 290, 293, 389, 392

boson field, 338, 401

boundary

   fundamental domain, 173, 187, 193, 197, 277, 281, 327

© Springer Science+Business Media New York 2016

A. Terras, *Harmonic Analysis on Symmetric Spaces—Higher Rank Spaces, Positive Definite Matrix Space and Generalizations*, DOI 10.1007/978-1-4939-3408-9